SUPPLEMENT TO THE INTERNATIONAL CODES®

2004

Second Printing

Publication Date: February 2005

Copyright © 2004
by
International Code Council, Inc.

ALL RIGHTS RESERVED. This 2004 International Supplement to the International Codes is a copyrighted work owned by the International Code Council, Inc. Without advance written permission from the copyright owner, no part of this book may be reproduced, distributed, or transmitted in any form or by any means, including, without limitations, electronic, optical or mechanical means (by way of example and not limitation, photocopying, or recording by or in an information storage retrieval system). For information on permission to copy material exceeding fair use, please contact: Publications, 4051 West Flossmoor Road, Country Club Hills, IL 60478-5795 [Phone: (800) 214-4321].

Trademarks: "International Code Council," the "International Code Council" logo are trademarks of the International Code Council, Inc.

PRINTED IN THE U.S.A

PREFACE

This document is the 2004 Supplement to the 2003 editions of the *International Building Code®, ICC Electrical Code™, International Energy Conservation Code®, International Existing Building Code®, International Fire Code®, International Fuel Gas Code®, International Mechanical Code®, International Plumbing Code®, International Private Sewage Disposal Code®, International Property Maintenance Code®, International Residential Code®, International Urban-Wildland Interface Code®* and the *International Zoning Code®*. It contains changes submitted in the 2003/2004 Code Development Cycle which were approved by the membership of the International Code Council. The *ICC Performance Code for Buildings and Facilities®* has not been changed from the 2003 edition.

This supplement is organized by code and includes all approved changes. Each section is identified, followed by the applicable code change number and the revised/added/deleted text. For example, Section 302.1.1.1 of the IBC (page IBC-1) was revised by Code Change FS19-03/04 in the 2003/2004 Code Development Cycle. Editorial revisions approved by the ICC Code Correlation Committee are indicated by "CCC."

This supplement is prepared in a form which facilitates adoption by reference and is convenient to use. A suggested form of an adoption ordinance is found on page v (page vii of the *ICC Performance Code*) of each of the *2003 International Codes*. Through the adoption of the above referenced ICC Codes and this document, jurisdictions will have the benefit of the latest developments in building regulations.

TABLE OF CONTENTS

2004 Supplement to the International Building Code IBC-1

2004 Supplement to the ICC Electrical Code . . . . ICC EC-1

2004 Supplement to the International Energy Conservation Code IECC-1

2004 Supplement to the International Existing Building Code IEBC-1

2004 Supplement to the International Fire Code IFC-1

2004 Supplement to the International Fuel Gas Code IFGC-1

2004 Supplement to the International Mechanical Code IMC-1

2004 Supplement to the International Plumbing Code IPC-1

2004 Supplement to the International Private Sewage Disposal Code IPSDC-1

2004 Supplement to the International Property Maintenance Code IPMC-1

2004 Supplement to the International Residential Code IRC-1

2004 Supplement to the International Urban-Wildland Interface Code IUWIC-1

2004 Supplement to the International Zoning Code IZC-1

Errata to the 2003 International Codes ERR-1

Errata to the 2004 Supplement ERR-18

*The ICC Performance Code for Buildings and Facilities has not been changed from the 2003 edition.

INTERNATIONAL BUILDING CODE®

2004 SUPPLEMENT

INTERNATIONAL BUILDING CODE 2004 SUPPLEMENT

SECTION 202
DEFINITIONS

Change the definition of "Approved" to read as shown: (G18-03/04)

Approved. Acceptable to the code official or authority having jurisdiction.

Add new definition of "Day box" to read as shown: (F163-03/04)

DAY BOX. See Section 307.2.1.

Delete the definition of "Flame Resistance": (G85-03/04)

Add new definitions to read as shown: (S2-03/04)

WALL, LOAD-BEARING. Any wall meeting either of the following classifications:

1. Any metal or wood stud wall that supports more than 100 pounds per linear foot (1459 N/m) of vertical load in addition to its own weight.

2. Any masonry or concrete wall that supports more than 200 pounds per linear foot (2919 N/m) of vertical load in addition to its own weight.

WALL, NONLOAD-BEARING. Any wall that is not a load-bearing wall.

CHAPTER 3
USE AND OCCUPANCY CLASSIFICATION

Table 302.1.1 Change to read as shown: (G27-03/04; G28-03/04; G29-03/04)

TABLE 302.1.1
INCIDENTAL USE AREAS

ROOM OR AREA	SEPARATION[a]
Rooms with boilers where the largest piece of equipment is over 15 psi and 10 horsepower	1 hour or provide automatic fire-extinguishing system
Hydrogen cut-off rooms, not classified as Group H.	1 hour in Group B, F, H, M, S and U occupancies. 2 hours in Group A, E, I and R occupancies.
Stationary lead-acid battery systems having a liquid capacity of more than 100 gallons used for facility standby power, emergency power or uninterrupted power supplies	1-hour in Group B, F, M, S and U occupancies. 2-hours in Group A, E, I and R occupancies.

(Portions of table not shown do not change)

Section 302.1.1.1 Change to read as shown: (FS19-03/04)

302.1.1.1 Separation. Where Table 302.1.1 requires a fire-resistance-rated separation, the incidental use area shall be separated from the remainder of the building with a fire barrier. Where Table 302.1.1 permits an automatic fire-extinguishing system without a fire barrier, the incidental use area shall be separated by construction capable of resisting the passage of smoke. The partitions shall extend from the floor to the underside of the fire-resistance-rated floor/ceiling assembly or fire-resistance-rated roof/ceiling assembly above or to the underside of the floor or roof sheathing, deck or slab above. Doors shall be self-closing or automatic-closing upon detection of smoke. Doors shall not have air transfer openings and shall not be undercut in excess of the clearance permitted in accordance with NFPA 80.

Section 302.3.2 Change to read as shown: (G27-03/04)

302.3.2 Separated uses. Each portion of the building shall be individually classified as to use and shall be completely separated from adjacent areas by fire barriers having a fire-resistance rating determined in accordance with Table 302.3.2 for uses being separated. Each fire area shall comply with this code based on the use of that space. Each fire area shall comply with the height limitations based on the use of that space and the type of construction classification. In each story, the building area shall be such that the sum of the ratios of the floor area of each use divided by the allowable area for each use shall not exceed one.

Exception: Except for Group H and I-2 areas, where the building is equipped throughout with an automatic sprinkler system, installed in accordance with Section 903.3.1.1, the fire-resistance ratings in Table 302.3.3 shall be reduced by 1 hour but to not less than 1 hour and to not less than that required for floor construction according to the type of construction.

Section 307.1 Change to read as shown and combine with Section 307.9 to make one section: (G27-03/04; F193-03/04)

307.1 High-Hazard Group H. High-Hazard Group H occupancy includes, among others, the use of a building or structure, or a portion thereof, that involves the manufacturing, processing, generation or storage of materials that constitute a physical or health hazard in quantities in excess of quantities allowed in control areas constructed and located as required in Section 414. Hazardous uses are classified in Groups H-1, H-2, H-3, H-4 and H-5 and shall be in accordance with this section, the requirements of Section 415 and the *International Fire Code*.

Exceptions: The following shall not be classified in Group H, but shall be classified in the occupancy that they most nearly resemble.

1. Buildings and structures that contain not more than the maximum allowable quantities per control area of hazardous materials as shown in Tables 307.7(1) and 307.7(2) provided that such buildings are maintained in accordance with the *International Fire Code*.

2. Buildings utilizing control areas in accordance with Section 414.2 that contain not more than the maximum allowable quantities per control area of hazardous materials as shown in Tables 307.7(1) and 307.7(2).

3. Buildings and structures occupied for the application of flammable finishes, provided that such buildings or areas conform to the requirements of Section 416 and the *International Fire Code*.

4. Wholesale and retail sales and storage of flammable and combustible liquids in mercantile occupancies conforming to the *International Fire Code*.

5. Closed systems housing flammable or combustible liquids or gases utilized for the operation of machinery or equipment.

6. Cleaning establishments that utilize combustible liquid solvents having a flash point of 140°F (60°C) or higher in closed systems employing equipment listed by an approved testing agency, provided that this occupancy is separated from all other areas of the building by 1-hour fire-resistance-rated fire barriers.

7. Cleaning establishments which utilize a liquid solvent having a flash point at or above 200°F (93°C).

8. Liquor stores and distributors without bulk storage.

9. Refrigeration systems.

10. The storage or utilization of materials for agricultural purposes on the premises.

11. Stationary batteries utilized for facility emergency power, uninterrupted power supply or telecommunication facilities provided that the batteries are provided with safety venting caps and ventilation is provided in accordance with the *International Mechanical Code*.

12. Corrosives shall not include personal or household products in their original packaging used in retail display or commonly used building materials.

13. Buildings and structures occupied for aerosol storage shall be classified as Group S-1, provided that such buildings conform to the requirements of the *International Fire Code*.

14. Display and storage of nonflammable solid and nonflammable or noncombustible liquid hazardous materials in quantities not exceeding the maximum allowable quantity per control area in Group M or S occupancies complying with Section 414.2.4.

15. The storage of black powder, smokeless propellant and small arms primers in Groups M and R-3 and special industrial explosive devices in Groups B, F, M and S, provided such storage conforms to the quantity limits and requirements prescribed in the *International Fire Code*.

Section 307.1.1 Add new section to read as shown: (F193-03/04)

307.1.1 Hazardous materials. Hazardous materials in any quantity shall conform to the requirements of this code, including Section 414, and the *International Fire Code*.

Section 307.2 Add new definition to read as shown: (F163-03/04)

DAY BOX. A portable magazine designed to hold explosive materials constructed in accordance with the requirements for a Type 3 magazine as defined and classified in Chapter 33 of the *International Fire Code*.

Section 307.2 Change definition to read as shown: (F187-03/04)

WATER-REACTIVE MATERIAL. A material that explodes; violently reacts; produces flammable, toxic or other hazardous gases; or evolves enough heat to cause auto-ignition or ignition of combustibles upon exposure to water or moisture. Water-reactive materials are subdivided as follows:

Class 3. Materials that react explosively with water without requiring heat or confinement.

Class 2. Materials that react violently with water or have the ability to boil water. Materials that produce flammable, toxic or other hazardous gases, or evolve enough heat to cause autoignition or ignition of combustibles upon exposure to water or moisture.

Class 1. Materials that react with water with some release of energy, but not violently.

Table 307.7(1) Change footnote e to read as shown: (F163-03/04)

TABLE 307.7(1)
MAXIMUM ALLOWABLE QUANTITY PER CONTROL AREA OF HAZARDOUS MATERIALS POSING A PHYSICAL HAZARD

(No change to current table body)
a. through d. (No change to current text)
e. Maximum allowable quantities shall be increased 100 percent when stored in approved storage cabinets, day boxes, gas cabinets, exhausted enclosures or safety cans. Where Note d also applies, the increase for both notes shall be applied accumulatively.
f. through n. (No change to current text)

Section 307.9 Relocate to be part of Section 307.1 (F193-03/04)

Section 308.3 Change to read as shown: (G42-03/04)

308.3 Group I-2. This occupancy shall include buildings and structures used for medical, surgical, psychiatric, nursing or custodial care of more than five persons who are not capable of self-preservation. This group shall include, but not be limited to, the following:

Hospitals
Nursing homes (both intermediate-care facilities and skilled nursing facilities)
Mental hospitals
Detoxification facilities

A facility such as the above with five or fewer persons shall be classified as Group R-3 or shall comply with the *International Residential Code* in accordance with Section 101.2.

Section 311.2 Change to read as shown: (G46-03/04)

311.2 Moderate-hazard storage, Group S-1. Buildings occupied for storage uses which are not classified as Group S-2 including, but not limited to, storage of the following:

Aerosols, Levels 2 and 3
Aircraft repair hangar
Bags; cloth, burlap and paper
Bamboos and rattan
Baskets
Belting; canvas and leather
Books and paper in rolls or packs
Boots and shoes
Buttons, including cloth covered, pearl or bone
Cardboard and cardboard boxes
Clothing, woolen wearing apparel
Cordage
Dry boat storage (indoor)
Furniture
Furs
Glues, mucilage, pastes and size
Grains
Horns and combs, other than celluloid
Leather
Linoleum
Lumber
Motor vehicle repair garages complying with the maximum allowable quantities of hazardous materials listed in Table 307.7(1) (see Section 406.6)
Photo engravings
Resilient flooring
Silks
Soaps
Sugar
Tires, bulk storage of
Tobacco, cigars, cigarettes and snuff
Upholstery and mattresses
Wax candles

CHAPTER 4
SPECIAL DETAILED REQUIREMENTS BASED ON USE AND OCCUPANCY

Section 402.4.5.1 Change to read as shown: (E1-03/04)

402.4.5.1 Exit passageways. Where exit passageways provide a secondary means of egress from a tenant space, doors to the exit passageway shall be 1-hour fire doors. Such doors shall be self-closing and be so maintained or shall be automatic-closing by smoke detection.

Section 402.4.6 Change to read as shown: (G50-03/04)

402.4.6 Service areas fronting on exit passageways. Mechanical rooms, electrical rooms, building service areas and service elevators are permitted to open directly into exit passageways provided the exit passageway is separated from such rooms with not less than 1-hour fire-resistance-rated fire barriers and 1-hour opening protectives.

Section 403.2 Change to read as shown: (FS40-03/04)

403.2 Automatic sprinkler system. Buildings and structures shall be equipped throughout with an automatic sprinkler system in accordance with Section 903.3.1.1 and a secondary water supply where required by Section 903.3.5.2.

Exceptions: An automatic sprinkler system shall not be required in spaces or areas of:

1. Open parking garages in accordance with Section 406.3.

2. Telecommunications equipment buildings used exclusively for telecommunications equipment, associated electrical power distribution equipment, batteries and standby engines, provided that those spaces or areas are equipped throughout with an automatic fire detection system in accordance with Section 907.2 and are separated from the remainder of the building by fire barriers consisting of not less than 1-hour fire-resistance-rated walls and 2-hour fire-resistance-rated floor/ceiling assemblies.

Section 403.3.1 Change to read as shown: (G55-03/04)

403.3.1 Type of construction. The following reductions in the minimum construction type allowed in Table 601 shall be allowed as provided in Section 403.3:

1. For buildings not greater than 420 feet (128 m) in height, Type IA construction shall be allowed to be reduced to Type IB.

2. In other than Groups F-1, M and S-1, Type IB construction shall be allowed to be reduced to Type IIA.

3. The height and area limitations of the reduced construction type shall be allowed to be the same as for the original construction type.

Section 403.3.2 Change to read as shown: (G55-03/04)

403.3.2 Shaft enclosures. For buildings not greater than 420 feet (128 m) in height, the required fire-resistance rating of the fire barrier walls enclosing vertical shafts, other than exit enclosures and elevator hoistway enclosures, shall be reduced to 1 hour where automatic sprinklers are installed within the shafts at the top and at alternate floor levels.

Section 404.1 Delete current text and substitute as shown: (G59-03/04)

404.1 General. The provisions of this section shall apply to buildings or structures containing vertical openings defined herein as atriums.

Exceptions:

1. This section is not applicable to floor openings meeting the requirements of Section 707.2, exception 2, 7, 8 or 9.

2. This section is not applicable to open stairs meeting the requirements of Section 1019.1, Exception 8 or 9.

3. Floor openings in Group H occupancies must be enclosed.

Section 404.3 Change to read as shown: (G27-03/04)

404.3 Automatic sprinkler protection. An approved automatic sprinkler system shall be installed throughout the entire building.

Exceptions:

1. That area of a building adjacent to or above the atrium need not be sprinklered provided that portion of the building is separated from the atrium portion by a 2-hour fire-resistance-rated fire barrier.

2. Where the ceiling of the atrium is more than 55 feet (16 764 mm) above the floor, sprinkler protection at the ceiling of the atrium is not required.

Section 404.4 Change to read as shown: (G59-03/04; G60-03/04)

404.4 Smoke control. A smoke control system shall be installed in accordance with Section 909.

Exception: Smoke control is not required for atriums that connect only two stories.

Section 404.5 Change to read as shown: (G61-03/04)

404.5 Enclosure of atriums. Atrium spaces shall be separated from adjacent spaces by a 1-hour fire barrier wall.

Exceptions:

1. A glass wall forming a smoke partition where automatic sprinklers are spaced 6 feet (1829 mm) or less along both sides of the separation wall, or on the room side only if there is not a walkway on the atrium side, and between 4 inches and 12 inches (102 mm and 305 mm) away from the glass and so designed that the entire surface of the glass is wet upon activation of the sprinkler system without obstruction. The glass shall be installed in a gasketed frame so that the framing system deflects without breaking (loading) the glass before the sprinkler system operates.

2. A glass-block wall assembly in accordance with Section 2110 and having a $^3/_4$-hour fire protection rating.

3. The adjacent spaces of any three floors of the atrium shall not be required to be separated from the atrium where such spaces are included in computing the atrium volume for the design of the smoke control system.

Section 405.4.2 Change to read as shown: (G62-03/04; FS72-03/04)

405.4.2 Smoke barrier penetration. The compartments shall be separated from each other by a smoke barrier in accordance with Section 709. Penetrations between the two compartments shall be limited to plumbing and electrical piping and conduit that are firestopped in accordance with Section 712. Doorways shall be protected by door assemblies installed in accordance with NFPA 105, and Section 715.3.3. Where provided, each compartment shall have an air supply and an exhaust system independent of the other compartments.

Section 405.4.3 Change to read as shown: (G62-03/04)

405.4.3 Elevators. Where elevators are provided, each compartment shall have direct access to an elevator. Where an elevator serves more than one compartment, an elevator lobby shall be provided and shall be separated from each compartment by a smoke barrier in accordance with Section 709. Doors shall be gasketed, have a drop sill, and be automatic-closing by smoke detection installed in accordance with Section 907.10.

Section 406.2.4 Change to read as shown: (G67-03/04)

406.2.4 Vehicle barriers. Parking areas shall be provided with exterior or interior walls or vehicle barriers, except at pedestrian or vehicular accesses, designed in accordance with Section 1607.7. Vehicle barriers not less than 2 feet (607 mm) high shall be placed at the ends of drive lanes, and at the end of parking spaces where the difference in adjacent floor elevation is greater than 1 foot (305 mm).

> **Exception:** Vehicle storage compartments in a mechanical access parking garage.

Section 406.2.5 Change to read as shown: (G68-03/04)

406.2.5 Ramps. Vehicle ramps shall not serve as an exit element. Vehicle ramps that are utilized for vertical circulation as well as for parking shall not exceed a slope of 1:15 (6.67 percent).

Section 406.3.6 Change to read as shown: (G69-03/04)

406.3.6 Area and height increases. The allowable area and height of open parking garages shall be increased in accordance with the provisions of this section. Garages with sides open on three-fourths of the building perimeter are permitted to be increased by 25 percent in area and one tier in height. Garages with sides open around the entire building perimeter are permitted to be increased 50 percent in area and one tier in height. For a side to be considered open under the above provisions, the total area of openings along the side shall not be less than 50 percent of the interior area of the side at each tier, and such openings shall be equally distributed along the length of the tier.

Allowable tier areas in Table 406.3.5 shall be increased for open parking garages constructed to heights less than the table maximum. The gross tier area of the garage shall not exceed that permitted for the higher structure. At least three sides of each such larger tier shall have continuous horizontal openings not less than 30 inches (762 mm) in clear height extending for at least 80 percent of the length of the sides, and no part of such larger tier shall be more than 200 feet (60 960 mm) horizontally from such an opening. In addition, each such opening shall face a street or yard accessible to a street with a width of at least 30 feet (9144 mm) for the full length of the opening, and standpipes shall be provided in each such tier.

Open parking garages of Type II construction, with all sides open, shall be unlimited in allowable area where the height does not exceed 75 feet (22 860 mm). For a side to be considered open, the total area of openings along the side shall not be less than 50 percent of the interior area of the side at each tier, and such openings shall be equally distributed along the length of the tier. All portions of tiers shall be within 200 feet (60 960 mm) horizontally from such openings.

Section 406.5.2.1 Add new text to read as shown: (F154-03/04)

406.5.2.1 Canopies used to support gaseous hydrogen systems. Canopies which are used to shelter dispensing operations where flammable compressed gases are located on the roof of the canopy shall be in accordance with the following:

1. The canopy shall meet or exceed Type I construction requirements.

2. Operations located under canopies shall be limited to refueling only.

3. The canopy shall be constructed in a manner that prevents the accumulation of hydrogen gas.

Section 407.3 Change to read as shown: (G73-03/04)

407.3 Corridor walls. Corridor walls shall be constructed as smoke partitions in accordance with Section 710.

Section 407.6 Change to read as shown: (F111-03/04)

407.6. Automatic fire detection. Corridors in nursing homes (both intermediate care and skilled nursing facilities), detoxification facilities and spaces permitted to be open to the corridors by Section 407.2 shall be equipped with an automatic fire detection system. Hospitals shall be equipped with smoke detection as required in Section 407.2.

> **Exceptions:**
>
> 1. Corridor smoke detection is not required where patient sleeping units are provided with smoke detectors that comply with UL 268. Such

detectors shall provide a visual display on the corridor side of each patient sleeping unit and an audible and visual alarm at the nursing station attending each unit.

2. Corridor smoke detection is not required where patient sleeping unit doors are equipped with automatic door-closing devices with integral smoke detectors on the unit sides installed in accordance with their listing, provided that the integral detectors perform the required alerting function.

Section 408.3.6 Change to read as shown: (E1-03/04)

408.3.6 Exit enclosures. One of the required exit enclosures in each building shall be permitted to have glazing installed in doors and interior walls at each landing level providing access to the enclosure, provided that the following conditions are met:

1. The exit enclosure shall not serve more than four floor levels.
2. Exit doors shall not be less than ¾-hour fire doors complying with Section 715.3.
3. The total area of glazing at each floor level shall not exceed 5,000 square inches (3.23 m^2) and individual panels of glazing shall not exceed 1,296 square inches (0.84 m^2).
4. The glazing shall be protected on both sides by an automatic fire sprinkler system. The sprinkler system shall be designed to wet completely the entire surface of any glazing affected by fire when actuated.
5. The glazing shall be in a gasketed frame and installed in such a manner that the framing system will deflect without breaking (loading) the glass before the sprinkler system operates.
6. Obstructions, such as curtain rods, drapery traverse rods, curtains, drapes or similar materials shall not be installed between the automatic sprinklers and the glazing.

Section 408.6 Change to read as shown: (G75-03/04)

408.6 Smoke barrier. Occupancies in Group I-3 shall have smoke barriers complying with Section 709 to divide every story occupied by residents for sleeping, or any other story having an occupant load of 50 or more persons, into at least two smoke compartments.

Exception: Spaces having direct exit to one of the following, provided that the locking arrangement of the doors involved complies with the requirements for doors at the smoke barrier for the use condition involved:

1. A public way.
2. A building separated from the resident housing area by a 2-hour fire-resistance-rated assembly or 50 feet (15 240 mm) of open space.
3. A secured yard or court having a holding space 50 feet (15 240 mm) from the housing area that provides 6 square feet (0.56 m^2) or more of refuge area per occupant, including residents, staff and visitors.

Section 408.7.1 Change to read as shown: (E3-03/04)

408.7.1 Occupancy Conditions 3 and 4. Each sleeping area in Occupancy Conditions 3 and 4 shall be separated from the adjacent common spaces by a smoke-tight partition where the travel distance from the sleeping area through the common space to the corridor exceeds 50 feet (15 240 mm).

Section 408.7.2 Change to read as shown: (E3-03/04)

408.7.2 Occupancy Condition 5. Each sleeping area in Occupancy Condition 5 shall be separated from adjacent sleeping areas, corridors and common spaces by a smoke-tight partition. Additionally, common spaces shall be separated from the corridor by a smoke-tight partition.

Section 410.2 Change definition of "Stage" to read as shown: (G77-03/04)

STAGE. A space within a building utilized for entertainment or presentations, which includes overhead hanging curtains, drops, scenery or stage effects other than lighting and sound.

Section 410.3.5 Change to read as shown: (G82-03/04)

410.3.5 Proscenium curtain. Where a proscenium wall is required to have a fire-resistance rating, the stage opening shall be provided with a fire curtain of approved material or an approved water curtain complying with Section 903.3.1.1. The fire curtain shall be designed and installed to intercept hot gases, flames and smoke, and to prevent a glow from a severe fire on the stage from showing on the auditorium side for a period of 20 minutes. The closing of the fire curtain from the full open position shall be accomplished in less than 30 seconds, with the last 8 feet (2438 mm) of travel requiring 5 or more seconds for full closure.

Section 410.3.6 Change to read as shown: (G85-03/04)

410.3.6 Scenery. Combustible materials used in sets and scenery shall meet the fire propagation performance criteria of NFPA 701, in accordance with Section 805 and the *International Fire Code*. Foam plastics and materials containing foam plastics shall comply with Section 2603 and the *International Fire Code*.

Section 410.5.1 Change to read as shown: (G27-03/04)

410.5.1 Separation from stage. Where the stage height is greater than 50 feet (15 240 mm), the stage shall be separated from dressing rooms, scene docks, property rooms, workshops, storerooms and compartments appurtenant to the stage and other parts of the building by a fire barrier with not less than a 2-hour fire-resistance rating with approved opening protectives. For stage heights of 50 feet (15 240 mm) or less, the required stage separation shall be a fire barrier with not less a 1-hour fire-resistance rating with approved opening protectives.

Section 410.5.2 Change to read as shown: (G27-03/04)

410.5.2 Separation from each other. Dressing rooms, scene docks, property rooms, workshops, storerooms and compartments appurtenant to the stage shall be separated from each other by fire barriers with not less than a 1-hour fire-resistance rating with approved opening protective.

Section 410.5.3 Opening protectives. Delete without substitution and renumber subsequent section: (G87-03/04)

Section 410.6 Change to read as shown: (F196-03/04; F197-03/04)

410.6 Automatic sprinkler system. Stages shall be equipped with an automatic fire-extinguishing system in accordance with Chapter 9. Sprinklers shall be installed under the roof and gridiron and under all catwalks and galleries over the stage. Sprinklers shall be installed in dressing rooms, performer lounges, shops and storerooms accessory to such stages.

Exceptions:

1. Sprinklers are not required under stage areas less than 4 feet (1219 mm) in clear height utilized exclusively for storage of tables and chairs, provided the concealed space is separated from the adjacent spaces by not less than $^5/_8$-inch (15.9 mm) Type X gypsum board.

2. Sprinklers are not required for stages 1,000 square feet (93 m^2) or less in area and 50 feet (15 240 mm) or less in height where curtains, scenery or other combustible hangings are not retractable vertically. Combustible hangings shall be limited to a single main curtain, borders, legs and a single backdrop.

3. Sprinklers are not required within portable orchestra enclosures on stages.

Section 412.2.3 Change to read as shown: (G89-03/04)

412.2.3 Floor surface. Floors shall be graded and drained to prevent water or fuel from remaining on the floor. Floor drains shall discharge through an oil separator to the sewer or to an outside vented sump.

Exception: Aircraft hangars with individual lease spaces not exceeding 2,000 square feet (186 m^2) each in which servicing, repairing or washing is not conducted, and in which fuel is not dispensed, shall have floors that are graded toward the door, but shall not require a separator.

Section 412.2.4 Change to read as shown: (G90-03/04)

412.2.4 Heating equipment. Heating equipment shall be placed in another room separated by 2-hour fire-resistance-rated construction. Entrance shall be from the outside or by means of a vestibule providing a two-doorway separation.

Exceptions:

1. Unit heaters and vented infrared radiant heating equipment suspended at least 10 feet (3048 mm) above the upper surface of wings or engine enclosures of the highest aircraft that are permitted to be housed in the hangar and at least 8 feet (2438 mm) above the floor in shops, offices and other sections of the hangar communicating with storage or service areas.

2. A single interior door shall be allowed, provided the sources of ignition in the appliances are at least 18 inches (457 mm) above the floor.

Section 412.3.2 Change to read as shown: (G50-03/04)

412.3.2 Fire separation. A hangar shall not be attached to a dwelling unless separated by a fire barrier having a fire-resistance rating of not less than 1 hour. Such separation shall be continuous from the foundation to the underside of the roof and unpierced except for doors leading to the dwelling unit. Doors into the dwelling unit must be equipped with self-closing devices and conform to the requirements of Section 715 with at least a 4-inch (102 mm) noncombustible raised sill. Openings from a hanger directly into a room used for sleeping purposes shall not be permitted.

Section 412.5.3 Change to read as shown: (G91-03/04)

412.5.3 Size. The landing area for helicopters less than 3,500 pounds (1588 kg) shall be a minimum of 20 feet (6096 mm) in length and width. The landing area shall be surrounded on all sides by a clear area having a minimum average width at roof level of 15 feet (4572 mm) but with no width less than 5 feet (1524 mm).

Section 412.5.5 Change to read as shown: (G92-03/04)

412.5.5 Means of egress. The means of egress from heliports and helistops shall comply with the provisions of Chapter 10. Landing areas located on buildings or structures shall have two or more means of egress. For landing areas less than 60 feet (18 288 mm) in length, or less than 2,000 square feet (186 m^2) in area, the second means of egress may be a fire escape or ladder leading to the floor below.

2004 SUPPLEMENT TO THE IBC

Section 414.2.1 Change to read as shown: (G94-03/04)

414.2.1 Construction requirements. Control areas shall be separated from each other by fire barriers constructed in accordance with Section 706.

Table 414.2.2 Change table heading to read as shown and delete current footnote b: (F165-03/04)

TABLE 414.2.2
DESIGN AND NUMBER OF CONTROL AREAS

Floor Level	Percentage of the Maximum Allowable Quantity Per Control Area[a]	Number of Control Areas Per Floor	Fire-resistance Rating for Fire Barriers in Hours [b]

(Portions of table not shown do not change)

a. (No change to current text)
b. Fire barriers shall include walls and floors as necessary to provide separation from other portions of the building.

Section 414.2.3 Change to read as shown: (G94-03/04)

414.2.3 Fire-resistance rating requirements. The required fire-resistance rating for fire barrier assemblies shall be in accordance with Table 414.2.2. The floor construction of the control area, and the construction supporting the floor of the control area, shall have a minimum 2-hour fire-resistance rating.

Sections 414.6.1, 414.6.1.1, 414.6.1.2 and 414.6.1.3 Change to read as shown: (F198-03/04)

414.6.1 Weather protection. Where weather protection is provided for sheltering outdoor hazardous material storage or use areas, such storage or use shall be considered outdoor storage or use when the weather protection structure complies with Sections 414.6.1.1 through 414.6.1.3.

414.6.1.1 Walls. Walls shall not obstruct more than one side of the structure.

> **Exception:** Walls shall be permitted to obstruct portions of multiple sides of the structure provided that the obstructed area does not exceed 25 percent of the structure's perimeter.

414.6.1.2 Separation distance. The distance from the structure to buildings, lot lines, public ways or means of egress to a public way shall not be less than the distance required for an outside hazardous material storage or use area without weather protection.

414.6.1.3 Noncombustible construction. The overhead structure shall be of approved noncombustible construction with a maximum area of 1,500 square feet (140 m^2).

> **Exception:** The increases permitted by Section 506 apply.

Section 415.2 Add new definition to read as shown: (F183-03/04)

PHYSIOLOGICAL WARNING THRESHOLD LEVEL. A concentration of airborne contaminants, normally expressed in parts per million (ppm) or milligrams per cubic meter, that represents the concentration at which persons can sense the presence of the contaminant due to odor, irritation or other quick-acting physiological response. When used in conjunction with the Permissible Exposure Limit (PEL) the physiological warning threshold levels are those consistent with the classification system used to establish the PEL. See the definition of Permissible Exposure Limit (PEL).

Section 415.3.1 Change to read as shown: (F199-03/04)

415.3.1 Group H occupancy minimum fire separation distance. Regardless of any other provisions, buildings containing Group H occupancies shall be set back a minimum fire separation distance as set forth in Items 1 through 4 below. Distances shall be measured from the walls enclosing the occupancy to lot lines, including those on a public way. Distances to assumed lot lines established for the purposes of determination of exterior wall and opening protection are not to be used to establish the minimum fire separation distance for buildings on sites where explosives are manufactured or used when separation is provided in accordance with the quantity distance tables specified for explosive materials in the *International Fire Code*.

1. Group H-1. Not less than 75 feet (22 860 mm) and not less than required by the *International Fire Code*.

   Exceptions:

   1. Fireworks manufacturing buildings separated in accordance with NFPA 1124.

   2. Buildings containing the following materials when separated in accordance with Table 415.3.1:

      2.1. Organic peroxides, unclassified detonable.

      2.2. Unstable reactive materials Class 4.

      2.3. Unstable reactive materials, Class 3 detonable.

      2.4. Detonable pyrophoric materials.

2. Group H-2. Not less than 30 feet (9144 mm) where the area of the occupancy exceeds 1,000 square feet (93 m^2) and it is not required to be located in a detached building.

3. Groups H-2 and H-3. Not less than 50 feet (15 240 mm) where a detached building is required (see Table 415.3.2).

4. Groups H-2 and H-3. Occupancies containing materials with explosive characteristics shall be separated as required by the *International Fire Code*. Where separations are not specified, the distances required shall not be less than the distances required by Table 415.3.1.

Section 415.6 Change to read as shown: (F132-03/04)

415.6 Smoke and heat venting. Smoke and heat vents complying with Section 910 shall be installed in occupancies classified as Group H-2 or H-3, having more than 15,000 square feet (1394 m^2) in single floor area.

> **Exception:** Buildings of noncombustible construction containing only noncombustible materials.

Section 415.7.1.2 Change to read as shown: (G27-03/04)

415.7.1.2 Grinding rooms. Every room or space occupied for grinding or other operations that produce combustible dusts shall be enclosed with fire barriers that have not less than a 2-hour fire-resistance rating where the area is not more than 3,000 square feet (279 m^2), and not less than a 4-hour fire-resistance rating where the area is greater than 3,000 square feet (279 m^2).

Section 415.7.3.4.1 Change to read as shown: (G27-03/04)

415.7.3.4.1 Fire separation assemblies. Separation of the attached structures shall be provided by fire barriers having a fire-resistance rating of not less than 1 hour and shall not have openings. Fire barriers between attached structures occupied only for the storage of LP-gas are permitted to have fire doors that comply with Section 715. Such fire barriers shall be designed to withstand a static pressure of at least 100 pounds per square foot (psf) (4788 Pa), except where the building to which the structure is attached is occupied by operations or processes having a similar hazard.

Section 415.7.3.5.2 Change to read as shown: (G27-03/04)

415.7.3.5.2 Common construction. Walls and floor/ceiling assemblies common to the room and to the building within which the room is located shall be fire barriers with not less than a 1 hour fire-resistance rating and without openings. Common walls for rooms occupied only for storage of LP-gas are permitted to have opening protectives complying with Section 715. The walls and ceilings shall be designed to withstand a static pressure of at least 100 psf (4788 Pa).

> **Exception:** Where the building, within which the room is located, is occupied by operations or processes having a similar hazard.

Section 415.8.3 Change to read as shown: (G50-03/04)

415.8.3 Separation—highly toxic solids and liquids. Highly toxic solids and liquids not stored in approved hazardous materials storage cabinets shall be isolated from other hazardous materials storage by a fire barrier having a fire-resistance rating of not less than 1 hour.

Section 415.9.2.2 Change to read as shown: (E4-03/04)

415.9.2.2 Separation. Fabrication areas, whose sizes are limited by the quantity of hazardous materials allowed by Table 415.9.2.1.1, shall be separated from each other, from corridors, and from other parts of the building by not less than 1-hour fire barriers.

> **Exceptions:**
> 1. Doors within such fire barrier walls, including doors to corridors, shall be only self-closing fire assemblies having a fire-protection rating of not less than ¾ hour.
> 2. Windows between fabrication areas and corridors are permitted to be fixed glazing listed and labeled for a fire protection rating of at least ¾ hour in accordance with Section 715.

Section 415.9.2.6 Change to read as shown: (F144-03/04)

415.9.2.6 Ventilation. Mechanical exhaust ventilation shall be provided throughout the fabrication area at the rate of not less than 1 cubic foot per minute per square foot (0.044 L/S/m^2) of floor area. The exhaust air duct system of one fabrication area shall not connect to another duct system outside that fabrication area within the building.

A ventilation system shall be provided to capture and exhaust gases, fumes and vapors at workstations.

Two or more operations at a workstation shall not be connected to the same exhaust system where either one or the combination of the substances removed could constitute a fire, explosion or hazardous chemical reaction within the exhaust duct system.

Exhaust ducts penetrating occupancy separations shall be contained in a shaft of equivalent fire-resistance-rated construction. Exhaust ducts shall not penetrate fire walls.

Fire dampers shall not be installed in exhaust ducts.

Section 415.9.2.7 Change to read as shown: (E4-03/04)

415.9.2.7 Transporting hazardous production materials to fabrication areas. Hazardous production materials shall be transported to fabrication areas through enclosed piping or tubing systems that comply with Section 415.9.6.1, through service corridors complying with Section 415.9.4, or in corridors as permitted in the exception to Section 415.9.3.

2004 SUPPLEMENT TO THE IBC

The handling or transporting of hazardous production materials within service corridors shall comply with the *International Fire Code*.

Section 415.9.3 Change to read as shown: (E4-03/04)

415.9.3 Corridors. Corridors shall comply with Chapter 10 and shall be separated from fabrication areas as specified in Section 415.9.2.2. Corridors shall not contain HPM and shall not be used for transporting such materials, except through closed piping systems as provided in Section 415.9.6.3.

> **Exception:** Where existing fabrication areas are altered or modified, HPM is allowed to be transported in existing corridors, subject to the following conditions:
>
> 1. Corridors. Corridors adjacent to the fabrication area where the alteration work is to be done shall comply with Section 1016 for a length determined as follows:
>
>   1.1 The length of the common wall of the corridor and the fabrication area; and
>
>   1.2. For the distance along the corridor to the point of entry of HPM into the corridor serving that fabrication area.
>
> 2. Emergency alarm system. There shall be an emergency telephone system, a local manual alarm station or other approved alarm-initiating device within corridors at not more than 150-foot (45 720 mm) intervals and at each exit and doorway. The signal shall be relayed to an approved central, proprietary or remote station service or the emergency control station and shall also initiate a local audible alarm.
>
> 3. Pass-throughs. Self-closing doors having a fire-protection rating of not less than 1 hour shall separate pass-throughs from existing corridors. Pass-throughs shall be constructed as required for the corridors, and protected by an approved automatic fire-extinguishing system.

Section 415.9.4.2 Change to read as shown: (E4-03/04)

415.9.4.2 Use conditions. Service corridors shall be separated from corridors as required by Section 415.9.2.2. Service corridors shall not be used as a required corridor.

Section 415.9.6.3 Change to read as shown: (G50-03/04; FS53-03/04; E4-03/04)

415.9.6.3 Installations in corridors and above other occupancies. The installation of hazardous production material piping and tubing within the space defined by the walls of corridors and the floor or roof above or in concealed spaces above other occupancies shall be in accordance with Section 415.9.6.2 and the following conditions:

1. Automatic sprinklers shall be installed within the space unless the space is less than 6 inches (152 mm) in the least dimension.

2. Ventilation not less than six air changes per hour shall be provided. The space shall not be used to convey air from any other area.

3. Where the piping or tubing is used to transport HPM liquids, a receptor shall be installed below such piping or tubing. The receptor shall be designed to collect any discharge or leakage and drain it to an approved location. The 1-hour enclosure shall not be used as part of the receptor.

4. HPM supply piping and tubing and HPM nonmetallic waste lines shall be separated from the corridor and from occupancies other than Group H-5 by fire barriers that have a fire-resistance rating of not less than 1 hour. Where gypsum wallboard is used, joints on the piping side of the enclosure are not required to be taped, provided the joints occur over framing members. Access openings into the enclosure shall be protected by approved fire-protection-rated assemblies.

5. Readily accessible manual or automatic remotely activated fail-safe emergency shutoff valves shall be installed on piping and tubing other than waste lines at the following locations:

    5.1. At branch connections into the fabrication area.

    5.2. At entries into corridors.

> **Exception:** Transverse crossings of the corridors by supply piping that is enclosed within a ferrous pipe or tube for the width of corridor need not comply with Items 1 through 5.

Section 415.9.7 Change to read as shown: (F183-03/04)

415.9.7 Continuous gas detection systems. A continuous gas detection system shall be provided for HPM gases when the physiological warning threshold level of the gas is at a higher level than the accepted permissible exposure limit (PEL) for the gas and for flammable gases in accordance with this section.

Section 415.9.7.1.4 Change to read as shown: (E4-03/04)

415.9.7.1.4 Corridors. When gases are transported in piping placed within the space defined by the walls of a corridor, and the floor or roof above the corridor, a continuous gas-detection system shall be provided where piping is located and in the corridor.

> **Exception:** A continuous gas-detection system is not required for occasional transverse crossings of the corridors by supply piping that is enclosed in a ferrous pipe or tube for the width of the corridor.

Section 415.9.9 Change to read as shown: (F146-03/04)

415.9.9 Emergency control station. An emergency control station shall be provided on the premises at an approved location outside of the fabrication area, and shall be continuously staffed by trained personnel. The emergency control station shall receive signals from emergency equipment and alarm and detection systems. Such emergency equipment and alarm and detection systems shall include, but not be limited to, the following where such equipment or systems are required to be provided either in this chapter or elsewhere in this code:

1. Automatic fire sprinkler system alarm and monitoring systems.
2. Manual fire alarm systems.
3. Emergency alarm systems.
4. Continuous gas-detection systems.
5. Smoke detection systems.
6. Emergency power system.
7. Automatic detection and alarm systems for pyrophoric liquids and Class 3 water reactive liquids required in Section 1805.2.2.4 of the *International Fire Code*.
8. Exhaust ventilation flow alarm devices for pyrophoric liquids and Class 3 water reactive liquids cabinet exhaust ventilation systems required in Section 1805.2.2.4 of the *International Fire Code*.

Section 415.9.10.1 Change to read as shown: (F146-03/04)

415.9.10.1 Required electrical systems. Emergency power shall be provided for electrically operated equipment and connected control circuits for the following systems:

1. HPM exhaust ventilation systems.
2. HPM gas cabinet ventilation systems.
3. HPM exhausted enclosure ventilation systems.
4. HPM gas room ventilation systems.
5. HPM gas detection systems.
6. Emergency alarm systems.
7. Manual fire alarm systems.
8. Automatic sprinkler system monitoring and alarm systems.
9. Automatic alarm and detection systems for pyrophoric liquids and Class 3 water reactive liquids required in Section 1805.2.2.4.
10. Flow alarm switches for pyrophoric liquids and Class 3 water reactive liquids cabinet exhaust ventilation systems required in Section 1805.2.2.4.
11. Electrically operated systems required elsewhere in this code or in the *International Fire Code* applicable to the use, storage or handling of HPM.

Sections 415.9.11.1 and 415.9.11.2 Change to read as shown: (F144-03/04)

415.9.11.1 Exhaust ducts for HPM. An approved automatic sprinkler system shall be provided in exhaust ducts conveying gases, vapors, fumes, mists or dusts generated from HPM in accordance with this section and the *International Mechanical Code*.

415.9.11.2 Metallic and noncombustible nonmetallic exhaust ducts. An approved automatic sprinkler system shall be provided in metallic and noncombustible nonmetallic exhaust ducts when all of the following conditions apply:

1. Where the largest cross-sectional diameter is equal to or greater than 10 inches (254 mm).
2. The ducts are within the building.
3. The ducts are conveying flammable gases, vapors or fumes.

Section 416.2 Change to read as shown: (G27-03/04)

416.2 Spray rooms. Spray rooms shall be enclosed with fire barriers with not less than a 1-hour fire-resistance rating. Floors shall be waterproofed and drained in an approved manner.

Section 416.3 Change to read as shown: (F142-03/04)

416.3 Spraying areas. Spraying areas shall be ventilated with an exhaust system to prevent the accumulation of flammable mist or vapors in accordance with the *International Mechanical Code*. Where such areas are not separately enclosed, noncombustible spray curtains shall be provided to restrict the spread of flammable vapors.

Section 416.3.1 Change to read as shown: (F142-03/04)

416.3.1 Surfaces. The interior surfaces of spraying areas shall be smooth and continuous without edges, and shall be so constructed to permit the free passage of exhaust air from all parts of the interior and to facilitate washing and cleaning, and shall be so designed to confine residues within the spraying area. Aluminum shall not be used.

2004 SUPPLEMENT TO THE IBC

Section 416.4 Change to read as shown: (F142-03/04)

416.4 Fire protection. An automatic fire-extinguishing system shall be provided in all spray, dip and immersing areas and storage rooms, and shall be installed in accordance with Chapter 9.

CHAPTER 5
GENERAL BUILDING HEIGHTS AND AREAS

Section 503.2 Party walls. Delete and relocate to Section 705.1.1: (FS24-03/04)

Table 503 Change to read as shown: (G97-03/04)

TABLE 503
ALLOWABLE HEIGHT AND BUILDING AREAS

Height limitations shown as stories and feet above grade plane. Area limitations as determined by the definition of "Area, building", per floor.

		Type of Construction	
		Type 1	
		A	B
Group	Hgt (ft) Hgt (S)	UL	160
H-5	S A	4 UL	4 UL

(Portions of table not shown remain unchanged)

Section 503.1.1 Basements. Delete without substitution and renumber subsequent subsections: (G98-03/04)

Section 504 title and Section 504.1 Change to read as shown: (G102-03/04)

SECTION 504
HEIGHT

504.1 General. The height permitted by Table 503 shall be increased in accordance with this section.

> **Exception:** The height of one-story aircraft hangars, aircraft paint hangars and buildings used for the manufacturing of aircraft shall not be limited if the building is provided with an automatic fire-extinguishing system in accordance with Chapter 9 and is entirely surrounded by public ways or yards not less in width than one and one-half times the height of the building.

Section 506.1 Delete current equations 5-1 and 5-2 and substitute as shown: (G104/03/04)

506.1 General, 506.2 Frontage increase.

For MULTI-STORY BUILDINGS

$A_a = \{A_t + [A_t \times I_f] + [A_t \times 2]\}$ (Equation 5 – 1 ms)

Where,

$I_f = [F / P - 0.25] W / 30$ (Equation 5 – 2 ms)

For SINGLE-STORY BUILDINGS

$A_a = A_t + [A_t \times I_f] + [A_t \times 3]$ (Equation 5-1 ss)

Where,

$I_f = [F / P - 0.25] W / 30$ (Equation 5-2 ss)

Section 506.4.1 Add new section to read as shown: (G106-03/04)

506.4.1 Mixed Occupancies: In buildings of mixed occupancy, the allowable area per floor (A_a) shall be based on the most restrictive provisions for each occupancy when the mixed occupancies are treated according to 302.3.1. When the occupancies are treated according to section 302.3.2 as separated occupancies, the maximum total floor area for a building shall be such that the sum of the ratios for each such area on all floors as calculated according to section 302.3.2 shall not exceed 2 for two story buildings and 3 for buildings three stories or higher.

Section 507.1 Add new section to read as shown: (G107-03/04)

507.1 General. The area of buildings of the occupancies and configurations specified herein shall not be limited.

(Renumber subsequent sections)

Section 507.1.10 Add new section to read as shown: (G107-03/04)

507.1.10 Covered mall buildings and anchor stores. The area of covered mall buildings and anchor stores not exceeding three stories in height that comply with Section 402.6 shall not be limited.

Section 507.2 Change to read as shown: (G112-03/04)

507.2 Sprinklered, one story. The area of a one-story, Group B, F, M or S building or a one-story Group A-4 building, of other than Type V construction shall not be limited when the building is provided with an automatic sprinkler system throughout in accordance with Section 903.3.1.1, and is surrounded and adjoined by public ways or yards not less than 60 feet (18 288 mm) in width.

> **Exceptions:**
> 1. Buildings and structures of Type I and II construction for rack storage facilities which do not

have access by the public shall not be limited in height provided that such buildings conform to the requirements of Section 507.1, Section 903.3.1.1 and NFPA 230.

2. The automatic sprinkler system shall not be required in areas occupied for indoor participant sports, such as tennis, skating, swimming and equestrian activities, in occupancies in Group A-4, provided that:

   2.1. Exit doors directly to the outside are provided for occupants of the participant sports areas, and

   2.2. The building is equipped with a fire alarm system with manual fire alarm boxes installed in accordance with Section 907.

Section 507.4 Change to read as shown: (G116-03/04; FS53-03/04)

507.4 Reduced open space. The permanent open space of 60 feet (18 288 mm) required in Sections 507.1, 507.2, 507.3, 507.5, and 507.9 shall be permitted to be reduced to not less than 40 feet (12 192 mm) provided the following requirements are met:

1. The reduced open space shall not be allowed for more than 75 percent of the perimeter of the building.

2. The exterior wall facing the reduced open space shall have a minimum fire-resistance rating of 3 hours.

3. Openings in the exterior wall, facing the reduced open space, shall have opening protectives with a minimum fire-protection rating of 3 hours.

Section 507.6 Change to read as shown: (G119-03/04; G120-03/04)

507.6 Group H occupancies. Group H-2, H-3 and H-4 occupancies shall be permitted in unlimited area buildings containing Groups F and S occupancies, in accordance with Sections 507.2 and 507.3 and the limitations of this section. The aggregate floor area of the Group H occupancies located at the perimeter of the unlimited area building shall not exceed 10 percent of the area of the building nor the area limitations for the Group H occupancies as specified in Table 503 as modified by Section 506.2, based upon the percentage of the perimeter of each Group H fire area that fronts on a street or other unoccupied space. The aggregate floor area of Group H occupancies not located at the perimeter of the building shall not exceed 25 percent of the area limitations for the Group H occupancies as specified in Table 503. Group H fire areas shall be separated from the rest of the unlimited area building and from each other in accordance with Table 302.3.2. For two-story unlimited area buildings the Group H fire areas shall not be located above the first story unless permitted by the allowable height in stories and feet as set forth in Table 503 based on the type of construction of the unlimited area building.

Section 508.2 Change to read as shown: (G121-03/04; G122-03/04; G123-03/04)

508.2 Group S-2 enclosed or open parking garage with Group A, B, M, R or S above. A basement and/or the first story above grade plane of a building shall be considered as a separate and distinct building for the purpose of determining area limitations, continuity of fire walls, limitation of number of stories and type of construction, when all of the following conditions are met:

1. The basement and/or the first story above grade plane is of Type IA construction and is separated from the building above with a horizontal assembly having a minimum 3-hour fire-resistance rating.

2. Shaft, stairway, ramp or escalator enclosures through the horizontal assembly shall have not less than a 2-hour fire-resistance rating with opening protectives in accordance with Table 715.3.

   Exception: Where the enclosure walls below the horizontal assembly have not less than a 3-hour fire-resistance rating with opening protectives in accordance with Table 715.3, the enclosure walls extending above the horizontal assembly shall be permitted to have a 1-hour fire-resistance rating provided:

   1. The building above the horizontal assembly is not required to be of Type I construction;

   2. The enclosure connects less than four stories, and

   3. The enclosure opening protectives above the horizontal assembly have a minimum 1-hour fire-protection rating.

3. The building above the horizontal assembly contains only Group A having an assembly room with an occupant load of less than 300, or Group B, M, R or S; and

4. The building below the horizontal assembly is a Group S-2 enclosed or open parking garage, used for the parking and storage of private motor vehicles.

   Exceptions:

   1. Entry lobbies, mechanical rooms and similar uses incidental to the operation of the building shall be permitted.

2. Group A having assembly rooms having an aggregate occupant load of less than 300 or Group B or M shall be permitted in addition to those uses incidental to the operation of the building (including storage areas), provided that the entire structure below the horizontal assembly is protected throughout by an approved automatic sprinkler system.

3. The maximum building height in feet shall not exceed the limits set forth in Table 503 for the least restrictive type of construction involved.

Section 508.7.1 Change to read as shown: (G50-03/04)

508.7.1 Fire separation. Fire barriers between the parking occupancy and the upper occupancy shall correspond to the required fire-resistance rating prescribed in Table 302.3.2 for the uses involved. The type of construction shall apply to each occupancy individually, except that structural members, including main bracing within the open parking structure, which is necessary to support the upper occupancy, shall be protected with the more restrictive fire-resistance-rated assemblies of the groups involved as shown in Table 601. Means of egress for the upper occupancy shall conform to Chapter 10 and shall be separated from the parking occupancy by fire barriers having at least a 2-hour fire-resistance rating as required by Section 706, with self-closing doors complying with Section 715. Means of egress from the open parking garage shall comply with Section 406.3.

CHAPTER 6
TYPES OF CONSTRUCTION

Table 602 Change first column to read as shown: (G127-03/04)

TABLE 602
FIRE- RESISTANCE RATING REQUIREMENTS FOR EXTERIOR WALLS BASED ON FIRE SEPARATION DISTANCE [a]

FIRE SEPARATION DISTANCE = X (Feet)
X < 5 [c]
5 ≤ X < 10
10 ≤ X < 30
X ≥ 30

(Portions of table not shown, including footnotes, do not change)

Section 602.2 Change to read as shown: (G129-03/04)

602.2 Types I and II. Types I and II construction are those types of construction in which the building elements listed in Table 601 are of noncombustible materials, except as permitted in Section 603 and elsewhere in this code.

Section 602.4 Change to read as shown: (G130-03/04)

602.4 Type IV. Type IV construction (Heavy Timber, HT) is that type of construction in which the exterior walls are of noncombustible materials and the interior building elements are of solid or laminated wood without concealed spaces. The details of Type IV construction shall comply with the provisions of this section. Fire-retardant-treated wood framing complying with Section 2303.2 shall be permitted within exterior wall assemblies with a 2-hour rating or less. Minimum solid sawn nominal dimensions are required for structures built using Type IV construction (Heavy Timber). For Glued Laminated members the equivalent net finished width and depths corresponding to the minimum nominal width and depths of solid sawn lumber are required as specified in Table 602.4.

Table 602.4 Add new table to read as shown: (G130-03/04)

TABLE 602.4
WOOD MEMBER SIZE

Minimum Nominal Solid Sawn Size		Minimum Glued Laminated Net Size	
Width, in.	Depth, in.	Width, in.	Depth, in.
8	8	6¾	8¼
6	10	5	10½
6	8	5	8¼
6	6	5	6
4	6	3	6⅞

For SI: 1 inch = 25.4 mm

Section 603.1 Change to read as shown: (FS29-03/04; FS93-03/04)

603.1 Allowable materials. Combustible materials shall be permitted in buildings of Type I or II construction in the following applications and in accordance with Sections 603.1.1 through 603.1.3:

1. Fire-retardant-treated wood shall be permitted in:

    1.1. Nonbearing partitions where the required fire-resistance rating is 2 hours or less.

    1.2. Nonbearing exterior walls where no fire rating is required.

    1.3. Roof construction as permitted in Table 601, Note c, Item 3.

2. Thermal and acoustical insulation, other than foam plastics, having a flame spread index of not more than 25.

   Exceptions:

   1. Insulation placed between two layers of noncombustible materials without an intervening airspace shall be allowed to have a flame spread index of not more than 100.

   2. Insulation installed between a finished floor and solid decking without intervening airspace shall be allowed to have a flame spread index of not more than 200.

3. Foam plastics in accordance with Chapter 26.

4. Roof coverings that have an A, B or C classification.

5. Interior floor finish and interior finish, trim and millwork such as doors, door frames, window sashes and frames.

6. Where not installed over 15 feet (4572 mm) above grade, show windows, nailing or furring strips, wooden bulkheads below show windows, their frames, aprons and show cases.

7. Finished flooring applied directly to the floor slab or to wood sleepers that are fireblocked in accordance with Section 717.2.7.

8. Partitions dividing portions of stores, offices or similar places occupied by one tenant only and which do not establish a corridor serving an occupant load of 30 or more shall be permitted to be constructed of fire-retardant-treated wood, 1-hour fire-resistance-rated construction or of wood panels or similar light construction up to 6 feet (1829 mm) in height.

9. Platforms as permitted in Section 410.

10. Combustible exterior wall coverings, balconies, bay or oriel windows, or similar appendages in accordance with Chapter 14.

11. Blocking such as for handrails, millwork, cabinets, and window and door frames.

12. Light-transmitting plastics as permitted by Chapter 26.

13. Mastics and caulking materials applied to provide flexible seals between components of exterior wall construction.

14. Exterior plastic veneer installed in accordance with Section 2605.2.

15. Nailing or furring strips as permitted by Section 803.4.

16. Heavy timber as permitted by Note c, Item 2, to Table 601 and Sections 602.4.7 and 1406.3.

17. Aggregates, component materials and admixtures as permitted by Section 703.2.2.

18. Sprayed cementitious and mineral fiber fire-resistance-rated materials installed to comply with Section 1704.11.

19. Materials used to protect penetrations in fire-resistance-rated assemblies in accordance with Section 712.

20. Materials used to protect joints in fire-resistance-rated assemblies in accordance with Section 713.

21. Materials allowed in the concealed spaces of buildings of Type I and II construction in accordance with Section 717.5.

22. Materials exposed within plenums complying with Section 602 of the *International Mechanical Code*.

CHAPTER 7
FIRE-RESISTANCE RATED CONSTRUCTION

Section 702.1 Change the following definitions to read as shown: (FS2-03/04)

COMBINATION FIRE/SMOKE DAMPER. A listed device installed in ducts and air transfer openings designed to close automatically upon the detection of heat and to also resist the passage of air and smoke. The device is installed to operate automatically, controlled by a smoke detection system, and where required, is capable of being positioned from a fire command center.

SMOKE DAMPER. A listed device installed in ducts and air transfer openings that is designed to resist the passage of air and smoke. The device is installed to operate automatically, controlled by a smoke detection system, and where required, is capable of being positioned from a fire command center.

Table 704.8 Change to read as shown: (FS12-03/04)

TABLE 704.8
MAXIMUM AREA OF EXTERIOR WALL OPENINGS [a]

Classification of Opening	Fire Separation Distance (feet)		
	0 to 3[e,h]	Greater than 3 to 5[b,f]	Greater than 5 to 10[b,d,f]
Unprotected	Not Permitted	Not Permitted [b]	10%
Protected	Not Permitted	15%	25%

2004 SUPPLEMENT TO THE IBC

(Portions of table, including footnotes, not shown do not change)

Section 704.9 Change to read as shown: (FS15-03/04)

704.9 Vertical separation of openings. Openings in exterior walls in adjacent stories shall be separated vertically to protect against fire spread on the exterior of the buildings where the openings are within 5 feet (1524 mm) of each other horizontally and the opening in the lower story is not a protected opening with a fire protection rating of not less than ¾ hour. Such openings shall be separated vertically at least 3 feet (914 mm) by spandrel girders, exterior walls or other similar assemblies that have a fire-resistance rating of at least 1 hour or by flame barriers that extend horizontally at least 30 inches (762 mm) beyond the exterior wall. Flame barriers shall also have a fire-resistance rating of at least 1 hour. The unexposed surface temperature limitations specified in ASTM E 119 shall not apply to the flame barriers or vertical separation unless otherwise required by the provisions of this code.

Exceptions:

1. This section shall not apply to buildings that are three stories or less in height.
2. This section shall not apply to buildings equipped throughout with an automatic sprinkler system in accordance with Section 903.3.1.1 or 903.3.1.2.
3. Open parking garages.

Section 704.10 Change to read as shown: (FS15-03/04)

704.10 Vertical exposure. For buildings on the same lot, opening protectives having a fire-protection rating of not less than ¾ hour shall be provided in every opening that is less than 15 feet (4572 mm) vertically above the roof of an adjoining building or adjacent structure that is within a horizontal fire separation distance of 15 feet (4572 mm) of the wall in which the opening is located.

Exception: Opening protectives are not required where the roof construction has a fire-resistance rating of not less than 1 hour for a minimum distance of 10 feet (3048 mm) from the adjoining building and the entire length and span of the supporting elements for the fire-resistance-rated roof assembly has a fire-resistance rating of not less than 1 hour.

Section 704.11 Change to read as shown: (FS19-03/04)

704.11 Parapets. Parapets shall be provided on exterior walls of buildings.

Exceptions: A parapet need not be provided on an exterior wall where any of the following conditions exist:

1. The wall is not required to be fire-resistance rated in accordance with Table 602 because of fire separation distance.

2. The building has an area of not more than 1,000 square feet (93 m^2) on any floor.

3. Walls that terminate at roofs of not less than 2-hour fire-resistance-rated construction or where the roof, including the deck or slab and supporting construction, is constructed entirely of noncombustible materials.

4. One-hour fire-resistance-rated exterior walls that terminate at the underside of the roof sheathing, deck or slab, provided:

    4.1. Where the roof/ceiling framing elements are parallel to the walls, such framing and elements supporting such framing shall not be of less than 1-hour fire-resistance-rated construction for a width of 4 feet (1220 mm) measured from the interior side of the wall for Groups R and U and 10 feet (3048 mm) for other occupancies.

    4.2. Where roof/ceiling framing elements are not parallel to the wall, the entire span of such framing and elements supporting such framing shall not be of less than 1-hour fire-resistance-rated construction.

    4.3. Openings in the roof shall not be located within 5 feet (1524 mm) of the 1-hour fire-resistance-rated exterior wall for Groups R and U and 10 feet (3048 mm) for other occupancies.

    4.4. The entire building shall be provided with not less than a Class B roof covering.

5. In occupancies of Groups R-2 and R-3 as applicable in Section 101.2, both provided with a Class C roof covering, the exterior wall shall be permitted to terminate at the roof sheathing or deck in Type III, IV and V construction provided:

    5.1. The roof sheathing or deck is constructed of approved noncombustible materials or of fire-retardant-treated wood, for a distance of 4 feet (1220 mm); or

    5.2. The roof is protected with 0.625-inch (15.88 mm) Type X gypsum board directly beneath the underside of the roof sheathing or deck, supported by a minimum of nominal 2-inch (51 mm) ledgers attached to the sides of the roof framing members, for a minimum distance of 4 feet (1220 mm).

6. Where the wall is permitted to have at least 25 percent of the exterior wall areas containing unprotected openings based on fire separation distance as determined in accordance with Section 704.8.

Section 704.12 Change to read as shown: (FS15-03/04; FS21-03/04)

704.12 Opening protection. Windows in exterior walls required to have protected openings in accordance with other sections of this code or determined to be protected in accordance with Section 704.3 or 704.8 shall comply with Section 715.4. Other openings required to be protected with fire door or shutter assemblies in accordance with other sections of this code or determined to be protected in accordance with Section 704.3 or 704.8 shall comply with Section 715.3.

Exception: Opening protectives are not required where the building is protected throughout by an automatic sprinkler system and the exterior openings are protected by an approved water curtain using automatic sprinklers approved for that use. The sprinklers and the water curtain shall be installed in accordance with NFPA 13 and shall have an automatic water supply and fire department connection.

Section 705.1 Change to read as shown: (FS24-03/04)

705.1 General. Each portion of a building separated by one or more fire walls that comply with the provisions of this section shall be considered a separate building. The extent and location of such fire walls shall provide a complete separation. Where a fire wall also separates occupancies that are required to be separated by a fire barrier wall, the most restrictive requirements of each separation shall apply.

Section 705.1.1 Add new section to read as shown (Relocated from Section 503.2): (FS24-03/04)

705.1.1 Party walls. Any wall located on a lot line between adjacent buildings, which is used or adapted for joint service between the two buildings, shall be constructed as a fire wall in accordance with Section 705. Party walls shall be constructed without openings and shall create separate buildings.

Section 705.6.1 Change to read as shown: (FS15-03/04)

705.6.1 Stepped buildings. Where a fire wall serves as an exterior wall for a building and separates buildings having different roof levels, such wall shall terminate at a point not less than 30 inches (762 mm) above the lower roof level, provided the exterior wall for a height of 15 feet (4572 mm) above the lower roof is not less than 1-hour fire-resistance-rated construction from both sides with openings protected by fire assemblies having a fire protection rating of not less than ¾ hour.

Exceptions: Where the fire wall terminates at the underside of the roof sheathing, deck or slab of the lower roof, provided:

1. The lower roof assembly within 10 feet (3048 mm) of the wall has not less than a 1-hour fire-resistance rating and the entire length and span of supporting elements for the rated roof assembly has a fire-resistance rating of not less than 1 hour.

2. Openings in the lower roof shall not be located within 10 feet (3048 mm) of the fire wall.

Section 706.1 Change to read as shown: (G94-03/04)

706.1 General. Fire barriers used for separation of shafts, exit enclosures, exit passageways, horizontal exits, incidental use areas, control areas, to separate different occupancies, to separate a single occupancy into different fire areas, or to separate other areas where a fire barrier is required elsewhere in this code or the *International Fire Code*, shall comply with this section.

Section 706.2.1 Add new text to read as shown: (FS28-03/04)

706.2.1 Fire-resistance-rated glazing. Fire-resistance-rated glazing when tested in accordance with ASTM E119 and complying with the requirements of Section 706 shall be permitted. Fire-resistance-rated glazing shall bear a label or other identification showing the name of the manufacturer, the test standard and the identifier "W-XXX" where the "XXX" is the fire-resistance rating in minutes. Such label or identification shall be issued by an approved agency and shall be permanently affixed.

Section 706.3.6 Add new section to read as shown and renumber subsequent sections: (G94-03/04)

706.3.6 Control areas. Fire barriers separating control areas shall have a fire-resistance rating of not less than that required in Section 414.2.3.

Section 706.4 Change to read as shown: (FS19-03/04; FS29-03/04)

706.4 Continuity of fire barrier walls. Fire barrier walls shall extend from the top of the floor/ceiling assembly below to the underside of the floor or roof sheathing, slab or deck above and shall be securely attached thereto. These walls shall be continuous through concealed spaces such as the space above a suspended ceiling. The supporting construction for fire barrier walls shall be protected to afford the required fire-resistance rating of the fire barrier supported except for 1-hour fire-resistance-rated incidental use area separations as required by Table 302.1.1 in buildings of Type IIB, IIIB and

VB construction. Hollow vertical spaces within the fire barrier wall shall be fireblocked in accordance with Section 717.2 at every floor level.

Exceptions:

1. The maximum required fire-resistance rating for assemblies supporting fire barriers separating tank storage as provided for in Section 415.7.2.1 shall be 2 hours, but not less than required by Table 601 for the building construction type.

2. Shaft enclosure shall be permitted to terminate at a top enclosure complying with Section 707.12.

Section 706.6 Change to read as shown: (FS30-03/04)

706.6 Exterior walls. Where exterior walls serve as a part of a required fire-resistance-rated enclosure or separation, such walls shall comply with the requirements of Section 704 for exterior walls and the fire-resistance-rated enclosure or separation requirements shall not apply.

Exception: Exterior walls required to be fire-resistance rated in accordance with Section 1013.5.1 for exterior egress balconies, Section 1019.1.4 for exit enclosures and Section 1022.6 for exterior exit ramps and stairways.

Section 706.7 Change to read as shown: (FS31-03/04)

706.7 Openings. Openings in a fire barrier wall shall be protected in accordance with Section 715. Openings shall be limited to a maximum aggregate width of 25 percent of the length of the wall, and the maximum area of any single opening shall not exceed 120 square feet (11 m^2). Openings in exit enclosures and exit passageways shall also comply with Section 1019.1.1 and 1020.4, respectively.

Exceptions:

1. Openings shall not be limited to 120 square feet (11 m^2) where adjoining fire areas are equipped throughout with an automatic sprinkler system in accordance with Section 903.3.1.1.

2. Fire doors serving an exit enclosure.

3. Openings shall not be limited to 120 square feet (11 m^2) or an aggregate width of 25 percent of the length of the wall where the opening protective assembly has been tested in accordance with ASTM E119 and has a minimum fire-resistance rating not less than the fire-resistance rating of the wall.

Section 706.8.1 Change to read as shown: (FS31-03/04)

706.8.1 Prohibited penetrations. Penetrations into an exit enclosure or an exit passageway shall be allowed only when permitted by Section 1019.1.2 or 1020.5, respectively.

Section 707.2 Change to read as shown: (FS32-03/04; FS33-03/04)

707.2 Shaft enclosure required. Openings through a floor/ceiling assembly shall be protected by a shaft enclosure complying with this section.

Exceptions:

1. A shaft enclosure is not required for openings totally within an individual residential dwelling unit and connecting four stories or less.

2. A shaft enclosure is not required in a building equipped throughout with an automatic sprinkler system in accordance with Section 903.3.1.1 for an escalator opening or stairway that is not a portion of the means of egress protected according to Item 2.1 or 2.2:

    2.1. Where the area of the floor opening between stories does not exceed twice the horizontal projected area of the escalator or stairway and the opening is protected by a draft curtain and closely spaced sprinklers in accordance with NFPA 13. In other than Groups B and M, this application is limited to openings that do not connect more than four stories.

    2.2. Where the opening is protected by approved power-operated automatic shutters at every floor penetrated. The shutters shall be of noncombustible construction and have a fire-resistance rating of not less than 1.5 hours. The shutter shall be so constructed as to close immediately upon the actuation of a smoke detector installed in accordance with Section 907.11 and shall completely shut off the well opening. Escalators shall cease operation when the shutter begins to close. The shutter shall operate at a speed of not more than 30 feet per minute (152.4 mm/s) and shall be equipped with a sensitive leading edge to arrest its progress where in contact with any obstacle, and to continue its progress on release therefrom.

3. A shaft enclosure is not required for penetrations by pipe, tube, conduit, wire, cable, and vents protected in accordance with Section 712.4.

4. A shaft enclosure is not required for penetrations by ducts protected in accordance with Section 712.4. Grease ducts shall be protected in accordance with the *International Mechanical Code*.

5. A shaft enclosure is not required for floor openings complying with the provisions for covered malls or atriums.

6. A shaft enclosure is not required for approved masonry chimneys, where annular space protection is provided at each floor level in accordance with Section 717.2.5.

7. In other than Groups I-2 and I-3, a shaft enclosure is not required for a floor opening or an air transfer opening that complies with the following:

    7.1. Does not connect more than two stories.

    7.2. Is not part of the required means of egress system except as permitted in Section 1019.1.

    7.3. Is not concealed within the building construction.

    7.4. Is not open to a corridor in Group I and R occupancies.

    7.5. Is not open to a corridor on nonsprinklered floors in any occupancy.

    7.6. Is separated from floor openings and air transfer openings serving other floors by construction conforming to required shaft enclosures.

    7.7. Is limited to the same smoke compartment.

8. A shaft enclosure is not required for automobile ramps in open parking garages and enclosed parking garages constructed in accordance with Sections 406.3 and 406.4, respectively.

9. A shaft enclosure is not required for floor openings between a mezzanine and the floor below.

10. A shaft enclosure is not required for joints protected by a fire-resistant joint system in accordance with Section 713.

11. Where permitted by other sections of this code.

12. A shaft enclosure shall not be required for floor openings created by unenclosed stairs or ramps in accordance with exception 8 or 9 in Section 1019.1.

Section 707.5 Change to read as shown: (FS19-03/04; FS29-03/04)

707.5 Continuity. Shaft enclosure walls shall extend from the top of the floor/ceiling assembly below to the underside of the floor or roof sheathing, slab or deck above and shall be securely attached thereto. These walls shall be continuous through concealed spaces such as the space above a suspended ceiling. The supporting construction shall be protected to afford the required fire-resistance rating of the element supported. Hollow vertical spaces within the shaft enclosure construction wall shall be fireblocked in accordance with Section 717.2 at every floor level.

Section 707.11 Change to read as shown: (FS40-03/04)

707.11 Enclosure at the bottom. Shafts that do not extend to the bottom of the building or structure shall:

1. Be enclosed at the lowest level with construction of the same fire-resistance rating as the lowest floor through which the shaft passes, but not less than the rating required for the shaft enclosure;

2. Terminate in a room having a use related to the purpose of the shaft. The room shall be separated from the remainder of the building by a fire barrier having a fire-resistance rating and opening protectives at least equal to the protection required for the shaft enclosure; or

3. Be protected by approved fire dampers installed in accordance with their listing at the lowest floor level within the shaft enclosure.

Exceptions:

1. The fire-resistance-rated room separation is not required provided there are no openings in or penetrations of the shaft enclosure to the interior of the building except at the bottom. The bottom of the shaft shall be closed off around the penetrating items with materials permitted by Section 717.3.1 for draftstopping, or the room shall be provided with an approved automatic fire suppression system.

2. A shaft enclosure containing a refuse chute or laundry chute shall not be used for any other purpose and shall terminate in a room protected in accordance with Section 707.13.4.

3. The fire-resistance-rated room separation and the protection at the bottom of the shaft are not required provided there are no combustibles in the shaft and there are no openings or other penetrations through the shaft enclosure to the interior of the building.

Section 707.12 Change to read as shown: (FS19-03/04)

707.12 Enclosure at the top. A shaft enclosure that does not extend to the underside of the roof sheathing, deck or slab of the building shall be enclosed at the top with construction of the same fire-resistance rating as the topmost floor penetrated by the shaft, but not less than the fire-resistance rating required for the shaft enclosure.

2004 SUPPLEMENT TO THE IBC

Section 707.13.1 Change to read as shown: (E3-03/04)

707.13.1 Refuse and laundry chute enclosures. A shaft enclosure containing a refuse or laundry chute shall not be used for any other purpose and shall be enclosed in accordance with Section 707.4. Openings into the shaft, including those from access rooms and termination rooms, shall be protected in accordance with this section and Section 715. Openings into chutes shall not be located in corridors. Opening protectives shall be self-closing or automatic-closing upon the actuation of a smoke detector installed in accordance with Section 907.10, except that heat-activated closing devices shall be permitted between the shaft and the termination room.

Section 707.13.3 Change to read as shown: (FS40-03/04)

707.13.3 Refuse and laundry chute access rooms. Access openings for refuse and laundry chutes shall be located in rooms or compartments enclosed by a fire barrier that has a fire-resistance rating of not less than 1 hour. Openings into the access rooms shall be protected by opening protectives having a fire protection rating of not less than ¾ hour and shall be self-closing or automatic-closing upon the detection of smoke.

Section 707.13.4 Change to read as shown: (FS40-03/04)

707.13.4 Termination room. Refuse and laundry chutes shall discharge into an enclosed room separated from the remainder of the building by a fire barrier that has a fire-resistance rating of not less than 1 hour. Openings into the termination room shall be protected by opening protectives having a fire protection rating of not less than ¾ hour and shall be self-closing or automatic-closing upon the detection of smoke. Refuse chutes shall not terminate in an incinerator room. Refuse and laundry rooms that are not provided with chutes need comply only with Table 302.1.1.

Section 707.14.1 Change to read as shown: (FS46-03/04; FS48-03/04)

707.14.1 Elevator lobby. An elevator lobby shall be provided at each floor where an elevator shaft enclosure connects more than three stories. The lobby shall separate the elevator shaft enclosure doors from each floor by fire partitions and the required opening protection. Elevator lobbies shall have at least one means of egress complying with Chapter 10 and other provisions within this code.

Exceptions:

1. In office buildings, separations are not required from a street-floor elevator lobby provided the entire street floor is equipped with an automatic sprinkler system in accordance with Section 903.3.1.1.

2. Elevators not required to be located in a shaft in accordance with Section 707.2.

3. Where additional doors are provided in accordance with Section 3002.6. Such doors shall be tested in accordance with UL 1784 without an artificial bottom seal.

4. In other than Group I-3, and buildings having occupied floors located more than 75 feet above the lowest level of fire department vehicle access, lobby separation is not required where the building is protected by an automatic sprinkler system installed in accordance with Section 903.3.1.1 or 903.3.1.2.

5. Smoke partitions shall be permitted to separate the elevator lobby at each floor where the building equipped throughout with an automatic sprinkler system installed in accordance with Section 903.3.1.1 or 903.3.1.2.

6. Elevator lobbies are not required provided that the elevator shaft enclosure is pressurized in accordance with Section 909.20.5.

Section 708.1 Change to read as shown: (G50-03/04)

708.1 General. The following wall assemblies shall comply with this section.

1. Walls separating dwelling units in the same building.

2. Walls separating sleeping units in occupancies in Group R-1, hotel occupancies, R-2 and I-1.

3. Walls separating tenant spaces in covered mall buildings as required by Section 402.7.2.

4. Corridor walls as required by Section 1016.1.

5. Elevator lobby separation.

6. Residential aircraft hangars.

Section 708.4 Change to read as shown: (FS19-03/04; FS29-03/04; E3-03/04)

708.4 Continuity. Fire partitions shall extend from the top of the floor assembly below to the underside of the floor or roof sheathing, slab or deck above or to the fire-resistance-rated floor/ceiling or roof/ceiling assembly above, and shall be securely attached thereto. If the partitions are not continuous to the sheathing, deck or slab, and where constructed of combustible construction, the space between the ceiling and the sheathing, deck or slab above shall be fireblocked or draftstopped in accordance with Sections 717.2 and 717.3 at the partition line. The supporting construction shall be protected to afford the required fire-resistance rating of the wall supported, except for tenant and sleeping unit separation walls and corridor walls in buildings of Type IIB, IIIB and VB construction.

Exceptions:

1. The wall need not be extended into the crawl space below where the floor above the crawl space has a minimum 1-hour fire-resistance rating.

2. Where the room-side fire-resistance-rated membrane of the corridor is carried through to the underside of the floor or roof sheathing deck or slab of a fire-resistance-rated floor or roof above, the ceiling of the corridor shall be permitted to be protected by the use of ceiling materials as required for a 1-hour fire-resistance-rated floor or roof system.

3. Where the corridor ceiling is constructed as required for the corridor walls, the walls shall be permitted to terminate at the upper membrane of such ceiling assembly.

4. The fire partition separating tenant spaces in a mall, complying with Section 402.7.2, is not required to extend beyond the underside of a ceiling that is not part of a fire-resistance-rated assembly. A wall is not required in attic or ceiling spaces above tenant separation walls.

5. Fireblocking or draftstopping is not required at the partition line in Group R-2 buildings that do not exceed four stories in height provided the attic space is subdivided by draftstopping into areas not exceeding 3,000 square feet (279 m^2) or above every two dwelling units, whichever is smaller.

6. Fireblocking or draftstopping is not required at the partition line in buildings equipped with an automatic sprinkler system installed throughout in accordance with Section 903.3.1.1 or 903.3.1.2 provided that automatic sprinklers are installed in combustible floor/ceiling and roof/ceiling spaces.

Section 709.4 Change to read as shown: (FS19-03/04)

709.4 Continuity. Smoke barriers shall form an effective membrane continuous from outside wall to outside wall and from floor slab to the underside of the floor or roof sheathing, deck or slab above, including continuity through concealed spaces, such as those found above suspended ceilings, and interstitial structural and mechanical spaces. The supporting construction shall be protected to afford the required fire-resistance rating of the wall or floor supported in buildings of other than Type IIB, IIIB or VB construction.

Exception: Smoke barrier walls are not required in interstitial spaces where such spaces are designed and constructed with ceilings that provide resistance to the passage of fire and smoke equivalent to that provided by the smoke barrier walls.

Section 709.5 Change to read as shown: (FS53-03/04)

709.5 Openings. Openings in a smoke barrier shall be protected in accordance with Section 715.

Exception: In Group I-2, where such doors are installed across corridors, a pair of opposite-swinging doors without a center mullion shall be installed having vision panels with fire-protection-rated glazing materials in fire-protection-rated frames, the area of which shall not exceed that tested. The doors shall be close fitting within operational tolerances, and shall not have undercuts, louvers or grilles. The doors shall have head and jamb stops, astragals or rabbets at meeting edges and automatic-closing devices. Positive-latching devices are not required.

Section 710.4 Change to read as shown: (FS19-03/04)

710.4 Continuity. Smoke partitions shall extend from the floor to the underside of the floor or roof sheathing, deck or slab above or to the underside of the ceiling above where the ceiling membrane is constructed to limit the transfer of smoke.

Section 711.4 Change to read as shown: (FS19-03/04)

711.4 Continuity. Assemblies shall be continuous without openings, penetrations or joints except as permitted by this section and Sections 707.2, 712.4 and 713. Skylights and other penetrations through a fire-resistance-rated roof deck or slab are permitted to be unprotected, provided that the structural integrity of the fire-resistance-rated roof construction is maintained. Unprotected skylights shall not be permitted in roof construction required to be fire-resistance rated in accordance with Section 704.10. The supporting construction shall be protected to afford the required fire-resistance rating of the horizontal assembly supported.

Section 712.3.3 Change to read as shown: (FS32-03/04)

712.3.3 Ducts and air transfer openings. Penetrations of fire-resistance-rated walls by ducts that are not protected with dampers shall comply with Sections 712.2 through 712.3.1. Ducts and air transfer openings that are protected with dampers shall comply with Section 716.

Section 712.4.4 Change to read as shown: (FS32-03/04)

712.4.4 Ducts and air transfer openings. Penetrations of horizontal assemblies by ducts that are not protected with dampers shall comply with Section 712.2 and Sections 712.4 through 712.4.3.2. Ducts and air transfer openings that are protected with dampers shall comply with Section 716.

Section 712.4.6 Change to read as shown: (FS60-03/04)

712.4.6 Floor fire doors. Floor fire doors used to protect openings in fire-resistance-rated floors shall be tested in accordance with NFPA 288, and shall achieve a fire-resistance rating not less than the assembly being penetrated. Floor fire doors shall be labeled by an approved agency.

2004 SUPPLEMENT TO THE IBC

Section 713.5 Add new section to read as shown: (FS65-03/04)

713.5 Spandrel wall. Height and fire-resistance requirements for curtain wall spandrels shall comply with Section 704.9. Where section 704.9 does not require a fire-resistance-rated spandrel wall, the requirements of Section 713.4 shall still apply to the intersection between the spandrel wall and the floor.

Table 715.3 Change to read as shown: (FS69-03/04)

**TABLE 715.3
FIRE DOOR AND FIRE SHUTTER
FIRE PROTECTION RATINGS**

Type of Assembly	Required Assembly Rating (hours)	Minimum Fire door and Fire Shutter Assembly Rating (hours)
Fire partitions:		
Corridor walls	1	1/3[b]
	0.5	1/3[b]
Other fire partitions	1	3/4
	0.5	1/3

(Portions of table not shown do not change)

Section 715.3.3 Change to read as shown: (FS72-03/04; FS75-03/04)

715.3.3 Door assemblies in corridors and smoke barriers. Fire door assemblies required to have a minimum fire protection rating of 20 minutes where located in corridor walls or smoke barrier walls having a fire-resistance rating in accordance with Table 715.3 shall be tested in accordance with NFPA 252 or UL 10C without the hose stream test. If a 20-minute fire door assembly contains glazing material, the glazing material in the door itself shall have a minimum fire protection rating of 20 minutes and be exempt from the hose stream test. Glazing material in any other part of the door assembly, including transom lites and sidelites, shall be tested in accordance with NFPA 257, including the hose stream test, in accordance with Section 715.4. Fire door assemblies shall also meet the requirements for a smoke- and draft-control door assembly tested in accordance with UL 1784 with an artificial bottom seal installed across the full width of the bottom of the door assembly during the test. The air leakage rate of the door assembly shall not exceed 3.0 cfm per square foot (0.01524 m^3/s . m^2) of door opening at 0.10 inch (24.9 Pa) of water for both the ambient temperature and elevated temperature tests. Louvers shall be prohibited. Installation of smoke doors shall be in accordance with NFPA 105.

Exceptions:

1. Viewports that require a hole not larger than 1 inch (25 mm) in diameter through the door, have at least a 0.25-inch-thick (6.4 mm) glass disc and the holder is of metal that will not melt out where subject to temperatures of 1,700°F (927°C).

2. Corridor door assemblies in occupancies of Group I-2 shall be in accordance with Section 407.3.1.

3. Unprotected openings shall be permitted for corridors in multitheater complexes where each motion picture auditorium has at least one-half of its required exit or exit access doorways opening directly to the exterior or into an exit passageway.

Section 715.3.4 Change to read as shown: (E1-03/04)

715.3.4 Doors in exit enclosures and exit passageways. Fire door assemblies in exit enclosures and exit passageways shall have a maximum transmitted temperature end point of not more than 450°F (250°C) above ambient at the end of 30 minutes of standard fire test exposure.

Exception: The maximum transmitted temperature end point is not required in buildings equipped throughout with an automatic sprinkler system installed in accordance with Section 903.3.1.1 or 903.3.1.2.

Section 715.3.5.1 Change to read as shown: (FS31-03/04)

715.3.5.1 Fire door labeling requirements. Fire doors shall be labeled showing the name of the manufacturer, the name of the third-party inspection agency, the fire protection rating and, where required for fire doors in exit enclosures and exit passageways by Section 715.3.4, the maximum transmitted temperature end point. Smoke and draft control doors complying with UL 1784 shall be labeled as such. Labels shall be approved and permanently affixed. The label shall be applied at the factory or location where fabrication and assembly are performed.

Section 715.3.6.3 Change to read as shown: (FS78-03/04)

715.3.6.3 Labeling. Fire-protection-rated glazing shall bear a label or other identification showing the name of the manufacturer, the test standard, and information required in Section 715.4.9.1, that shall be issued by an approved agency and shall be permanently affixed.

Section 715.3.6.3.1 Add new section to read as shown: (FS78-03/04)

715.3.6.3.1 Identification. For fire protection-rated glazing, the label shall bear the following four-part identification: "D – H or NH – T or NT – XXX". "D" indicates that the glazing shall be used in fire door assemblies and that the glazing meets the fire resistance requirements of the test standard. "H" shall indicate that the glazing meets the hose stream requirements of the test standard. "NH" shall indicate that the glazing does not meet the hose stream requirements of the test. "T" shall indicate that the glazing meets the temperature requirements of Section 715.3.4.1. "NT" shall indicate that the glazing does not meet the temperature requirements of Section 715.3.4.1. The placeholder "XXX" shall specify the fire protection rating period, in minutes.

Section 715.3.7.3 Change to read as shown: (E3-03/04)

715.3.7.3 Smoke-activated doors. Automatic-closing fire doors installed in the following locations shall be automatic-closing by the actuation of smoke detectors installed in accordance with Section 907.10 or by loss of power to the smoke detector or hold-open device. Fire doors that are automatic-closing by smoke detection shall not have more than a 10-second delay before the door starts to close after the smoke detector is actuated.

1. Doors installed across a corridor.
2. Doors that protect openings in horizontal exits, exits or corridors required to be of fire-resistance-rated construction.
3. Doors that protect openings in walls required to be fire-resistance rated by Table 302.1.1.
4. Doors installed in smoke barriers in accordance with Section 709.5.
5. Doors installed in fire partitions in accordance with Section 708.6.
6. Doors installed in a fire wall in accordance with Section 705.8.

Section 715.4 Change to read as shown: (FS15-03/04)

715.4 Fire-protection-rated glazing. Glazing in fire window assemblies shall be fire-protection-rated in accordance with this section and Table 715.4. Glazing in fire door assemblies shall comply with Section 715.3.6. Fire-protection-rated glazing shall be tested in accordance with and shall meet the acceptance criteria of NFPA 257. Fire-protection-rated glazing shall also comply with NFPA 80. Openings in nonfire-resistance-rated exterior wall assemblies that require protection in accordance with Section 704.3, 704.8, 704.9 or 704.10 shall have a fire-protection rating of not less than ¾ hour.

Exceptions:

1. Wired glass in accordance with Section 715.4.3.
2. Fire-protection-rated glazing in 0.5-hour fire-resistance-rated partitions is permitted to have an 0.33-hour fire protection rating.

Table 715.4 Change to read as shown: (FS15-03/04)

**TABLE 715.4
FIRE WINDOW ASSEMBLY
FIRE PROTECTION RATINGS**

Type of Assembly	Required Assembly Rating (hours)	Minimum Fire Window Assembly Rating (hours)
Interior walls:		
Fire walls	All	NP[a]
Fire barriers	4	3
	3	3
	2	1 ½
	1	¾
Smoke barriers and fire partitions	1	¾
Exterior walls	>1	1 ½
	1	¾
Party wall	All	NP

(No change to footnote)

Section 715.4.8 Exterior fire window assemblies. Delete without substitution and renumber subsequent sections: (FS15-03/04)

Section 715.4.9 Change to read as shown: (FS80-03/04)

715.4.9 Labeling requirements. Fire-protection-rated glazing shall bear a label or other identification showing the name of the manufacturer, the test standard, and information required in Section 715.4.9.1, that shall be issued by an approved agency and shall be permanently affixed.

Section 715.4.9.1 Add new section to read as shown: (FS80-03/04)

715.4.9.1 Identification. For fire protection-rated glazing, the label shall bear the following two-part identification: "OH – XXX". "OH" indicates that the glazing met both the fire-resistance and the hose-stream requirements of NFPA 257 and is permitted to be used in openings. "XXX" represents the fire-protection rating period, in minutes, that was tested.

Section 716.1.1 Change to read as shown: (FS32-03/04)

716.1.1 Ducts without dampers. Ducts that penetrate fire-resistance-rated assemblies and are not required by this section to have dampers shall comply with the requirements of Section 712.

Section 716.2 Change to read as shown: (FS82-03/04)

716.2 Installation. Fire dampers, smoke dampers, combination fire/smoke dampers and ceiling radiation dampers located within air distribution and smoke control systems shall be installed in accordance with the

2004 SUPPLEMENT TO THE IBC

requirements of this section, the manufacturer's installation instructions and their listing.

Section 716.3.2.1 Change to read as shown: (FS84-03/04)

716.3.2.1 Smoke damper actuation methods. The smoke damper shall close upon actuation of a listed smoke detector or detectors installed in accordance with Section 907.10 and one of the following methods, as applicable:

1. Where a damper is installed within a duct, a smoke detector shall be installed in the duct within 5 feet (1524 mm) of the damper with no air outlets or inlets between the detector and the damper. The detector shall be listed for the air velocity, temperature and humidity anticipated at the point where it is installed. Other than in mechanical smoke control systems, dampers shall be closed upon fan shutdown where local smoke detectors require a minimum velocity to operate.

2. Where a damper is installed above smoke barrier doors in a smoke barrier, a spot-type detector listed for releasing service shall be installed on either side of the smoke barrier door opening.

3. Where a damper is installed within an unducted opening in a wall, a spot-type detector listed for releasing service shall be installed within 5 feet (1524 mm) horizontally of the damper.

4. Where a damper is installed in a corridor wall or ceiling, the damper shall be permitted to be controlled by a smoke detection system installed in the corridor.

5. Where a total-coverage smoke detector system is provided within areas served by a heating, ventilation and air-conditioning (HVAC) system, dampers shall be permitted to be controlled by the smoke detection system.

Section 716.4 Change to read as shown: (FS85-03/04)

716.4 Access and identification. Fire and smoke dampers shall be provided with an approved means of access, large enough to permit inspection and maintenance of the damper and its operating parts. The access shall not affect the integrity of fire-resistance-rated assemblies. The access openings shall not reduce the fire-resistance rating of the assembly. Access points shall be permanently identified on the exterior by a label having letters not less than 0.5 inch (12.7 mm) in height reading: FIRE/SMOKE DAMPER, SMOKE DAMPER or FIRE DAMPER. Access doors in ducts shall be tight fitting and suitable for the required duct construction.

Section 716.5.2 Change to read as shown: (FS87-03/04)

716.5.2 Fire barriers. Duct and air transfer openings of fire barriers shall be protected with approved fire dampers installed in accordance with their listing.

Exception: Fire dampers are not required at penetrations of fire barriers where any of the following apply:

1. Penetrations are tested in accordance with ASTM E 119 as part of the fire-resistance-rated assembly.

2. Ducts are used as part of an approved smoke control system in accordance with Section 909 and where the use of a fire damper would interfere with the operation of the smoke control system.

3. Such walls are penetrated by ducted HVAC systems, have a required fire-resistance rating of 1 hour or less, are in areas of other than Group H and are in buildings equipped throughout with an automatic sprinkler system in accordance with Section 903.3.1.1 or 903.3.1.2. For the purposes of this exception, a ducted HVAC system shall be a duct system for conveying supply, return or exhaust air as part of the structure's HVAC system. Such a duct system shall be constructed of sheet steel not less than 26 gage thickness and shall be continuous from the air-handling appliance or equipment to the air outlet and inlet terminals.

Section 716.5.3.1 Change to read as shown: (FS90-03/04)

716.5.3.1 Penetrations of shaft enclosures. Shaft enclosures that are permitted to be penetrated by ducts and air transfer openings shall be protected with approved fire and smoke dampers installed in accordance with their listing.

Exceptions:

1. Fire dampers are not required at penetrations of shafts where:

    1.1. Steel exhaust subducts extended at least 22 inches (559 mm) vertically in exhaust shafts provided there is a continuous airflow upward to the outside, or
    1.2. Penetrations are tested in accordance with ASTM E 119 as part of the rated assembly, or
    1.3. Ducts are used as part of an approved smoke control system designed and installed in accordance with Section 909, and where the fire damper will interfere with the operation of the smoke control system, or
    1.4. The penetrations are in parking garage exhaust or supply shafts that are separated from other building shafts by not less than 2-hour fire-resistance-rated construction.

2. In Group B occupancies, equipped throughout with an automatic sprinkler system in accordance with Section 903.3.1.1, smoke dampers are not required at penetrations of shafts where bathroom

and toilet room exhaust openings with steel exhaust subducts, having a wall thickness of at least 0.019 inch (0.48 mm) that extend at least 22 inches (559 mm) vertically and the exhaust fan at the upper terminus, powered continuously in accordance with the provisions of Section 909.11, maintains airflow upward to the outside.

3. Smoke dampers are not required at penetration of exhaust or supply shafts in parking garages that are separated from other building shafts by not less than 2-hour fire-resistance-rated construction.

4. Smoke dampers are not required at penetrations of shafts where ducts are used as part of an approved mechanical smoke control system designed in accordance with Section 909 and where the smoke damper will interfere with the operation of the smoke control system.

Section 716.5.4 Change to read as shown: (FS32-03/04; FS91-03/04)

716.5.4 Fire partitions. Ducts and air transfer openings that penetrate fire partitions shall be protected with approved fire dampers installed in accordance with their listing.

Exceptions: In occupancies other than Group H, fire dampers are not required where any of the following apply:

1. The partitions are tenant separation or corridor walls in buildings equipped throughout with an automatic sprinkler system in accordance with Section 903.3.1.1 or 903.3.1.2 and the duct is protected as a through penetration in accordance with Section 712.

2. The duct system is constructed of approved materials in accordance with the *International Mechanical Code* and the duct penetrating the wall meets all of the following minimum requirements:

    2.1. The duct shall not exceed 100 square inches (0.06 m^2).

    2.2. The duct shall be constructed of steel a minimum of 0.0217 inch (0.55 mm) in thickness.

    2.3. The duct shall not have openings that communicate the corridor with adjacent spaces or rooms.

    2.4. The duct shall be installed above a ceiling.

    2.5. The duct shall not terminate at a wall register in the fire-resistance-rated wall.

    2.6. A minimum 12-inch-long (0.30 m) by 0.060-inch-thick (1.52 mm) steel sleeve shall be centered in each duct opening. The sleeve shall be secured to both sides of the wall and all four sides of the sleeve with minimum 1$^1/_2$-inch by 1$^1/_2$-inch by 0.060-inch (0.038 m by 0.038 m by 1.52 mm) steel retaining angles. The retaining angles shall be secured to the sleeve and the wall with No. 10 (M5) screws. The annular space between the steel sleeve and wall opening shall be filled with rock (mineral) wool batting on all sides.

Section 716.6.1 Change to read as shown: (FS32-03/04)

716.6.1 Through penetrations. In occupancies other than Groups I-2 and I-3, a duct constructed of approved materials in accordance with the *International Mechanical Code* that penetrates a fire-resistance-rated floor/ceiling assembly that connects not more than two stories is permitted without shaft enclosure protection provided a fire damper is installed at the floor line or the duct is protected in accordance with Section 712.4. For air transfer openings, see exception 7 to Section 707.2.

Exceptions: A duct is permitted to penetrate three floors or less without a fire damper at each floor provided it meets all of the following requirements.

1. The duct shall be contained and located within the cavity of a wall and shall be constructed of steel not less than 0.019 inch (0.48 mm) (26 gage) in thickness.

2. The duct shall open into only one dwelling unit or sleeping unit and the duct system shall be continuous from the unit to the exterior of the building.

3. The duct shall not exceed 4-inch (102 mm) nominal diameter and the total area of such ducts shall not exceed 100 square inches (0.065 m^2) in any 100 square feet (9.3 m^2) of floor area.

4. The annular space around the duct is protected with materials that prevent the passage of flame and hot gases sufficient to ignite cotton waste where subjected to ASTM E 119 time-temperature conditions under a minimum positive pressure differential of 0.01 inch (2.49 Pa) of water at the location of the penetration for the time period equivalent to the fire-resistance rating of the construction penetrated

5. Grille openings located in a ceiling of a fire-resistance-rated floor/ceiling or roof/ceiling assembly shall be protected with a ceiling radiation damper in accordance with Section 716.6.2.

Section 716.6.2 Change to read as shown: (FS83-03/04)

716.6.2 Membrane penetrations. Duct systems constructed of approved materials in accordance with *the International*

Mechanical Code that penetrate the ceiling membrane of a fire-resistance-rated floor/ceiling or roof/ceiling assembly shall be protected with one of the following:

1. A fire-resistance-rated shaft enclosure in accordance with Sections 707 and 712.4.

2. An approved ceiling radiation damper installed at the ceiling line where the duct system penetrates the ceiling of a fire-resistance-rated floor/ceiling or roof/ceiling assembly.

3. An approved ceiling radiation damper installed at the ceiling line where a diffuser with no duct attached penetrates the ceiling of a fire-resistance-rated floor/ceiling or roof/ceiling assembly.

Section 716.6.2.1 Add new section to read as shown: (FS83-03/04)

716.6.2.1 Ceiling radiation dampers. Ceiling radiation dampers shall be tested in accordance with UL 555C and installed in accordance with the manufacturer's installation instructions and listing. Ceiling radiation dampers are not required where either of the following apply:

1. ASTM E 119 fire tests have shown that ceiling radiation dampers are not necessary in order to maintain the fire-resistance rating of the assembly.

2. Where exhaust duct penetrations are protected in accordance with Section 712.4.2 and the exhaust ducts are located within the cavity of a wall, and do not pass through another dwelling unit or tenant space.

Section 716.6.3 Change to read as shown: (FS92-03/04; FS93-03/04)

716.6.3 Nonfire-resistance-rated floor assemblies. Duct systems constructed of approved materials in accordance with the *International Mechanical Code* that penetrate nonfire-resistance-rated floor assemblies shall be protected by any of the following methods:

1. A fire-resistance-rated shaft enclosure that meets the requirements of Section 712.4.3.

2. The duct connects not more than two stories, and the annular space around the penetrating duct is protected with an approved noncombustible material to resist the free passage of flame and the products of combustion.

3. The duct connects not more than three stories, and the annular space around the penetrating duct is protected with an approved noncombustible material to resist the free passage of flame and the products of combustion, and a fire damper is installed at each floor line.

Exception: Fire dampers are not required in ducts within individual residential dwelling units.

Section 717.1 Change to read as shown: (FS93-03/04)

717.1 General. Fireblocking and draftstopping shall be installed in combustible concealed locations in accordance with this section. Fireblocking shall comply with Section 717.2. Draftstopping in floor/ceiling spaces and attic spaces shall comply with Sections 717.3 and 717.4, respectively. The permitted use of combustible materials in concealed spaces of buildings of Type I or II construction shall be limited to the applications indicated in Section 717.5.

Section 717.2.6 Change to read as shown: (FS94-03/04)

717.2.6 Architectural trim. Fireblocking shall be installed within concealed spaces of exterior wall finish and other exterior architectural elements where permitted to be of combustible construction in Section 1406 or where erected with combustible frames, at maximum intervals of 20 feet (6096 mm), and so that there will be no open space exceeding 100 square feet (9.3 m^3). Where wood furring strips are used, they shall be of approved wood of natural decay resistance or preservative-treated wood. If noncontinuous, such elements shall have closed ends, with at least 4 inches (102 mm) of separation between sections.

Exceptions:

1. Fireblocking of cornices is not required in single-family dwellings, as applicable in Section 101.2. Fireblocking of cornices of a two-family dwelling as applicable in Section 101.2 is required only at the line of dwelling unit separation.

2. Fireblocking shall not be required where installed on noncombustible framing and the face of the exterior wall finish exposed to the concealed space is covered by one of the following materials:

    2.1. Aluminum having a minimum thickness of 0.019 inch (0.5 mm).

    2.2. Corrosion-resistant steel having a base metal thickness not less than 0.016 inch (0.4 mm) at any point.

    2.3. Other approved noncombustible materials.

Section 717.5 Change to read as shown: (FS93/03/04)

717.5 Combustible materials in concealed spaces in Type I or II construction. Combustible materials shall not be permitted in concealed spaces of buildings of Type I or II construction.

Exceptions:

1. Combustible materials in accordance with Section 603.

2. Combustible materials exposed within plenums complying with Section 602 of the *International Mechanical Code*.

3. Class A interior finish materials classified in accordance with Section 803.

4. Combustible piping within partitions or shaft enclosures installed in accordance with the provisions of this code.

5. Combustible piping within concealed ceiling spaces installed in accordance with the *International Mechanical Code* and the *International Plumbing Code*.

Section 720.1.2 Change to read as shown: (FS97-03/04)

720.1.2 Unit masonry protection. Where required, metal ties shall be embedded in bed joints of unit masonry for protection of steel columns. Such ties shall be as set forth in Table 720.1(1) or be equivalent thereto.

Table 720.1(3) Change entry in first column to read as shown: (FS105-03/04)

TABLE 720.1(3)
MINIMUM PROTECTION FOR FLOOR
AND ROOF SYSTEMS [a, q]

Floor or Roof Construction
22. Wood joists, wood I-joists, floor trusses and flat or pitched roof trusses spaced a maximum 24" o.c. with ½" wood structural panels with exterior glue applied at right angles to top of joist or top chord of trusses with 8d nails. The wood structural panel thickness shall not be less than nominal ½" less than required by Chapter 23.

(Portions of table not shown, including footnotes, do not change)

Section 721.5.2.2 Change to read as shown: (FS106-03/04)

721.5.2.2 Spray-applied fire-resistant materials. The provisions in this section apply to structural steel beams and girders protected with spray-applied fire-resistant materials. Larger or smaller beam and girder shapes shall be permitted to be substituted for beams specified in approved unrestrained or restrained fire-resistant assemblies provided that the thickness of the fire-resistant material is adjusted in accordance with the following expression:

$$h_2 = \left[\frac{W_1/D_1 + 0.60}{W_2/D_2 + 0.60} \right] h_1 \qquad \text{(Equation 7-17)}$$

where:
h = Thickness of spray-applied fire-resistant material in inches.
W = Weight of the structural steel beam or girder in pounds per linear foot.
D = Heated perimeter of the structural steel beam or girder in inches.

Subscript 1 refers to the beam and fire-resistant material thickness in the fire-resistance-rated assembly.

Subscript 2 refers to the substitute beam or girder and the required thickness of fire-resistant material.

Section 721.5.2.2.1 Change to read as shown: (FS106-03/04)

721.5.2.2.1 Minimum thickness. The use of Equation 7-17 is subject to the following conditions.

1. The weight-to-heated-perimeter ratio for the substitute beam or girder (W_2/D_2) shall not be less than 0.37.

2. The thickness of fire protection materials calculated for the substitute beam or girder (T_1) shall not be less than ⅜ inch (9.5 mm).

3. The unrestrained or restrained beam rating shall not be less than 1 hour.

4. When used to adjust the material thickness for a restrained beam, the use of this procedure is limited to steel sections classified as compact in accordance with the AISC Specification for Structural Steel Buildings, (AISC-LRFD).

Section 721.7 Other reference documents. Delete without substitution: (FS107-03/04)

CHAPTER 8
INTERIOR FINISHES

Section 801.1.2 Change to read as shown: (G85-03/04; F43-03/04)

801.1.2 Decorative materials and trim. Decorative materials and trim shall be restricted by combustibility and the flame propagation performance criteria of NFPA 701, in accordance with Section 805.

Section 801.2.2 Change to read as shown: (FS109-03/04)

801.2.2 Foam plastics. Foam plastics shall not be used as interior finish or trim except as provided in Section 2603.4 or 2604. This section shall apply both to exposed foam plastics and to foam plastics used in conjunction with a textile or vinyl facing or cover.

Section 802.1 Delete the definition of "Flame Resistance" (G85-03/04)

Section 802.1 Change definitions of "Flame Spread Index" and "Smoke-Developed Index" to read as shown: (FS110-03/04)

FLAME SPREAD INDEX. A comparative measure, expressed as a dimensionless number, derived from visual measurements of the spread of flame versus time for a material tested in accordance with ASTM E 84.

SMOKE-DEVELOPED INDEX. A comparative measure, expressed as a dimensionless number, derived from measurements of smoke obscuration versus time for a material tested in accordance with ASTM E 84.

Table 803.5 Change to read as shown: (E1-03/04; E3-03/04)

TABLE 803.5
INTERIOR WALL AND CEILING FINISH REQUIREMENTS BY OCCUPANCY[k]

GROUP	SPRINKLERED		NONSPRINKLERED	
	Exit enclosures and exit passageways[a,b]	Corridors	Exit enclosures and exit passageways[a,b]	Corridors

(Portions of headings and table entries not shown do not change)

For SI: (No change)

a. (No change to current text)
b. In exit enclosures of buildings less than three stories in height of other than Group I-3, Class B interior finish for nonsprinklered buildings and Class C interior finish for sprinklered buildings shall be permitted.
c. (No change to current text)
d. Lobby areas in Group A-1, A-2 and A-3 occupancies shall not be less than Class B materials.
e. through f. (No change to current text)
g. Class B material is required where the building exceeds two stories.
h. and i. (No change to current text)
j. Class B materials shall be permitted as wainscotting extending not more than 48 inches above the finished floor in corridors.
k. (No change to current text)
l. Applies when the exit enclosures, exit passageways, corridors, or rooms and enclosed spaces are protected by a sprinkler system installed in accordance with Section 903.3.1.1 or 903.3.1.2.

Section 804.5 Change to read as shown: (E1-03/04; E3-03/04)

804.5 Interior floor finish requirements. In all occupancies, interior floor finish in exit enclosures, exit passageways, corridors and rooms or spaces not separated from corridors by full-height partitions extending from the floor to the underside of the ceiling shall withstand a minimum critical radiant flux as specified in Section 804.5.1.

Section 804.5.1 Change to read as shown: (E1-03/04; E3-03/04)

804.5.1 Minimum critical radiant flux. Interior floor finish in exit enclosures, exit passageways and corridors shall not be less than Class I in Groups I-2 and I-3 and not less than Class II in Groups A, B, E, H, I-4, M, R-1, R-2 and S. In all other areas, the interior floor finish shall comply with the DOC FF-1 "pill test" (CPSC 16 CFR, Part 1630).

> **Exception:** Where a building is equipped throughout with an automatic sprinkler system in accordance with Section 903.3.1.1, Class II materials are permitted in any area where Class I materials are required and materials complying with DOC FF-1 "pill test" (CPSC 16 CFR, Part 1630) are permitted in any area where Class II materials are required.

Section 805.1 Change to read as shown: (F43-03/04; F47-03/04)

805.1 General. In occupancies of Groups A, E, I and R-1 and dormitories in Group R-2, curtains, draperies, hangings and other decorative materials suspended from walls or ceilings shall meet the flame propagation performance criteria of NFPA 701 in accordance with Section 805.2 or be noncombustible.

In Groups I-1 and I-2, combustible decorations shall meet the flame propagation performance criteria of NFPA 701 unless the decorations, such as photographs and paintings, are of such limited quantities that a hazard of fire development or spread is not present. In Group I-3, combustible decorations are prohibited.

Fixed or movable walls and partitions, paneling, wall pads and crash pads, applied structurally or for decoration, acoustical correction, surface insulation or other purposes shall be considered interior finish if they cover 10 percent or more of the wall or of the ceiling area, and shall not be considered decorations or furnishings.

Section 805.1.2 Change to read as shown: (F43-03/04)

805.1.2 Combustible decorative materials. The permissible amount of decorative materials meeting the flame propagation performance criteria of NFPA 701 shall not exceed 10 percent of the aggregate area of walls and ceilings.

> **Exception:** In auditoriums of Group A, the permissible amount of decorative material meeting the flame propagation performance criteria of NFPA 701 shall not exceed 50 percent of the aggregate area of walls and ceiling where the building is equipped throughout with an

automatic sprinkler system in accordance with Section 903.3.1.1, and where the material is installed in accordance with Section 803.3.

Section 805.2 Change to read as shown: (F43-03/04; F48-03/04)

805.2 Acceptance criteria and reports. Where required by Section 805.1, decorative materials shall be tested by an approved agency and meet the flame propagation performance criteria of NFPA 701, or such materials shall be noncombustible. Reports of test results shall be prepared in accordance with NFPA 701 and furnished to the fire code official upon request.

CHAPTER 9
FIRE PROTECTION SYSTEMS

Section 903.2 Change to read as shown: (FS40-03/04)

903.2 Where required. Approved automatic sprinkler systems in new buildings and structures shall be provided in the locations described in this section.

> **Exception:** Spaces or areas in telecommunications buildings used exclusively for telecommunications equipment, associated electrical power distribution equipment, batteries and standby engines, provided those spaces or areas are equipped throughout with an automatic fire alarm system and are separated from the remainder of the building by fire barriers consisting of not less than 1 hour fire-resistance-rated walls and 2-hour fire-resistance-rated floor/ceiling assemblies.

Section 903.2.1.2 Change to read as shown: (F58-03/04)

903.2.1.2 Group A-2. An automatic sprinkler system shall be provided for Group A-2 occupancies where one of the following conditions exists:

1. The fire area exceeds 5,000 square feet (464.5 m^2);
2. The fire area has an occupant load of 100 or more; or
3. The fire area is located on a floor other than the level of exit discharge.

Section 904.11.1 Change to read as shown: (F88-03/04)

904.11.1 Manual system operation. A manual actuation device shall be located at or near a means of egress from the cooking area, a minimum of 10 feet (3048 mm) and a maximum of 20 feet (6096 mm) from the kitchen exhaust system. The manual actuation device shall be installed not more than 48 inches (1200 mm), nor less than 42 inches (1067 mm) above the floor and shall clearly identify the hazard protected. The manual actuation shall require a maximum force of 40 pounds (178 N) and a maximum movement of 14 inches (356 mm) to actuate the fire suppression system.

> **Exception:** Automatic sprinkler systems shall not be required to be equipped with manual actuation means.

Section 905.3.3 Change to read as shown: (F97-03/04)

905.3.3 Covered mall buildings. A covered mall building shall be equipped throughout with a standpipe system where required by Section 905.3.1. Covered mall buildings not required to be equipped with a standpipe system by Section 905.3.1 shall be equipped with Class I hose connections connected to a system sized to deliver 250 gallons per minute (946.4 L/min) at the most hydraulically remote outlet. Hose connections shall be provided at each of the following locations:

1. Within the mall at the entrance to each exit passageway or corridor.
2. At each floor-level landing within enclosed stairways opening directly on the mall.
3. At exterior public entrances to the mall.

Section 905.3.4 Change to read as shown: (F98-03/04)

905.3.4 Stages. Stages greater than 1,000 square feet in area (93 m^2) shall be equipped with a Class III wet standpipe system with 1.5-inch and 2.5-inch (38 mm and 64 mm) hose connections on each side of the stage.

> **Exception:** Where the building or area is equipped throughout with an automatic sprinkler system, a 1.5 inch (38 mm) hose connection shall be installed and shall be allowed to be supplied from the automatic sprinkler system and shall have a flow rate of not less than 100 gpm (329 L/m) at a minimum residual pressure of 65 psi (448.2 kPa), added to the sprinkler flow demand, at the most hydraulically remote hose connection.

Section 905.3.7 Add new section to read as shown: (F99-03/04)

905.3.7 Marinas and boatyards. Marinas and boatyards shall be equipped throughout with standpipe systems in accordance with NFPA 303.

Section 905.4 Change to read as shown: (F100-03/04)

905.4 Location of Class I standpipe hose connections. Class I standpipe hose connections shall be provided in all of the following locations:

1. In every required stairway, a hose connection shall be provided for each floor level above or below grade. Hose connections shall be located at an intermediate floor level landing between floors, unless otherwise approved by the building official.
2. On each side of the wall adjacent to the exit opening of a horizontal exit.

Exception: Where floor areas adjacent to a horizontal exit are reachable from exit stairway hose connections by a 30-foot (9144 mm) hose stream from a nozzle attached to 100 feet (30,480 mm) of hose, a hose connection shall not be required at the horizontal exit.

3. In every exit passageway at the entrance from the exit passageway to other areas of a building.

4. In covered mall buildings, adjacent to each exterior public entrance to the mall and adjacent to each entrance from an exit passageway or exit corridor to the mall.

5. Where the roof has a slope less than four units vertical in 12 units horizontal (33.3-percent slope), each standpipe shall be provided with a hose connection located either on the roof or at the highest landing of stairways with stair access to the roof. An additional hose connection shall be provided at the top of the most hydraulically remote standpipe for testing purposes.

6. Where the most remote portion of a nonsprinklered floor or story is more than 150 feet (45 720 mm) from a hose connection or the most remote portion of a sprinklered floor or story is more than 200 feet (60 960 mm) from a hose connection, the building official is authorized to require that additional hose connections be provided in approved locations.

Section 907.2.6 Change to read as shown: (F111-03/04)

907.2.6 Group I. A manual fire alarm system shall be installed in Group I occupancies. An electrically supervised, automatic smoke detection system shall be provided in accordance with Sections 907.2.6.1 and 907.2.6.2.

Exception: Manual fire alarm boxes in patient sleeping areas of Group I-1 and I-2 occupancies shall not be required at exits if located at all nurses' control stations or other constantly attended staff locations, provided such stations are visible and continuously accessible and that travel distances required in Section 907.3.1 are not exceeded.

Section 907.2.6.1 Add new section to read as shown: (F111-03/04)

907.2.6.1 Group I-1. Corridors, habitable spaces other than sleeping rooms and kitchens and waiting areas that are open to corridors shall be equipped with an automatic smoke detection system.

Exceptions:

1. Smoke detection in habitable spaces is not required where the facility is equipped throughout with an automatic sprinkler system.

2. Smoke detection is not required for exterior balconies.

(Renumber subsequent sections)

Section 907.2.6.1 Change to read as shown: (F111-03/04)

907.2.6.1 Group I-2. Corridors in nursing homes (both intermediate care and skilled nursing facilities), detoxification facilities and spaces permitted to be open to the corridors by Section 407.2 shall be equipped with an automatic fire detection system. Hospitals shall be equipped with smoke detection as required in Section 407.2.

Exceptions:

1. Corridor smoke detection is not required in smoke compartments that contain patient sleeping rooms where patient sleeping units are provided with smoke detectors that comply with UL 268. Such detectors shall provide a visual display on the corridor side of each patient sleeping unit and an audible and visual alarm at the nursing station attending each unit.

2. Corridor smoke detection is not required in smoke compartments that contain patient sleeping rooms where patient sleeping unit doors are equipped with automatic door-closing devices with integral smoke detectors on the unit sides installed in accordance with their listing, provided that the integral detectors perform the required alerting function.

Section 907.2.9 Change to read as shown: (F114-03/04)

907.2.9 Group R-2. A manual fire alarm system shall be installed in Group R-2 occupancies where:

1. Any dwelling unit or sleeping unit is located three or more stories above the lowest level of exit discharge;

2. Any dwelling unit or sleeping unit is located more than one story below the highest level of exit discharge of exits serving the dwelling unit or sleeping unit; or

3. The building contains more than 16 dwelling units or sleeping units.

Exceptions:

1. A fire alarm system is not required in buildings not over two stories in height where all dwelling units or sleeping units and contiguous attic and crawl spaces are separated from each other and public or common areas by at least 1-hour fire partitions and each dwelling unit or sleeping unit has an exit directly to a public way, exit court or yard.

2. Manual fire alarm boxes are not required throughout the building when the following conditions are met:

2.1. The building is equipped throughout with an automatic sprinkler system in accordance with Section 903.3.1.1 or Section 903.3.1.2; and

2.2. The notification appliances will activate upon sprinkler flow.

3. A fire alarm system is not required in buildings that do not have interior corridors serving dwelling units and are protected by an approved automatic sprinkler system installed in accordance with Section 903.3.1.1 or 903.3.1.2, provided that dwelling units either have a means of egress door opening directly to an exterior exit access that leads directly to the exits or are served by open-ended corridors designed in accordance with Section 1022.6, Exception 4.

Sections 907.2.12.2, 907.2.12.2.1 and 907.2.12.2.2 Change to read as shown: (F120-03/04)

907.2.12.2 Emergency voice/alarm communication system. The operation of any automatic fire detector, sprinkler water-flow device or manual fire alarm box shall automatically sound an alert tone followed by voice instructions giving approved information and directions for a general or staged evacuation on a minimum of the alarming floor, the floor above and the floor below in accordance with the building's fire safety and evacuation plans required by Section 404. Speakers shall be provided throughout the building by paging zones. As a minimum, paging zones shall be provided as follows:

1. Elevator groups.
2. Exit stairways.
3. Each floor.
4. Areas of refuge as defined in Section 1002.

Exception: In Group I-1 and I-2 occupancies, the alarm shall sound in a constantly attended area and a general occupant notification shall be broadcast over the overhead page.

907.2.12.2.1 Manual override. A manual override for emergency voice communication shall be provided on a selective and all-call basis for all paging zones.

907.2.12.2.2 Live voice messages. The emergency voice/alarm communication system shall also have the capability to broadcast live voice messages through paging zones on a selective and all-call basis.

Section 908.3 Change to read as shown: (F183-03/04)

908.3 Gas detection system. A gas detection system shall be provided to detect the presence of gas at or below the permissible exposure limit (PEL) or ceiling limit of the gas for which detection is provided. The system shall be capable of monitoring the discharge from the treatment system at or below one-half the IDLH limit.

Exception: A gas detection system is not required for toxic gases when the physiological warning threshold level for the gas is at a level below the accepted PEL for the gas.

Section 909.4.6 Change to read as shown: (FS117-03/04)

909.4.6 Duration of operation. All portions of active or passive smoke control systems shall be capable of continued operation after detection of the fire event for a period of not less than either 20 minutes or 1.5 times the calculated egress time, whichever is less.

Section 909.5.2 Change to read as shown: (FS53-03/04)

909.5.2 Opening protection. Openings in smoke barriers shall be protected by automatic-closing devices actuated by the required controls for the mechanical smoke control system. Door openings shall be protected by door assemblies complying with Section 715.3.3.

Exceptions:

1. Passive smoke control systems with automatic-closing devices actuated by spot-type smoke detectors listed for releasing service installed in accordance with Section 907.10.

2. Fixed openings between smoke zones that are protected utilizing the airflow method.

3. In Group I-2, where such doors are installed across corridors, a pair of opposite-swinging doors without a center mullion shall be installed having vision panels with fire-protection-rated glazing materials in fire-protection-rated frames, the area of which shall not exceed that tested. The doors shall be close fitting within operational tolerances and shall not have undercuts, louvers or grilles. The doors shall have head and jamb stops, astragals or rabbets at meeting edges, and automatic-closing devices. Positive-latching devices are not required.

4. Group I-3.

5. Openings between smoke zones with clear ceiling heights of 14 feet (4267 mm) or greater and bank-down capacity of greater than 20 minutes as determined by the design fire size.

Section 909.8.1 Change to read as shown: (FS118-03/04; FS119-03/04)

909.8.1 Exhaust rate. The height of the lowest horizontal surface of the accumulating smoke layer shall be maintained at least 6 feet (1829 mm) above any walking surface which forms a portion of a required egress system within the smoke zone. The required exhaust rate for the zone shall be the largest of the calculated plume mass flow rates for the possible plume configurations. Provisions shall be made for natural or mechanical supply of air from outside or adjacent smoke zones to make up for the air exhausted. Makeup airflow rates, when measured at the potential fire location, shall not increase the smoke production rate beyond the capabilities of the smoke control system. The temperature of the makeup air shall be such that it does not expose temperature-sensitive fire protection systems beyond their limits.

Section 909.8.3 Balcony spill plumes. Delete without substitution: (FS120-03/04)

Section 909.8.4 Window plumes. Delete without substitution: (FS121-03/04)

Section 909.9 Change to read as shown: (FS122-03/04)

909.9 Design fire. The design fire shall be based on a rational analysis performed by the registered design professional and approved by the building official. The design fire shall be based on the analysis in accordance with Section 909.4 and this section.

Section 909.9.2 Change to read as shown: (FS122-03/04)

909.9.2 Separation distance. Determination of the design fire shall include consideration of the type of fuel, fuel spacing and configuration.

Section 909.16 Change to read as shown: (F125-03/04)

909.16 Fire-fighter's smoke control panel. A fire-fighter's smoke control panel for fire department emergency response purposes only shall be provided and shall include manual control or override of automatic control for mechanical smoke control systems. The panel shall be located in a fire command center complying with Section 509 in high rise buildings or buildings with smoke protected assembly seating. In all other buildings, the fire-fighter's smoke control panel shall be installed in an approved location adjacent to the fire alarm control panel. The fire-fighter's smoke control panel shall comply with Sections 909.16.1 through 909.16.3.

Section 909.20.4.1 Change to read as shown: (FS124-03/04)

909.20.4.1 Vestibule doors. The door assembly from the building into the vestibule shall be a fire door complying with Section 715.3.3. The door assembly from the vestibule to the stairway shall have not less than a 20-minute fire protection rating and meet the requirements for a smoke door assembly in accordance with Section 715.3.3. The door shall be installed in accordance with NFPA 105.

Section 910.1 Change to read as shown: (F126-03/04)

910.1 General. Where required by this code or otherwise installed, smoke and heat vents, or mechanical smoke exhaust systems, and draft curtains shall conform to the requirements of this section.

Exceptions:

1. Frozen food warehouses used solely for storage of Class I and Class II commodities where protected by an approved automatic sprinkler system.

2. Where areas of buildings are equipped with early suppression fast-response (ESFR) sprinklers, automatic smoke and heat vents shall not be required within these areas.

Section 910.2 Change to read as shown: (F127-03/04)

910.2 Where required. Smoke and heat vents shall be installed in the roofs of one-story buildings or portions thereof occupied for the uses set forth in Sections 910.2.1 through 910.2.4.

Section 910.2.2 Change to read as shown: (F132-03/04)

910.2.2 Group H. Buildings and portions thereof used as a Group H-2 or H-3 occupancy having over 15,000 square feet (1394 m^2) in single floor area.

Exception: Buildings of noncombustible construction containing only noncombustible materials.

Table 910.3 Change table to read as shown: (F134-03/04; F135-03/04)

TABLE 910.3
REQUIREMENTS FOR DRAFT CURTAINS AND
SMOKE AND HEAT VENTS[a]

OCCUPANCY GROUP AND COMMODITY CLASSIFICATION	VENT AREA TO FLOOR AREA RATIO[c]
Groups F-1 and S-1	(No change)

Table 910.3 (continued)

High-piled Storage (see Section 910.2.3 I-IV (Option 1)	(No change)
High-piled Storage (see Section 910.2.3 I-IV (Option 2)	
High-piled Storage (see Section 910.2.3 High Hazard (Option 1)	
High-piled Storage (see Section 910.2.3) High Hazard (Option 2)	

(Portions of the table not shown do not change)
a. and b. (No change to current text)
c. Where draft curtains are not required, the vent area to floor area ratio shall be calculated based on a minimum draft curtain depth of 6 feet (Option 1).

Section 910.3.1 Add new section to read as shown: (F127-03/04)

Section 910.3.1 Design. Smoke and heat vents shall be listed and labeled.

(Renumber subsequent sections)

Section 910.3.1 Change to read as shown: (F127-03/04)

910.3.1 Vent operation. Smoke and heat vents shall be approved and shall be capable of being operated by approved automatic and manual means. Automatic operation of smoke and heat vents shall conform to the provisions of this section.

CHAPTER 10
MEANS OF EGRESS

Section 1002.1 Change definition of "Accessible Means of Egress" to read as shown: (E6-03/04)

ACCESSIBLE MEANS OF EGRESS. A continuous and unobstructed way of egress travel from any accessible point in a building or facility to a public way.

Section 1003.2 Change to read as shown: (E8-03/04)

1003.2 Ceiling height. The means of egress shall have a ceiling height of not less than 7 feet 6 inches (2286 mm).

Exceptions:

1. Sloped ceilings in accordance with Section 1208.2.
2. Ceilings of dwelling units and sleeping units within residential occupancies in accordance with Section 1208.2.

3. Allowable projections in accordance with Section 1003.3.
4. Stair headroom in accordance with Section 1009.2.
5. Door height in accordance with Section 1008.1.1.

Section 1004.1 Change to read as shown: (E9-03/04)

1004.1 Design occupant load. In determining means of egress requirements, the number of occupants for whom means of egress facilities shall be provided shall be determined in accordance with this section. Where occupants from accessory areas egress through a primary space, the calculated occupant load for the primary space shall include the total occupant load of the primary space plus the number of occupants egressing through it from the accessory area.

Section 1004.1.1 Actual number. Delete section without substitution: (E9-03/04) (Renumber subsequent sections and tables)

Section 1004.1.2 Change to read as shown: (E9-03/04)

1004.1.2 Areas without fixed seating. The number of occupants computed at the rate of one occupant per unit of area as prescribed in Table 1004.1.1. For areas without fixed seating, the occupant load shall not be less than that number determined by dividing the floor area under consideration by the occupant per unit of area factor assigned to the occupancy as set forth in Table 1004.1.1. Where an intended use is not listed in Table 1004.1.1, the building official shall establish a use based on a listed use that most nearly resembles the intended use.

Exception: Where approved by the building official, the actual number of occupants for whom each occupied space, floor or building is designed, although less than those determined by calculation, shall be permitted to be used in the determination of the design occupant load.

Section 1004.1.3 Number by combination. Delete section without substitution: (E9-03/04)

Section 1004.2 Change to read as shown: (E9-03/04)

1004.2 Increased occupant load. The occupant load permitted in any building or portion thereof is permitted to be increased from that number established for the occupancies in Table 1004.1.1 provided that all other requirements of the code are also met based on such modified number and the occupant load shall not exceed one occupant per 5 square feet (0.47 m^2) of occupiable floor space. Where required by the building official, an approved aisle, seating or fixed equipment diagram substantiating any increase in occupant load shall be submitted. Where required by the building official, such diagram shall be posted.

Section 1004.7 Change to read as shown: (E12-03/04)

1004.7 Fixed seating. For areas having fixed seats and aisles, the occupant load shall be determined by the number of fixed seats installed therein. The occupant load for areas in which fixed seating is not installed, such as waiting spaces and wheelchair spaces, shall be determined in accordance with Section 1004.1.2 and added to the number of fixed seats.

For areas having fixed seating without dividing arms, the occupant load shall not be less than the number of seats based on one person for each 18 inches (457 mm) of seating length.

The occupant load of seating booths shall be based on one person for each 24 inches (610 mm) of booth seat length measured at the backrest of the seating booth.

Section 1006.3 Change to read as shown: (E3-03/04; E14-03/04)

1006.3 Illumination emergency power. The power supply for means of egress illumination shall normally be provided by the premise's electrical supply.

In the event of power supply failure, an emergency electrical system shall automatically illuminate the following areas:

1. Aisles and unenclosed egress stairways in rooms and spaces that require two or more means of egress.
2. Corridors, exit enclosures and exit passageways in buildings required to have two or more exits.
3. Exterior egress components at other than the level of exit discharge until exit discharge is accomplished for buildings required to have two or more exits.
4. Interior exit discharge elements, as permitted in Section 1023.1, in buildings required to have two or more exits.
5. The portion of the exterior exit discharge immediately adjacent to exit discharge doorways in buildings required to have two or more exits.

The emergency power system shall provide power for a duration of not less than 90 minutes and shall consist of storage batteries, unit equipment or an on-site generator. The installation of the emergency power system shall be in accordance with Section 2702.

Section 1007.1 Change to read as shown: (E16-03/04)

1007.1 Accessible means of egress required. Accessible means of egress shall comply with this section. Accessible spaces shall be provided with not less than one accessible means of egress. Where more than one means of egress is required by Section 1014.1 or 1018.1 from any accessible space, each accessible portion of the space shall be served by not less than two accessible means of egress.

Exceptions:

1. Accessible means of egress are not required in alterations to existing buildings.
2. One accessible means of egress is required from an accessible mezzanine level in accordance with Section 1007.3, 1007.4 or 1007.5.
3. In assembly spaces with sloped floors, one accessible means of egress is required from a space where the common path of travel of the accessible route for access to the wheelchair spaces meets the requirements in Section 1024.9.

Section 1007.2 Change to read as shown: (E17-03/04)

1007.2 Continuity and components. Each required accessible means of egress shall be continuous to a public way and shall consist of one or more of the following components:

1. Accessible routes complying with Section 1104.
2. Stairways within exit enclosures complying with Sections 1007.3 and 1019.1.
3. Elevators complying with Section 1007.4.
4. Platform lifts complying with Section 1007.5.
5. Horizontal exits.
6. Smoke barriers.
7. Ramps complying with Section 1010.

Exceptions:

1. Where the exit discharge is not accessible, an exterior area for assisted rescue must be provided in accordance with Section 1007.8.
2. Where the exit stairway is open to the exterior, the accessible means of egress shall include either an area of refuge in accordance with Section 1007.6 or an exterior area for assisted rescue in accordance with Section 1007.8.

Section 1008.1.1 Change to read as shown: (E20-03/04)

1008.1.1 Size of doors. The minimum width of each door opening shall be sufficient for the occupant load thereof and shall provide a clear width of not less than 32 inches (813 mm). Clear openings of doorways with swinging doors shall be measured between the face of the door and the stop, with the door open 90 degrees (1.57 rad). Where this section requires a minimum clear width of 32 inches (813 mm) and a door opening includes two door leaves without a mullion, one leaf shall provide a clear opening width of 32 inches (813 mm). The maximum width of a swinging door leaf shall be 48 inches (1219 mm) nominal. Means of egress doors in an occupancy in Group I-2 used for the movement of beds shall provide a clear width not less than 41½ inches (1054 mm). The height of doors shall not be less than 80 inches (2032 mm).

Exceptions:

1. The minimum and maximum width shall not apply to door openings that are not part of the required means of egress in occupancies in Groups R-2 and R-3 as applicable in Section 101.2.

2. Door openings to resident sleeping units in occupancies in Group I-3 shall have a clear width of not less than 28 inches (711 mm).

3. Door openings to storage closets less than 10 square feet (0.93 m^2) in area shall not be limited by the minimum width.

4. Width of door leafs in revolving doors that comply with Section 1008.1.3.1 shall not be limited.

5. Door openings within a dwelling unit or sleeping unit shall not be less than 78 inches (1981 mm) in height.

6. Exterior door openings in dwelling units and sleeping units, other than the required exit door, shall not be less than 76 inches (1930 mm) in height.

7. Interior egress doors within a dwelling unit or sleeping unit which is not required to be an Accessible unit, Type A unit or Type B unit.

8. Door openings required to be accessible within Type B dwelling units shall have a minimum clear width of 31$^3/_4$ inches (806 mm).

Section 1008.1.2 Change to read as shown: (E22-03/04)

1008.1.2 Door swing. Egress doors shall be side-hinged swinging.

Exceptions:

1. Private garages, office areas, factory and storage areas with an occupant load of 10 or less.

2. Group I-3 occupancies used as a place of detention.

3. Doors within or serving a single dwelling unit in Groups R-2 and R-3 as applicable in Section 101.2.

4. In other than Group H occupancies, revolving doors complying with Section 1008.1.3.1.

5. In other than Group H occupancies, horizontal sliding doors complying with Section 1008.1.3.3 are permitted in a means of egress.

6. Power-operated doors in accordance with Section 1008.1.3.1.

7. Doors serving a bathroom within an individual sleeping unit in Group R-1

Doors shall swing in the direction of egress travel where serving an occupant load of 50 or more persons or a Group H occupancy.

The opening force for interior side-swinging doors without closers shall not exceed a 5-pound (22 N) force. For other side-swinging, sliding and folding doors, the door latch shall release when subjected to a 15-pound (67 N) force. The door shall be set in motion when subjected to a 30-pound (133 N) force. The door shall swing to a full-open position when subjected to a 15-pound (67 N) force. Forces shall be applied to the latch side.

Section 1008.1.6 Change exception to read as shown: (E25-03/04; E26-03/04)

1008.1.6 Thresholds. Thresholds at doorways shall not exceed 0.75 inch (19.1 mm) in height for sliding doors serving dwelling units or 0.5 inch (12.7 mm) for other doors. Raised thresholds and floor level changes greater than 0.25 inch (6.4 mm) at doorways shall be beveled with a slope not greater than one unit vertical in two units horizontal (50-percent slope).

Exception: The threshold height shall be limited to 7 ¾ inches (197 mm) where the occupancy is Group R-2 or R-3 as applicable in Section 101.2, the door is an exterior door that is not a component of the required means of egress, the door, other than an exterior storm or screen door does not swing over the landing or step, and the doorway is not on an accessible route as required by Chapter 11.

Section 1008.1.8.5 Change to read as shown: (E28-03/04)

1008.1.8.5 Unlatching. The unlatching of any door or leaf shall not require more than one operation.

Exceptions:

1. Places of detention or restraint.

2004 SUPPLEMENT TO THE IBC

2. Where manually operated bolt locks are permitted by Section 1008.1.8.4.

3. Doors with automatic flush bolts as permitted by Section 1008.1.8.3, Exception 3.

4. Doors from individual dwelling units and guestrooms of Group R occupancies as permitted by Section 1008.1.8.3, Exception 4.

Section 1008.1.8.7 Change to read as shown: (E29-03/04)

1008.1.8.7 Stairway doors. Interior stairway means of egress doors shall be openable from both sides without the use of a key or special knowledge or effort.

Exceptions:

1. Stairway discharge doors shall be openable from the egress side and shall only be locked from the opposite side.

2. This section shall not apply to doors arranged in accordance with Section 403.12.

3. In stairways serving not more than four stories, doors are permitted to be locked from the side opposite the egress side, provided they are openable from the egress side and capable of being unlocked simultaneously without unlatching upon a signal from the fire command station, if present, or a signal by emergency personnel from a single location inside the main entrance to the building.

Section 1008.1.9 Change to read as shown: (E31-03/04; E33-03/04; E34-03/04)

1008.1.9 Panic and fire exit hardware. Where panic and fire exit hardware is installed, it shall comply with the following:

1. The actuating portion of the releasing device shall extend at least one-half of the door leaf width.

2. The maximum unlatching force shall not exceed 15 pounds (67 N).

Each door in a means of egress from a Group A or E occupancy having an occupant load of more than 50 and any Group H-1, H-2, H-3 or H-5 occupancy, shall not be provided with a latch or lock unless it is panic hardware or fire exit hardware.

Exception: A main exit of a Group A use in compliance with Section 1008.1.8.3, Item 2.

Electrical rooms with equipment rated 1200 amperes or more and over 6 feet (1.9 m) wide that contain overcurrent devices, switching devices, or control devices, with exit access doors must be equipped with panic hardware and doors must swing in the direction of egress.

If balanced doors are used and panic hardware is required, the panic hardware shall be the push-pad type and the pad shall not extend more then one-half the width of the door measured from the latch side.

Section 1008.2.1 Change to read as shown: (E35-03/04)

1008.2.1 Stadiums. Panic hardware is not required on gates surrounding stadiums where such gates are under constant immediate supervision while the public is present, and further provided that safe dispersal areas based on 3 square feet (0.28 m^2) per occupant are located between the fence and enclosed space. Such required safe dispersal areas shall not be located less than 50 feet (15 240 mm) from the enclosed space. See Section 1023.6 for means of egress from safe dispersal areas.

Section 1009.1 Change to read as shown: (E37-03/04)

1009.1 Stairway width. The width of stairways shall be determined as specified in Section 1005.1, but such width shall not be less than 44 inches (1118 mm). See Section 1007.3 for accessible means of egress stairways.

Exceptions:

1. Stairways serving an occupant load of 50 or less shall have a width of not less than 36 inches (914 mm).

2. Spiral stairways as provided for in Section 1009.9.

3. Aisle stairs complying with Section 1024.

4. Where an incline platform lift or stairway chairlift is installed on stairways serving occupancies in Group R-3, or within dwelling units in occupancies in Group R-2, both as applicable in Section 101.2, a clear passage width not less than 20 inches (508 mm) shall be provided. If the seat and platform can be folded when not in use, the distance shall be measured from the folded position.

Section 1009.3 Change to read as shown: (E40-03/04)

1009.3 Stair treads and risers. Stair riser heights shall be 7 inches (178 mm) maximum and 4 inches (102 mm) minimum. Stair tread depths shall be 11 inches (279 mm) minimum. The riser height shall be measured vertically between the leading edges of adjacent treads. The tread depth shall be measured horizontally between the vertical planes of the foremost projection of adjacent treads and at right angle to the tread's leading edge. Winder treads shall have a minimum tread depth of 11 inches (279 mm) measured at a right angle to the tread's leading edge at a point 12 inches (305 mm) from the side where the treads are narrower and a minimum tread depth of 10 inches (254 mm).

Exceptions:

1. Alternating tread devices in accordance with Section 1009.10.

2. Spiral stairways in accordance with Section 1009.9.

3. Aisle stairs in assembly seating areas where the stair pitch or slope is set, for sightline reasons, by the slope of the adjacent seating area in accordance with Section 1024.11.2.

4. In occupancies in Group R-3, as applicable in Section 101.2, within dwelling units in occupancies in Group R-2, as applicable in Section 101.2, and in occupancies in Group U, which are accessory to an occupancy in Group R-3, as applicable in Section 101.2, the maximum riser height shall be 7.75 inches (197 mm) and the minimum tread depth shall be 10 inches (254 mm), the minimum winder tread depth at the walk line shall be 10 inches (254 mm), and the minimum winder tread depth shall be 6 inches (152 mm). A nosing not less than 0.75 inch (19.1 mm) but not more than 1.25 inches (32 mm) shall be provided on stairways with solid risers where the tread depth is less than 11 inches (279 mm).

5. See the *International Existing Building Code* for the replacement of existing stairways.

Section 1009.3.1 Add new section to read as shown: (E40-03/04)

1009.3.1 Winder treads. Winder treads are not permitted in means of egress stairways except within a dwelling unit.

Exceptions:

1. Circular stairways in accordance with Section 1009.7

2. Spiral stairways in accordance with Section 1009.9

(Renumber subsequent sections)

Section 1009.3.1 Change to read as shown: (E40-03/04)

1009.3.1 Dimensional uniformity. Stair treads and risers shall be of uniform size and shape. The tolerance between the largest and smallest riser height or between the largest and smallest tread depth shall not exceed 0.375 inch (9.5 mm) in any flight of stairs. The greatest winder tread depth at the 12-inch (305 mm) walk line within any flight of stairs shall not exceed the smallest by more than 0.375 inch (9.5 mm) measured at a right angle to the treads leading edge

Exceptions:

1. Nonuniform riser dimensions of aisle stairs complying with Section 1024.11.2.

2. Consistently shaped winders, complying with Section 1009.3, differing from rectangular treads in the same stairway flight.

Where the bottom or top riser adjoins a sloping public way, walkway or driveway having an established grade and serving as a landing, the bottom or top riser is permitted to be reduced along the slope to less than 4 inches (102 mm) in height with the variation in height of the bottom or top riser not to exceed one unit vertical in 12 units horizontal (8-percent slope) of stairway width. The nosings or leading edges of treads at such nonuniform height risers shall have a distinctive marking stripe, different from any other nosing marking provided on the stair flight. The distinctive marking stripe shall be visible in descent of the stair and shall have a slip-resistant surface. Marking stripes shall have a width of at least 1 inch (25 mm) but not more than 2 inches (51 mm).

Section 1009.5.2 Change to read as shown: (E46-03/04)

1009.5.2 Outdoor conditions. Outdoor stairways and outdoor approaches to stairways shall be designed so that water will not accumulate on walking surfaces.

Section 1009.5.3 Relocated from Section 1019.1.5. Change to read as shown: (E96-03/04)

1009.5.3 Enclosures under stairways. The walls and soffits within enclosed usable spaces under enclosed and unenclosed stairways shall be protected by 1-hour fire-resistance-rated construction, or the fire-resistance rating of the stairway enclosure, whichever is greater. Access to the enclosed space shall not be directly from within the stair enclosure.

Exception: Spaces under stairways serving and contained within a single residential dwelling unit in Group R-2 or R-3, as applicable in Section 101.2, shall be permitted to be protected on the enclosed side with ½ inch (12.7 mm) gypsum board.

There shall be no enclosed usable space under exterior exit stairways unless the space is completely enclosed in 1-hour fire-resistance-rated construction. The open space under exterior stairways shall not be used for any purpose.

Section 1009.8 Winders. Delete without substitution: (E40-03/04)

Section 1009.12.2 Add new section to read as shown: (E51-03/04)

1009.12.2 Protection at roof hatch openings. Where the roof hatch opening providing the required access is located within 10 feet (3049 mm) of the roof edge, such roof access or roof edge shall be protected by guards installed in accordance with the provisions of Section 1012.

Section 1010.2 Change to read as shown: (E52-03/04)

1010.2 Slope. Ramps used as part of a means of egress shall have a running slope not steeper than one unit vertical in 12 units horizontal (8-percent slope). The slope of other pedestrian ramps shall not be steeper than one unit vertical in eight units horizontal (12.5-percent slope).

Exception: Aisle ramp slope in occupancies of Group A shall comply with Section 1024.11.

Section 1010.6.3 Change to read as shown: (E53-03/04)

1010.6.3 Length. The landing length shall be 60 inches (1525 mm) minimum.

Exceptions:

1. Landings in nonaccessible Group R-2 and R-3 individual dwelling units, as applicable in Section 101.2, are permitted to be 36 inches (914 mm) minimum.

2. Where the ramp is not a part of an accessible route, the length of the landing shall not be required to be more than 48 inches (1220 mm) in the direction of travel.

Section 1010.7 Change to read as shown: (E1-03/04)

1010.7 Ramp construction. All ramps shall be built of materials consistent with the types permitted for the type of construction of the building; except that wood handrails shall be permitted for all types of construction. Ramps used as an exit shall conform to the applicable requirements of Sections 1019.1 and 1019.1.1 through 1019.1.3 for exit enclosures.

Section 1010.7.2 Change to read as shown: (E46-03/04)

1010.7.2 Outdoor conditions. Outdoor ramps and outdoor approaches to ramps shall be designed so that water will not accumulate on walking surfaces.

Section 1010.9 Change to read as shown: (E54-03/04)

1010.9 Edge protection. Edge protection complying with Section 1010.9.1, 1010.9.2 or 1010.9.3 shall be provided on each side of ramp runs and at each side of ramp landings.

Exceptions:

1. Edge protection is not required on ramps not required to have handrails, provided they have flared sides that comply with the ICC A117.1 curb ramp provisions.

2. Edge protection is not required on the sides of ramp landings serving an adjoining ramp run or stairway.

3. Edge protection is not required on the sides of ramp landings having a vertical dropoff of not more than 0.5 inch (13 mm) within 10 inches (254 mm) horizontally of the required landing area.

Section 1010.9.3 Add new section to read as shown: (E54-03/04)

1010.9.3 Extended floor or ground surface. The floor or ground surface of the ramp run or landing shall extend 12 inches (305 mm) minimum beyond the inside face of a handrail complying with Section 1009.11.

Section 1011.1 Change to read as shown: (E3-03/04)

1011.1 Where required. Exits and exit access doors shall be marked by an approved exit sign readily visible from any direction of egress travel. Access to exits shall be marked by readily visible exit signs in cases where the exit or the path of egress travel is not immediately visible to the occupants. Exit sign placement shall be such that no point in a corridor is more than 100 feet (30 480 mm) or the listed viewing distance for the sign, whichever is less, from the nearest visible exit sign.

Exceptions:

1. Exit signs are not required in rooms or areas which require only one exit or exit access.

2. Main exterior exit doors or gates which obviously and clearly are identifiable as exits need not have exit signs where approved by the building official.

3. Exit signs are not required in occupancies in Group U and individual sleeping units or dwelling units in Group R-1, R-2 or R-3.

4. Exit signs are not required in sleeping areas in occupancies in Group I-3.

5. In occupancies in Groups A-4 and A-5, exit signs are not required on the seating side of vomitories or openings into seating areas where exit signs are provided in the concourse that are readily apparent from the vomitories. Egress lighting is provided to identify each vomitory or opening within the seating area in an emergency.

Section 1011.5.1 Change to read as shown: (E59-03/04)

1011.5.1 Graphics. Every exit sign and directional exit sign shall have plainly legible letters not less than 6 inches (152 mm) high with the principal strokes of the letters not less than 0.75 inch (19.1 mm) wide. The word "EXIT" shall have letters having a width not less than 2 inches (51 mm) wide except the letter "I," and the minimum spacing between letters shall not be less than 0.375 inch (9.5 mm). Signs larger than the minimum established in this section shall have letter widths, strokes and spacing in proportion to their height.

The word "EXIT" shall be in high contrast with the background and shall be clearly discernible when the exit sign illumination means is or is not energized. If a chevron directional indicator is provided as part of the exit sign, the construction shall be such that the direction of the chevron directional indicator cannot be readily changed.

Section 1012.5 Change to read as shown: (E51-03/04; E67-03/04)

1012.5 Mechanical equipment. Guards shall be provided where appliances, equipment, fans, roof hatch openings or other components that require service are located within 10 feet (3048 mm) of a roof edge or open side of a walking surface and such edge or open side is located more than 30 inches (762 mm) above the floor, roof or grade below. The guard shall be constructed so as to prevent the passage of a 21 inch (533 mm) diameter sphere. The guard shall extend not less than 30 inches beyond each end of such appliance, equipment, fan or component.

Section 1012.6 Add new section to read as shown: (E51-03/04)

1012.6 Roof access. Guards shall be provided where the roof hatch opening is located within 10 feet (3048 mm) of a roof edge or open side of a walking surface and such edge or open side is located more than 30 inches (762 mm) above the floor, roof or grade below. The guard shall be constructed so as to prevent the passage of a 21 inch (533 mm) diameter sphere.

Section 1013.2 Change to read as shown: (E68-03/04)

1013.2 Egress through intervening spaces. Egress through intervening spaces shall be controlled in accordance with this section.

1. Egress from a room or space shall not pass through adjoining or intervening rooms or areas, except where such adjoining rooms or areas are accessory to the area served; are not a high-hazard occupancy and provide a discernible path of egress travel to an exit.

   Exception: Means of egress are not prohibited through adjoining or intervening rooms or spaces in a Group H occupancy when the adjoining or intervening rooms or spaces are the same or a lesser hazard occupancy group.

2. Egress shall not pass through kitchens, storage rooms, closets or spaces used for similar purposes.

   Exception: Means of egress are not prohibited through a kitchen area serving adjoining rooms constituting part of the same dwelling unit or sleeping unit.

3. An exit access shall not pass through a room that can be locked to prevent egress.

4. Means of egress from dwelling units or sleeping areas shall not lead through other sleeping areas, toilet rooms or bathrooms.

Section 1013.2.2 Change to read as shown: (E3-03/04)

1013.2.2 Group I-2. Habitable rooms or suites in Group I-2 occupancies shall have an exit access door leading directly to a corridor.

Exceptions:

1. Rooms with exit doors opening directly to the outside at ground level.

2. Patient sleeping rooms are permitted to have one intervening room if the intervening room is not used as an exit access for more than eight patient beds.

3. Special nursing suites are permitted to have one intervening room where the arrangement allows for direct and constant visual supervision by nursing personnel.

4. For rooms other than patient sleeping rooms, suites of rooms are permitted to have one intervening room if the travel distance within the suite to the exit access door is not greater than 100 feet (30 480 mm) and are permitted to have two intervening rooms where the travel distance within the suite to the exit access door is not greater than 50 feet (15 240 mm).

Suites of sleeping rooms shall not exceed 5,000 square feet (465 m^2). Suites of rooms, other than patient sleeping rooms, shall not exceed 10,000 square feet (929 m^2). Any patient sleeping room, or any suite that includes patient sleeping rooms, of more than 1,000 square feet (93 m^2) shall have at least two exit access doors remotely located from each other. Any room or suite of rooms, other than patient sleeping rooms, of more than 2,500 square feet (232 m^2) shall have at least two access doors remotely located from each other. The travel distance between any point in a Group I-2 occupancy and an exit access door in the room shall not exceed 50 feet (15 240 mm). The travel distance between any point in a suite of sleeping rooms and an exit access door of that suite shall not exceed 100 feet (30 480 mm).

Section 1013.5 Change to read as shown: (E46-03/04)

1013.5 Egress balconies. Balconies used for egress purposes shall conform to the same requirements as corridors for width, headroom, dead ends and projections.

Section 1016.4 Change to read as shown: (E3-03/04)

1016.4 Air movement in corridors. Corridors shall not serve as supply, return, exhaust, relief or ventilation air ducts or plenums.

Exceptions:

1. Use of a corridor as a source of makeup air for exhaust systems in rooms that open directly onto

such corridors, including toilet rooms, bathrooms, dressing rooms, smoking lounges and janitor closets, shall be permitted provided that each such corridor is directly supplied with outdoor air at a rate greater than the rate of makeup air taken from the corridor.

2. Where located within a dwelling unit, the use of corridors for conveying return air shall not be prohibited.

3. Where located within tenant spaces of 1,000 square feet (93 m^2) or less in area, utilization of corridors for conveying return air is permitted.

Section 1018.1 Change to read as shown: (E89-03/04)

1018.1 Minimum number of exits. All rooms and spaces within each story shall be provided with and have access to the minimum number of approved independent exits required by Table 1018.1 based on the occupant load of the story, except as modified in Section 1014.1 or 1018.2. For the purposes of this chapter, occupied roofs shall be provided with exits as required for stories. The required number of exits from any story, basement or individual space shall be maintained until arrival at grade or the public way.

Table 1018.1 Change entire table to read as shown: (E89-03/04)

**TABLE 1018.1
MINIMUM NUMBER OF EXITS
FOR OCCUPANT LOAD**

OCCUPANT LOAD (persons per story)	MINIMUM NUMBER OF EXITS (per story)
1-500	2
501-1,000	3
More than 1,000	4

Section 1019.1 Change to read as shown: (E90-03/04; E91-03/04; E93-03/04)

1019.1 Enclosures required. Interior exit stairways and interior exit ramps shall be enclosed with fire barriers constructed in accordance with Section 706. Exit enclosures shall have a fire-resistance rating of not less than 2 hours where connecting four stories or more and not less than 1 hour where connecting less than four stories. The number of stories connected by the shaft enclosure shall include any basements but not any mezzanines. An exit enclosure shall not be used for any purpose other than means of egress.

Exceptions:

1. In other than Group H and I occupancies, a stairway is not required to be enclosed when serving an occupant load of less than 10 either not more than one story above, or one story below, the level of exit discharge, but not both.

2. Exits in buildings of Group A-5 where all portions of the means of egress are essentially open to the outside need not be enclosed.

3. Stairways serving and contained within a single residential dwelling unit or sleeping unit in occupancies in Group R-1, R-2 or R-3 are not required to be enclosed.

4. Stairways that are not a required means of egress element are not required to be enclosed where such stairways comply with Section 707.2.

5. Stairways in open parking structures which serve only the parking structure are not required to be enclosed.

6. Stairways in occupancies in Group I-3 as provided for in Section 408.3.6 are not required to be enclosed.

7. Means of egress stairways as required by Section 410.5.4 are not required to be enclosed.

8. In other than occupancy Groups H and I, a maximum of 50 percent of egress stairways serving one adjacent floor are not required to be enclosed, provided at least two means of egress are provided from both floors served by the unenclosed stairways. Any two such interconnected floors shall not be open to other floors.

9. In other than occupancy Groups H and I, interior egress stairways serving only the first and second stories of a building equipped throughout with an automatic sprinkler system in accordance with Section 903.3.1.1 are not required to be enclosed, provided at least two means of egress are provided from both floors served by the unenclosed stairways. Such interconnected stories shall not be open to other stories.

Section 1019.1.3 Change to read as shown: (FS53-03/04)

1019.1.3 Ventilation. Equipment and ductwork for exit enclosure ventilation shall comply with one of the following items:

1. Such equipment and ductwork shall be located exterior to the building and shall be directly connected to the exit enclosure by ductwork enclosed in construction as required for shafts.

2. Where such equipment and ductwork is located within the exit enclosure, the intake air shall be taken directly from the outdoors and the exhaust air shall be discharged directly to the outdoors, or such air shall

be conveyed through ducts enclosed in construction as required for shafts.

3. Where located within the building, such equipment and ductwork shall be separated from the remainder of the building, including other mechanical equipment, with construction as required for shafts.

In each case, openings into the fire-resistance-rated construction shall be limited to those needed for maintenance and operation and shall be protected by opening protectives in accordance with Section 715 for shaft enclosures.

Exit enclosure ventilation systems shall be independent of other building ventilation systems.

Section 1019.1.4 Change to read as shown: (E1-03/04)

1019.1.4 Exit enclosure exterior walls. Exterior walls of an exit enclosure shall comply with the requirements of Section 704 for exterior walls. Where nonrated walls or unprotected openings enclose the exterior of the stairway and the walls or openings are exposed by other parts of the building at an angle of less than 180 degrees (3.14 rad), the building exterior walls within 10 feet (3048 mm) horizontally of a nonrated wall or unprotected opening shall be constructed as required for a minimum 1-hour fire-resistance rating with ¾-hour opening protectives. This construction shall extend vertically from the ground to a point 10 feet (3048 mm) above the topmost landing of the stairway or to the roof line, whichever is lower.

Section 1019.1.5 Enclosures under stairways. Relocated to be Section 1009.5.3): (E96-03/04)

Section 1019.1.7 Change to read as shown: (E1-03/04)

1019.1.7 Stairway floor number signs. A sign shall be provided at each floor landing in interior exit enclosures connecting more than three stories designating the floor level, the terminus of the top and bottom of the stair enclosure and the identification of the stair. The signage shall also state the story of, and the direction to the exit discharge and the availability of roof access from the stairway for the fire department. The sign shall be located 5 feet (1524 mm) above the floor landing in a position that is readily visible when the doors are in the open and closed positions.

Section 1024.3 Change to read as shown: (E105-03/04)

1024.3 Assembly other exits. In addition to having access to a main exit, each level of an occupancy in Group A having an occupant load greater than 300 shall be provided with additional means of egress that shall provide an egress capacity for at least one-half of the total occupant load served by that level and comply with Section 1014.2.

Exception: In assembly occupancies where there is no well-defined main exit or where multiple main exits are provided, exits shall be permitted to be distributed around the perimeter of the building provided that the total width of egress is not less than 100 percent of the required width.

Section 1024.5.1 Change to read as shown: (E1-03/04)

1024.5.1 Enclosure of balcony openings. Interior stairways and other vertical openings shall be enclosed in an exit enclosure as provided in Section 1019.1, except that stairways are permitted to be open between the balcony and the main assembly floor in occupancies such as theaters, churches and auditoriums. At least one accessible means of egress is required from a balcony or gallery level containing accessible seating locations in accordance with Section 1007.3 or 1007.4.

CHAPTER 11
ACCESSIBILITY

Section 1103.1 Change to read as shown: (E110-03/04)

1103.1 Where required. Sites, buildings, structures, facilities, elements and spaces, temporary or permanent, shall be accessible to persons with physical disabilities.

Section 1103.2 Change to read as shown: (E110-03/04)

1103.2 General exceptions. Sites, buildings, structures, facilities, elements and spaces shall be exempt from this chapter to the extent specified in this section.

Section 1104.3 Change exceptions to read as shown: (E114-03/04; E116-03/04)

1104.3 Connected spaces. When a building, or portion of a building, is required to be accessible, an accessible route shall be provided to each portion of the building, to accessible building entrances connecting accessible pedestrian walkways and the public way.

Exception: In assembly areas with fixed seating required to be accessible, an accessible route shall not be required to serve fixed seating where wheelchair spaces or designated aisle seats required to be on an accessible route are not provided.

Section 1104.5 Change to read as shown: (E116-03/04)

1104.5 Location. Accessible routes shall coincide with or be located in the same area as a general circulation path. Where the circulation path is interior, the accessible route shall also be interior. Where only one accessible route is provided, the accessible route shall not pass through kitchens, storage rooms, restrooms, closets or similar spaces.

Exceptions:

1. Accessible routes from parking garages contained within and serving Type B dwelling units are not required to be interior.

2. A single accessible route is permitted to pass through a kitchen or storage room in an Accessible, Type A or Type B dwelling unit.

Section 1105.1.2 Add exception to read as shown: (E120-03/04)

1105.1.2 Entrances from tunnels or elevated walkways. Where direct access is provided for pedestrians from a pedestrian tunnel or elevated walkway to a building or facility, at least one entrance to the building or facility from each tunnel or walkway shall be accessible.

Exception: Where the entrance serves stories containing only dwelling units and sleeping units intended to be occupied as a residence, the entrance is required to be accessible only if the story contains required Accessible units, required Type A units, or is required by Section 1107.7.1.4 to contain Type B units.

Section 1106.6 Change exceptions to read as shown: (E119-03/04)

1106.6 Location. Accessible parking spaces shall be located on the shortest accessible route of travel from adjacent parking to an accessible building entrance. Accessible parking spaces shall be dispersed among the various types of parking facilities provided. In parking facilities that do not serve a particular building, accessible parking spaces shall be located on the shortest route to an accessible pedestrian entrance to the parking facility. Where buildings have multiple accessible entrances with adjacent parking, accessible parking spaces shall be dispersed and located near the accessible entrances.

Exceptions:

1. In multilevel parking structures, van-accessible parking spaces are permitted on one level.

2. Parking spaces shall be permitted to be located in different parking facilities if substantially equivalent or greater accessibility is provided in terms of distance from an accessible entrance or entrances, parking fee and user convenience.

Section 1107.7.1 Change to read as shown: (E120-03/04)

1107.7.1 Buildings without elevator service. Where no elevator service is provided in a building, only the dwelling and sleeping units that are located on stories indicated in Sections 1107.7.1.1 through 1107.7.1.4 are required to be Type A and Type B units. The number of Type A units shall be determined in accordance with Section 1107.6.2.1.1.

Sections 1107.7.1.3 and 1107.7.1.4 Add new sections to read as shown: (E120-03/04)

1107.7.1.3 Additional stories with entrances through fire walls. Where an entrance is provided to a story of a building from an accessible story of an adjacent building by an opening in a fire wall, all dwelling units and sleeping units intended to be occupied as a residence on that story shall be Type B units, provided that the planned finished floor elevation within 5 feet (1524 mm) of each side of the door does not include a change in level in excess of 12 inches (305 mm).

1107.7.1.4 Additional stories with entrances from bridges or elevated walkways. Where an entrance is provided to a story of a building from an accessible story of an adjacent building by a bridge or elevated walkway, all dwelling units and sleeping units intended to be occupied as a residence on that story shall be Type B units, provided the slope between the planned finished floor elevation at the building entrance and the planned finish floor elevation at the bridge or elevated walkway connection to the adjacent building is 10 percent or less.

Section 1108.2.3 Integration. Delete without substitution: (E122-03/04)

Section 1108.2.4 Combine with Section 1108.2.4.1 to read as shown: (E122-03/04)

1108.2.4 Dispersion of wheelchair spaces in multilevel assembly seating areas. In multilevel assembly seating areas, wheelchair spaces shall be provided on the main floor level and on one of each two additional floor or mezzanine levels. Wheelchair spaces shall be provided in each luxury box, club box and suite within assembly facilities.

Exceptions:

1. In multilevel assembly spaces utilized for worship services, where the second floor or mezzanine level contains 25 percent or less of the total seating capacity, wheelchair spaces shall be permitted to all be located on the main level.

2. In multilevel assembly seating where the second floor or mezzanine level provides 25 percent or less of the total seating capacity and 300 or fewer seats, wheelchair spaces shall be permitted to all be located on the main level.

Section 1108.2.4.1 Multilevel assembly seating areas. Deleted (see Section 1108.2.4) (E122-03/04)

Section 1108.2.5 Companion seats. Delete without substitution: (E122-03/04)

Section 1108.4.2 Change to read as shown: (E126-03/04)

1108.4.2 Holding cells. Central holding cells and court-floor holding cells shall comply with Sections 1108.4.2.1 and 1108.4.2.2.

Section 1108.4.3.1 Change exception to read as shown: (E127-03/04)

1108.4.3.1 Cubicles and counters. At least 5 percent, but no fewer than one, of cubicles shall be accessible on both the visitor and detainee sides. Where counters are provided, at least one shall be accessible on both the visitor and detainee sides.

> **Exception:** This requirement shall not apply to the detainee side of cubicles or counters at noncontact visiting areas not serving Accessible unit holding cells.

Section 1109.2 Change to read as shown (current exception 4 deleted): (E129-03/04)

1109.2 Toilet and bathing facilities. Toilet rooms and bathing facilities shall be accessible. Where a floor level is not required to be connected by an accessible route, the only toilet rooms or bathing facilities provided within the facility shall not be located on the inaccessible floor. At least one of each type of fixture, element, control or dispenser in each accessible toilet room and bathing facility shall be accessible.

> **Exceptions:**
>
> 1. In toilet rooms or bathing facilities accessed only through a private office, not for common or public use, and intended for use by a single occupant, any of the following alternatives are allowed:
>   1.1. Doors are permitted to swing into the clear floor space provided the door swing can be reversed to meet the requirements in ICC A117.1,
>   1.2. The height requirements for the water closet in ICC A117.1 are not applicable,
>   1.3. Grab bars are not required to be installed in a toilet room, provided that reinforcement has been installed in the walls and located so as to permit the installation of such grab bars, and
>   1.4. The requirement for height, knee and toe clearance shall not apply to a lavatory.
> 2. This section is not applicable to toilet and bathing facilities that serve dwelling units or sleeping units that are not required to be accessible by Section 1107.
> 3. Where multiple single-user toilet rooms or bathing facilities are clustered at a single location and contain fixtures in excess of the minimum required number of plumbing fixtures, at least 5 percent, but not less than one room for each use at each cluster, shall be accessible.
> 4. Where no more than one urinal is provided in a toilet room or bathing facility, the urinal is not required to be accessible.
> 5. Toilet rooms that are part of critical-care or intensive-care patient sleeping rooms are not required to be accessible.

Section 1109.3 Change to read as shown: (current exception 2 deleted): (E135-03/04)

1109.3 Sinks. Where sinks are provided, at least 5 percent, but not less than one, provided in accessible spaces shall comply with ICC A117.1.

> **Exception:** Mop or service sinks are not required to be accessible.

Section 1109.5 Change to read as shown: (E136-03/04)

1109.5 Drinking fountains. Where drinking fountains are provided on an exterior site, on a floor, or within a secured area, the drinking fountains shall be provided in accordance with Sections 1109.5.1 and 1109.5.2.

Sections 1109.5.1 and 1109.5.2 Add new sections to read as shown: (E136-03/04)

1109.5.1 Minimum number. No fewer than two drinking fountains shall be provided. One drinking fountain shall comply with the requirements for people who use a wheelchair and one drinking fountain shall comply with the requirements for standing persons.

> **Exception.** A single drinking fountain that complies with the requirements for people who use a wheelchair and standing persons shall be permitted to be substituted for two separate drinking fountains.

1109.5.2 More than the minimum number. Where more than the minimum number of drinking fountains specified in Section 1109.5.1 are provided, 50 percent of the total number of drinking fountains provided shall comply with the requirements for persons who use a wheelchair and 50 percent of the total number of drinking fountains provided shall comply with the requirements for standing persons.

Exception. Where 50 percent of the drinking fountains yields a fraction, 50 percent shall be permitted to be rounded up or down provided that the total number of drinking fountains complying with this section equals 100 percent of drinking fountains.

Section 1109.8.3 Change to read as shown: (E140-03/04)

1109.8.3 Coat hooks and shelves. Where coat hooks and shelves are provided in toilet rooms or toilet compartments, or in dressing, fitting or locker rooms, at least one of each type shall be accessible and shall be provided in accessible toilet rooms without toilet compartments, accessible toilet compartments and accessible dressing, fitting and locker rooms.

Section 1109.10 Change to read as shown: (E121-03/04)

1109.10 Assembly area seating. Assembly areas with fixed seating shall comply with Section 1108.2 for accessible seating and assistive listening devices.

Section 1109.12.2 Change to read as shown (delete exception): (E142-03/04)

1109.12.2 Check-out aisles. Where check-out aisles are provided, accessible check-out aisles shall be provided in accordance with Table 1109.12.2. Where check-out aisles serve different functions, at least one accessible check-out aisle shall be provided for each function. Where checkout aisles serve different functions, accessible check-out aisles shall be provided in accordance with Table 1109.12.2 for each function. Where check-out aisles are dispersed throughout the building or facility, accessible check-out aisles shall also be dispersed. Traffic control devices, security devices and turnstiles located in accessible check-out aisles or lanes shall be accessible.

Section 1109.15 Stairways. Delete section without substitution: (E143-03/04)

CHAPTER 14
EXTERIOR WALLS

Section 1403.2 Change to read as shown: (FS128-03/04)

1403.2 Weather protection. Exterior walls shall provide the building with a weather-resistant exterior wall envelope. The exterior wall envelope shall include flashing, as described in Section 1405.3. The exterior wall envelope shall be designed and constructed in such a manner as to prevent the accumulation of water within the wall assembly by providing a water-resistive barrier behind the exterior veneer, as described in Section 1404.2 and a means for draining water that enters the assembly to the exterior. Protection against condensation in the exterior wall assembly shall be provided in accordance with the *International Energy Conservation Code*.

Exceptions:

1. A weather-resistant exterior wall envelope shall not be required over concrete or masonry walls designed in accordance with Chapters 19 and 21, respectively.

2. Compliance with the requirements for a means of drainage, and the requirements of Sections 1405.2 and 1405.3, shall not be required for an exterior wall envelope that has been demonstrated through testing to resist wind-driven rain, including joints, penetrations and intersections with dissimilar materials, in accordance with ASTM E 331 under the following conditions:

    2.1. Exterior wall envelope test assemblies shall include at least one opening, one control joint, one wall/eave interface and one wall sill. All tested openings and penetrations shall be representative of the intended end-use configuration.

    2.2. Exterior wall envelope test assemblies shall be at least 4 feet by 8 feet (1219 mm by 2438 mm) in size.

    2.3. Exterior wall envelope assemblies shall be tested at a minimum differential pressure of 6.24 pounds per square foot (psf) (0.297 kN/m^2).

    2.4. Exterior wall envelope assemblies shall be subjected to a minimum test exposure duration of 2 hours.

The exterior wall envelope design shall be considered to resist wind-driven rain where the results of testing indicate that water did not penetrate control joints in the exterior wall envelope, joints at the perimeter of openings or intersections of terminations with dissimilar materials.

Section 1403.6 Change to read as shown: (FS129-03/04)

1403.6 Flood resistance. For buildings in flood hazard areas as established in Section 1612.3, exterior walls extending below the design flood elevation shall be resistant to water damage. Wood shall be pressure-preservative treated in accordance with AWPA U1 for the species, product and end use using a preservative listed in AWPA standards P1/13, P2, P3, P5, P8 or P9 or decay-resistant heartwood of redwood, black locust or cedar.

Section 1404.2 Change to read as shown: (FS131-03/04)

1404.2 Water-resistive barrier. A minimum of one layer of No. 15 asphalt felt, complying with ASTM D 226 for Type 1 felt, shall be attached to the studs or sheathing, with flashing as described in Section 1405.3, in such a manner as to provide a continuous water-resistive barrier behind the exterior wall veneer.

Section 1405.4 Change to read as shown: (FS136-03/04)

1405.4 Wood veneers. Wood veneers on exterior walls of buildings of Type I, II, III and IV construction shall be not less than 1-inch (25 mm) nominal thickness, 0.438-inch (11.1 mm) exterior hardboard siding or 0.375-inch (9.5 mm) exterior-type wood structural panels or particleboard and shall conform to the following:

1. The veneer shall not exceed three stories in height, measured from the grade plane. Where fire-retardant-treated wood is used, the height shall not exceed four stories.

2. The veneer is attached to or furred from a noncombustible backing that is fire-resistance rated as required by other provisions of this code.

3. Where open or spaced wood veneers (without concealed spaces) are used, they shall not project more than 24 inches (610 mm) from the building wall.

Section 1405.12.2 Add new section to read as shown: (FS138-03/04)

1405.12.2 Window sills. In occupancy Group R, one-and two family and multiple single family dwellings, where the rough opening for the sill portion of an operable window is located more than 72 inches above the grade or other surface below, the rough opening for the sill, or lowest part of the operable portion of the window, shall be a minimum of 24 inches above the finished floor of the room in which the window is located.

Exception. Windows whose openings will not allow a 4 inch diameter sphere to pass through the opening when the opening is in its largest opened position.

Section 1406.2.4 Change to read as shown: (FS94-03/04)

1406.2.4 Fireblocking. Where the combustible exterior wall covering is furred from the wall and forms a solid surface, the distance between the back of the covering and the wall shall not exceed 1.625 inches (41 mm) and the space thereby created shall be fireblocked in accordance with Section 717.

Section 1407.1.1 Add new section to read as shown: (FS140-03/04)

1407.1.1 Plastic core. The plastic core of the MCM shall not contain foam plastic insulation as defined in Section 2602.1.

CHAPTER 15
ROOF ASSEMBLIES AND ROOFTOP STRUCTURES

Section 1504.3.2 Add exception to read as shown: (FS142-03/04)

1504.3.2 Metal panel roof systems. Metal panel roof systems through fastened or standing seam shall be tested in accordance with UL 580 or ASTM E 1592.

Exception: Metal roofs constructed of cold-formed steel, where the roof deck acts as the roof covering and provides both weather protection and support for structural loads, shall be permitted to be designed and tested in accordance with the applicable referenced structural design standard in Section 2209.1.

Section 1504.7 Change to read as shown: (FS144-03/04)

1504.7 Impact resistance. Roof coverings installed on low-slope roofs (roof slope < 2:12) in accordance with Section 1507 shall resist impact damage based on the results of tests conducted in accordance with ASTM D 3746, ASTM D 4272, CGSB 37-GP-52M or the Resistance to Foot Traffic Test (Section 5.5) of FM 4470.

Section 1504.8 Add new section and table to read as shown: (S1-03/04)

1504.8 Gravel and stone. Gravel or stone shall not be used on the roof of a building located in a hurricane-prone region as defined in Section 1609.2, or on any other building with a mean roof height exceeding that permitted by Table 1504.8 based on the exposure category and basic wind speed at the building site.

TABLE 1504.8
MAXIMUM ALLOWABLE MEAN ROOF HEIGHT PERMITTED FOR BUILDINGS WITH GRAVEL OR STONE ON THE ROOF IN AREAS OUTSIDE A HURRICANE-PRONE REGION

BASIC WIND SPEED FROM FIGURE 1609 (mph)[b]	MAXIMUM MEAN ROOF HEIGHT (ft)[a,d] Exposure category[c]		
	B	C	D
85	170	60	30
90	110	35	15
95	75	20	NP
100	55	15	NP
105	40	NP	NP
110	30	NP	NP
115	20	NP	NP
120	15	NP	NP
Greater than 120	NP	NP	NP

For SI: 1 foot = 304.8 mm; 1 mile per hour = 0.447 m/s

a. Mean roof height in accordance with Section 1609.2.
b. For intermediate values of basic wind speed, the height associated with the next higher value of wind speed shall be used, or direct interpolation is permitted.
c. Exposure category determined from Section 1609.4 when Section 1609.6 is used to determine design wind pressures. When the provisions of Section 6 of ASCE 7 are used to determine the design wind pressures, the exposure category shall be the most restrictive of all the categories determined for the building.
d. NP = gravel and stone not permitted for any roof height.

Section 1507.2.3 Change to read as shown: (FS148-03/04)

1507.2.3 Underlayment. Unless otherwise noted, required underlayment shall conform to ASTM D 226, Type I, ASTM D 4869, Type I, or ASTM D6757.

Section 1507.2.7 Change to read as shown: (FS150-03/04)

1507.2.7 Attachment. Asphalt shingles shall have the minimum number of fasteners required by the manufacturer and Section 1504.1. Asphalt shingles shall be secured to the roof with not less than four fasteners per strip shingle or two fasteners per individual shingle. Where the roof slope exceeds 20 units vertical in 12 units horizontal (166-percent slope), special methods of fastening are required. For roofs located where the basic wind speed in accordance with Figure 1609 is 110 mph (49 m/s) or greater, special methods of fastening are required. Special fastening methods shall be tested in accordance with ASTM D3161, Class F. In these areas asphalt shingle wrappers shall bear a label indicating compliance with ASTM D3161, Class F.

Section 1507.2.8 Change to read as shown: (FS151-03/04)

1507.2.8 Underlayment application. For roof slopes from two units vertical in 12 units horizontal (17-percent slope), up to four units vertical in 12 units horizontal (33-percent slope), underlayment shall be two layers applied in the following manner. Apply a minimum 19-inch-wide (483 mm) strip of underlayment felt parallel with and starting at the eaves, fastened sufficiently to hold in place. Starting at the eave, apply 36-inch-wide (914 mm) sheets of underlayment overlapping successive sheets 19 inches (483 mm) and fastened sufficiently to hold in place. Distortions in the underlayment shall not interfere with the ability of the shingles to seal. For roof slopes of four units vertical in 12 units horizontal (33-percent slope) or greater, underlayment shall be one layer applied in the following manner. Underlayment shall be applied shingle fashion, parallel to and starting from the eave and lapped 2 inches (51 mm), fastened sufficiently to hold in place. Distortions in the underlayment shall not interfere with the ability of the shingles to seal.

Section 1507.2.9.2 Change to read as shown: (FS153-03/04; FS155-03/04)

1507.2.9.2 Valleys. Valley linings shall be installed in accordance with the manufacturer's instructions before applying shingles. Valley linings of the following types shall be permitted:

1. For open valleys (valley lining exposed) lined with metal, the valley lining shall be at least 16 inches (406 mm) wide and of any of the corrosion-resistant metals in Table 1507.2.9.2.

2. For open valleys, valley lining of two plies of mineral-surfaced roll roofing complying with ASTM D3909 or ASTM D6380 shall be permitted. The bottom layer shall be 18 inches (457 mm) and the top layer a minimum of 36 inches (914 mm) wide.

3. For closed valleys (valleys covered with shingles), valley lining of one ply of smooth roll roofing complying with ASTM D6380, Class S Type III, Class M Type II or ASTM D3909 and at least 36 inches (914 mm) wide or types as described in Items 1 and 2 above shall be permitted. Specialty underlayment shall comply with ASTM D 1970.

Section 1507.3.3 Change to read as shown: (FS157-03/04)

1507.3.3 Underlayment. Unless otherwise noted, required underlayment shall conform to: ASTM D 226, Type II; ASTM D 2626 or ASTM D 6380, Class M mineral-surfaced roll roofing.

Section 1507.3.5 Change to read as shown: (FS158-03/04)

1507.3.5 Concrete tile. Concrete roof tile shall comply with ASTM C1492.

Delete Table 1507.3.5 Transverse Breaking Strength of Concrete Roof Tile without substitution: (FS158-03/04)

Section 1507.3.9 Change to read as shown: (FS159-03/04)

1507.3.9 Flashing. At the juncture of the roof vertical surfaces, flashing and counterflashing shall be provided in accordance with the manufacturer's installation instructions, and where of metal, shall not be less than 0.019-inch (0.48 mm) (No. 26 galvanized sheet gage) corrosion-resistant. The valley flashing shall extend at least 11 inches (279 mm) from the centerline each way and have a splash diverter rib not less than 1 inch (25 mm) high at the flow line formed as part of the flashing. Sections of flashing shall have an end lap of not less than 4 inches (102 mm). For roof slopes of three units vertical in 12 units horizontal (25-percent slope) and over, the valley flashing shall have a 36-inch-wide (914 mm) underlayment of either one layer of Type I underlayment running the full length of the valley, or a self-adhering polymer-modified bitumen sheet complying with ASTM D1970, in addition to other required underlayment. In areas where the average daily temperature in January is 25°F (-4°C) or less or where there is a possibility of ice forming along the eaves causing a backup of water, the metal valley flashing underlayment shall be solid cemented to the roofing underlayment for slopes under seven units vertical in 12 units horizontal (58-percent slope) or install self-adhering polymer-modified bitumen sheet.

Section 1507.4.2 Change to read as shown: (FS161-03/04)

1507.4.2 Deck slope. Minimum slopes for metal roof panels shall comply with the following:

1. The minimum slope for lapped, nonsoldered seam metal roofs without applied lap sealant shall be three units vertical in 12 units horizontal (25-percent slope).

2. The minimum slope for lapped, nonsoldered seam metal roofs with applied lap sealant shall be one-half unit vertical in 12 units horizontal (4-percent slope). Lap sealants shall be applied in accordance with the approved manufacturer's installation instructions.

3. The minimum slope for standing seam of roof systems shall be one-quarter unit vertical in 12 units horizontal (2-percent slope).

Table 1507.4.3 Add new entries as shown: (FS160-03/04; FS162-03/04)

TABLE 1507.4.3
METAL ROOF COVERINGS

Roof Covering Type	Standard Application Rate/Thickness
Aluminum Alloy-coated steel	ASTM A875 GF60
Aluminum-coated steel	ASTM A463 T2 65
Galvanized steel	ASTM A653 G-90 zinc-coated

(Portions of table not shown do not change)

Section 1507.4.4 Change to read as shown: (FS164-03/04)

1507.4.4 Attachment. Metal roof panels shall be secured to the supports in accordance with the approved manufacturer's fasteners. In the absence of manufacturer recommendations, the following fasteners shall be used:

1. Galvanized fasteners shall be used for steel roofs.

2. 300 series stainless-steel fasteners shall be used for copper roofs.

3. Stainless-steel fasteners are acceptable for all types of metal roofs.

Section 1507.5.3 Change to read as shown: (FS165-03/04)

1507.5.3 Underlayment. Underlayment shall comply with ASTM D226, Type I or ASTM D4869. In areas where the average daily temperature in January is 25°F (-4°C) or less or where there is a possibility of ice forming along the eaves causing a backup of water, an ice barrier that consists of at least two layers of underlayment cemented

2004 SUPPLEMENT TO THE IBC

together or of a self-adhering polymer-modified bitumen sheet, shall be used in lieu of normal underlayment and extend from the eave's edge to a point at least 24 inches (610 mm) inside the exterior wall line of the building.

Exception: Detached accessory structures that contain no conditioned floor area.

Section 1507.5.6 Change to read as shown: (FS166-03/04)

1507.5.6 Flashing. Roof valley flashing shall be of corrosion-resistant metal of the same material as the roof covering or shall comply with the standards in Table 1507.4.3. The valley flashing shall extend at least 8 inches (203 mm) from the centerline each way and shall have a splash diverter rib not less than 0.75 inch (19.1 mm) high at the flow line formed as part of the flashing. Sections of flashing shall have an end lap of not less than 4 inches (102 mm). In areas where the average daily temperature in January is 25°F (-4°C) or less or where there is a possibility of ice forming along the eaves causing a backup of water, the metal valley flashing shall have a 36-inch-wide (914 mm) underlayment directly under it consisting of either one layer of underlayment running the full length of the valley, or a self-adhering polymer-modified bitumen sheet complying with ASTM D1970, in addition to underlayment required for metal roof shingles. The metal valley flashing underlayment shall be solid cemented to the roofing underlayment for roof slopes under seven units vertical in 12 units horizontal (58-percent slope) or install self-adhering polymer-modified bitumen sheet.

Section 1507.6.3 Change to read as shown: (FS165-03/04)

1507.6.3 Underlayment. Underlayment shall comply with ASTM D226, Type I or ASTM D4869. In areas where the average daily temperature in January is 25°F (-4°C) or less or where there is a possibility of ice forming along the eaves causing a backup of water, an ice barrier that consists of at least two layers of underlayment cemented together or of a self-adhering polymer-modified bitumen sheet, shall extend from the eave's edge to a point at least 24 inches (610 mm) inside the exterior wall line of the building.

Exception: Detached accessory structures that contain no conditioned floor area.

Section 1507.6.4 Change to read as shown: (FS167-03/04)

1507.6.4 Material standards. Mineral-surfaced roll roofing shall conform to ASTM D3909 or ASTM D6380.

Section 1507.8.3 Change to read as shown: (FS165-03/04)

1507.8.3 Underlayment. Underlayment shall comply with ASTM D 226, Type I or ASTM D4869. In areas where the average daily temperature in January is 25°F (-4°C) or less or where there is a possibility of ice forming along the eaves causing a backup of water, an ice barrier that consists of at least two layers of underlayment cemented together or of a self-adhering polymer-modified bitumen sheet shall extend from the eave's edge to a point at least 24 inches (610 mm) inside the exterior wall line of the building.

Exception: Detached accessory structures that contain no conditioned floor area.

Section 1507.8.7 Change to read as shown: (FS168-03/04)

1507.8.7 Flashing. At the juncture of the roof and vertical surfaces, flashing and counterflashing shall be provided in accordance with the manufacturer's installation instructions, and where of metal, shall not be less than 0.019-inch (0.48 mm) (No. 26 galvanized sheet gage) corrosion-resistant metal. The valley flashing shall extend at least 11 inches (279 mm) from the centerline each way and have a splash diverter rib not less than 1 inch (25 mm) high at the flow line formed as part of the flashing. Sections of flashing shall have an end lap of not less than 4 inches (102 mm). For roof slopes of three units vertical in 12 units horizontal (25-percent slope) and over, the valley flashing shall have a 36-inch-wide (914 mm) underlayment of either one layer of Type I underlayment running the full length of the valley, or a self-adhering polymer-modified bitumen sheet complying with ASTM D1970, in addition to other required underlayment. In areas where the average daily temperature in January is 25°F (-4°C) or less or where there is a possibility of ice forming along the eaves causing a backup of water, the metal valley flashing underlayment shall be solid cemented to the roofing underlayment for slopes under seven units vertical in 12 units horizontal (58-percent slope) or install self-adhering polymer-modified bitumen sheet.

Section 1507.9.3 Change to read as shown: (FS165-03/04)

1507.9.3 Underlayment. Underlayment shall comply with ASTM D226, Type I or ASTM D4869. In areas where the average daily temperature in January is 25°F (-4°C) or less or where there is a possibility of ice forming along the eaves causing a backup of water, an ice barrier that consists of at least two layers of underlayment cemented together or a self-adhering polymer-modified bitumen sheet shall extend from the edge of the eave to a point at least 24 inches (610 mm) inside the exterior wall line of the building.

Exception: Detached accessory structures that contain no conditioned floor area.

Table 1507.9.5 Change to read as shown: (FS169-03/04)

TABLE 1507.9.5
WOOD SHAKE MATERIAL REQUIREMENTS

Material	Minimum Grades	Applicable Grading Rules
Preservative-treated taper sawn shakes of Southern pine treated in accordance with AWPA Standard U1 (Commodity Specification A, Use Category 3B and Section 5.6)	1 or 2	TFS

(Portions of table and notes not shown do not change)

Section 1507.9.8 Change to read as shown: (FS170-03/04)

1507.9.8 Flashing. At the juncture of the roof and vertical surfaces, flashing and counterflashing shall be provided in accordance with the manufacturer's installation instructions, and where of metal, shall not be less than 0.019-inch (0.48 mm) (No. 26 galvanized sheet gage) corrosion-resistant metal. The valley flashing shall extend at least 11 inches (279 mm) from the centerline each way and have a splash diverter rib not less than 1 inch (25 mm) high at the flow line formed as part of the flashing. Sections of flashing shall have an end lap of not less than 4 inches (102 mm). For roof slopes of 3 units vertical in 12 units horizontal (25-percent slope) and over, the valley flashing shall have a 36-inch-wide (914 mm) underlayment of either one layer of Type I underlayment running the full length of the valley, or a self-adhering polymer-modified bitumen sheet complying with ASTM D1970, in addition to other required underlayment. In areas where the average daily temperature in January is 25°F (-4°C) or less or where there is a possibility of ice forming along the eaves causing a backup of water, the metal valley flashing underlayment shall be solid cemented to the roofing underlayment for slopes under seven units vertical in 12 units horizontal (58-percent slope) or install self-adhering polymer-modified bitumen sheet.

Section 1507.12.2 Change to read as shown: (FS171-03/04)

1507.12.2 Material standards. Thermoset single-ply roof coverings shall comply with ASTM D 4637, ASTM D 5019 or CGSB 37-GP-52M.

Section 1507.13.2 Change to read as shown: (FS172-03/04; FS173-03/04)

1507.13.2 Material standards. Thermoplastic single-ply roof coverings shall comply with ASTM D4434, ASTM D6754, ASTM D6878 or CGSB 37-GP-54M.

Section 1507.15.2 Change to read as shown: (FS174-03/04)

1507.15.2 Material standards. Liquid-applied roof coatings shall comply with ASTM C836, ASTM C957, ASTM D1227 or ASTM D3468, ASTM D6083 or ASTM D6694.

Section 1509.4 Change to read as shown: (FS177-03/04)

1509.4 Cooling towers. Cooling towers in excess of 250 square feet (23.2 m^2) in base area or in excess of 15 feet (4572 mm) high where located on building roofs more than 50 feet (15 240 mm) high shall be of noncombustible construction. Cooling towers shall not exceed one-third of the supporting roof area.

Exception: Drip boards and the enclosing construction of wood not less than 1 inch (25 mm) nominal thickness, provided the wood is covered on the exterior of the tower with noncombustible material.

Section 1510.3 Change to read as shown: (FS179-03/04)

1510.3 Recovering versus replacement. New roof coverings shall not be installed without first removing all existing layers of roof coverings where any of the following conditions occur:

1. Where the existing roof or roof covering is water soaked or has deteriorated to the point that the existing roof or roof covering is not adequate as a base for additional roofing.

2. Where the existing roof covering is wood shake, slate, clay, cement or asbestos-cement tile.

3. Where the existing roof has two or more applications of any type of roof covering.

Exceptions:

1. Complete and separate roofing systems, such as standing-seam metal roof systems, that are designed to transmit the roof loads directly to the building's structural system and that do not

rely on existing roofs and roof coverings for support, shall not require the removal of existing roof coverings.

2. Metal panel, metal shingle, and concrete and clay tile roof coverings shall be permitted to be installed over existing wood shake roofs when applied in accordance with Section 1510.4.

3. The application of a new protective coating over an existing spray polyurethane foam roofing system shall be permitted without tear-off of existing roof coverings.

CHAPTER 16
STRUCTURAL DESIGN

Section 1602 DEFINITIONS. Delete the following definitions: (S2-03/04)

BASE SHEAR.

BASIC SEISMIC-FORCE-RESISTING SYSTEMS.
Bearing wall system.
Building frame system.
Dual system.
Inverted pendulum system.
Moment-resisting frame system.
Shear wall-frame interactive system.

BOUNDARY MEMBERS.
Boundary element.

CANTILEVERED COLUMN SYSTEM.

COLLECTOR ELEMENTS.

CONFINED REGION.

DEFORMABILITY.
High deformability element.
Limited deformable element.
Low deformability element.

DEFORMATION.
Limit deformation.
Ultimate deformation.

ELEMENT.
Ductile element.
Limited ductile element.
Nonductile element.

EQUIPMENT SUPPORT.

FLEXIBLE EQUIPMENT CONNECTIONS.

FRAME.
Braced frame.
Concentrically braced frame (CBF).
Eccentrically braced frame (EBF)
Ordinary concentrically braced frame (OCBF).
Special concentrically braced frame (SCBF).
Moment frame.

JOINT.

P-DELTA EFFECT.

SHALLOW ANCHORS.

SHEAR PANEL.

SHEAR WALL.

SPACE FRAME.

SPECIAL TRANSVERSE REINFORCEMENT.

WALL, LOAD BEARING.

WALL, NONLOAD BEARING.

Section 1602 Change definition of Notations to read as shown: (S3-03/04)

NOTATIONS
$F =$ Load due to fluids with well-defined pressures and maximum heights.
$H =$ Load due to lateral earth pressures, ground water pressure or pressure of bulk materials.

Delete the Notation P = Ponding Load

(Notations not shown remain unchanged)

Section 1603.1 Change to read as shown: (S4-03/04)

1603.1 General. Construction documents shall show the size, section and relative locations of structural members with floor levels, column centers and offsets fully dimensioned. The design loads and other information pertinent to the structural design required by Sections 1603.1.1 through 1603.1.8 shall be clearly indicated on the construction documents for parts of the building or structure.

Exception: Construction documents for buildings constructed in accordance with the conventional light-frame construction provisions of Section 2308 shall indicate the following structural design information:

1. Floor and roof live loads.

2. Ground snow load, P_g.

3. Basic wind speed (3-second gust), miles per hour (mph) (km/hr) and wind exposure.

4. Seismic design category and site class.

5. Flood design data, if located in flood hazard areas established in 1612.3.

Section 1603.1.6 Change to read as shown: (S5-03/04; S6-03/04)

1603.1.6 Flood design data. For buildings located in whole or in part in flood hazard areas as established in Section 1612.3, the documentation pertaining to design, if required in Section 1612.5, shall be included and the following information, referenced to the datum on the community's Flood Insurance Rate Map (FIRM), shall be shown, regardless of whether flood loads govern the design of the building:

1. In flood hazard areas not subject to high-velocity wave action, the elevation of proposed lowest floor, including basement.

2. In flood hazard areas not subject to high-velocity wave action, the elevation to which any nonresidential building will be dry floodproofed.

3. In flood hazard areas subject to high-velocity wave action, the proposed elevation of the bottom of the lowest horizontal structural member of the lowest floor, including basement.

Section 1604.8.2 Change to read as shown: (S8-03/04)

1604.8.2 Concrete and masonry walls. Concrete and masonry walls shall be anchored to floors, roofs and other structural elements that provide lateral support for the wall. Such anchorage shall provide a positive direct connection capable of resisting the horizontal forces specified in this chapter but not less than a minimum strength design horizontal force of 280 plf (4.10 kN/m) of wall, substituted for "*E*" in the load combinations of Section 1605.2 or 1605.3. Walls shall be designed to resist bending between anchors where the anchor spacing exceeds 4 feet (1219 mm). Required anchors in masonry walls of hollow units or cavity walls shall be embedded in a reinforced grouted structural element of the wall. See Sections 1609.6.2.2 for wind design requirements and see Sections 1620 and 1621 for earthquake design requirements.

Section 1605.2.2 Change to read as shown: (S3-03/04)

1605.2.2 Other loads. Where *F*, *H* or *T* is to be considered in design, the load combinations of Section 2.3.2 of ASCE 7 shall be used, except the factor f_2 on snow load, S, in combination 5 shall be in accordance with Section 1605.2.1. Where F_a is to be considered in design, the load combinations of Section 2.3.3 of ASCE 7 shall be used.

Section 1605.3.1.1 Change to read as shown: (S9-03/04)

1605.3.1.1 Load reduction. It is permitted to multiply the combined effect of two or more variable loads by 0.75 and add to the effect of dead load. The combined load used in design shall not be less than the sum of the effects of dead load and any one of the variable loads.

Increases in allowable stresses specified in the appropriate materials section of this code or referenced standard shall not be used with the load combinations of Section 1605.3.1 except that a duration of load increase shall be permitted in accordance with Chapter 23.

Section 1605.3.1.2 Change to read as shown: (S3-03/04)

1605.3.1.2 Other loads. Where *F*, *H* or *T* are to be considered in design, the load combinations of Section 2.4.1 of ASCE 7 shall be used. Where F_a is to be considered in design, the load combinations of Section 2.4.2 of ASCE 7 shall be used.

Section 1605.3.2.1 Change to read as shown: (S3-03/04)

1605.3.2.1. Other loads. Where F, H or T are to be considered in design, 1.0 times each applicable load shall be added to the combinations specified in Section 1605.3.2.

Section 1605.5 Change to read as shown: (S11-03/04)

1605.5 Heliports and helistops. Heliport and helistop landing areas shall be designed for the following loads, combined in accordance with Section 1605:

1. Dead load, *D*, plus the gross weight of the helicopter, D_h, plus snow load, *S*.

2. Dead load, *D*, plus two single concentrated impact loads, *L*, approximately 8 feet (2438 mm) apart applied anywhere on the touchdown pad (representing each of the helicopter's two main landing gear, whether skid type or wheeled type), having a magnitude of 0.75 times the gross weight of the helicopter. Both loads acting together total 1.5 times the gross weight of the helicopter.

3. Dead load, *D*, plus a uniform live load, *L*, of 100 psf (4.79 kN/m^2).

2004 SUPPLEMENT TO THE IBC

Table 1607.1 Change to read as shown: (S14-03/04)

TABLE 1607.1
MINIMUM UNIFORMLY DISTRIBUTED LIVE LOADS AND MINIMUM CONCENTRATED LIVE LOADS[g]

OCCUPANCY OR USE	UNIFORM (psf)	CONCENTRATED (lbs.)
27. Residential One- and two-family dwellings		
Uninhabitable attics without storage[i]	10	
Uninhabitable attics with limited storage[i, j, k]	20	

(Portions of table not shown do not change)

a. through h. (No change to current text)

i. Attics without storage are those where the maximum clear height between joist and rafter is less than 42 inches, or where there are not 2 or more adjacent trusses with the same web configuration capable of containing a rectangle 42 inches high by 2 feet wide, or greater, located within the plane of the truss. For attics without storage, this live load need not be assumed to act concurrently with any other live load requirements.

j. For attics with limited storage and constructed with trusses, this live load need only be applied to those portions of the bottom chord where two or more adjacent trusses with the same web configuration contain a rectangle 42 inches high or greater by 2 feet wide or greater, located within the plane of the truss. The rectangle shall fit between the top of the bottom chord and the bottom of any other truss member, provided that each of the following criteria is met:
   i. The attic area is accessible by a pull-down stairway or framed opening in accordance with Section 1209.2; and
   ii. The truss shall have a bottom chord pitch less than 2:12.
   iii Bottom chords of trusses shall be designed for the greater of actual imposed dead loads or 10 psf, uniformly distributed over the entire span.

k. Attic spaces served by a fixed stair shall be designed to support the minimum live load specified for habitable attics and sleeping rooms.

Section 1607.9.2 Change to read as shown: (S12-03/04)

1607.9.2 Alternate floor live load reduction. As an alternative to Section 1607.9.1, floor live loads are permitted to be reduced in accordance with the following provisions. Such reductions shall apply to slab systems, beams, girders, columns, piers, walls and foundations.

1. A reduction shall not be permitted in Group A occupancies.

2. A reduction shall not be permitted when the live load exceeds 100 psf (4.79 kN/m^2) except that the design live load for columns may be reduced by 20 percent.

3. For live loads not exceeding 100 psf (4.79 kN/m^2), the design live load for any structural member supporting 150 square feet (13.94 m^2) or more is permitted to be reduced in accordance with the following equation:

$R = r(A - 150)$ **(Equation 16-22)**

For SI: $R = r(A - 13.94)$

Such reduction shall not exceed the smallest of:

1. 40 percent for horizontal members,

2. 10 psf (0.48 KN/m^2 for horizontal members in passenger vehicle garages

3. 60 percent for vertical members, nor

4. R as determined by the following equation.

$R = 23.1(1 + D/L_o)$ **(Equation 16-23)**

where:
A = Area of floor or roof supported by the member, square feet (m^2).
D = Dead load per square foot (m^2) of area supported.
L_o = Unreduced live load per square foot (m^2) of area supported.
R = Reduction in percent.
r = Rate of reduction equal to 0.08 percent for floors.

Section 1609.1.4 Change to read as shown: (S17-03/04)

1609.1.4 Protection of openings. In wind-borne debris regions, glazing that receives positive external pressure in the lower 60 feet (18 288 mm) in buildings shall be assumed to be openings unless such glazing is impact resistant or protected with an impact-resistant covering meeting the requirements of an approved impact-resisting standard or ASTM E1996 and of ASTM E1886 referenced therein as follows:

1. Glazed openings located within 30 feet (9144 mm) of grade shall meet the requirements of the Large Missile Test of ASTM E1996.
2. Glazed openings located more than 30 feet (9144 mm) above grade shall meet the provisions of the Small Missile Test of ASTM E1996.

Exceptions:

1. Wood structural panels with a minimum thickness of 7/16 inch (11.1 mm) and maximum panel span of 8 feet (2438 mm) shall be permitted for opening protection in one- and two-story buildings. Panels shall be precut so that they shall be attached to the framing surrounding the opening containing the product with the glazed opening. Panels shall be secured with the attachment hardware provided. Attachments shall be designed to resist the components and cladding loads determined in accordance with the provisions of Section 1609.6.1.2. Attachment in accordance with Table 1609.1.4 is permitted for buildings with a mean roof height of 33 feet (10,058 mm) or less where wind speeds do not exceed 130 mph (57.2 m/s).
2. Buildings in Category I as defined in Table 1604.5, including production greenhouses as defined in Section 1608.3.3.

Table 1609.1.4 Change entire table to read as shown: (S17-03/04)

TABLE 1609.1.4
WIND-BORNE DEBRIS PROTECTION FASTENING SCHEDULE FOR WOOD STRUCTURAL PANELS[a,b,c,d]

FASTENER TYPE	FASTENER SPACING (inches)		
	Panel Span ≤ 4 feet	4 feet < Panel Span ≤ 6 feet	6 feet < Panel Span ≤ 8 feet
No. 6 Screws	16	12	9
No. 8 Screws	16	16	12

For SI: 1 inch = 25.4 mm, 1 foot = 304.8 mm, 1 pound = 4.4 N. 1 mile per hour = 0.44 m/s.

a. This table is based on a maximum wind speed (3-second gust) of 130 mph and mean roof height of 33 feet or less.
b. Fasteners shall be installed at opposing ends of the wood structural panel. Fasteners shall be located a minimum of 1 inch from the edge of the panel.
c. Fasteners shall be long enough to penetrate through the exterior wall covering and a minimum of 1¾ inch into wood wall framing and a minimum of 1¼ inch into concrete block or concrete or into steel framing by at least three threads. Fasteners shall be located a minimum of 2½ inch from the edge of concrete block or concrete.
d. Where screws are attached to masonry or masonry/stucco, they shall be attached utilizing vibration-resistant anchors having a minimum withdrawal capacity of 490 pounds.

Section 1609.7.2 Change to read as shown: (S21-03/04; S22-03/04)

1609.7.2 Roof coverings. Roof coverings shall comply with Section 1609.7.1.

Exception: Rigid tile roof coverings that are air permeable and installed over a roof deck complying with Section 1609.7.1 are permitted to be designed in accordance with Section 1609.7.3.

Asphalt shingles installed over a roof deck complying with IBC Section 1609.7.1 shall be tested to determine the resistance of the sealant to uplift forces using ASTM D6381.

Asphalt shingles installed over a roof deck complying with Section 1609.7.1 are permitted to be designed using UL2390 to determine appropriate uplift and force coefficients applied to the shingle.

Section 1612.1 Change to read as shown: (S24-03/04)

1612.1 General. Within flood hazard areas as established in Section 1612.3, all new construction of buildings, structures and portions of buildings and structures, including substantial improvement and restoration of substantial damage to buildings and structures, shall be designed and constructed to resist the effects of flood hazards and flood loads. For buildings that are located in more than one flood hazard area, the provisions associated with the most restrictive flood hazard area shall apply.

Section 1613 Change, add or delete the following definitions as shown: (S2-03/04)

BASE SHEAR. Total design lateral force or shear at the base.

BASIC SEISMIC FORCE-RESISTING-SYSTEMS

Bearing wall system. A structural system without a complete vertical load-carrying space frame. Bearing walls or bracing elements provide support for substantial vertical loads. Seismic lateral force resistance is provided by shear walls or braced frames.

Building frame system. A structural system with an essentially complete space frame providing support for vertical loads. Seismic lateral force resistance is provided by shear walls or braced frames.

Dual system. A structural system with an essentially complete space frame providing support for vertical loads. Seismic lateral force resistance is provided by shear walls and/or braced frames together with supplemental moment frames.

Inverted pendulum system. A structure with a large portion of its mass concentrated at the top; therefore, having essentially one degree of freedom in horizontal translation. Seismic lateral force resistance is provided by the columns acting as cantilevers.

Moment resisting frame system. A structural system with an essentially complete space frame providing support for vertical loads. Seismic lateral force resistance is provided by moment frames.

Shear wall-frame interactive system. A structural system which uses combinations of shear walls and frames designed to resist seismic lateral forces in proportion to their rigidities, considering interaction between shear walls and frames on all levels. Support of vertical loads is provided by the same shear walls and frames.

CANTILEVERED COLUMN SYSTEM. A structural system relying on column elements that cantilever from a fixed base and have minimal rotational resistance capacity at the top with lateral forces applied essentially at the top and are used for lateral resistance.

DEFORMABILITY. The ratio of the ultimate deformation to the limit deformation.

High-deformability element. An element whose deformability is not less than 3.5 when subjected to four fully reversed cycles at the limit deformation.

Limited-deformability element. An element that is neither a low deformability or a high deformability element.

Low-deformability element. An element whose deformability is 1.5 or less.

DEFORMATION

Limit deformation. Two times the initial deformation that occurs at a load equal to 40 percent of the maximum strength.

Ultimate deformation. The deformation at which failure occurs and which shall be deemed to occur if the sustainable load reduces to 80 percent or less of the maximum strength.

Delete the definition of "DISPLACEMENT RESTRAINT SYSTEM."

Delete the definition of "EFFECTIVE DAMPING."

Delete the definition of "EFFECTIVE STIFFNESS."

ELEMENT

Ductile element. An element capable of sustaining large cyclic deformations beyond the attainment of its nominal strength without any significant loss of strength.

Limited ductile element. An element that is capable of sustaining moderate cyclic deformations beyond the attainment of nominal strength without significant loss of strength.

Nonductile element. An element having a mode of failure that results in an abrupt loss of resistance when the element is deformed beyond the deformation corresponding to the development of its nominal strength. Nonductile elements cannot reliably sustain significant deformation beyond that attained at their nominal strength.

EQUIPMENT SUPPORT. Those structural members or assemblies of members or manufactured elements, including braces, frames, lugs, snubbers, hangers or saddles, that transmit gravity load and operating load between the equipment and the structure.

FLEXIBLE EQUIPMENT CONNECTIONS. Those connections between equipment components that permit rotational and/or transactional movement without degradation of performance.

FRAME

Braced frame. An essentially vertical truss, or its equivalent, of the concentric or eccentric type that is provided in a building frame system or dual system to resist lateral forces.

Concentrically braced frame (CBF). A braced frame in which the members are subjected primarily to axial forces. Concentrically braced frames are categorized as either ordinary concentrically braced frames (OCBF) or special concentrically braced frames (SCBF).

Eccentrically braced frame (EBF). A diagonally braced frame in which at least one end of each brace frames into a beam a short distance from a beam-column connection or from another diagonal brace.

Moment frame. A frame in which lateral forces are resisted primarily by the development of flexure in beams, columns and their connections. Moment frames are categorized as "intermediate moment frames" (IMF), "ordinary moment frames" (OMF), and "special moment frames" (SMF).

Delete the definition of "ISOLATION INTERFACE."

Delete the definition of "ISOLATION SYSTEM."

Delete the definition of "ISOLATOR UNIT."

P-DELTA EFFECT. The second-order effect on shears, axial forces and moments of frame members induced by axial loads on a laterally displaced building frame.

SHALLOW ANCHORS. Shallow anchors are those with embedment length-to-diameter ratios of less than 8.

SHEAR PANEL. A floor, roof or wall component sheathed to act as a shear wall or diaphragm.

Delete the definition of "SHEAR WALL FRAME INTERACTIVE SYSTEM."

SPACE FRAME. A structure composed of interconnected members, other than bearing walls, that is capable of supporting vertical loads and that also may provide resistance to seismic lateral forces.

(Definitions not shown are unchanged)

Section 1614.1 Change to read as shown: (S25-03/04)

1614.1 Scope. Every structure, and portion thereof, shall as a minimum, be designed and constructed to resist the effects of earthquake motions and assigned a seismic design category as set forth in Section 1616.3. Structures determined to be in Seismic Design Category A need only comply with Section 1616.4.

Exceptions:

1. Structures designed in accordance with the provisions of Sections 9.1 through 9.6, 9.13 and 9.14 of ASCE 7 shall be permitted.

2. Detached one- and two-family dwellings, as applicable in Section 101.2, in Seismic Design Categories A, B and C, or located where the mapped short-period spectral response acceleration, S_S, is less than 0.4 g, are exempt from the requirements of Sections 1613 through 1622.

3. The seismic-force-resisting system of wood frame buildings that conform to the provisions of Section 2308 are not required to be analyzed as specified in Section 1616.1.

4. Agricultural storage structures intended only for incidental human occupancy are exempt from the requirements of Sections 1613 through 1623.

5. Structures located where mapped short-period spectral response acceleration, S_S, determined in accordance with Section 1615.1, is less than or equal to 0.15 g and where the mapped spectral response acceleration at 1-second period, S_1, determined in accordance with Section 1615.1, is less than or equal to 0.04 g shall be categorized as Seismic Design Category A. Seismic Design Category A structures need only comply with Section 1616.4.

Section 1615.1.5 Change to read as shown: (S26-03/04)

1615.1.5 Site classification for seismic design. Site classification for Site Class C, D or E shall be determined from Table 1615.1.5.

The notations presented below apply to the upper 100 feet (30 480 mm) of the site profile. Profiles containing distinctly different soil and/or rock layers shall be subdivided into those layers designated by a number that ranges from 1 to n at the bottom where there is a total of n distinct layers in the upper 100 feet (30 480 mm). The symbol, i, then refers to any one of the layers between 1 and n.

where:

v_{si} = The shear wave velocity in feet per second (m/s).
d_i = The thickness of any layer between 0 and 100 feet (30 480 mm).

where:

$$\bar{V}_s = \frac{\sum_{i=1}^{n} d_i}{\sum_{i=1}^{n} \frac{d_i}{v_{si}}} \qquad \text{(Equation 16-44)}$$

$$\sum_{i=1}^{n} d_i = 100 \text{ feet (30 480 mm)}$$

N_i is the Standard Penetration Resistance (ASTM D 1586) not to exceed 100 blows/foot (305 mm) as directly measured in the field without corrections. When refusal is met for a rock layer, N_i shall be taken as 100 blows/foot (305 mm).

$$\bar{N} = \frac{\sum_{i=1}^{n} d_i}{\sum_{i=1}^{n} \frac{d_i}{N_i}} \qquad \text{(Equation 16-45)}$$

where N_i and d_i in Eq. 16-45 are for cohesionless soil, cohesive soil, and rock layers.

2004 SUPPLEMENT TO THE IBC

$$\overline{N}_{ch} = \frac{d_s}{\sum_{i=1}^{m} \frac{d_i}{N_i}}$$ (Equation 16-46)

where:

$$\sum_{i=1}^{m} d_i = d_s$$

Use d_i and N_i for cohesionless soils layers only in Equation 16-46.

d_s = The total thickness of cohesionless soil layers in the top 100 feet (30 480 mm).
m = The number of cohesionless soil layers in the top 100 feet (30 480 mm).
s_{ui} = The undrained shear strength in psf (kPa), not to exceed 5,000 psf (240 kPa), ASTM D 2166 or D 2850.

$$\overline{s}_u = \frac{d_c}{\sum_{i=1}^{k} \frac{d_i}{s_{ui}}}$$ (Equation 16-47)

where:

$$\sum_{i=1}^{k} d_i = d_c$$

d_c = The total thickness of cohesive soil layers in the top 100 feet (30 480 mm).
k = The number of cohesive soil layers in the top 100 feet (30 480 mm).
PI = The plasticity index, ASTM D 4318.
w = The moisture content in percent, ASTM D 2216.

The shear wave velocity for rock, Site Class B, shall be either measured on site or estimated by a geotechnical engineer or engineering geologist/seismologist for competent rock with moderate fracturing and weathering. Softer and more highly fractured and weathered rock shall either be measured on site for shear wave velocity or classified as Site Class C.

The hard rock category, Site Class A, shall be supported by shear wave velocity measurements either on site or on profiles of the same rock type in the same formation with an equal or greater degree of weathering and fracturing. Where hard rock conditions are known to be continuous to a depth of 100 feet (30 480 mm), surficial shear wave velocity measurements are permitted to be extrapolated to assess v_s.

The rock categories, Site Classes A and B, shall not be used if there is more than 10 feet (3048 mm) of soil between the rock surface and the bottom of the spread footing or mat foundation.

Section 1616.5 Change to read as shown: (S27-03/04)

1616.5 Building configuration. Buildings shall be classified as regular or irregular based on the criteria in Section 9.5.2.3 of ASCE 7.

Delete the following sections and tables without substitution: (S27-03/04)

1616.5.1 Building configuration (for use in the simplified analysis procedure of Section 1617.5).

1615.5.1.1 Plan irregulatiry.

1615.5.1.2 Vertical irregularity.

Table 1616.5.1.1 Plan Structural Irregularities

Table 1616.5.1.2 Vertical Structural Irregularities.

Section 1616.6 Change to read as shown: (S27-03/04)

1616.6 Analysis procedures. A structural analysis conforming to one of the types permitted in Section 9.5.2.5.1 of ASCE 7 shall be made for all structures. The analysis shall form the basis for determining the seismic forces, E and E_m, to be applied in the load combinations of Section 1605 and shall form the basis for determining the design drift as required by Section 9.5.2.8 of ASCE 7.

Exception: Structures assigned to Seismic Design Category A.

Section 1616.6.1 Simplified analysis. Delete section without substitution: (S27-03/04)

Section 1617.1 Change to read as shown: (S27-03/04)

1617.1 Seismic load effect E and E_m. The seismic load effect, E, for use in the basic load combinations of Sections 1605.2 and 1605.3 shall be determined from Section 9.5.2.7 of ASCE 7. The maximum seismic load effect, E_m, for use in the special seismic load combination of Section 1605.4 shall be the special seismic load determined from Section 9.5.2.7.1 of ASCE 7.

Delete the following sections without substitution: (S27-03/04)

1617.1.1 Seismic load effects, E and E_m (for use in the simplified analysis procedure of Section 1617.5).

1617.1.1.1 Seismic load effect, E.

1617.1.1.2 Maximum seismic load effect, E_m.

Section 1617.2 Redundancy and Section 1617.2.1 ASCE 7, Sections 9.5.2.4.2 and 9.5.4.3. Delete without substitution: (S28-03/04)

Sections 1617.2.2, 1617.2.2.1, 1617.2.2.2 Renumber as 1617.2, 1617.2.1 and 1617.2.2 and revise to read as shown: (S28-03/04)

1617.2. Redundancy. A redundancy coefficient, ρ, shall be assigned to each structure in accordance with this section.

1617.2.1 Seismic Design Category A, B or C. For structures assigned to Seismic Design Category A, B or C (see Section 1616), the value of the redundancy coefficient ρ is 1.0.

1617.2.2 Seismic Design Category D, E or F. For structures in Seismic Design Category D, E or F (see Section 1616), the redundancy coefficient, ρ, shall be taken as the largest of the values of, ρ_i, calculated at each story "i" of the structure in accordance with Equation 16-54, as follows:

$$\rho_i = 2 - \frac{20}{r_{max_i}\sqrt{A_i}}$$ (Equation 16-54)

For SI:

$$\rho_i = 2 - \frac{6.1}{r_{max_i}\sqrt{A_i}}$$

where:

r_{maxi} = The ratio of the design story shear resisted by the most heavily loaded single element in the story to the total story shear, for a given direction of loading.

r_{maxi} = For braced frames, the value is equal to the horizontal force component in the most heavily loaded brace element divided by the story shear.

r_{maxi} = For moment frames, r_{maxi} shall be taken as the maximum of the sum of the shears in any two adjacent columns in a moment frame divided by the story shear. For columns common to two bays with moment-resisting connections on opposite sides at the level under consideration, it is permitted to use 70 percent of the shear in that column in the column shear summation.

r_{maxi} = For shear walls, r_{maxi}, shall be taken as the maximum value of the product of the shear in the wall or wall pier and $10/l_w$ ($3.3/l_w$ for SI), divided by the story shear, where l_w is the length of the wall or wall pier in feet (m). In light-framed construction, the value of the ratio of $10/l_w$ need not be greater than 1.0.

r_{maxi} = For dual systems, r_{maxi}, shall be taken as the maximum value defined above, considering all lateral-load-resisting elements in the story. The lateral loads shall be distributed to elements based on relative rigidities considering the interaction of the dual system. For dual systems, the value of ρ need not exceed 80 percent of the value calculated above.

A_i = The floor area in square feet (m²) of the diaphragm level immediately above the story.

For a story with a flexible diaphragm immediately above r_{maxi} shall be permitted to be calculated from an analysis that assumes rigid diaphragm behavior and ρ need not exceed 1.25.

The value, ρ, shall not be less than 1.0, and need not exceed 1.5.

Calculation of r_{maxi} need not consider the effects of accidental torsion and any dynamic amplification of torsion required by Section 9.5.5.5.2 of ASCE 7.

For structures with seismic-force-resisting systems in any direction comprised solely of special moment frames, the seismic-force-resisting system shall be configured such that the value of ρ calculated in accordance with this section does not exceed 1.25 for structures assigned to Seismic Design Category D, and does not exceed 1.1 for structures assigned to Seismic Design Category E or F.

Exception: The calculated value of ρ is permitted to exceed these limits when the design story drift, Δ, does not exceed Δ_a/ρ for any story where Δ_a is the allowable story drift from Section 1617.3.

The value ρ shall be permitted to be taken equal to 1.0 in the following circumstances:

1. When calculating displacements for dynamic amplification of torsion in Section 9.5.5.5.2 of ASCE 7.

2. When calculating deflections, drifts and seismic shear forces related to Sections 9.5.5.7.1 and 9.5.5.7.2 of ASCE 7.

3. For design calculations required by Section 1620, 1621 or 1622.

For structures with vertical combinations of seismic-force-resisting systems, the value, ρ, shall be determined independently for each seismic-force-resisting system. The redundancy coefficient of the lower portion shall not be less than the following:

$$\rho_L = \frac{R_L \rho_u}{R_u}$$

2004 SUPPLEMENT TO THE IBC

where:

ρ_L = ρ of lower portion
R_L = R of lower portion
ρ_u = ρ of upper portion
R_u = R of upper portion

Section 1617.3 Change to read as shown: (S27-03/04)

1617.3 Deflection and drift limits. The provisions given in Section 9.5.2.8 of ASCE 7 shall be used.

Delete the following sections and tables without substitution: (S27-03/04)

1617.3.1 Deflection and drift limits (for use in the simplified analysis procedure of Section 1617.5).

Table 1617.3.1 Allowable Story Drift, Δ_a (inches)[a].

1617.4 Equivalent lateral force procedure for seismic design of buildings.

1617.5 Simplified analysis procedure for seismic design of buildings.

1617.5.1 Seismic base shear.

1617.5.2 Vertical distribution.

1617.5.3 Horizontal distribution.

1617.5.4 Design drift.

Section 1617.6, Table 1617.6, Sections 1617.6.1, 1617.6.2, 1617.6.3. Change to read as shown: (S27-03/04)

1617.6 Seismic-force-resisting systems. The provisions given in Section 9.5.2.2 of ASCE 7 shall be used except as modified in Sections 1617.6.1 through 1617.6.3. In addition, the systems identified in Table 1617.6 shall use the provisions of Section 9.5.2.2 of ASCE 7 with the parameters identified in Table 1617.6 and subject to the limitations contained therein.

**TABLE 1617.6
DESIGN COEFFICIENTS AND FACTORS FOR
BASIC SEISMIC-FORCE-RESISTING SYSTEMS**

BASIC SEISMIC FORCE-RESISTING SYSTEMS	RESPONSE MODIFICATION COEFFICIENT, R	SYSTEM OVER-STRENGTH FACTOR Ω[b]	DEFLECTION AMPLIFICATION FACTOR, C_d	SYSTEM LIMITATIONS AND BUILDING HEIGHT LIMITATIONS (FT) BY SEISMIC DESIGN CATEGORY AS DETERMINED IN SECTION 1616.3[a]				
				A & B	C	D	E	F
1. Bearing Wall Systems								
Ordinary plain prestressed masonry shear walls	1½	2½	1¼	NL	NP	NP	NP	NP
Intermediate prestressed masonry shear walls	2½	2½	2½	NL	35	NP	NP	NP
Special prestressed masonry shear walls	4½	2½	3½	NL	35	35	35	35
2. Building Frame Systems								
Ordinary plain prestressed masonry shear walls	1½	2½	1¼	NL	NP	NP	NP	NP
Intermediate prestressed masonry shear walls	3	2½	2½	NL	35	NP	NP	NP
Special prestressed masonry shear walls	4½	2½	4	NL	35	35	35	35

For SI: 1 foot = 304.8 mm

a. NL = Not limited and NP = Not permitted
b. The tabulated value of the overstrength factor Ω_o is permitted to be reduced by subtracting ½ for structures with flexible diaphragms but shall not be taken as less than 2.0 for any structure.

(Portions of table and footnotes not shown do not change)

1617.6.1 ASCE 7, Table 9.5.2.2. Modify Table 9.5.2.2 as follows:

1. Bearing wall systems: Ordinary reinforced masonry shear walls shall use a response modification coefficient of 2½. Light-framed walls sheathed with wood structural panels rated for shear resistance or steel sheets shall use a response modification coefficient of 6½.

2. Building frame systems: Ordinary reinforced masonry shear walls shall use a response modification coefficient of 3. Light-framed walls sheathed with wood structural panels rated for shear resistance or steel sheets shall use a response modification coefficient of 7.

3. Dual systems with intermediate moment frames capable of resisting at least 25 percent of prescribed seismic forces. Special steel concentrically braced frames shall use a deflection amplification factor of 4.

1617.6.2 ASCE 7, Section 9.5.2.2.2.1. Modify Section 9.5.2.2.2.1 by adding an exception 3 as follows:

3. The following two-stage static analysis procedure is permitted to be used for structures having a flexible upper portion supported on a rigid lower portion where both portions of the structure considered separately can be classified as being regular, the average story stiffness of the lower portion is at least 10 times the average story stiffness of the upper portion and the period of the entire structure is not greater than 1.1 times the period of the upper portion considered as a separate structure fixed at the base.

   3.1 The flexible upper portion shall be designed as a separate structure using the appropriate values of R and ρ.

   3.2 The rigid lower portion shall be designed as a separate structure using the appropriate values of R and ρ. The reactions from the upper portion shall be those determined from the analysis of the upper portion amplified by the ratio of the R/ρ of the upper portion over R/ρ of the lower portion. This ratio shall not be less than 1.0.

1617.6.3 ASCE 7, Section 9.5.2.2.4.3. Modify Section 9.5.2.2.4.3 by changing exception to read as follows:

Exception: Reinforced concrete frame members not designed as part of the seismic-force-resisting system and slabs shall comply with Section 31.11 of Ref.9.9-1.

Delete the following sections and table without substitution: (S27-03/04)

1617.6.2 Seismic-force-resisting systems (for use in the Simplified analysis procedure of Section 1617.5).

1617.6.2.1 Dual systems.

1617.6.2.2 Combination along the same axis.

1617.6.2.3 Combinations of framing systems.

1617.6.2.3.1 Combination framing factor.

1617.6.2.3.2 Combination framing detailing requirements.

1617.6.2.4 System limitations for Seismic Design Category D, E or F.

1617.6.2.4.1 Limited building height.

1617.6.2.4.2 Interaction effects.

1617.6.2.4.3 Deformational compatibility.

1617.6.2.4.4 Special moment frame.

Table 1617.6.2 Design Coefficients and Factors for Basic Seismic-Force-Resisting Systems.

Section 1620.1 Change to read as shown: (S27-03/04; S29-03/04)

1620.1 Structural component design and detailing. The design and detailing of the components of the seismic-force-resisting system shall comply with the requirements of Section 9.5.2.6 of ASCE 7, except as modified in Sections 1620.1.1 through 1620.1.4, in addition to the nonseismic requirements of this code.

Add new section to read as shown: (S29-03/04)

1620.1.1 ASCE 7, Section 9.5.2.6.1. Modify Section 9.5.2.6.1 of ASCE 7 to read as follows:
  9.5.2.6.1 Seismic Design Category A. Structures assigned to Seismic Design Category A shall conform to the requirements of Sections 1616.4.2 and 1616.4.3 of the *International Building Code*.

(Renumber subsequent sections)

Delete the following sections without substitution: (S27-03/04)

1620.2 Structural component design and detailing (for use in the simplified analysis procedure of Section 1617.5).

2004 SUPPLEMENT TO THE IBC

1620.2.1 Second order load effects.

1620.2.2 Openings.

1620.2.3 Discontinuities in vertical system.

1620.2.4 Connections.

1620.2.5 Diaphragms.

1620.2.6 Collector elements.

1620.2.7 Bearing walls and shear walls.

1620.2.8 Inverted pendulum-type structures.

1620.2.9 Elements supporting discontinuous walls or frames.

1620.2.10 Direction of seismic load.

1620.3 Seismic Design Category C.

1620.3.1 Anchorage of concrete or masonry walls.

1620.3.2 Direction of seismic load.

1620.4 Seismic Design Category D.

1620.4.1 Plan or vertical irregularities.

1620.4.2 Vertical seismic forces.

1620.4.3 Diaphragms.

1620.4.4 Collector elements.

1620.4.5 Building separations.

1620.4.6 Anchorage of concrete or masonry walls to flexible diaphragms.

1620.5 Seismic Design Category E or F.

1620.5.1 Plan or vertical irregularities.

CHAPTER 17
STRUCTURAL TESTS AND SPECIAL INSPECTIONS

Section 1703.4.1 Change to read as shown: (S32-03/04)

1703.4.1 Research and investigation. Sufficient technical data shall be submitted to the building official to substantiate the proposed use of any material or assembly. If it is determined that the evidence submitted is satisfactory proof of performance for the use intended, the building official shall approve the use of the material or assembly subject to the requirements of this code. The costs, reports and investigations required under these provisions shall be paid by the permit applicant.

Section 1704.9 Pier foundations. Delete without substitution: (S37-03/04)

Section 1704.10 Wall panels and veneers. Delete without substitution: (S37-03/04)

Section 1707.5 Add new section to read as shown: (S37-03/04)

1707.5 Pier foundations. Special inspection is required for pier foundations for buildings assigned to Seismic Design Category C, D, E or F in accordance with Section 1616.3. Periodic special inspection is required during drilling of the piers and placement of reinforcement, and continuous special inspection is required during placement of the concrete.

(Renumber subsequent sections)

Section 1707.8.1 Change to read as shown: (S39-03/04)

1707.8.1 Component inspection. Special inspection is required for the installation of the following components, where the component has a Component Importance Factor greater than 1.0 in accordance with Section 9.6.1.5 of ASCE 7.

1. Equipment using combustible energy sources.

2. Electrical motors, transformers, switchgear unit substations and motor control centers.

3. Reciprocating and rotating-type machinery.

4. Piping distribution systems 3 inches (76 mm) and larger.

5. Tanks, heat exchangers and pressure vessels.

Section 1707.8.2 Change to read as shown: (S39-03/04)

1707.8.2 Component and attachment testing. The component manufacturer shall test or analyze the component and the component mounting system or anchorage for the design forces in Chapter 16 for those components having a Component Importance Factor greater than 1.0 in accordance with Chapter 16. The manufacturer shall submit a certificate of compliance for review and acceptance by the registered design professional responsible for the design, and for approval by the building official. The basis of certification shall be by test on a shaking table, by three-dimensional shock tests, by an analytical method using dynamic characteristics and forces from Chapter 16 or by more rigorous analysis. The special inspector shall inspect the component and verify that the label, anchorage or mounting conforms to the certificate of compliance.

Section 1707.8.3 Change to read as shown: (S39-03/04)

1707.8.3 Component manufacturer certification. Each manufacturer of equipment to be placed in a building assigned to Seismic Design Category E or F, in accordance with Chapter 16, where the equipment has a Component Importance Factor greater than 1.0 in accordance with Chapter 16, shall maintain an approved quality control program. Evidence of the quality control program shall be permanently identified on each piece of equipment by a label.

Section 1715.1.1 Change to read as shown: (S42-03/04)

1715.1 Test standards for joist hangers and connectors.

1715.1.1 Test standards for joist hangers. The vertical load-bearing capacity, torsional moment capacity and deflection characteristics of joist hangers shall be determined in accordance with ASTM D 1761, using lumber having a specific gravity of 0.49 or greater, but not greater than 0.55, as determined in accordance with AFPA NDS for the joist and headers.

Exception: The joist length shall not be required to exceed 24 inches (610 mm).

Section 1715.1.2 Change to read as shown: (S42-03/04)

1715.1.2 Vertical load capacity for joist hangers. The vertical load capacity for the joist hanger shall be determined by testing a minimum of three joist hanger assemblies as specified in ASTM D 1761. If the ultimate vertical load for any one of the tests varies more than 20 percent from the average ultimate vertical load, at least three additional tests shall be conducted. The allowable vertical load of the joist hanger shall be the lowest value determined from the following:

1. The lowest ultimate vertical load for a single hanger from any test divided by three (where three tests are conducted and each ultimate vertical load does not vary more than 20 percent from the average ultimate vertical load).

2. The average ultimate vertical load for a single hanger from all tests divided by three (where six or more tests are conducted).

3. The average from all tests of the vertical loads which produce a vertical movement of the joist with respect to the header of 0.125 inch (3.2 mm).

4. The sum of the allowable design loads for nails or other fasteners utilized to secure the joist hanger to the wood members and allowable bearing loads that contribute to the capacity of the hanger.

5. The allowable design load for the wood members forming the connection.

Section 1715.1.3 Change to read as shown: (S42-03/04)

1715.1.3 Torsional moment capacity for joist hangers. The torsional moment capacity for the joist hanger shall be determined by testing at least three joist hanger assemblies as specified in ASTM D 1761. The allowable torsional moment of the joist hanger shall be the average torsional moment at which the lateral movement of the top or bottom of the joist with respect to the original position of the joist is 0.125 inch (3.2 mm).

Section 1715.1.4 Change to read as shown: (S42-03/04)

1715.1.4 Design value modifications for joist hangers. Allowable design values for joist hangers that are determined by Item 4 or 5 in Section 1715.1.2 shall be permitted to be modified by the appropriate duration of loading factors as specified in AFPA NDS but shall not exceed the direct loads as determined by Item 1, 2 or 3 in Section 1715.1.2. Allowable design values determined by Item 1, 2 or 3 in Section 1715.1.2 shall not be modified by duration of loading factors.

CHAPTER 18
SOILS AND FOUNDATIONS

1801.2.1 Change to read as shown: (S43-03/04)

1801.2.1 Foundation design for seismic overturning. Where the foundation is proportioned using the strength design load combinations of Section 1605.2, the seismic overturning moment shall be computed in accordance with the equivalent lateral force method or the modal analysis method.

Section 1803.3 Change to read as shown: (S44-03/04)

1803.3 Site grading. The ground immediately adjacent to the foundation shall be sloped away from the building at a slope of not less than one unit vertical in 20 units horizontal (5-percent slope) for a minimum distance of 10 feet (3048 mm) measured perpendicular to the face of the wall. If physical obstructions or lot lines prohibit 10 feet (3048 mm) of horizontal distance, a 5-percent slope shall be provided to an approved alternate method of diverting water away from the foundation. Swales used for this purpose shall be sloped a minimum of 2 percent where located within 10 feet (3048 mm) of the building foundation. Impervious surfaces within 10 feet (3048 mm) of the building foundation shall be sloped a minimum of 2 percent away from the building.

Exception: Where climatic or soil conditions warrant, the slope of the ground away from the building foundation is permitted to be reduced to not less than one unit vertical in 48 units horizontal (2-percent slope).

The procedure used to establish the final ground level adjacent to the foundation shall account for additional settlement of the backfill.

Section 1803.4 Change to read as shown: (S45-03/04; S46-03/04; S47-03/04)

1803.4 Grading and fill in flood hazard areas. In flood hazard areas established in Section 1612.3, grading and/or fill shall not be approved:

1. Unless such fill is placed, compacted, and sloped to minimize shifting, slumping and erosion during the rise and fall of flood water and, as applicable, wave action.

2. In floodways, unless it has been demonstrated through hydrologic and hydraulic analyses performed by a registered design professional in accordance with standard engineering practice that the proposed grading or fill, or both, will not result in any increase in flood levels during the occurrence of the design flood.

3. In flood hazard areas subject to high velocity wave action, unless such fill is conducted and/or placed to avoid diversion of water and waves toward any building or structure.

4. Where design flood elevations are specified but floodways have not been designated, unless it has been demonstrated that the cumulative effect of the proposed flood hazard area encroachment, when combined with all other existing and anticipated flood hazard area encroachment, will not increase the design flood elevation more than one foot (305 mm) at any point.

Section 1804.2 Change to read as shown: (S48-03/04)

1804.2 Presumptive load-bearing values. The maximum allowable foundation pressure, lateral pressure or lateral sliding-resistance values for supporting soils near the surface shall not exceed the values specified in Table 1804.2 unless data to substantiate the use of a higher value are submitted and approved.

Presumptive load-bearing values shall apply to materials with similar physical characteristics and dispositions.

Mud, organic silt, organic clays, peat or unprepared fill shall not be assumed to have a presumptive load-bearing capacity unless data to substantiate the use of such a value are submitted.

Exception: A presumptive load-bearing capacity is permitted to be used where the building official deems the load-bearing capacity of mud, organic silt or unprepared fill is adequate for the support of lightweight and temporary structures.

Section 1805.2.1 Change to read as shown: (S49-03/04)

1805.2.1 Frost protection. Except where otherwise protected from frost, foundation walls, piers and other permanent supports of buildings and structures shall be protected from frost by one or more of the following methods:

1. Extending below the frost line of the locality;

2. Constructing in accordance with ASCE 32; or

3. Erecting on solid rock.

Exception: Free-standing buildings meeting all of the following conditions shall not be required to be protected:

1. Classified in Importance Category I, in accordance with Table 1604.5;

2. Area of 600 square feet (56 m^2) or less for light-frame construction or 400 square feet (37 m^2) or less for other than light-frame construction; and

3. Eave height of 10 feet (3048 mm) or less.

Footings shall not bear on frozen soil unless such frozen condition is of a permanent character.

Section 1805.5.1.1 Change to read as shown: (S50-03/04)

1805.5.1.1 Thickness at top of foundation wall. The thickness of foundation walls shall not be less than the thickness of the wall supported, except that foundation walls of at least 8-inch (203 mm) nominal width are permitted to support brick-veneered frame walls and 10-inch-wide (254 mm) cavity walls provided the requirements of Section 1805.5.1.2 are met. Corbeling of masonry shall be in accordance with Section 2104.2. Where an 8-inch (203 mm) wall is corbeled, the top corbel shall not extend higher than the bottom of the floor framing and shall be a full course of headers at least 6 inches (152 mm) in length or the top course bed joint shall be tied to the vertical wall projection. The tie shall be W2.8 (4.8 mm) and spaced at a maximum horizontal distance of 36 inches (914 mm); the hollow space behind the corbelled masonry shall be filled with mortar or grout.

Section 1805.4.5 Change to read as shown: (S51-03/04)

1805.4.5 Timber footings. Timber footings are permitted for buildings of Type V construction and as otherwise approved by the building official. Such footings shall be treated in accordance with AWPA U1 (Commodity Specification A, Use Category 4B). Treated timbers are not required where placed entirely below permanent water level, or where used as capping for wood piles that project above the water level over submerged or marsh lands. The compressive stresses perpendicular to grain in untreated timber footings supported upon treated piles shall not exceed 70 percent of the allowable stresses for the species and grade of timber as specified in the AFPA NDS.

Section 1805.4.6 Change to read as shown: (S51-03/04)

1805.4.6 Wood foundations. Wood foundation systems shall be designed and installed in accordance with AFPA Technical Report No. 7. Lumber and plywood shall be treated in accordance with AWPA U1 (Commodity Specification A. Use Category 4B and Section 5.2) and shall be identified in accordance with Section 2303.1.8.1.

Section 1805.7.1 Change to read as shown: (S51-03/04)

1805.7.1 Limitations. The design procedures outlined in this section are subject to the following limitations:

1. The frictional resistance for structural walls and slabs on silts and clays shall be limited to one-half of the normal force imposed on the soil by the weight of the footing or slab.

2. Posts embedded in earth shall not be used to provide lateral support for structural or nonstructural materials such as plaster, masonry or concrete unless bracing is provided that develops the limited deflection required.

Wood poles shall be treated in accordance with AWPA U1 for sawn timber posts (Commodity Specification A, Use Category 4B) and for round timber posts (Commodity Specification B, Use Category 4B).

Section 1807.3.1 Change to read as shown: (S52-03/04)

1807.3.1 Floors. Floors required to be waterproofed shall be of concrete, designed and constructed to withstand the hydrostatic pressures to which the floors will be subjected.

Waterproofing shall be accomplished by placing a membrane of rubberized asphalt, butyl rubber, fully adhered/fully bonded HDPE or polyolefin composite membrane, or not less than 6-mil (0.006 inch; 0.152 mm) polyvinyl chloride with joints lapped not less than 6 inches (152 mm) or other approved materials under the slab. Joints in the membrane shall be lapped and sealed in accordance with the manufacturer's installation instructions.

Section 1809.1.2 Change to read as shown: (S51-03/04)

1809.1.2 Preservative treatment. Timber piles used to support permanent structures shall be treated in accordance with this section unless it is established that the tops of the untreated timber piles will be below the lowest ground-water level assumed to exist during the life of the structure. Preservative and minimum final retention shall be in accordance with AWPA U1 (Commodity Specification E, Use Category 4C) for round timber piles and AWPA U1 (Commodity Specification A, Use Category 4B) for sawn timber piles. Preservative-treated timber piles shall be subject to a quality control program administered by an approved agency. Pile cutoffs shall be treated in accordance with AWPA M4.

CHAPTER 19
CONCRETE

Section 1908.1.5 Add new section to read as shown: (S53-03/04)

1908.1.5 ACI 318, Section 21.9.5.3. Modify ACI 318, Section 21.9.5.3, by adding a second paragraph to read as follows:

21.9.5.3 - Structural truss elements, struts, ties, diaphragm chords, and collector elements with compressive stresses exceeding $0.2f'_c$ at any section shall have transverse reinforcement, as given in 21.4.4.1 through 21.4.4.3, over the length of the element. The special transverse reinforcement is allowed to be discontinued at a section where the calculated compressive strength is less than $0.15f'_c$. Stresses shall be calculated for the factored forces using a linearly elastic model and gross-section properties of the elements considered.

Where design forces are amplified by the overstrength factor, Ω_o, as required by Section 1620.2.6 of the International Building Code, the limit of $0.2f'_c$ shall be increased to $0.5f'_c$ and the limit of $0.15f'_c$ shall be increased to $0.4f'_c$.

(Renumber subsequent sections)

CHAPTER 21
MASONRY

Section 2103.2 Change to read as shown: (S54-03/04)

2103.2 Clay or shale masonry units. Clay or shale masonry units shall conform to the following standards:

2004 SUPPLEMENT TO THE IBC

ASTM C 34 for structural clay load-bearing wall tile; ASTM C 56 for structural clay nonload-bearing wall tile; ASTM C 62 for building brick (solid masonry units made from clay or shale); ASTM C 1088 for solid units of thin veneer brick; ASTM C 126 for ceramic-glazed structural clay facing tile, facing brick and solid masonry units; ASTM C 212 for structural clay facing tile; ASTM C 216 for facing brick (solid masonry units made from clay or shale); ASTM C 652 for hollow brick (hollow masonry units made from clay or shale) and ASTM C 1405 for glazed brick (single-fired solid brick units).

> **Exception:** Structural clay tile for nonstructural use in fireproofing of structural members and in wall furring shall not be required to meet the compressive strength specifications. The fire-resistance rating shall be determined in accordance with ASTM E119 and shall comply with the requirements of Table 602.

Section 2103.9.2 Electrically conductive dry set mortars. Delete section without substitution: (S57-03/04)

Section 2104.1.2.5 Add new section to read as shown: (S55-03/04)

2104.1.2.5 Grouted masonry. Between grout pours, a horizontal construction joint shall be formed by stopping all wythes at the same elevation and with the grout stopping a minimum of 1½ inches (38 mm) below a mortar joint, except at the top of the wall. Where bond beams occur, the grout pour shall be stopped a minimum of ½ inch (12.7 mm) below the top of the masonry.

(Renumber subsequent sections)

Section 2109.8.4.6 Change to read as shown: (S56-03/04)

2109.8.4.6 Exterior finish. Exterior walls constructed of unstabilized adobe units shall have their exterior surface covered with a minimum of two coats of portland cement plaster having a minimum thickness of $^3/_4$ inch (19.1 mm) and conforming to ASTM C 926. Lathing shall comply with ASTM C1063. Fasteners shall be spaced at 16 inches (406 mm) o.c. maximum. Exposed wood surfaces shall be treated with an approved wood preservative or other protective coating prior to lath application.

Section 2113.12 Change to read as shown: (RB259-03/04)

2113.12 Clay flue lining (installation). Clay flue liners shall be installed in accordance with ASTM C 1283 and extend from a point not less than 8 inches (203 mm) below the lowest inlet or, in the case of fireplaces, from the top of the smoke chamber to a point above the enclosing walls. The lining shall be carried up vertically, with a maximum slope no greater than 30 degrees (0.52 rad) from the vertical.

Clay flue liners shall be laid in medium-duty refractory mortar conforming to ASTM C199 with tight mortar joints left smooth on the inside and installed to maintain an air space or insulation not to exceed the thickness of the flue liner separating the flue liners from the interior face of the chimney masonry walls. Flue lining shall be supported on all sides. Only enough mortar shall be placed to make the joint and hold the liners in position.

CHAPTER 23
WOOD

Section 2302.1 Change the following definitions to read as shown: (S59-03/04; S69-03/04)

Delete definition of "Adjusted Shear Resistance."

PERFORATED SHEAR WALL SEGMENT. A section of shear wall with full-height sheathing that meets the height-to-width ratio limits of Section 2305.3.3.

Delete definition of "Unadjusted Shear Resistance."

(Definitions not shown are unchanged)

Section 2303.1.8 Change to read as shown: (S51-03/04)

2303.1.8 Preservative-treated wood. Lumber, timber, plywood, piles and poles supporting permanent structures required by Section 2304.11 to be preservative-treated shall conform to the requirements of the applicable AWPA U1 and M4, for the species, product, preservative and end use. Preservatives shall conform to AWPA P1/P13, P2, P5, P8 or P9. Lumber and plywood used in wood foundation systems shall conform to Chapter 18.

2004 SUPPLEMENT TO THE IBC

Table 2304.9.1 Change entire table to read as shown (S62-03/04)

TABLE 2304.9.1
FASTENING SCHEDULE

CONNECTION	FASTENING [a, m]	LOCATION
1. Joist to sill or girder	3 - 8d common (2½" x 0.131") 3 - 3" x 0.131" nails 3 - 3" 14 gage staples	toenail
2. Bridging to joist	2 - 8d common (2½" x 0.131") 2 - 3" x 0.131" nails 2 - 3" 14 gage staples	toenail each end
3. 1" x 6" subfloor or less to each joist	2 - 8d common (2½" x 0.131")	face nail
4. Wider than 1" x 6" subfloor to each joist	3 - 8d common (2½" x 0.131")	face nail
5. 2" subfloor to joist or girder	2 - 16d common (3½" x 0.162")	blind and face nail
6. Sole plate to joist or blocking	16d (3½" x 0.135") at 16" o.c. 3" x 0.131" nails at 8" o.c. 3" 14 gage staples at 12" o.c.	typical face nail
Sole plate to joist or blocking at braced wall panel	3 - 16d (3½" x 0.135") at 16" 4 - 3" x 0.131" nails at 16" 4 - 3" 14 gage staples per 16"	braced wall panel
7. Top plate to stud	2 - 16d common (3½" x 0.162") 3 - 3" x 0.131" nails 3 - 3" 14 gage staples	end nail
8. Stud to sole plate	4 - 8d common (2½" x 0.131") 4 - 3" x 0.131" nails 3 - 3 14 gage staples	toenail
	2 - 16d common (3½" x 0.162") 3 - 3" x 0.131" nails 3 - 3" 14 gage staples	end nail
9. Double studs	16d (3½" x 0.135") at 24" o.c. 3" x 0.131" nail at 8" o.c. 3" 14 gage staple at 8" o.c.	face nail
10. Double top plates	16d (3½" x 0.135") at 16" o.c. 3" x 0.131" nail at 12" o.c. 3" 14 gage staple at 12" o.c.	typical face nail
Double top plates	8-16d common (3½" x 0.162") 12-3" x 0.131" nails 12-3" 14 gage staples	lap splice
11. Blocking between joists or rafters to top plate	3 - 8d common (2½" x 131") 3 -3" x 0.13" nails 3 - 3" 14 gage staples	toenail
12. Rim joist to top plate	8d (2½" x 0.113") at 6" o.c. 3" x 0.131" nail at 6" o.c. 3" x 14 gage staple at 6" o.c.	toenail
13. Top plates, laps and intersections	2 - 16d common (3½" x 0.162") 3 -3" x 0.131" nails 3 -3" 14 gage staples	face nail
14. Continuous header, two pieces	16d common (3½" x 0.162")	16" o.c. along edge
15. Ceiling joists to plate	3 - 8d common (2½" x 0.131") 5 - 3" x 0.131" nails 5 - 3" 14 gage staples	toenail
16. Continuous header to stud	4 – 8d common (2 1/2" x 0.131")	toenail

IBC-65

2004 SUPPLEMENT TO THE IBC

TABLE 2304.9.1 (continued)

CONNECTION	FASTENING[a, m]	LOCATION
17. Ceiling joists, laps over partitions (see Section 2308.10.4.1, Table 2308.10.4.1)	3 - 16d common (3½ x 0.162") minimum, Table 2308.10.4.1 4 - 3" x 0.131" nails 4 - 3 14 gage staples	face nail
18. Ceiling joists to parallel rafters (see Section 2308.10.4.1, Table 2308.104.1)	3 - 16d common (3½" x 0.162") minimum, Table 2308.10.4.1 4 - 3" x 0.131" nails 4 - 3" 14 gage staples	face nail
19. Rafter to plate (see Section 2308.10.1, Table 2308.10.1)	3 - 8d common (2½" x 0.131") 3 - 3" x 0.131" nails 3 - 3" 14 gage staples	toenail
20. 1" diagonal brace to each stud and plate	2 - 8d common (2½" x 0.131") 2 - 3" x 0.131" nails 3 - 3" 14 gage staples	face nail
21. 1" x 8" sheathing to each bearing wall	3 - 8d common (2½" x 0.131")	face nail
22. Wider than 1" x 8" sheathing to each bearing	3 8d common (2½" x 0.131")	face nail
23. Built-up corner studs	16d common (3½" x 0.162") 3" x 0.131" nails 3" 14 gage staples	24" o.c. 16" o.c. 16" o.c.
24. Built-up girder and beams	20d common (4" x 0.192") 32" o.c. 3" x 0.131" nail at 24" o.c. 3" 14 gage staple at 24" o.c. 2 - 20d common (4" x 0.192") 3 - 3 " x 0.131" nails 3 - 3" 14 gage staples	face nail at top and bottom staggered on opposite sides face nail at ends and at each splice
25. 2" planks	16d common (3½" x 0.162")	at each bearing
26. Collar tie to rafter	3 - 10d common (3" x 0.148") 4 - d" x 0.131" nails 4 - 3" 14 gage staples	face nail
27. Jack rafter to hip	3 - 10d common (3" x 0.148") 4 - 3" x 0.131" nails 4 - 3" 14 gage staples	toenail
	2 - 16d common (3½" x 0.162") 3 - 3" x 0.131" nails 3 - 3" 14 gage staples	face nail
28. Roof rafter to 2-by ridge beam	2 - 16d common (3½" x 0.162") 3 - 3" x 0.131" nails 3 - 3" 14 gage staples	toenail
	2-16d common (3½" x 0.162") 3 - 3" x 0.131" nails 3 - 3" 14 gage staples	face nail
29. Joist to band joist	3 - 16d common (3½" x 0.162") 4 - 3" x 0.131 nails 4 - 3" 14 gage staples	face nail
30. Ledger strip	3 - 16d common (3½" x 0.162") 4 - 3" x 0.131" nails 4 - 3" 14 gage staples	face nail

TABLE 2304.9.1 (continued)

CONNECTION	FASTENING[a, m]		LOCATION
31. Wood structural panels and particleboard: [b] Subfloor, roof and wall sheathing (to framing): Single Floor (combination subfloor-underlayment to framing):	½" and less 19/32" to ¾" ⅞" to 1" 1⅛ to 1-¼" ¾" and less ⅞" to 1" 1⅛" to 1-¼"	6d [c, l] 2 ⅜" x 0.113" nail [n] 1 ¾" 16 gage [o] 8d [d] or 6d [e] 2 ⅜" x 0.113" nail [p] 2" 16 gage [p] 8d [c] 10d [d] or 8d [d] 6d [e] 8d [e] 10d [d] or 8d[e]	
32. Panel siding (to framing)	½" or less ⅝"	6d [f] 8d [f]	
33. Fiberboard sheathing: [g]	 ½" 25/32"	No. 11 gage roofing Nail [h] 6d common nail (2" x 0.113") No. 16 gage staple [i] No. 11 gage roofing Nail [h] 8d common nail (2½" x 0.131") No. 16 gage staple [i]	
34. Interior paneling	¼" ⅜"	4d [j] 6d [k]	

For SI: 1 inch = 25.4 mm

a. and b. (No change to current text)

c. Common or deformed shank. (6d - 2" x 0.113"; 8d - 2½" x 0.131"; 10d - 3" x 0.148")

d. Common. (6d - 2" x 0.113"; 8d - 2½" x 0.131"; 10d - 3" x 0.148")

e. Deformed shank. (6d - 2" x 0.113"; 8d - 2½" x 0.131"; 10d - 3" x 0.148")

f. Corrosion-resistant siding (6d - 1⅞" x 0.106"; 8d - 2⅜" x 0.128") or casing (6d - 2" x 0.099"; 8d - 2½" x 0.113") nail.

g. through i. (No change to current text)

j. Casing (1½" x 0.080") or finish (1½" x 0.072") nails spaced 6 inches on panel edges, 12 inches at intermediate supports.

k. (No change to current text)

l. For roof sheathing applications, 8d nails (2½" x 0.113") are the minimum required for wood structural panels.

m. through p.(No change to current text)

Section 2304.9.5 Change to read as shown: (S61-03/04)

2304.9.5 Fasteners in preservative-treated and fire-retardant-treated wood. Fasteners for preservative-treated and fire-retardant-treated wood shall be of hot dipped zinc-coated galvanized steel, stainless steel, silicon bronze or copper. The coating weights for zinc-coated fasteners shall be in accordance with ASTM A 153. Fastenings for wood foundations shall be as required in AF&PA Technical Report No. 7.

Section 2304.11.2 Change to read as shown: (S51-03/04; S64-03/04)

2304.11.2 Wood used above ground. Wood used above ground in the locations specified in Sections 2304.11.2.1 through 2304.11.2.6, 2304.11.3 and 2304.11.5 shall be naturally durable wood or preservative-treated wood using water-borne preservatives, in accordance with AWPA U1 (Commodity Specifications A or F) for above-ground use.

Section 2304.11.3 Change to read as shown: (S65-03/04)

2304.11.3 Laminated timbers. The portions of glued-laminated timbers that form the structural supports of a building or other structure and are exposed to weather and not fully protected from moisture by a roof, eave or similar covering shall be pressure treated with preservative, or be manufactured from naturally durable or preservative-treated wood.

Section 2304.11.4 Change to read as shown: (S51-03/04)

2304.11.4 Wood in contact with the ground or fresh water. Wood used in contact with the ground (exposed earth) in the locations specified in Sections 2304.11.4.1 and 2304.11.4.2 shall be naturally durable (species for both decay and termite resistance) or preservative-treated using water-borne preservatives in accordance with AWPA U1 (Commodity Specifications A or F) for soil or fresh water use.

Section 2304.11.6 Change to read as shown: (S51-03/04)

2304.11.6 Termite protection. In geographical areas where hazard of termite damage is known to be very heavy, wood floor framing shall be of naturally durable species (termite resistant) or preservative-treated in accordance with AWPA U1 for the species, product preservative and end use, or provided with approved methods of termite protection.

Section 2304.11.7 Change to read as shown: (S51-03/04)

2304.11.7 Wood used in retaining walls and cribs. Wood installed in retaining or crib walls shall be preservative-treated in accordance with AWPA U1 (Commodity Specifications A or F) for soil and fresh water use.

Section 2304.12 Delete and substitute as shown: (S66-03/04)

2304.12 Long-term loading. Wood members supporting concrete, masonry, or similar materials, shall be checked for the effects of long-term loading using the provisions of the NDS. The total deflection, including the effects of long-term loading, shall be limited in accordance with Section 1604.3.1 for these supported materials.

> **Exception:** Horizontal wood members supporting masonry or concrete nonstructural floor or roof surfacing not more than 4 inches (102mm) thick need not be checked for long-term loading.

Section 2305.1.5 Change to read as shown: (S67-03/04)

2305.1.5 Wood members resisting horizontal seismic forces contributed by masonry and concrete walls. Wood shear walls, diaphragms, horizontal trusses and other members shall not be used to resist horizontal seismic forces contributed by masonry or concrete walls in structures over one story in height.

> **Exceptions:**
> 1. Wood floor and roof members are permitted to be used in horizontal trusses and diaphragms to resist horizontal seismic forces contributed by masonry or concrete walls provided such forces do not result in torsional force distribution through the truss or diaphragm.
> 2. Wood structural-panel-sheathed shear walls are permitted to be used to provide resistance to seismic forces contributed by masonry or concrete construction in two-story structures of masonry or concrete walls, provided the following requirements are met:
>   2.1. Story-to-story wall heights shall not exceed 12 feet (3658 mm).
>   2.2. Diaphragms shall not be designed to transmit lateral forces by rotation. Diaphragms shall not cantilever past the outermost supporting shear wall.
>   2.3. Combined deflections of diaphragms and shear walls shall not permit story drift of supported masonry or concrete walls to exceed the limit of Section 1617.3.
>   2.4. Wood structural panel sheathing in diaphragms shall have unsupported edges blocked. Wood structural panel sheathing for both stories of shear walls shall have unsupported edges blocked and, for the lower story, shall have a minimum thickness of $^{15}/_{32}$ inch (11.9 mm).
>   2.5. There shall be no out-of-plane horizontal offsets between the first and second stories of wood structural panel shear walls.

Section 2305.1.6 Add new section to read as shown: (S68-03/04)

2305.1.6 Wood members resisting seismic forces from non-structural concrete or masonry. Wood members shall be permitted to resist horizontal seismic forces from non-structural concrete, masonry veneer or concrete floors.

Section 2305.2.2 Change to read as shown: (S71-03/04)

2305.2.2 Deflection. Permissible deflection shall be that deflection up to which the diaphragm and any attached distributing or resisting element will maintain its structural integrity under design load conditions, such that the resisting element will continue to support design loads without danger to occupants of the structure. Calculations for diaphragm deflection shall account for the usual bending and shear components as well as any other factors, such as nail deformation, which will contribute to deflection.

The deflection (Δ) of a blocked wood structural panel diaphragm uniformly nailed throughout is permitted to be calculated by using the following formula. If not uniformly nailed, the constant 0.188 (For SI: 1/1627) in the third term must be modified accordingly.

$$\Delta = \frac{5vL^3}{8EAb} + \frac{vL}{4Gt} + 0.188Le_n + \frac{\Sigma(\Delta_c X)}{2b}$$

(Equation 23-1)

For SI: $\Delta = \dfrac{0.052L^3}{EAb} + \dfrac{vL}{4Gt} + \dfrac{Le_n}{1627} + \dfrac{\Sigma(\Delta_c X)}{2b}$

where:
- A = Area of chord cross section, in square inches (mm²).
- b = Diaphragm width, in feet (mm).
- E = Elastic modulus of chords, in pounds per square inch (N/mm²).
- e_n = Nail or staple deformation, in inches (mm). [See Table 2305.2.2(1)]
- Gt = Panel rigidity through the thickness, in pounds per inch (N/mm) of panel width or depth. [See Table 2305.2.2(2)]
- L = Diaphragm length, in feet (mm).
- v = Maximum shear due to design loads in the direction under consideration, in pounds per linear foot (plf) (N/mm).
- Δ = The calculated deflection, in inches (mm).
- $\Sigma(\Delta_c X)$ = Sum of individual chord-splice values on both sides of the diaphragm, each multiplied by its distance to the nearest support.

Table 2305.2.2(1) Add new table to read as shown: (S71-03/04)

TABLE 2305.2.2(1)
"e_n VALUES (INCHES) FOR USE IN CALCULATING DIAPHRAGM DEFLECTION DUE TO FASTENER SLIP (STRUCTURAL I)[a,d]

LOAD PER FASTENER[c] (pounds)	FASTENER DESIGNATIONS[b]			
	6d	8d	10d	14-Ga staple x 2 inches long
60	0.012	0.008	0.006	0.011
80	0.020	0.012	0.010	0.018
100	0.030	0.018	0.013	0.028
120	0.045	0.023	0.018	0.04
140	0.068	0.031	0.023	0.053
160	0.102	0.041	0.029	0.068
180	---	0.056	0.037	---
200	---	0.074	0.470	---
220	---	0.096	0.060	---
240	---	---	0.077	---

For SI: 1 inch = 25.4 mm, 1 foot = 305 mm, 1 pound = 4.448 N.

a. Increase "e_n" values 20 percent for plywood grades other than Structural I.
b. Nail values apply to common wire nails or staples identified.
c. Load per fastener = maximum shear per foot divided by the number of fasteners per foot at interior panel edges.
d. Decrease e_n values 50 percent for seasoned lumber (moisture content < 19%).

2004 SUPPLEMENT TO THE IBC

Table 2305.2.2(2) Add new table to read as shown: (S71-03/04)

TABLE 2305.2.2(2)
VALUES OF Gt FOR USE IN CALCULATING DEFLECTION OF WOOD STRUCTURAL PANEL SHEAR WALLS AND DIAPHRAGMS

| PANEL TYPE | SPAN RATING | VALUES OF Gt (lb/in. panel depth or width) ||||||||
| | | OTHER |||| STRUCTURAL I ||||
		3-ply Plywood	4-ply Plywood	5-ply Plywood[a]	OSB	3-ply Plywood	4-ply Plywood	5-ply Plywood[a]	OSB
Sheathing	24/0	25000	32500	37500	77500	32500	42500	41500	77500
	24/16	27000	35000	40500	83500	35000	45500	44500	83500
	32/16	27000	35000	40500	83500	35000	45500	44500	83500
	40/20	28500	37000	43000	88500	37000	48000	47500	88500
	48/24	31000	40500	46500	96000	40500	52500	51000	96000
Single Floor	16 o.c.	27000	35000	40500	83500	35000	45500	44500	83500
	20 o.c.	28000	36500	42000	87000	36500	47500	46000	87000
	24 o.c.	30000	39000	45000	93000	39000	50500	49500	93000
	32 o.c.	36000	47000	54000	110000	47000	61000	59500	110000
	48 o.c.	50500	65500	76000	155000	65500	85000	83500	155000

| | Thickness (in.) | Other ||| Structural I |||
		A-A, A-C	Marine	All Other Grades	A-A, A-C	Marine	All Other Grades
Sanded Plywood	¼	24000	31000	24000	31000	31000	31000
	11/32	25500	33000	25500	33000	33000	33000
	38054	26000	34000	26000	34000	34000	34000
	15/32	38000	49500	38000	49500	49500	49500
	½	38500	50000	38500	50000	50000	50000
	19/32	49000	63500	49000	63500	63500	63500
	⅝	49500	64500	49500	64500	64500	64500
	23/32	50500	65500	50500	65500	65500	65500
	¾	51000	66500	51000	66500	66500	66500
	⅞	52500	68500	52500	68500	68500	68500
	1	73500	95500	73500	95500	95500	95500
	1⅛	75000	97500	75000	97500	97500	97500

For SI: 1 inch = 25.4 mm, 1 pound/inch = 0.1751 N/mm.

a. Applies to plywood with 5 or more layers; for 5 ply/3 layer plywood, use values for 4 ply.

Section 2305.3.2 Change to read as shown: (S70-03/04; S71-03/04)

2305.3.2 Deflection. Permissible deflection shall be that deflection up to which the shear wall and any attached distributing or resisting element will maintain its structural integrity under design load conditions, i.e., continue to support design loads without danger to occupants of the structure.

The deflection (Δ) of a blocked wood structural panel shear wall uniformly fastened throughout is permitted to be calculated by the use of the following formula:

$$\Delta = \frac{8vh^3}{Eab} + \frac{vh}{Gt} + 0.75he_n + d_a\frac{h}{b}$$ (Equation 23-2)

For SI: $\Delta = \frac{vh^3}{3EAb} + \frac{vh}{Gt} + \frac{he_n}{407.6} + d_a\frac{h}{b}$

where:
- A = Area of boundary element cross section in square inches (mm²) (vertical member at shear wall boundary).
- b = Wall width, in feet (mm).
- d_a = Vertical elongation of overturning anchorage (including fastener slip, device elongation, anchor rod elongation, etc.) at the design shear load (v).
- E = Elastic modulus of boundary element (vertical member at shear wall boundary), in pounds per square inch (N/mm²).
- e_n = Nail or staple deformation, in inches (mm). [See Table 2305.2.2(2)].

Gt = Panel rigidity through the thickness, in pounds per inch (N/mm) of panel width or depth. [See Table 2305.2.2(2)]
h = Wall height, in feet (mm).
v = Maximum shear due to design loads at the top of the wall, in pounds per linear foot (N/mm).
Δ = The calculated deflection, in inches (mm).

Section 2305.3.3 Change to read as shown: (S59-03/04)

2305.3.3 Shear wall aspect ratios. Size and shape of shear walls, perforated shear wall segments within perforated shear walls and wall piers within shear walls with openings designed for force transfer around openings shall be limited as set forth in Table 2305.3.3. The height, *h*, and the width, *w*, shall be determined in accordance with Sections 2305.3.4 through 2305.3.4.2 and 2305.3.5 through 2305.3.5.2, respectively.

Table 2305.3.3 Change footnote to read as shown: (S69-03/04)

**Table 2305.3.3
MAXIMUM SHEAR WALL
DIMENSION RATIOS**

(No change to table entries)

a. (No change to current text)
b. Ratio shown is for unblocked construction. Height-to-width ratio is permitted to be 2:1 where the wall is installed as blocked construction in accordance with Section 2306.4.5.1.2.

Section 2305.3.4 Change to read as shown: (S59-03/04)

2305.3.4 Shear wall height definition. The height of a shear wall, *h*, shall be defined as:

1. The maximum clear height from top of foundation to bottom of diaphragm framing above; or
2. The maximum clear height from top of diaphragm to bottom of diaphragm framing above [see Figure 2305.3.4(a)].

Figure 2305.3.4 Change figure titles to read as shown: (S59-03/04)

No change to illustration.

(a) Height-to-Width Ratio for Shear Walls and Perforated Shear Walls
(b) Height-to-Width Ratio with Design for Force Transfer Around Openings

Sections 2305.3.4.1 and 2305.3.4.2 Add new sections to read as shown: (S59-03/04)

2305.3.4.1 Perforated shear wall segment height definition. The height of a perforated shear wall segment, *h*, shall be defined as specified in Section 2305.3.4 for shear walls.

2305.3.4.2 Force transfer shear wall pier height definition. The height, *h*, of a wall pier in a shear wall with openings designed for force transfer around openings shall be defined as the clear height of the pier at the side of an opening [see Figure 2305.3.4(b)].

Section 2305.3.5, 2305.3.5.1 Change to read as shown: (S59-03/04)

2305.3.5 Shear wall width definition. The width of a shear wall, *w*, shall be defined as the sheathed dimension of the shear wall in the direction of application of force [see Figure 2305.3.4(a)].

2305.3.5.1 Perforated shear wall segment width definition. The width of a perforated shear wall segment, *w*, shall be defined as the width of full-height sheathing adjacent to openings in the perforated shear wall [see Figure 2305.3.4(a)].

Section 2305.3.5.2 Add new section to read as shown: (S59-03/04)

2305.3.5.2 Force transfer shear wall pier width definition. The width, *w*, of a wall pier in a shear wall with openings designed for force transfer around openings shall be defined as the sheathed width of the pier at the side of an opening [see Figure 2305.3.4(b)].

Section 2305.3.7.1 Change to read as shown: (S59-03/04)

2305.3.7.1 Force transfer around openings. Where shear walls with openings are designed for force transfer around the openings, the limitations of Table 2305.3.3 shall apply to the overall shear wall including openings and to each wall pier at the side of an opening. Design for force transfer shall be based on a rational analysis. Detailing of boundary elements around the opening shall be provided in accordance with the provisions of this section [see Figure 2305.3.4(b)].

Section 2305.3.7.2 Change to read as shown: (S59-03/04)

2305.3.7.2 Perforated shear walls. The provisions of Section 2305.3.7.2 shall be permitted to be used for the design of perforated shear walls. For the determination of the height and width of perforated shear wall segments, see Sections 2305.3.4.1 and 2305.3.5.1, respectively.

Section 2305.3.7.2.2 Change to read as shown: (S59-03/04)

2305.3.7.2.2 Perforated shear wall resistance. The resistance of a perforated shear wall shall be calculated in accordance with the following:

1. The percent of full-height sheathing shall be calculated as the sum of the widths of perforated shear wall segments divided by the total width of the perforated shear wall including openings.

2. The maximum opening height shall be taken as the maximum opening clear height. Where areas above and below an opening remain unsheathed, the height of opening shall be defined as the height of the wall.

3. The unadjusted shear resistance shall be the allowable shear set forth in Table 2306.4.1 for height-to-width ratios of perforated shear wall segments that do not exceed 2:1 for seismic forces and 3½:1 for other than seismic forces. For seismic forces, where the height-to-width ratio of any perforated shear wall segment used in the calculation of the sum of the widths of perforated shear wall segments, $\sum L_i$, is greater than 2:1 but not exceeding 3½:1, the unadjusted shear resistance shall be multiplied by 2 w/h.

4. The adjusted shear resistance shall be calculated by multiplying the unadjusted shear resistance by the shear resistance adjustment factors of Table 2305.3.7.2. For intermediate percentages of full-height sheathing, the values in Table 2305.3.7.2 are permitted to be interpolated.

5. The perforated shear wall resistance shall be equal to the adjusted shear resistance times the sum of the widths of the perforated shear wall segments.

Section 2305.3.7.2.4 Change to read as shown: (S59-03/04)

2305.3.7.2.4 Uplift anchorage at perforated shear wall ends. Anchorage for uplift forces due to overturning shall be provided at each end of the perforated shear wall. The uplift anchorage shall conform to the requirements of Section 2305.3.6, except that for each story the minimum tension chord uplift force, T, shall be calculated in accordance with the following:

$$T = \frac{Vh}{C_o \sum L_i}$$ (Equation 23-3)

where:
- T = Tension chord uplift force, pounds (N)
- V = Shear force in perforated shear wall, pounds (N)
- h = Perforated shear wall height, feet (mm)
- C_o = Shear resistance adjustment factor from Table 2305.3.7.2
- $\sum L_i$ = Sum of widths of perforated shear wall segments, feet (mm).

Section 2305.3.7.2.5 Change to read as shown: (S59-03/04)

2305.3.7.2.5 Anchorage for in-plane shear. The unit shear force, v, transmitted into the top of a perforated shear wall, out of the base of the perforated shear wall at full height sheathing, and into collectors connecting shear wall segments, shall be calculated in accordance with the following:

$$v = \frac{V}{C_o \sum L_i}$$ (Equation 23-4)

where:
- v = Unit shear force, pounds per lineal feet (N/m)
- V = Shear force in perforated shear wall, pounds (N)
- C_o = Shear resistance adjustment factor from Table 2305.3.7.2
- $\sum L_i$ = Sum of widths of perforated shear wall segments, feet (mm).

Section 2305.3.10 Change to read as shown: (S83-03/04)

2305.3.10 Sill plate size and anchorage in Seismic Design Category D, E or F. Anchor bolts for shear walls shall include steel plate washers, a minimum of ¼ inch by 3 inches by 3 inches (6.4 mm by 76 mm by 76 mm) in size, between the sill plate and nut. The hole in the plate washer is permitted to be diagonally slotted with a width of up to 3/16 inch (4.76 mm) larger than the bolt diameter and a slot length not to exceed 1¾ inches (44 mm), provided a standard cut washer is placed between the plate washer and the nut. Sill plates resisting a design load greater than 490 plf (LRFD)(7154N/m) or 350 plf (ASD)(5110N/m) shall not be less than a 3-inch (76 mm) nominal member. Where a single 3-inch (76 mm) nominal sill plate is used, 2- 20d box end nails shall be substituted for 2- 16d common end nails found in Line 8 of Table 2304.9.1.

Exception: In shear walls where the design load is less than 840 plf (LRFD) (12 264 N/m) or 600 plf (ASD) (8760 N/m), the sill plate is permitted to be a 2-inch (51 mm) nominal member if the sill plate is anchored by two times the number of bolts required by design and $3/_{16}$ inch by 2 inch by 2 inch (4.76 mm by 51 mm by 51 mm) plate washers are used.

Section 2306.3.1 Change to read as shown: (S74-03/04)

2306.3.1 Shear capacities modifications. The allowable shear capacities in Table 2306.3.1 and Table 2306.3.2 for horizontal wood structural panel diaphragms shall be increased 40 percent for wind design.

2004 SUPPLEMENT TO THE IBC

Table 2306.3.1 Change table to read as shown: (S62-03/04; S75-03/04; S76-03/04)

Table 2306.3.1
ALLOWABLE SHEAR (POUNDS PER FOOT) FOR WOOD STRUCTURAL PANEL DIAPHRAGMS
WITH FRAMING OF DOUGLAS-FIR-LARCH, OR SOUTHERN PINE[a] FOR WIND OR SEISMIC LOADING

PANEL GRADE	COMMON NAIL SIZE OR STAPLE[f] LENGTH AND GAGE	MINIMUM FASTENER PENETRATION IN FRAMING (INCHES)	MINIMUM NOMINAL PANEL THICKNESS (inch)	MINIMUM NOMINAL WIDTH OF FRAMING MEMBERS AT ADJOINING PANEL EDGES AND BOUNDARIES[g] (INCHES)
Structural 1 Grades	6d [e] (2" x 0.113")	1 ¼	No changes to this column	No changes to this column
	1½ 16 Gage	1		
	8d (2-½" x 0.131")	1 ⅜		
	1½ 16 Gage	1		
	10d [d] (3" x 0.148")	1 ½		
	1½ 16 Gage	1		
Sheathing, single floor and other grades covered in DOC PS 1 and PS 2	6d [e] (2" x 0.113")	1 ¼		
	1½ 16 Gage	1		
	6d [e] (2" x 0.113")	1 ¼		
	8d (2-1/2" x 0.131")	1 ⅜		
Sheathing, single floor and other grades covered in DOC PS 1 and PS2 (continued)	1½ 16 Gage	1		
	8d (2-1/2" x 0.131")	1 ⅜		
	1½ 16 Gage	1		
	8d (2-1/2" x 0.131")	1 ⅜		
	10d [d] (3" x 0.148")	1 ½		
	1½ 16 Gage	1		
	10d [d] (3" x 0.148")	1 ½		
	1¾ 16 Gage	1		

(Portions of table and headings not shown do not change)

a. through e. (No change to current text)
f. Staples shall have a minimum crown width of 7/16 inch, and shall be installed with their crowns parallel to the long dimension of the framing members.
g. The minimum nominal width of framing members not located at boundaries or adjoining panel edges shall be 2 inches.

2004 SUPPLEMENT TO THE IBC

Table 2306.3.2 Change table heading and add footnote f to read as shown: (S75-03/04; S76-03/04)

TABLE 2306.3.2
ALLOWABLE SHEAR IN POUNDS PER FOOT FOR HORIZONTAL BLOCKED DIAPHRAGMS UTILIZING MULTIPLE ROWS OF FASTENERS (HIGH LOAD DIAPHRAGMS) WITH FRAMING OF DOUGLAS FIR-LARCH OR SOUTHERN PINE[a] FOR WIND OR SEISMIC LOADING[b]

PANEL GRADE[c]	COMMON NAIL SIZE OR STAPLE[f] GAGE	MINIMUM FASTENER PENETRATION IN FRAMING (inches)	MINIMUM NOMINAL PANEL THICKNESS (inch)	MINIMUM NOMINAL WIDTH OF FRAMING MEMBER AT ADJOINING PANEL EDGES AND BOUNDARIES[a]

(Column headings and table contents not shown do not change)

a. For framing of other species: (1) Find specific gravity for species of framing lumber in AFPA and National Design Specification. (2) For staples, find shear value from table above for Structural I panels (regardless of actual grade) and multiply value by 0.82 for species with specific gravity of 0.42 or greater, or from 0.65 for all other species. (3) For nails, find shear value from table above for nail size of actual grade and multiply value by the following adjustment factor: Specific Gravity Adjustment Factor = [1- (0.5 - SG)], where SG = Specific Gravity of the framing lumber. This adjustment factor shall not be greater than 1.
b. Fastening along intermediate framing members: Space fasteners a maximum of 12 inches on center, except 6 inches on center for spans greater than 32 inches.
c. and d. (No change to current text)
e. The minimum nominal depth of framing members shall be inches nominal. The minimum nominal width of framing members not located at boundaries or adjoining panel edges shall be 2 inches.
f. Staples shall have a minimum crown width of 7/16 inch, and shall be installed with their crowns parallel to the long dimension of the framing members.

2004 SUPPLEMENT TO THE IBC

Table 2306.4.1 Change to read as shown: (S62-03/04; S75-03/04; S77-03/04)

TABLE 2306.4.1
ALLOWABLE SHEAR (POUNDS PER FOOT) FOR WOOD STRUCTURAL PANEL SHEAR WALLS WITH FRAMING OF DOUGLAS-FIR-LARCH, OR SOUTHERN PINE [a] FOR WIND OR SEISMIC LOADING [b, h, i, j]

PANEL GRADE	MINIMUM NOMINAL PANEL THICKNESS (inch)	MINIMUM FASTENER PENETRATION IN FRAMING (inches)	PANELS APPLIED DIRECT TO FRAMING – NAIL (common or galvanized box) or staple size [k]	Fastener spacing at panel edges (inches) 6	4	3	2 [e]	PANELS APPLIED OVER ½ or 5/8 GYPSUM SHEATHING – NAIL (common or galvanized box) or staple size [k]	Fastener spacing at panel edges (inches) 6	4	3	2 [e]
Structural I Sheathing	5/16	1 1/4	6d (2" x 0.113" common, 2" x 0.099" galvanized box)	200	300	390	510	8d (2½" x 0.131" common, 2-½" x 0.113" galvanized box)	200	300	390	510
		1	1 ½ 16 Gage	165	245	325	415	2 16 Gage	125	185	245	315
	3/8	1 3/8	8d (2-1/2" x 0.131" common, 2-1/2" x 0.113" galvanized box)	230[d]	360[d]	460[d]	610[d]	10d (3" x 0.148" common, 3" x 0.128" galvanized box)	280	430	550[f]	730
		1	1 ½ 16 Gage	155	235	315	400	2 16 Gage	155	235	310	400
	7/16	1 3/8	8d (2-1/2" x 0.131" common, 2-1/2" x 0.113" galvanized box)	255[d]	395[d]	505[d]	670[d]	10d (3" x 0.148" common, 3" x 0.128" galvanized box)	280	430	550[f]	730
		1	1 ½ 16 Gage	170	260	345	440	2 16 Gage	155	235	310	400
	15/32	1 3/8	8d (2-1/2" x 0.131" common, 2-1/2" x 0.113" galvanized box)	280	430	550	730	10d (3" x 0.148" common, 3" x 0.128" galvanized box)	280	430	550[f]	730
		1	1 ½ 16 Gage	185	280	375	475	2 16 Gage	155	235	300	400
		1 1/2	10d (3" x 0.148" common, 3" x 0.128" galvanized box)	340	510	665[f]	870	10d (3" x 0.148" common, 3" x 0.128" galvanized box)	—	—	—	—
Sheathing, plywood siding[g] except Group 5 species	15/16 or 1/4[c]	1 1/4	6d (2" x 0.113" common, 2" x 0.099" galvanized box)	180	270	350	450	8d (2-1/2" x 0.131" common, 2-1/2" x 0.113" galvanized box)	180	270	350	450
		1	1 ½ 16 Gage	145	220	295	375	2 16 Gage	110	165	220	285
	3/8	1 1/4	6d (2" x 0.113" common, 2" x 0.099" galvanized box)	200	300	390	510	8d (2-1/2" x 0.131" common, 2-1/2" x 0.113" galvanized box)	200	300	390	510
		1 3/8	8d (2½" x 0.131" common, 2½" x 0.113" galvanized box)	220[d]	320[d]	410[d]	530[d]	10d (3" x 0.148" common, 3" x 0.128" galvanized box)	260	380	490[f]	640
		1	1 ½ 16 Gage	140	210	280	360	2 16 Gage	140	210	280	360

IBC-75

2004 SUPPLEMENT TO THE IBC

TABLE 2306.4.1 (continued)
ALLOWABLE SHEAR (POUNDS PER FOOT) FOR WOOD STRUCTURAL PANEL SHEAR WALLS WITH FRAMING OF DOUGLAS-FIR-LARCH, OR SOUTHERN PINE [a] FOR WIND OR SEISMIC LOADING [b, h, i, j]

PANEL GRADE	MINIMUM NOMINAL PANEL THICKNESS (inch)	MINIMUM FASTENER PENETRATION IN FRAMING (inches)	PANELS APPLIED DIRECT TO FRAMING – NAIL (common or galvanized box) or staple size [k]	6	4	3	2	PANELS APPLIED OVER ½ or 5/8 GYPSUM SHEATHING – NAIL (common or galvanized box) or staple size [k]	6	4	3	2
	7/16	1 3/8	8d (2½" x 0.131" common, 2½" x 0.113" galvanized box)	240[d]	350[d]	450[d]	585[d]	10d (3" x 0.148" common, 3" x 0.128" galvanized box)	260	380	490[f]	640
		1	1 ½ 16 Gage	155	230	310	395	2 16 Gage	140	210	280	360
	15/32	1 3/8	8d (2½" x 0.131" common, 2½" x 0.113" galvanized box)	260	380	490	640	10d (3" x 0.148" common, 3" x 0.128" galvanized box)	260	380	490[f]	640
		1 ½	10d (3" x 0.148" common, 3" x 0.128" galvanized box)	310	460	600[f]	770	—	—	—	—	—
		1	.1 ½ 16 Gage	170	255	335	430	2 16 Gage	140	210	280	360
	19/32	1 1/2	10d (3" x 0.148" common, 3" x 0.128" galvanized box)	340	510	665[f]	870	—	—	—	—	—
		1	1 3/4 16 Gage	185	280	375	475					
			Nail size (galvanized casing)					Nail Size (galvanized casing)				
	5/16[c]	1 1/4	6d (2" x 0.099")	140	210	275	360	8d (2 1/2" x 0.113")	140	210	275	360
	3/8	1 3/8	8d (2½" x 0.113")	160	240	310	410	10d (3" x 0.128")	160	240	310[f]	410

For SI: 1 inch = 25.4 mm, 1 pound per foot = 14.5939 N/m.

a. through c. (No change to current text)
d. Allowable shear values are permitted to be increased to values shown for 15/32-inch sheathing with same nailing provided (a) studs are spaced a maximum of 16 inches on center, or (b) if panels are applied with long dimension across studs.
e. (No change to current text)
f. Framing at adjoining panel edges shall be 3 inches nominal or wider, and nails shall be staggered where both of the following conditions are met: (1) 10d (3" x 0.148") nails having penetration into framing of more than 1 1/2 inches and (2) nails are spaced 3 inches on center.
g. (No change to current text)
h. Where panels are applied on both faces of a wall and nail spacing is less than 6 inches o.c. on either side, panel joints shall be offset to fall on different framing members. Or framing shall be 3-inch nominal or thicker at adjoining panel edges and nails on each side shall be staggered.
i. In Seismic Design Category D, E or F, where shear design values exceed 490 pounds per lineal foot (LRFD) or 350 pounds per lineal foot (ASD) all framing members receiving edge nailing from abutting panels shall not be less than a single 3-inch nominal member, or two 2-inch nominal members fastened together in accordance with Section 2307.1 (LRFD) or Section 2306.1 (ASD) to transfer the design shear value between framing members. Wood structural panel joint and sill plate nailing shall be staggered in all cases. See Section 2305.3.10 for sill plate size and anchorage requirements.
j. (No change to current text)
k. Staples shall have a minimum crown width of 7/16 inch, and shall be installed with their crowns parallel to the long dimension of the framing members.

2004 SUPPLEMENT TO THE IBC

Table 2306.4.3 Add new footnote b to read as shown: (S75-03/04)

TABLE 2306.4.3
ALLOWABLE SHEAR FOR PARTICLEBOARD SHEAR WALL SHEATHING [b]

(No change to table entries)

a. (No change to current text)
b. Galvanized nails shall be hot-dipped or tumbled.

Table 2306.4.5 Change table to read as shown: (S62-03/04; S75-03/04)

TABLE 2306.4.5
ALLOWABLE SHEAR FOR WIND OR SEISMIC FORCES FOR SHEAR WALLS OF LATH AND PLASTER OR GYPSUM BOARD WOOD FRAMED WALL ASSEMBLIES

Type of Material	Thickness of Material	Wall Construction	Fastener Spacing Maximum (inches)	Shear Value[a, e] (plf)	Minimum Fastener Size [c,d,j,k]
4. Gypsum board, gypsum veneer base, or water-resistant gypsum backing board	½"	Unblocked [f]	(No changes)	(No changes)	5d cooler (1⅝" x 0.086") or wallboard 0.120" Nail, min. ⅜" head, 1 ½" long 16 Gage Staple, 1½" long
		Unblocked [f]			
		Unblocked			
		Unblocked			
		Blocked [g]			
		Blocked [g]			
		Unblocked			No. 6-1 ¼" screws [i]
		Blocked [g]			
		Blocked [g]			
		Blocked [f, g]			
		Blocked [g]			
	⅝"	Unblocked [f]			6d cooler (1⅞" x 0.092") or wallboard
		Blocked [g]			0.120" Nail, min. ⅜" head, 1¾ long 16 Gage Staple, 1½" legs, 1⅝" long
		Blocked [g] Two-ply			Base ply-6d cooler (1⅞" x 0.092") or wallboard 1¾"' x 0.120" Nail, min. ⅜" head 1⅝" 16 Gage Galv. Staple Face ply-8d cooler (2⅜" x 0.113") or wallboard 0.120" Nail, min. ⅜" head, 2⅜" long 15 Gage Galv. Staple, 2¼" long
		Unblocked			No. 6-11/4" screws [i]
		Blocked [g]			

(Portions of table not shown do not change)

IBC-77

a. (No change to current text)
b. Applies to fastening at studs, top and bottom plates and blocking.
c. Alternate fasteners are permitted to be used if their dimensions are not less than the specified dimensions. Drywall screws are permitted to substitute for the 5d (1⅝" x 0.086"), and 6d (1⅞" x 0.092")(cooler) nails listed above, and No. 6-1¼ inch Type S or W screws for 6d (1⅞" x 0.092) (cooler) nails.
d. through f. (No change to current text)
g. All edges are blocked, and edge fastening is provided at all supports and all panel edges.
h. First number denotes fastener spacing at the edges; second number denotes fastener spacing at intermediate framing members.
i. (No change to current text)
j. Staples shall have a minimum crown width of 7/16 inch, measured outside the legs, and shall be installed with their crowns parallel to the long dimension of the framing members.
k. (No change to current text)

Section 2306.4.5.1.3 Change to read as shown: (S75-03/04)

2306.4.5.1.3 Fastening. Studs, top and bottom plates and blocking shall be fastened in accordance with Table 2304.9.1.

Section 2306.4.5.1.4 Change to read as shown: (S75-03/04)

2306.4.5.1.4 Fasteners. The size and spacing of fasteners shall be set forth in Table 2306.4.5. Fasteners shall be spaced not less than ⅜ inch (9.5 mm) from edges and ends of gypsum boards or sides of studs, blocking and top and bottom plates.

Table 2308.9.3(1) Change footnote c to read as shown: (S80-03/04)

**TABLE 2308.9.3(1)
BRACED WALL PANELS[a]**

(No change to table entries.)

a. and b. (No change to current text)
c. See Sections 2308.9.3.1 and 2308.9.3.2 for alternative braced panel requirement.
d. through f. (No change to current text)

Section 2308.9.3.2 Add new section to read as shown: (S80-03/04)

2308.9.3.2 Alternate bracing wall panel adjacent to a door or window opening. Any bracing required by Section 2308.9.3 is permitted to be replaced by the following when used adjacent to a door or window opening with a full-length header:

1. In one-story buildings, each panel shall have a length of not less than 16 inches (406 mm) and a height of not more than 10 feet (3048 mm). Each panel shall be sheathed on one face with a single layer of ⅜-inch-minimum-thickness (9.5 mm) wood structural panel sheathing nailed with 8d common or galvanized box nails in accordance with Figure 2308.9.3.2. The wood structural panel sheathing shall extend up over the solid sawn or glued-laminated header and shall be nailed in accordance with Figure 2308.9.3.2. A built-up header consisting of at least two 2 x 12s and fastened in accordance with item 24 of Table 2304.9.1 shall be permitted to be used. A spacer, if used, shall be placed on the side of the built-up beam opposite the wood structural panel sheathing. The header shall extend between the inside faces of the first full-length outer studs of each panel. The clear span of the header between the inner studs of each panel shall be not less than 6 feet (1829 mm) and not more than 18 feet (5486 mm) in length. A strap with an uplift capacity of not less than 1,000 pounds (4,400 N) shall fasten the header to the inner studs opposite the sheathing. One anchor bolt not less than ⅝-inch (15.9 mm) diameter and installed in accordance with Section 2308.6 shall be provided in the center of each sill plate. The studs at each end of the panel shall have a tie-down device fastened to the foundation with an uplift capacity of not less than 4,200 pounds (18,480 N).

Where a panel is located on one side of the opening, the header shall extend between the inside face of the first full-length stud of the panel and the bearing studs at the other end of the opening. A strap with an uplift capacity of not less than 1000 pounds (4,400 N) shall fasten the header to the bearing studs. The bearing studs shall also have a tie-down device fastened to the foundation with an uplift capacity of not less than 1,000 pounds (4,400N).

The tie-down devices shall be an embedded strap type, installed in accordance with the manufacturer's recommendations. The panels shall be supported directly on a foundation which is continuous across the entire length of the braced wall line. This foundation shall be reinforced with not less than one No. 4 bar top and bottom.

Where the continuous foundation is required to have a depth greater than 12 inches (305 mm), a minimum 12-inch-by-12-inch (305 mm by 305 mm) continuous footing or turned down slab edge is permitted at door openings in the braced wall line.

2004 SUPPLEMENT TO THE IBC

This continuous footing or turned down slab edge shall be reinforced with not less than one No. 4 bar top and bottom. This reinforcement shall be lapped not less than 15 inches (381 mm) with the reinforcement required in the continuous foundation located directly under the braced wall line.

2. In the first story of two-story buildings, each wall panel shall be braced in accordance with Item 1 above, except that each panel shall have a length of not less than 24 inches (610 mm)

Figure 2308.9.3.2 Add new figure as shown: (S80-03/04)

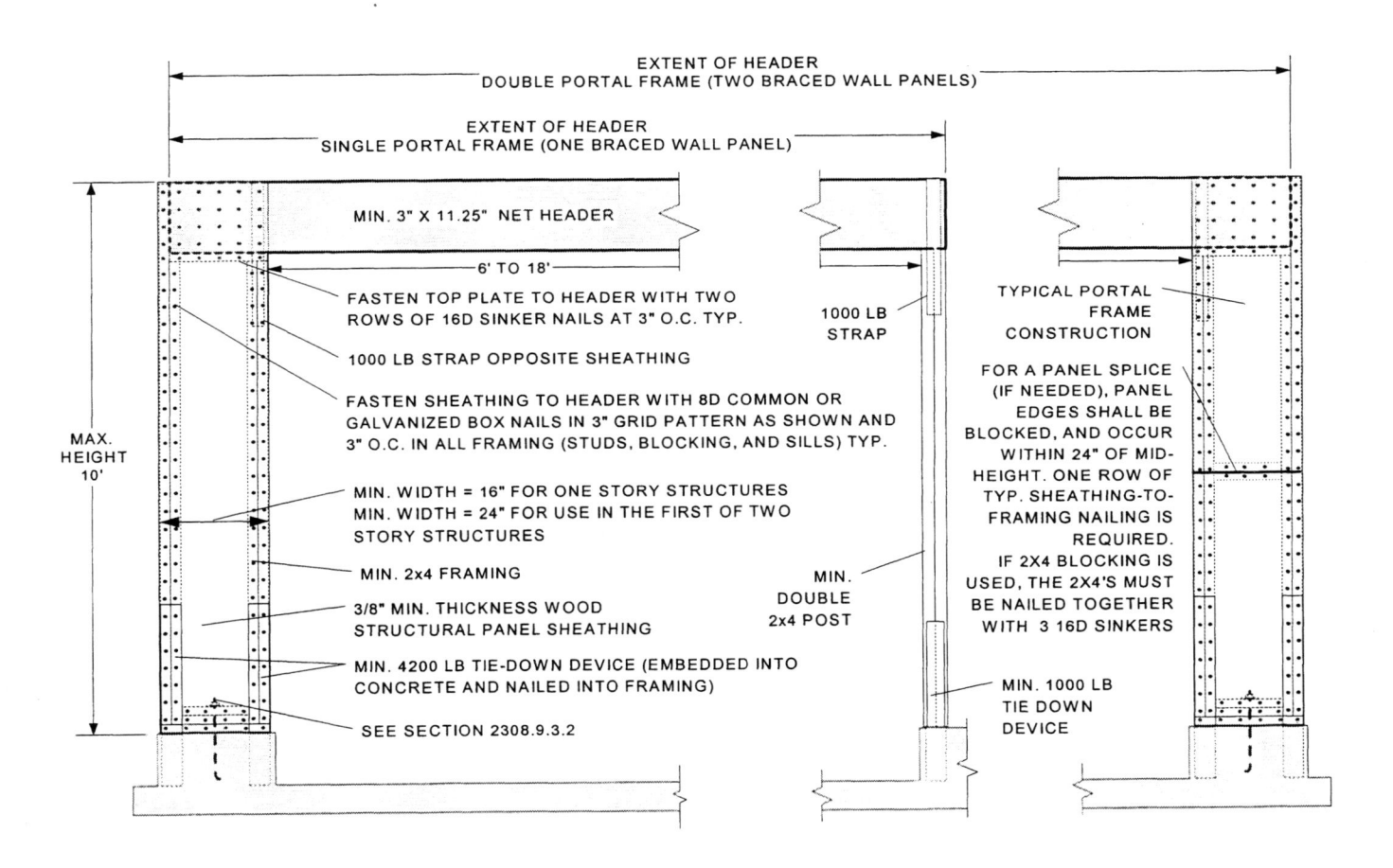

For SI: 1 foot = 304.8 mm; 1 inch = 25.4 mm; 1 pound = 4.448 N

FIGURE 2308.9.3.2
ALTERNATE BRACED WALL PANEL ADJACENT TO A DOOR OR WINDOW OPENING

2004 SUPPLEMENT TO THE IBC

Table 2308.10.4.1 Change to read as shown: (S62-03/04)

TABLE 2308.10.4.1
RAFTER TIE CONNECTIONS [g]

RAFTER SLOPE	TIE SPACING (inches)	NO SNOW LOAD				GROUND SNOW LOAD (pound per square foot)							
						30 pounds per square foot				50 pounds per square foot			
						Roof span (feet)							
		12	20	28	36	12	20	28	36	12	20	28	36
		Required number of 16d common (3-½" x 0.162") nails [a,b] per connection [c,d,e,f]												

(Portions of table not shown do not change)

a. 40d box (5" x 0.162") or l6d sinker (3¼" x 0.148") nails are permitted to be substituted for 16d common (3-½" x 0.162") nails.
b. though g. (No change to current text)

Section 2308.10.7 Add new section to read as shown: (S82-03/04)

2308.10.7 Engineered wood products. Prefabricated wood I-joists, structural glued-laminated timber and structural composite lumber shall not be notched or drilled unless allowed by the manufacturer's recommendations or unless the effects of such penetrations are considered in the design of the member by a registered design professional.

(Renumber subsequent sections)

Section 2308.12.8 Change to read as shown: (S83-03/04)

2308.12.8 Steel plate washers. Steel plate washers shall be placed between the foundation sill plate and the nut. Such washers shall be a minimum of ¼ inch by 3 inches by 3 inches (6.4 mm by 76 mm by 76 mm) in size. The hole in the plate washer is permitted to be diagonally slotted with a width of up to 3/16 inch (4.76 mm) larger than the bolt diameter and a slot length not to exceed 1¾ inches (44 mm), provided a standard cut washer is placed between the plate washer and the nut.

CHAPTER 24
GLASS AND GLAZING

Section 2404.1 Change to read as shown: (S84-03/04)

2404.1 Vertical glass. Glass sloped 15 degrees (0.26 rad) or less from vertical in windows, curtain and window walls, doors and other exterior applications shall be designed to resist the wind loads in Section 1609 for components and cladding. Glass in glazed curtain walls, glazed storefronts and glazed partitions shall meet the seismic requirements of ASCE 7, Section 9.6.2.10. The load resistance of glass under uniform load shall be determined in accordance with ASTM E1300.

The design of vertical glazing shall be based on the following equation:

$$F_{gw} \leq F_{ga} \quad \text{(Equation 24-1)}$$

where:

F_{gw} is the wind load on the glass computed in accordance with Section 1609 and F_{ga} is the short duration load on the glass as determined in accordance with ASTM E1300:

Table 2404.1 c_1 FACTORS FOR VERTICAL AND SLOPED GLASS [a] Delete table without substitution: (S84-03/04)

Section 2404.2 Change to read as shown: (S84-03/04)

2404.2 Sloped glass. Glass sloped more than 15 degrees (0.26 rad) from vertical in skylights, sunrooms, sloped roofs and other exterior applications shall be designed to resist the most critical of the following combinations of loads.

$$F_g = W_o - D \quad \text{(Equation 24-2)}$$

$$F_g = W_i + D + 0.5\,S \quad \text{(Equation 24-3)}$$

$$F_g = 0.5\,W_i + D + S \quad \text{(Equation 24-4)}$$

where:

D = Glass dead load psf (kN/m^2)

For glass sloped 30 degrees (0.52 rad) or less from horizontal,

$D = 13\, t_g$ (For SI: $0.0245\, t_g$)

For glass sloped more than 30 degrees (0.52 rad) from horizontal,

$D = 13\, t_g \cos \theta$ (For SI: $0.0245\, t_g \cos \theta$).

F_g = Total load, psf (kN/m²) on glass.
S = Snow load, psf (kN/m²) as determined in Section 1608.
t_g = Total glass thickness, inches (mm) of glass panes and plies.
W_i = Inward wind force, psf (kN/m²) as calculated in Section 1609.
W_o = Outward wind force, psf (kN/m²) as calculated in Section 1609.
θ = Angle of slope from horizontal.

Exception: Unit skylights shall be designed in accordance with Section 2405.5

The design of sloped glazing shall be based on the following equation:

$$F_g \leq F_{ga} \quad \text{(Equation 24-5)}$$

where F_g is the total load on the glass determined from the load combinations above and F_{ga} is the short duration load resistance of the glass as determined according to ASTM E1300 for Equations 24-2 and 24-3; or the long duration load resistance of the glass as determined according to ASTM E1300 for Equation 24-4.

Table 2404.2 c_2 FACTORS FOR SLOPED GLASS ª [For use with Figures 2404(1) through 2404(12)] Delete table without substitution: (S84-03/04)

Figures 2404(1) through 2404(12) Delete figures without substitution: (S84-03/04)

Sections 2404.3 through 2404.3.5 Add new sections to read as shown: (S84-03/04)

2404.3 Wired, patterned, and sandblasted glass.

2404.3.1 Vertical wired glass. Wired glass sloped 15 degrees (0.26 rad) or less from vertical in windows, curtain and window walls, doors and other exterior applications shall be designed to resist the wind loads in Section 1609 for components and cladding according to the following formula:

$$F_{gw} < 0.5\, F_{ge} \quad \text{(Equation 24-6)}$$

where F_{gw} is the wind load on the glass computed per Section 1609 and F_{ge} is the nonfactored load from ASTM E1300 using a thickness designation for monolithic glass that is not greater than the thickness of wired glass.

2404.3.2 Sloped wired glass. Wired glass sloped more than 15 degrees (0.26 rad) from vertical in skylights, sunspaces, sloped roofs and other exterior applications shall be designed to resist the most critical of the combinations of loads from Section 2404.2. For Equations 24-2 and 24-3:

$$F_g < 0.5\, F_{ge} \quad \text{(Equation 24-7)}$$

For Equation 24-4:

$$F_g < 0.3\, F_{ge} \quad \text{(Equation 24-8)}$$

where F_g is total load on the glass and F_{ge} is the nonfactored load from ASTM E1300.

2404.3.3 Vertical patterned glass. Patterned glass sloped 15 degrees (0.26 rad) or less from vertical in windows, curtain and window walls, doors and other exterior applications shall be designed to resist the wind loads in Section 1609 for components and cladding according to the following formula:

$$F_{gw} < 1.0\, F_{ge} \quad \text{(Equation 24-9)}$$

where F_{gw} is the wind load on the glass computed per Section 1609 and F_{ge} is the nonfactored load from ASTM E1300. The value for patterned glass shall be based on the thinnest part of the glass. Interpolation between nonfactored load charts in ASTM E1300 shall be permitted.

2404.3.4 Sloped patterned glass. Patterned glass sloped more than 15 degrees (0.26 rad) from vertical in skylights, sunspaces, sloped roofs and other exterior applications shall be designed to resist the most critical of the combinations of loads from Section 2404.2. For load Equations 24-2 and 24-3:

$$F_g < 1.0\, F_{ge} \quad \text{(Equation 24-10)}$$

For Equation 24-4:

$$F_g < 0.6\, F_{ge} \quad \text{(Equation 24-11)}$$

where F_g is total load on the glass and F_{ge} is the nonfactored load from ASTM E1300. The value for patterned glass shall be based on the thinnest part of the glass. Interpolation between the nonfactored load charts in ASTM E1300 shall be permitted.

2404.3.5 Vertical sandblasted glass. Sandblasted glass sloped 15 degrees (0.26 rad) or less from vertical in windows, curtain and window walls, doors, and other exterior applications shall be designed to resist the wind loads in Section 1609 for components and cladding according to the following formula:

$$F_g < 0.5\, F_{ge} \quad \text{(Equation 24-12)}$$

where F_g is total load on the glass and F_{ge} is the nonfactored load from ASTM E1300. The value for sandblasted glass is for moderate levels of sandblasting.

Section 2404.4 Add new section to read as shown: (S84-03/04)

2404.4 Other designs. For designs outside the scope of this section, an analysis or test data for the specific installation shall be prepared by a registered design professional.

Section 2406.1 Change to read as shown: (S85-03/04)

2406.1 Human Impact Loads. Individual glazed areas, including glass mirrors, in hazardous locations as defined in Section 2406.3 shall comply with Sections 2406.1.1 through 2406.1.4.

Section 2406.1.2 Wired glass. Delete section without substitution: (S85-03/04)

Section 2406.2 Change to read as shown: (S86-03/04)

2406.2 Identification of safety glazing. Except as indicated in Section 2406.2.1, each pane of safety glazing installed in hazardous locations shall be identified by a label specifying the labeler, whether the manufacturer or installer, and the safety glazing standard with which it complies, as well as the information specified in Section 2403.1. The label shall be acid etched, sand blasted, ceramic fired, laser burned, or an embossed mark, or shall be of a type that once applied, cannot be removed without being destroyed.

Exceptions:

1. For other than tempered glass, labels are not required, provided the building official approves the use of a certificate, affidavit or other evidence confirming compliance with this code.

2. Tempered spandrel glass is permitted to be identified by the manufacturer with a removable paper label.

Section 2409 Glass in Floors and Sidewalks (includes Sections 2409.1 through 2409.4) Delete sections without substitution: (S88-03/04)

CHAPTER 25
GYPSUM BOARD AND PLASTER

Section 2510.6 Add exception to read as shown: (S93-03/04)

2510.6 Weather-resistant barriers. Weather-resistant barriers shall be installed as required in Section 1404.2 and, where applied over wood-based sheathing, shall include a weather-resistant vapor-permeable barrier with a performance at least equivalent to two layers of Grade D paper.

Exception: Where the weather-resistant barrier that is applied over wood-based sheathing has a water resistance equal to or greater than that of 60-minute Grade D paper and is separated from the stucco by an intervening, substantially nonwater-absorbing layer or drainage space.

CHAPTER 26
PLASTIC

Section 2608.3 Change to read as shown: (FS183-03/04)

2603.8 Special approval. Foam plastic shall not be required to comply with the requirements of Sections 2603.4 through 2603.7, where specifically approved based on large-scale tests such as, but not limited to, NFPA 286 (with the acceptance criteria of Section 803.2), FM 4880, UL 1040 or UL 1715. Such testing shall be related to the actual end-use configuration and be performed on the finished manufactured foam plastic assembly in the maximum thickness intended for use. Foam plastics that are used as interior finish on the basis of special tests shall also conform to the flame spread requirements of Chapter 8. Assemblies tested shall include seams, joints and other typical details used in the installation of the assembly and shall be tested in the manner intended for use.

Section 2606.7.5 Change to read as shown: (EL3-03/04)

2606.7.5 Electrical luminaires. Light-transmitting plastic panels and light-diffuser panels that are installed in approved electrical lighting fixtures shall comply with the requirements of Chapter 8 unless the light-transmitting plastic panels conform to the requirements of Section 2606.7.2. The area of approved light-transmitting plastic materials that are used in required exits or corridors shall not exceed 30 percent of the aggregate area of the ceiling in which such panels are installed, unless the building is equipped throughout with an automatic sprinkler system in accordance with Section 903.3.1.1.

2004 SUPPLEMENT TO THE IBC

CHAPTER 29
PLUMBING SYSTEMS

Table 2902.1 Change entire table to read as shown: (P11-03/04; P14-03/04; P15-03/04)

TABLE 2902.1
MINIMUM NUMBER OF PLUMBING REQUIRED PLUMBING FIXTURES
(See Sections P2902.2 and P2902.2)

NO.	CLASSIFICATION	OCCUPANCY	DESCRIPTION	WATER CLOSETS (URINALS SEE SECTION 419.2 of the *International Plumbing Code*) MALE	WATER CLOSETS FEMALE	LAVATORIES MALE	LAVATORIES FEMALE	BATHTUBS/ SHOWERS	DRINKING FOUNTAIN (SEE SECTION 410.1 of the *International Plumbing Code*)	OTHER
1	Assembly (see Sections 2902.2, 2902.5 and 2902.6)	A-1 [d]	Theaters usually with fixed seats and other buildings for the performing arts and motion pictures	1 per 125	1 per 65	1 per 200		—	1 per 500	1 service sink
		A-2 [d]	Nightclubs, bars, taverns, dance halls and buildings for similar purposes	1 per 40	1 per 40	1 per 75		—	1 per 500	1 service sink
			Restaurants, banquet halls and food courts	1 per 75	1 per 75	1 per 200		—	1 per 500	1 service sink
		A-3 [d]	Auditoriums without permanent seating, art galleries, exhibition halls, museums, lecture halls, libraries, arcades and gymnasiums	1 per 125	1 per 65	1 per 200		—	1 per 500	1 service sink
			Passenger terminals and transportation facilities	1 per 500	1 per 500	1 per 750		—	1 per 1000	1 service sink
			Places of worship and other religious services. Churches without assembly halls.	1 per 150	1 per 75	1 per 200		—	1 per 1000	1 service sink

IBC-83

2004 SUPPLEMENT TO THE IBC

Table 2902.1 (continued)

NO.	CLASSIFICATION	OCCUPANCY	DESCRIPTION	WATER CLOSETS (URINALS SEE SECTION 419.2 of the *International Plumbing Code*) MALE	FEMALE	LAVATORIES MALE	FEMALE	BATHTUBS/ SHOWERS	DRINKING FOUNTAIN (SEE SECTION 410.1 of the *International Plumbing Code*)	OTHER
2	Business (see Sections 2902.2, 2902.4 and 2902.6)	B	Buildings for the transaction of business, professional services, other services involving merchandise, office buildings, banks, light industrial and similar uses.	1 per 25 for the first 50 and 1 per 50 for the remainder exceeding 50		1 per 40 for the first 80 and 1 per 80 for the remainder exceeding 80		See Section 411 of the *International Plumbing Code*	1 per 100	1 service sink
5	Institutional	I-4	Adult daycare and childcare	1 per 15		1 per 15		—	1 per 100	1 service sink

a. The fixtures shown are based on one fixture being the minimum required for the number of persons indicated or any fraction of the number of persons indicated. The number of occupants shall be determined by the this code.
b. Toilet facilities for employees shall be separate from facilities for inmates or patients.
c. A single-occupant toilet room with one water closet and one lavatory serving not more than two adjacent patient rooms shall be permitted where such room is provided with direct access from each patient room and with provisions for privacy.
d. The occupant load for seasonal outdoor seating and entertainment areas shall be included when determining the minimum number of facilities required.
e. For attached one- and two-family dwellings, one automatic clothes washer connection shall be required per 20 dwelling units.

Section 2902.2 Change exception to read as shown: (P16-03/04)

2902.2 Separate facilities. Where plumbing fixtures are required, separate facilities shall be provided for each sex.

Exceptions:

1. Separate facilities shall not be required for dwelling units and sleeping units.

2. Separate employee facilities shall not be required in occupancies in which 15 or fewer people are employed.

3. Separate facilities shall not be required in structures or tenant spaces with a total occupant load, including both employees and customers, of 15 or less.

4. Separate facilities shall not be required in mercantile occupancies in which the maximum occupant load is 50 or less.

Section 2902.6 Change to read as shown: (P19-03/04)

2902.6 Public facilities. Customers, patrons, and visitors shall be provided with public toilet facilities in structures and tenant spaces intended for public utilization. The accessible route to public facilities shall not pass through kitchens, storage rooms, closets or similar spaces. Public toilet facilities shall be located not more than one story above or below the space required to be provided with public toilet facilities and the path of travel to such facilities shall not exceed a distance of 500 feet (152 m).

CHAPTER 30
ELEVATORS AND CONVEYING SYSTEMS

Section 3002.4 Change to read as shown: (G142-03/04; G143-03/04)

3002.4 Elevator car to accommodate ambulance stretcher. Where elevators are provided in buildings four stories in height or more, at least one elevator shall be provided for fire department emergency access to all floors. Such elevator car shall be of such a size and arrangement to accommodate a 24-inch by 84-inch (610 mm by 2134 mm) ambulance stretcher in the horizontal, open position and shall be identified by the international symbol for emergency medical services (star of life). The symbol shall not be less than 3 inches (76 mm) high and shall be placed inside on both sides of the hoistway door frame.

Section 3006.4 Change to read as shown: (G50-03/04; FS53-03/04)

3006.4 Machine rooms and machinery spaces. Elevator machine rooms and machinery spaces shall be enclosed with fire barriers having a fire-resistance rating not less than the required rating of the hoistway enclosure served by the machinery. Openings shall be protected with assemblies having a fire-protection rating not less than that required for the hoistway enclosure doors.

CHAPTER 31
SPECIAL CONSTRUCTION

Section 3102.3 Change to read as shown: (G85-03/04)

3102.3 Type of construction. Noncombustible membrane structures shall be classified as Type IIB construction. Noncombustible frame or cable-supported structures covered by an approved membrane in accordance with Section 3102.3.1 shall be classified as Type IIB construction. Heavy timber frame-supported structures covered by an approved membrane in accordance with Section 3102.3.1 shall be classified as Type IV construction. Other membrane structures shall be classified as Type V construction.

> **Exception:** Plastic less than 30 feet (9144 mm) above any floor used in greenhouses, where occupancy by the general public is not authorized, and for aquaculture pond covers, is not required to meet the fire propagation performance criteria of NFPA 701.

Section 3102.3.1 Change to read as shown: (G85-03/04)

3102.3.1 Membrane and interior liner material. Membranes and interior liners shall be either noncombustible as set forth in Section 703.4, or meet the fire propagation performance criteria of NFPA 701 and the manufacturer's test protocol.

> **Exception:** Plastic less than 20 mil (0.5 mm) in thickness used in greenhouses, where occupancy by the general public is not authorized, and for aquaculture pond covers, is not required to meet the fire propagation performance criteria of NFPA 701.

Section 3102.6.1.1 Change to read as shown: (G85-03/04)

3102.6.1.1 Membrane. A membrane meeting the fire propagation performance criteria of NFPA 701 shall be permitted to be used as the roof or as a skylight on buildings of Type IIB, III, IV and V construction provided it is at least 20 feet (6096 mm) above any floor, balcony or gallery.

Section 3105.4 Change to read as shown: (G85-03/04)

3105.4 Canopy materials. Canopies shall be constructed of a rigid framework with an approved covering, that meets the fire propagation performance criteria of NFPA 701 or has a flame spread index not greater than 25 when tested in accordance with ASTM E 84.

Section 3109.5 Delete and substitute as follows: (G146-03/04)

3109.5 Entrapment avoidance. Suction outlets shall be designed to produce circulation throughout the pool or spa. Single-outlet systems, such as automatic vacuum cleaner systems, or other such multiple suction outlets whether isolated by valves or otherwise shall be protected against user entrapment.

Sections 3109.5.1 through 3109.5.4 Add new sections to read as follows: (G146-03/04)

3109.5.1 Suction fittings. All pool and spa suction outlets shall be provided with a cover that conforms to ANSI/ASME A112.19.8M, or a 12" x 12" drain grate or larger, or an approved channel drain system.

> **Exception:** Surface skimmers

3109.5.2 Atmospheric vacuum relief system required. All pool and spa single or multiple outlet circulation systems shall be equipped with an atmospheric vacuum relief should grate covers located therein become missing or broken. Such vacuum relief systems shall include at least one approved or engineered method of the type specified herein, as follows:

1. Safety vacuum release systems conforming to ANSI/ASME A112.19.17, or

2. Approved gravity drainage system.

3109.5.3 Dual drain separation. Single or multiple pump circulation systems shall be provided with a minimum of two suction outlets of the approved type. A minimum horizontal or vertical distance of 3 feet shall separate such outlets. These suction outlets shall be piped so that water is drawn through them simultaneously through a vacuum-relief-protected line to the pump or pumps.

3109.5.4 Pool cleaner fittings. Where provided, vacuum or pressure cleaner fitting(s) shall be located in an accessible position(s) at least 6 inches and not greater than 12 inches below the minimum operational water level or as an attachment to the skimmer(s).

CHAPTER 34
EXISTING STRUCTURES

Sections 3403.1 and 3403.1.1 Change to read as shown: (EB34-03/04)

3403.1 Existing buildings or structures. Additions or alterations to any building or structure shall conform with the requirements of the code for new construction. Additions or alterations shall not be made to an existing building or structure which will cause the existing building or structure to be in violation of any provisions of this code. An existing building plus additions shall comply with the height and area provisions of Chapter 5. Portions of the structure not altered and not affected by the alteration are not required to comply with the code requirements for a new structure.

3403.1.1 Flood hazard areas. For buildings and structures in flood hazard areas established in Section 1612.3, any additions, alterations or repairs that constitute substantial improvement of the existing structure, as defined in Section 1612.2, shall comply with the flood design requirements for new construction and all aspects of the existing structure shall be brought into compliance with the requirements for new construction for flood design.

Section 3409.4 Change to read as shown: (EB30-03/04)

3409.4 Additions. Provisions for new construction shall apply to additions. An addition that affects the accessibility to, or contains an area of primary function, shall comply with the requirements in Section 3409.6.

Section 3410.2.4.1 Add new section to read as shown: (EB34-03/04)

3410.2.4.1 Flood hazard areas. For existing buildings located in flood hazard areas established in Section 1612.3, if the alterations and repairs constitute substantial improvement of the existing building, the existing building shall be brought into compliance with the requirements for new construction for flood design.

Section 3410.6.6 Change to read as shown: (E1-03/04)

3410.6.6 Vertical openings. Evaluate the fire-resistance rating of exit enclosures, hoistways, escalator openings and other shaft enclosures within the building, and openings between two or more floors. Table 3410.6.6(1) contains the appropriate protection values. Multiply that value by the construction-type factor found in Table 3410.6.6(2). Enter the vertical opening value and its sign (positive or negative) in Table 3410.7 under Safety Parameter 3410.6.6, Vertical Openings, for fire safety, means of egress and general safety. If the structure is a one-story building, enter a value of 2. Unenclosed vertical openings that conform to the requirements of Section 707 shall not be considered in the evaluation of vertical openings.

2004 SUPPLEMENT TO THE IBC

CHAPTER 35
REFERENCED STANDARDS

Change, delete or add the following referenced standards to read as shown: (G112-03/04; G146; G152; FS60-03/04; FS72; FS107; FS124; FS129; FS143; FS148; FS150; FS153; FS155; FS157; FS158; F159; FS160; FS162; F165; F166; FS167; FS168; FS169; FS170; FS171; FS172; FS173; FS174; FS178; FS184; S21-03/04; S22-03/04; S51; S54; S56; S57-03/04; S84; S96; E146-03/04; E147; F99-03/04)
(STANDARDS NOT SHOWN REMAIN UNCHANGED)

AA

Aluminum Association Inc.
900 19th St NW, Suite 300
Washington, DC 20006

Standard reference number	Title	Referenced In code section number
ASM 35-00	Aluminum Sheet Metal Work in Building Construction (Fourth Edition)	2002.1

AISC

American Institute of Steel Construction
One East Wacker Drive, Suite 3100
Chicago, IL 60601

Standard reference number	Title	Referenced In code section number
335-89s1	Specification for Structural Steel Buildings - Allowable Stress Design and Plastic Design, including Supplement No. 1, April 15, 2002	1604.3.3, Table 1704.3, 2203.2, 2205.1
341-02	Seismic Provisions for Structural Steel Buildings, May 21, 2002	1602.1, 1707.2, 1708.4, 2205.2.1, 2205.2.2, 2205.3, 2205.3.1

AITC

American Institute of Timber Construction
7012 S. Revere Parkway, Suite 140
Englewood, CO 80112

Standard reference number	Title	Referenced In code section number
A 190.1-02	Structural Glued Laminated Timber	2303.1.3, 2306.1

ANSI

American National Standards Institute
25 West 43rd Street - 4th Floor
New York, NY 10036

Standard reference number	Title	Referenced In code section number
A42.2-71	DELETED	
A42.3-71	DELETED	
A108.7-92	DELETED	
A118.2-99	DELETED	

ASME

American Society of Mechanical Engineers
Three Park Avenue
New York, NY 10016-5990

Standard reference number	Title	Referenced In code section number
A17.1-2000	Safety Code for Elevators and Escalators - with A17.1a 2002 Addenda	1007.4, 1607.8.1, 3001.2, 3001.4, 3002.5, 3003.2, 3409.7.2
A18.1-2003	Safety Standard for Platform Lifts and Stairway Chairlifts	1007.5, 1109.1, 3409.7.3
A112.19.8M-1987	Suction Fittings for Use in Swimming Pools, Wading Pools, Spas, Hot Tubs and Whirlpool Bathing Appliances	3109.5.1
A112.19.17-1987	Manufacturers Safety Vacuum Release Systems (SVRS) for Residential and Commercial Swimming Pool, Spa, Hot Tub and Wading Pool	3109.5.2
B16.18-2001	Cast Copper Alloy Solder Joint Pressure Fittings	909.15.1
B16.22-2000	Wrought Copper and Copper Alloy Solder Joint Pressure Fittings	909.13.1
B31.3-2002	Process Piping	415.9.6.1

ASTM

ASTM International
100 Barr Harbor Drive
West Conshohocken, PA 19428-2859

Standard reference number	Title	Referenced In code section number
A 6/A 6M-02b	Specification for General Requirements for Rolled, Structural Steel Bars, Plates, Shapes, and Sheet Piling	Table 1704.3
A 36/A 36M-02	Specification for Carbon Structural Steel	1809.3.1, 2103.11.5
A 82-02	Specification for Steel Wire, Plain, for Concrete Reinforcement	2103.11.5, 2103.11.6
A 123/A 123M-02	Specification of Zinc (Hot-Dip Galvanized) Coating on Iron and Steel Products	2103.11.7.1
A 153-02	Specification for Zinc Coating (Hot Dip) on Iron and Steel Hardware	2103.11.7.1
A 185-02	Specification for Steel Welded Wire Reinforcement, Plain for Concrete	2103.11.4,
A 307-02	Specification for Carbon Steel Bolts and Studs, 60,000 psi Tensile Strength	1912.1, 2103.11.5
A 252-98 (2002)	Specification for Welded and Seamless Steel Pipe Piles	1809.3.1, 1810.6.1
A 416/A 416M-02	Specification for Steel Strand, Uncoated Seven-Wire for Prestressed Concrete	1809.2.3.1, 2103.11.6
A 421/A 421M-02	Specification for Uncoated Stress-Relieved Steel Wire for Prestressed Concrete	2103.11.6
A 496-02	Specification for Steel Wire, Deformed for Concrete Reinforcement	2103.11.3, 2103.11.4
A463M-02a/A463	Specification for Steel Sheet, Aluminum-Coated, by the Hot Dip Process	Table 1507.4.3
A 510-02	Specification for General Requirements for Wire Rods and Coarse Round Wire, Carbon Steel	2103.11.6
A 568/A 568M-02	Specification for Steel, Sheet, Carbon, and High-Strength, Low-Alloy, Hot-Rolled and Cold-Rolled, General Requirements for	Table 1704.3
A 572/A 572M-02	Specification for High-Strength Low-Alloy Columbium-Vanadium Structural Steel	1809.3.1
A 615/A 615M-02	Specification for Deformed and Plain Billet-Steel Bars for Concrete Reinforcement	2103.11.1, 2103.11.6
A 653/A 653M-02a	Specification for Steel Sheet, Zinc-Coated Galvanized or Zinc-Iron Alloy-Coated Galvanized by the Hot-Dip Process	Table 1507.4.3, 2211.2, 2211.2.2.1
A 706/A 706M-02	Specification for Low-Alloy Steel Deformed and Plain Bars for Concrete Reinforcement	1704.4.1, 1903.5.2, 1908.1.3
A 722/A 722M-98 (2003)	Specification for Uncoated High-Strength Steel Bar for Prestressing Concrete	2103.11.6, 2106.1.1.3.1

ASTM (continued)

A 755/A 755M-01	Specification for Steel Sheet, Metallic-Coated by the Hot-Dip Process and Prepainted by the Coil-Coating Process for Exterior Exposed Building Products	Table 1507.4.3
A 792/A 792M-02	Specification for Steel Sheet, 55% Aluminum-Zinc Alloy-Coated by the Hot-Dip Process	Table 1507.4.3, 2211.2.2, 2211.2.2.1
A 875/A 875M-02a	Standard Specification for Steel Sheet Zinc-5%, Aluminum Alloy-Coated by the Hot-Dip Process	Table 1507.4.3, 2211.2.2, 2211.2.2.1
A 884/A 884M-02	Specification for Epoxy-Coated Steel Wire and Welded Wire Fabric for Reinforcement	2103.11.7.2
A 899-91 (2002)	Specification for Steel Wire Epoxy-Coated	2103.11.7.2
A 951-02	Specification for Masonry Joint Reinforcement	2103.11.2
A 996/A 996M-02	Specification for Rail-Steel and Axle-Steel Deformed Bars for Concrete Reinforcement	2103.11.1, 2103.11.6
A 1008-03	Specification for Steel, Sheet, Cold-Rolled, Carbon, Structural, High-Strength Low-Alloy and High-Strength Low-Alloy with Improved Formability	2103.11.5
B42-02	Specification for Seamless Copper Pipe, Standard Sizes	909.13.1
B68-02	Specification for Seamless Copper Tube, Bright Annealed	909.13.1
B88-02	Specification for Seamless Copper Water Tube	909.13.1
B101-02	Specification for Lead-Coated Copper Sheet and Strip for Building Construction	Table 1507.4.3
B209-02a	Specification for Aluminum and Aluminum-Alloy Steel and Plate	Table 1507.4.3
B251-02	Specification for General Requirements for Wrought Seamless Copper and Copper-Alloy Tube	909.13.1
B280-02	Specification for Seamless Copper Tube for Air Conditioning and Refrigeration Field Service	909.13.1
B633-98e01	Specification for Electodeposited Coatings of Zinc on Iron and Steel	2211.2
C 27-98 (2002)	Specification for Standard Classification of Fireclay and High-Alumina Refractory Brick	2111.5, 2111.8
C 28/C 28M-00 e01	Specification for Gypsum Plasters	Table 2507.2
C 33-02a	Specification for Concrete Aggregates	Table 1904.2.1
C 39/C 39M-99	Test Method for Compressive Strength of Cylindrical Concrete Specimens	1905.6.3.2
C 42/C 42M-99	Test Method for Obtaining and Testing Drilled Cores and Sawed Beams of Concrete	1905.6.5.2
C 67-03	Test Methods of Sampling and Testing Brick and Structural Clay Tile	721.4.1.1.1, 1507.3.5, 2104.5, 2105.2.2.1.1, 2109.8.1.1
C 90-02a	Specification for Loadbearing Concrete Masonry Units	Table 721.3.2, 1805.5.2, 2103.1, 2105.2.2.1.2
C 91-03	Specification for Masonry Cement	Table 2103.7(1), Table 2507.2
C 140-02c	Test Method Sampling and Testing Concrete Masonry Units and Related Units	721.3.1.2, 1507.3.5, 2105.2.2.1.2
C 270-03	Specification for Mortar for Unit Masonry	2103.7, Table 2103.7(2)
C 315-02	Specification for Clay Flue Linings	2113.11.1, Table 2113.16(1), Table 2113.16(2)
C 331-03	Specification for Lightweight Aggregates for Concrete Masonry Units	721.3.1.4, 721.4.1.1.3
C 474-02	Test Methods for Joint Treatment Materials for Gypsum Board Construction	Table 2506.2
C 475/C 475M-02	Specification for Joint Compound and Joint Tape for Finishing Gypsum Wallboard	Table 2506.2
C 476-02	Specification for Grout for Masonry	2103.10, 2105.2.2.1.1, 2105.2.2.1.2
C516-02	Specification for Vermiculite Loose Fill Thermal Insulation	721.3.1.4, 721.4.1.1.3
C549-02	Specification for Perlite Loose Fill Insulation	Table 721.1(2), Table 721.1(3)
C 587-02	Specification for Gypsum Veneer Plaster	Table 2507.2
C 588/C 588M-01	Specification for Gypsum Base for Veneer Plasters	Table 2507.2
C 595-03	Specification for Blended Hydraulic Cements	1904.1, Table 1904.2.3
C 612-00a	Specification for Mineral Fiber Black and Board Thermal Insulation	
C 631-95a (2002)	Specification for Bonding Compounds for Interior Gypsum Plastering	Table 2507.2
C836-03	Specification for High-Solids Content, Cold Liquid-Applied Elastomeric Waterproofing Membrane for Use with Separate Wearing Course	1507.15.2
C 840-02	Specification for Application and Finishing of Gypsum Board	Table 2508.1, 2509.2
C 887-79a (2001)	Specification for Packaged, Dry, Combined Materials for Surface Bonding Mortar	1807.2.2, 2103.8

2004 SUPPLEMENT TO THE IBC

ASTM (continued)

C926-98a	Specification for Application of Portland Cement Based Plaster	2109.8.4.6, 2510.3, Table 2511.1, 2511.3, 2511.4, 2512.1, 2512.1.2, 2512.2, 2512.6, 2512.8.2, 2513.7, 2512.9
C931/C 931M-02	Specification for Exterior Gypsum Soffit Board	Table 2506.2
C956-97 (2002)e01	Specification for Installation of Cast-in-Place Reinforced Gypsum Concrete	1915.1
C960/C 960M-01	Specification for Predecorated Gypsum Board	Table 2506.2
C1019-02	Test Method for Sampling and Testing Grout	2105.2.2.1.1, 2105.2.2.1.2
C1029-02	Specification for Spray-Applied Rigid Cellular Polyurethane Thermal Insulation	1507.14.2
C1032-96 (2002)	Specification for Woven Wire Plaster Base	Table 2507.2
C1063-99	Specification for Installation of Lathing and Furring to Receive Interior and Exterior Portland Cement Based Plaster	2109.8.4.6, 2510.3, Table 2511.1, 2512.1.1
C1088-02	Specification for Thin Veneer Brick Units Made From Clay or Shale	2103.2
C1167-03	Specification for Clay Roof Tiles	1507.3.4, 1507.3.5
C1186-02	Specification for Flat Nonasbestos Fiber Cement Sheets	1404.10
C1283-02	Practice for Installing Clay Flue Lining	2113.12
C1314-02a	Test Method for Compressive Strength of Masonry Prisms	2105.2.2.2.2, 2105.3.1, 2105.3.2
C1328-03	Specification for Plastic (Stucco Cement)	Table 2507.2
C1329-03	Specification for Mortar Cement	Table 2103.7(1)
C1405-00a	Standard Specification for Glazed Brick (Single Fired, Solid Brick Units)	2103.2
C1492-03	Standard Specification for Concrete Roof Tile	1507.3.5
D224-89 (1996)	DELETED	
D225-02	Specification for Asphalt Shingles (Organic Felt) Surfaced with Mineral Granules	1507.2.5
D249-89 (1996)	DELETED	
D371-89 (1996)	DELETED	
D422-63 (2002)	Test Method for Particle-Size Analysis of Soils	1802.3.2
D1557-02	Test Method for Laboratory Compaction Characteristics of Soil Using Modified Effort (56,000 ft-lb/ft3 (2,700kN-m/m3))	1803.5
D1761-88 (2000) e01	Test Methods for Mechanical Fasteners in Wood	1715.1.1, 1715.1.2, 1715.1.3
D1970-01	Specification for Self-Adhering Polymer Modified Bituminous Sheet Materials Used as Steep Roof Underlayment for Ice Dam Protection	1507.2.4, 1507.2.9.2, 1507.3.9, 1507.5.6, 1507.8.7, 1507.9.8
D2850-03	Test Method for Unconsolidated, Undrained Triaxial Compression Test on Cohesive Soils	1615.1.5
D3161-03	Test Method for a Wind Resistance of Asphalt Shingles (Fan Induced Method)	1507.2.7
D3462-03	Specification for Asphalt Shingles Made From Glass Felt and Surfaced with Mineral Granules	1507.2.5
D3679-02a	Specification for Rigid Poly (Vinyl Chloride) (PVC) Siding	1404.9, 1405.13
D3737-02	Practice for Establishing Allowable Properties for Structural Glued Laminated Timber (Glulam)	2303.1.3
D3746-85 (2002)	Test Method for Impact Resistance of Bituminous Roofing Systems	1504.7
D3747-79 (2002)	Specification for Emulsified Asphalt Adhesive for Adhering Roof Insulation	Table 1507.10.2
D3909-97b	Specification for Asphalt Roll Roofing (Glass Felt) Surfaced with Mineral Granules	1507.2.9.2, 1507.5.4, Table 1507.10.2
D4869-02	Specification for Asphalt-Saturated (Organic Felt) Underlayment Used in Steep Slope Roofing	Table1507.2, 1507.2.3, 1507.5.3, 1507.6.3, 1507.8.3, 1507.9.3
D5019-96e01	Specification for Reinforced Non-Vulcanized Polymeric Sheet Used in Roofing Membrane	1507.12.2
D5055-02	Specification for Establishing and Monitoring Structural Capacities of Prefabricated Wood I-Joists	2303.1.2
D5516-02	Test Method of Evaluating the Flexural Properties of Fire-Retardant Treated Softwood Plywood Exposed to the Elevated Temperatures	2303.2.2.1
D5664-02	Test Methods for Evaluating the Effects of Fire-Retardant Treatments and Elevated Temperatures on Strength Properties of Fire-Retardant Treated Lumber	2303.2.2.2
D6222-02	Specification for Atactic Polypropylene (APP) Modified Bituminous Sheet Materials Using Polyester Reinforcement	1507.11.2

2004 SUPPLEMENT TO THE IBC

ASTM (continued)

D6223-02	Specification for Atactic Polypropylene (APP) Modified Bituminous Sheet Materials Using a Combination of Polyester and Glass Fiber Reinforcement	1507.11.2
D6305-02e01	Practice for Calculating Bending Strength Design Adjustment Factors for Fire-Retardant-Treated Plywood Roof Sheathing	2303.2.2.1
D6380-01^{E1}	Standard Specification for Asphalt Roll Roofing (Organic) Felt	1507.2.9.2, 1507.3.3, 1507.6.4
D6381-03	Standard Test Method for Measurement of Asphalt Shingle Mechanical Uplift Resistance	1609.7.2
D6694-01	Standard Specification for Liquid-Applied Silicone Coating Used in Spray Polyurethane Foam Roofing	1507.15.2
D6754-02	Standard Specification for Ketone Ethylene Ester Based Sheet Roofing	1507.13.2
D6757-02	Standard Specification for Inorganic Underlayment for Use with Steep Slope Roofing Products	1507.2.3
D6878-03	Standard Specification for Thermoplastic Polyolefin Based Sheet Roofing	1507.13.2
E 84-03	Test Method for Surface Burning Characteristics of Building Materials	402.10, 402.14.4, 406.5.2, 410.3.5.3, 703.4.2, 719.1, 719.4, 802.1, 803.1, 803.5, 803.6.1, 803.6.2, 1407.10, 1407.10.1, 2303.2, 2603.3, 2603.4.1.13, 2603.5.4, 2604.2.4, 2606.4, 3105.3
E90-02	Test Method for Laboratory Measurement of Airborne Sound Transmission Loss of Building Partitions and Elements	1207.2
E96-00e01	Test Method for Water Vapor Transmission of Materials	1203.2
E136-99e01	Test Method for Behavior of Materials in a Vertical Tube Furnace at 750 Degrees C	703.4.1
E328-02	Methods for Stress Relaxation for Materials and Structures	2103.11.6
E330-02	Test Method for Structural Performance of Exterior Windows, Doors, Skylights and Curtain Walls by Uniform Static Air Pressure Difference	1714.5.2
E814-02	Test Method of Fire Tests of Through-Penetration Firestops	702.1, 712.3.1.2, 712.4.1.2
E1300-03	Standard Practice for Determining Load Resistance of Glass in Buildings	2404.1, 2404.2, 2404.3.1, 2404.3.2, 2404.3.3, 2404.3.4, 2404.3.5
E1886-02	Test Method for Performance of Exterior Windows, Curtain Walls, Doors and Storm Shutters Impacted by Missiles and Exposed to Cyclic Pressure Differentials	1609.1.4
E1996-02	Specification for Performance of Exterior Windows, Curtain Walls, Doors and Storm Shutters Impacted by Windborne Debris in Hurricanes	702.1, 712.3
F1346-91 (2003)	Performance Specifications for Safety Covers and Labeling Requirements for All Covers for Swimming Pools, Spas and Hot Tubs	3104.9, 3109.4.1.8
F1667-02a	Specification for Driven Fasteners: Nails, Spikes, and Staples	Table 721.1(2), 721.1(2), Table 721.1(3), 1507.2.6, 2303.6, Table 2506.2

AWPA

American Wood-Preservers' Association
P.O. Box 5690
Grandbury, TX 76049

Standard reference number	Title	Referenced In code section number

C1-00, C2-01, C3-99, C4-99, C9-00, C14-99, C15-00, C16-00, C22-96, C23-00, C24-96, C28-99, C31-00, C33-00 (STANDARDS DELETED; REPLACE BY U1-02)

U1-02	USE CATEGORY SYSTEM: User Specification for Treated Wood except Section 7, Commodity Specification H	1403.6, 1505.6, Table 1507.9.5, 1805.4.5, 1805.4.6, 1805.7.1, 1809.1.2, 2303.1.8, 2304.11.2, 2304.11.4, 2304.11.6, 2304.11.7

AWS

American Welding Society
550 NW Le Jeune Road
Miami, FL 33126

Standard reference number	Title	Referenced In code section number
D1.1-2002	Structural Welding Code - Steel	Table 1704.3, 1704.3.1, 1708.4

BHMA

Builders Hardware Manufacturers' Association
355 Lexington Avenue, 17th Floor
New York, NY 10017

Standard reference number	Title	Referenced In code section number
A156.19-2002	Standard for Power Assist and Low Energy Power Operated Doors Power Assist and Low Energy Power Operated Doors	1008.1.3.2

DASMA

Door and Access Systems Manufacturer's
1300 Summer Avenue
Cleveland, OH 44115-2851

Standard reference number	Title	Referenced In code section number
107-98 (03)	Room Fire Test Standard for Garage Doors Using Foam Plastic Insulation	2603.4.1.9

GA

Gypsum Association
810 First Street, NE - Suite 510
Washington, D.C. 20002-4268

Standard reference number	Title	Referenced In code section number
GA 600-03	Fire Resistance Design Manual, 17th Edition	Table 721.1(1), Table 721.1(2), Table 721.1(3)

ICC

International Code Council
5203 Leesburg Pike, Suite 600
Falls Church, VA 22041

Standard reference number	Title	Referenced In code section number
A117.1-03	Accessible & Usable Buildings and Facilities	406.2.2, 907.9.1.3, 1007.6.5, 1010.1, 1010.6.5, 1010.9, 1011.3, 1101.2, 1102.1, 1103.2.13, 1106.6, 1107.2, 1109.2.2, 1109.3, 1109.4, 1109.8, 1109.15, 3001.3, 3409.5, 3409.7.2, 3409.7.3

2004 SUPPLEMENT TO THE IBC

NAAMM
National Association of Architectural Metal
8 South Michigan Avenue, Suite 1000
Chicago, IL 60603

Standard reference number	Title	Referenced In code section number
FP 1001-97	Guide Specifications for Design of Metal Flag Poles, Fourth Edition	1609.1.1

NIST
National Institute of Standards and Technology
U.S. Department of Commerce
100 Bureau Drive - Stop 3460
Gaithersburg, MD 20899-3460

Standard reference number	Title	Referenced In code section number
BMS 71-41	DELETED	
TRBM-44 - 46	DELETED	

(Since these are the only 2 standards referenced in the IBC for NIST, the entire listing, including the heading is deleted)

NFPA
National Fire Protection Association
1 Batterymarch Park
Quincy, MA 02269-9101

Standard reference number	Title	Referenced In code section number
13-02	Installation of Sprinkler Systems	704.12, 707.2, 903.3.1.1, 903.3.2, 903.3.5.1.1, 904.11, 907.8, 1621.3.10.1, 3104.5, 3104.9
13D-02	Installation of Sprinkler Systems in One- and Two-family Dwellings and Manufactured Homes	903.1.2, 903.3.1.3, 903.3.5.1.1
13R-02	Installation of Sprinkler Systems in Residential Occupancies Up to and Including Four Stories in Height	903.1.2, 903.3.1.2, 903.3.5.1.1, 903.3.5.1.2, 903.4
17-02	Dry Chemical Extinguishing Systems	904.6, 904.11
17A-02	Wet Chemical Extinguishing Systems	904.5, 904.11
40-01	Storage and Handling of Cellulose Nitrate Film	409.1
72-02	National Fire Alarm Code	505.4, 901.6, 903.4.1, 904.3.5, 907.2, 907.2.1, 907.2.1.1, 907.2.10, 907.2.10.4, 907.2.11.2, 907.2.11.3, 907.2.12.2.3, 907.2.12.3, 907.4, 907.5, 907.9.2, 907.10, 907.14, 907.16, 907.17, 911.1, 3006.5
105-03	Standard for the Installation of Smoke Door Assemblies	405.4.2, 715.3.3
110-02	Emergency and Standby Power Systems	2702.1
230-99	Standard for the Fire Protection of Storage	507.2, 909.20.4.1
231C-98	DELETED	
265-02	Methods of Fire Tests for Evaluating Room Fire Growth Contribution of Textile Coverings on Full Height Panels and Walls	803.6.1, 803.6.1.1, 803.6.1.2
268-01	Standard Test Method for Determining Ignitibility of Exterior Wall Assemblies Using a Radiant Heat Energy Source	1406.2.1, 1406.2.1.1, 1406.2.1.2, 2603.5.7
288-01	Standard Methods of Fire Tests of Floor Fire Door Assemblies In Fire-Resistance-Rated Floor Systems	712.4.6
303-00	Fire Protection Standards for Marinas and Boatyards	905.3.7
409-01	Aircraft Hangars	412.2.6, 412.4.5
704-01	Standard System for the Identification of the Hazards of Materials for Emergency Response	414.7.2, 415.2
1124-03	Manufacture, Transportation, and Storage of Fireworks and Pyrotechnic Articles	415.3.1

RMA

Rubber Manufacturers Association
1400 K. Street, N.W. #900
Washington, D.C. 20005

Standard reference number	Title	Referenced In code section number
RP-1--90	DELETED	
RP-2-90	DELETED	
RP-3-85	DELETED	

(Since these are the only 3 standards referenced in the IBC for RMA, the entire listing, including the heading is deleted)

RMI

Rack Manufacturers Institute
8720 Red Oak Boulevard, Suite 201
Charlotte, NC 28217

Standard reference number	Title	Referenced In code section number
RMI (2002)	Specification for Design, Testing and Utilization of Industrial Steel Storage Racks	2208.1

SJI

Steel Joist Institute
3127 10th Avenue, North
Myrtle Beach, SC 29577-6760

Standard reference number	Title	Referenced In code section number
JG-1.0 (2002)	Standard Specification for Joist Girders	1604.3.3, 2206
K-1.0 (2002)	Standard Specification for Open Web Steel Joists, K-Series	2206
LH/DLH-1.0 (2002)	Standard Specification for Longspan Steel Joists, LH Series and Deep Longspan Steel Joists, DLH Series	2206

SPRI

Single-Ply Roofing Institute
77 Rumford Avenue, Suite 3-B
Waltham, MA 02453

Standard reference number	Title	Referenced In code section number
ES-1-03	Wind Design Standard for Edge Systems Used with Low Slope Roofing Systems	1504.5
RP4-02	Wind Design Guide for Ballasted Single-ply Roofing Systems	1504.4

UL

Underwriters Laboratories, Inc.
333 Pfingsten Road
Northbrook, IL 60062-2096

Standard reference number	Title	Referenced In code section number
10B-1997	Fire Tests of Door Assemblies - with Revisions through October 2001	715.3.2

UL (continued)

14C-1999	Swinging Hardware for Standard Tin Clad Fire Doors Mounted Singly and in Pairs	715.3
103-2001	Factory-Built Chimneys, for Residential Type and Building Heating Appliances	717.2.5
555-1999	Fire Dampers - with Revisions through January 2002	716.3
555S-1999	Smoke Dampers - with Revisions through January 2002	716.3, 716.3.1.1
790-1998	Tests for Fire Resistance of Roof Covering Materials - with Revisions through July 1998	1505.1, 2603.6, 2610.2, 2610.3
864-1996	Control Units for Fire Protective Signaling Systems - with Revisions through August 2001	909.12
1256-2002	Fire Test of Roof Deck Construction	1508.1, 2603.3, 2603.4.1.5
1479-1994	Fire Tests of Through-Penetration Firestops - with Revisions through August 2000	712.3.1.2, 712.4.1.2
1715-1997	Fire Test of Interior Finish Material - with Revisions through October 2002	1407.10.2, 1407.10.3, 2603.4, 2603.8
1897-1998	Uplift Tests for Roof Covering Systems - with Revisions through November 2002	1504.3.1
2390-04	Test Method for Measuring the Wind Uplift Coefficients for Asphalt Shingles	1609.7.2

ULC

Underwriters Laboratories of Canada
7 Crouse Road
Toronto, Ontario, Canada M1R 3A9

Standard reference number	Title	Referenced In code section number
S102-2-1988	Standard Method of Test for Surface Burning Characteristics of Flooring, Floor Coverings, and Miscellaneous Materials and Assemblies - with 2000 Revisions	719.4

WRI

Wire Reinforcement Institute, Inc.
942 Main Street, Suite 300
Hartford, CT 06103

Standard reference number	Title	Referenced In code section number
WRI/CRSI-81	Design of Slab-On-Ground Foundations - with 1996 Update	1805.8.2

2004 SUPPLEMENT TO THE IBC

APPENDIX D
FIRE DISTRICTS

Section D102.2.8. Change to read as shown: (G85-03/04)

D102.2.8 Permanent canopies. Permanent canopies are permitted to extend over adjacent open spaces provided:

1. The canopy and its supports shall be of noncombustible material, fire-retardant-treated wood, Type IV construction, or of 1-hour fire-resistance-rated construction.

    Exception: Any textile covering for the canopy shall meet the fire propagation performance criteria of NFPA 701 after both accelerated water leaching and accelerated weathering.

2. Any canopy covering, other than textiles, shall have a flame spread index not greater than 25 when tested in accordance with ASTM E 84 in the form intended for use.

3. The canopy shall have at least one long side open.

4. The maximum horizontal width of the canopy shall not exceed 15 feet (4572 mm).

5. The fire resistance of exterior walls shall not be reduced.

APPENDIX E
SUPPLEMENTAL ACCESSIBILITY REQUIREMENTS

Section E109.3 Existing facilities: key stations. Delete section without substitution: (E149-03/04)

Section E109.3.1 Accessible route. Delete section without substitution: (E149-03/04)

Section E109.3.2 Platform and vehicle floor coordination. Delete section without substitution: (E149-03/04)

Section E109.3.3 Direct connections. Delete section without substitution: (E149-03/04)

Section E111 Qualified Historic Buildings and Facilities. Delete entire section without substitution: (E149-03/04)

Section E112 Referenced Standards. Delete the following referenced standards without substitution: (E149-03/04)

DOT 49/CFR Part 37   DELETED
DOT 49/CFR Part 38   DELETED
Y3.H626 2P   DELETED

APPENDIX G
FLOOD-RESISTANT CONSTRUCTION

Section G101.3 Change to read as shown: (S97-03/04)

G101.3 Scope. The provisions of this appendix shall apply to all proposed development in a flood hazard area established in Section 1612 of this code, including certain building work exempt from permit under Section 105.2.

Section G102.1 Change to read as shown: (S97-03/04)

G102.1 General. This appendix, in conjunction with the *International Building Code*, provides minimum requirements for development located in flood hazard areas including the subdivision of land; installation of utilities; placement and replacement of manufactured homes; new construction and repair, reconstruction, rehabilitation or additions to new construction; substantial improvement of existing buildings and structures, including restoration after damage; and certain building work exempt from permit under Section 105.2.

Section G103.1 Change to read as shown: (S97-03/04)

G103.1 Permit applications. The building official shall review all permit applications to determine whether proposed development sites will be reasonably safe from flooding. If a proposed development site is in a flood hazard area, all site development activities, including grading, filling, utility installation and drainage modification, all new construction and substantial improvements (including the placement of prefabricated buildings and manufactured homes), and certain building work exempt from permit under Section 105.2, shall be designed and constructed with methods, practices and materials that minimize flood damage and that are in accordance with this code and ASCE 24.

Section G103.4 Change to read as shown: (S98-03/04)

G103.4 Activities in riverine flood hazard areas. In riverine flood hazard areas where design flood elevations are specified but floodways have not been designated, the building official shall not permit any new construction, substantial improvement or other development, including fill, unless the applicant demonstrates that the cumulative effect of the proposed development, when combined with all other existing and anticipated flood hazard area encroachment, will not increase the design flood elevation more than 1 foot (305 mm) at any point within the community.

Section G801 Add new sections to read as shown: (S97-03/04)

SECTION G801
OTHER BUILDING WORK

G801.1 Detached accessory structures. Detached accessory structures shall be anchored to prevent flotation, collapse or lateral movement resulting from hydrostatic loads, including the effects of bouyancy, during conditions of the design flood. Fully enclosed accessory structures shall have flood openings to allow for the automatic entry and exit of flood waters.

G801.2 Fences. Fences in floodways that may block the passage of floodwaters, such as stockade fences and wire mesh fences, shall meet the requirement of Section G103.5.

G801.3 Oil derricks. Oil derricks located in flood hazard areas shall be designed in conformance to the flood loads in Section 1603.1.6 and Section 1612.

G801.4 Retaining walls, sidewalks and driveways. Retaining walls, sidewalks and driveways shall meet the requirements of Section 1803.4.

G801.5 Prefabricated swimming pools. Prefabricated swimming pools in floodways shall meet the requirement of Section G103.5

APPENDIX H
SIGNS

Section H106.1.1 Change to read as shown: (G151-03/04)

H106.1.1 Internally illuminated signs. Except as provided for in Sections 402.14 and 2611, where internally illuminated signs have sign facings of wood or approved plastic, the area of such facing section shall not be more than 120 square feet (11.16 m^2) and the wiring for electric lighting shall be entirely enclosed in the sign cabinet with a clearance of not less than 2 inches (51 mm) from the facing material. The dimensional limitation of 120 square feet (11.16 m^2) shall not apply to sign facing sections made from flame-resistant-coated fabric (ordinarily known as "flexible sign face plastic") that weighs less than 20 ounces per square yard (678 g/m^2) and which, when tested in accordance with NFPA 701, meets the fire propagation performance requirements of both Test 1 and Test 2, or which, when tested in accordance with an approved test method, exhibits an average burn time for 10 specimens of 2 seconds or less and a burning extent of 5.9 inches (150 mm) or less.

APPENDIX J
GRADING

Section J101.2 Change to read as shown: (S99-03/04)

J101.2 Flood hazard areas. The provisions of this chapter shall not apply to grading, excavation and earthwork construction, including fills and embankments, in floodways within flood hazard areas established in Section 1612.3, or in flood hazard areas where design flood elevations are specified but floodways have not been designated, unless it has been demonstrated through hydrologic and hydraulic analyses performed in accordance with standard engineering practice that the proposed work will not result in any increase in the level of the base flood.

International Code Council Electrical Code®

Administrative Provisions

2004 Supplement

ICC ELECTRICAL CODE
2004 SUPPLEMENT

CHAPTER 12
ELECTRICAL PROVISIONS

Sections 1202.6 and 1203.1.5 Change to read as shown: (EL3-03/04)

1202.6 Appliance access. Where appliances requiring access are installed in attics or underfloor spaces, a luminaire controlled by a switch located at the required passageway opening to such space and a receptacle outlet shall be provided at or near the appliance location.

1203.1.5 Luminaires. Every public hall, interior stairway, toilet rooms, kitchen, bathroom, laundry room, boiler room and furnace room shall be provided with at least one electric luminaire.

Section 1202.12 Change to read as shown: (M68-03/04)

1202.12 Stationary fuel cell power systems. Stationary fuel cell power systems having a power output not exceeding 10 MW, shall be tested in accordance with ANSI CSA American FC1 and shall be installed in accordance with the manufacturer's installation instructions and NFPA 853.

International Energy Conservation Code

2004 Supplement

2004 SUPPLEMENT TO THE IECC

As a result of the 2003/2004 Cycle, the International Energy Conservation Code (IECC) has been substantially revised and reformatted. As such, the ENTIRE IECC is reproduced here. The following is a cross reference between the 2003 edition and the 2004 supplement.

2003 IECC	2004 Supplement to the IECC
Chapter 1 Administration	**Chapter 1** Administration and Enforcement
Chapter 2 Definitions	**Chapter 2** Definitions
Chapter 3 Design Conditions	**Chapter 3** Climate Zones
Chapter 4 Residential Building Design by Systems Analyses	**Chapter 4** Residential Energy Efficiency
Chapter 5 Residential Building Design by Performance Approach	**Chapter 5** Deleted – Residential provisions located in Chapter 4
Chapter 6 Simplified Prescriptive Requirements for Detached One- and Two-Family Dwellings	**Chapter 6** Deleted – Residential provisions located in Chapter 4
Chapter 7 Building Design for All Commercial Buildings	**Chapter 7** Deleted – Commercial provisions located in Chapter 8
Chapter 8 Design Acceptable Commercial Practice	**Chapter 8** Building Design for Commercial Buildings
Chapter 9 Climate Maps	**Chapter 9** Deleted – Climate Information located in Chapter 3
Chapter 10 Reference Standards	**Chapter 10** Reference Standards
Appendix	**Appendix** Deleted

Delete Chapter 1 in its entirety and replace as shown: (EC48-03/04)

CHAPTER 1
ADMINISTRATION AND ENFORCEMENT

SECTION 101
SCOPE AND GENERAL REQUIREMENTS

101.1 Title. This code shall be known as the *International Energy Conservation Code* of [**NAME OF JURISDICTION**], and shall be cited as such. It is referred to herein as "this code."

101.2 Scope. This code applies to residential and commercial buildings.

> **Exception:** Existing buildings undergoing repair, alteration, or additions, and change of occupancy shall be permitted to comply with the *International Existing Building Code*.

101.3 Intent. This code shall regulate the design and construction of buildings for the effective use of energy. This code is intended to provide flexibility to permit the use of innovative approaches and techniques to achieve the effective use of energy. This code is not intended to abridge safety, health or environmental requirements contained in other applicable codes or ordinances.

101.4 Applicability.

101.4.1 Existing buildings. Except as specified in this chapter, this code shall not be used to require the removal, alteration or abandonment of, nor prevent the continued use and maintenance of, an existing building or building system lawfully in existence at the time of adoption of this code.

101.4.2 Historic buildings. Any building or structure that is listed in the State or National Register of Historic Places; designated as a historic property under local or state designation law or survey; certified as a contributing resource with a National Register listed or locally designated historic district; or with an opinion or certification that the property is eligible to be listed on the National or State Registers of Historic Places either individually or as a contributing building to a historic district by the State Historic Preservation Officer or the Keeper of the National Register of Historic Places, are exempt from this code.

101.4.3 Additions, alterations, renovations or repairs. Additions, alterations, renovations or repairs to an existing building, building system or portion thereof shall conform to the provisions of this code as they relate to new construction without requiring the unaltered portion(s) of the existing building or building system to comply with this code. Additions, alterations, renovations, or repairs shall not create an unsafe or hazardous condition or overload existing building systems.

> **Exceptions:** The following need not comply provided the energy use of the building is not increased.
>
> 1. Storm windows installed over existing fenestration.
> 2. Glass only replacements in an existing sash and frame.
> 3. Existing ceiling, wall or floor cavities exposed during construction provided that these cavities are filled with insulation.
> 4. Construction where the existing roof, wall or floor cavity is not exposed.

101.4.4 Change in occupancy. Buildings undergoing a change in occupancy that would result in an increase in demand for either fossil fuel or electrical energy shall comply with this code.

101.4.5 Mixed occupancy. Where a building includes both residential and commercial occupancies, each occupancy shall be separately considered and meet the applicable provisions of Chapter 4 for residential and Chapter 8 for commercial.

101.5 Compliance. Residential buildings shall meet the provisions of Chapter 4. Commercial buildings shall meet the provisions of Chapter 8.

101.5.1 Compliance materials. The code official shall be permitted to approve specific computer software, worksheets, compliance manuals and other similar materials that meet the intent of this code.

101.5.2 Low energy buildings. The following buildings, or portions thereof, separated from the remainder of the building by building thermal envelope assemblies complying with this code shall be exempt from the building thermal envelope provisions of this code.

1. Those with a peak design rate of energy usage less than 3.4 Btu/h·ft^2 (10.7 W/m^2) or 1.0 watt/ft^2 (10.7 W/m^2) of floor area for space conditioning purposes.
2. Those that do not contain conditioned space.

SECTION 102
MATERIALS, SYSTEMS AND EQUIPMENT

102.1 Identification. Materials, systems and equipment shall be identified in a manner that will allow a determination of compliance with the applicable provisions of this code.

102.1.1 Building thermal envelope insulation. An *R*-value identification mark shall be applied by the manufacturer to each piece of building thermal envelope insulation 12 inches (305 mm) or greater in width. Alternately, the insulation installers shall provide a certification listing the type, manufacturer and *R*-value of insulation installed in each element of the building thermal envelope. For blown or sprayed insulation, the initial installed thickness, settled thickness, settled *R*-value, installed density, coverage area and number of bags installed shall be listed on the certification. The insulation installer shall sign, date and post the certification in a conspicuous location on the job site.

102.1.1.1 Blown or sprayed roof/ceiling insulation. The thickness of blown in or sprayed roof/ceiling insulation shall be written in inches (mm) on markers that are installed at least one for every 300 ft^2 (28 m^2) throughout the attic space. The markers shall be affixed to the trusses or joists and marked with the minimum initial installed thickness with numbers a minimum of 1 inch (25 mm) in height. Each marker shall face the attic access opening.

102.1.2 Insulation mark installation. Insulating materials shall be installed such that the manufacturer's *R*-value mark is readily observable upon inspection.

102.1.3 Fenestration product rating. *U*-factors of fenestration products (windows, doors and skylights) shall be determined in accordance with NFRC 100 by an accredited, independent laboratory, and labeled and certified by the manufacturer. Products lacking such a labeled *U*-factor shall be assigned a default *U*-factor from Table 102.1.3(1) or 102.1.3(2). The solar heat gain coefficient (SHGC) of glazed fenestration products (windows, glazed doors and skylights) shall be determined in accordance with NFRC 200 by an accredited, independent laboratory, and labeled and certified by the manufacturer. Products lacking such a labeled SHGC shall be assigned a default SHGC from Table 102.1.3(3).

TABLE 102.1.3(1)
DEFAULT GLAZED FENESTRATION *U*-FACTOR

FRAME TYPE	SINGLE PANE	DOUBLE PANE	SKYLIGHT SINGLE	SKYLIGHT DOUBLE
Metal	1.20	0.80	2.00	1.30
Metal with Thermal Break	1.10	0.65	1.90	1.10
Non-Metal or Metal Clad	0.95	0.55	1.75	1.05
Glazed Block	0.60			

**TABLE 102.1.3(2)
DEFAULT DOOR U-FACTORS**

Door Type	U-factor
Uninsulated Metal	1.20
Insulated Metal	0.60
Wood	0.50
Insulated, non-metal edge, max 45% glazing, any glazing double pane	0.35

**TABLE 102.1.3(3)
DEFAULT GLAZED FENESTRATION SHGC**

Single Glazed		Double Glazed		Glazed Block
Clear	Tinted	Clear	Tinted	
0.7	0.6	0.6	0.5	0.6

102.2 Installation. All materials, systems and equipment shall be installed in accordance with the manufacturer's installation instructions and the *International Building Code*.

102.2.1 Protection of exposed foundation insulation. Insulation applied to the exterior of basement walls, crawlspace walls and the perimeter of slab-on-grade floors shall have a rigid, opaque and weather-resistant protective covering to prevent the degradation of the insulation's thermal performance. The protective covering shall cover the exposed exterior insulation and extend a minimum of 6 inches (153 mm) below grade.

102.3 Maintenance information. Maintenance instructions shall be furnished for equipment and systems that require preventive maintenance. Required regular maintenance actions shall be clearly stated and incorporated on a readily accessible label. The label shall include the title or publication number for the operation and maintenance manual for that particular model and type of product.

SECTION 103
ALTERNATE MATERIALS—METHOD OF CONSTRUCTION, DESIGN OR INSULATING SYSTEMS

103.1 General. This code is not intended to prevent the use of any material, method of construction, design or insulating system not specifically prescribed herein, provided that such construction, design or insulating system has been approved by the code official as meeting the intent of this code.

103.1.1 Above code programs. The code official or other authority having jurisdiction shall be permitted to deem a national, state or local energy efficiency program to exceed the energy efficiency required by this code. Buildings approved in writing by such an energy efficiency program shall be considered in compliance with this code.

SECTION 104
CONSTRUCTION DOCUMENTS

104.1 General. Construction documents and other supporting data shall be submitted in one or more sets with each application for a permit. The code official is authorized to require necessary construction documents to be prepared by a registered design professional.

> **Exception:** The code official is authorized to waive the requirements for construction documents or other supporting data if the code official determines they are not necessary to confirm compliance with this code.

104.2 Information on construction documents. Construction documents shall be drawn to scale upon suitable material. Electronic media documents are permitted to be submitted when approved by the code official. Construction documents shall be of sufficient clarity to indicate the location, nature and extent of the work proposed, and show in sufficient detail pertinent data and features of the building, systems and equipment as herein governed. Details shall include, but are not limited to, insulation materials and their *R*-values; fenestration *U*-factors and SHGCs; system and equipment efficiencies, types, sizes and controls; duct sealing, insulation and location; and air sealing details.

SECTION 105
INSPECTIONS

105.1 General. Construction or work for which a permit is required shall be subject to inspection by the code official.

105.2 Required approvals. No work shall be done on any part of the building beyond the point indicated in each successive inspection without first obtaining the written approval of the code official. No construction shall be concealed without being inspected and approved.

105.3 Final inspection. The building shall have a final inspection and not be occupied until approved.

105.4 Reinspection. A building shall be reinspected when determined necessary by the code official.

SECTION 106
VALIDITY

106.1 General. If a portion of this code is held to be illegal or void, such a decision shall not affect the validity of the remainder of this code.

SECTION 107
REFERENCED STANDARDS

107.1 General. The standards, and portions thereof, referred to in this code and listed in Chapter 10 shall be considered part of the requirements of this code to the extent of such reference.

107.2 Conflicting requirements. Where the provisions of this code and the referenced standards conflict, the provisions of this code shall take precedence.

Delete Chapter 2 in its entirety and replace as shown (EC48-03/04; EC5-03/04; EC6-03/04; EC31-03/04; EC37-03/04; EC40-03/04)

CHAPTER 2
DEFINITIONS

SECTION 201
GENERAL

201.1 Scope. Unless stated otherwise, the following words and terms in this code shall have the meanings indicated in this chapter.

201.2 Interchangeability. Words used in the present tense include the future; words in the masculine gender include the feminine and neuter; the singular number includes the plural and the plural includes the singular.

201.3 Terms defined in other codes. Terms that are not defined in this code but are defined in the *International Building Code*, ICC *Electrical Code, International Fire Code, International Fuel Gas Code, International Mechanical Code, International Plumbing Code*, or the *International Residential Code* shall have the meanings ascribed to them in those codes.

201.4 Terms not defined. Terms not defined by this chapter shall have ordinarily accepted meanings such as the context implies.

SECTION 202
GENERAL DEFINITIONS

ABOVE GRADE WALL. A wall more than 50 percent above grade and enclosing conditioned space. This includes between-floor spandrels, peripheral edges of floors, roof and basement knee walls, dormer walls, gable end walls, walls enclosing a mansard roof, and skylight shafts.

ACCESSIBLE. Admitting close approach as a result of not being guarded by locked doors, elevation or other effective means (see "Readily accessible").

ADDITION. An extension or increase in the conditioned space floor area or height of a building or structure.

ALTERATION. Any construction or renovation to an existing structure other than repair or addition that requires a permit. Also, a change in a mechanical system that involves an extension, addition or change to the arrangement, type or purpose of the original installation that requires a permit.

APPROVED. Acceptable to the code official.

AUTOMATIC. Self-acting, operating by its own mechanism when actuated by some impersonal influence, as, for example, a change in current strength, pressure, temperature or mechanical configuration (see "Manual").

BASEMENT WALL. A wall 50 percent or more below grade and enclosing conditioned space.

BUILDING. Any structure used or intended for supporting or sheltering any use or occupancy.

BUILDING THERMAL ENVELOPE. The basement walls, exterior walls, floor, roof, and any other building element that enclose conditioned space. This boundary also includes the boundary between conditioned space and any exempt or unconditioned space.

CODE OFFICIAL. The officer or other designated authority charged with the administration and enforcement of this code, or a duly authorized representative.

COMMERCIAL BUILDING. For this code, all buildings that are not included in the definition of Residential Buildings.

CONDITIONED FLOOR AREA. The horizontal projection of the floors associated with the conditioned space.

CONDITIONED SPACE. An area or room within a building being heated or cooled, containing uninsulated ducts, or with a fixed opening directly into an adjacent conditioned space.

CRAWLSPACE WALL. The opaque portion of a wall that encloses a crawl space and is partially or totally below grade.

DUCT. A tube or conduit utilized for conveying air. The air passages of self-contained systems are not to be construed as air ducts.

DUCT SYSTEM. A continuous passageway for the transmission of air that, in addition to ducts, includes duct fittings, dampers, plenums, fans and accessory air-handling equipment and appliances.

DWELLING UNIT. A single unit providing complete independent living facilities for one or more persons, including permanent provisions for living, sleeping, eating, cooking and sanitation.

ECONOMIZER, AIR. A duct and damper arrangement and automatic control system that allows a cooling system to supply outside air to reduce or eliminate the need for mechanical cooling during mild or cold weather.

ECONOMIZER, WATER. A system where the supply air of a cooling system is cooled indirectly with water that is itself cooled by heat or mass transfer to the environment without the use of mechanical cooling.

ENERGY ANALYSIS. A method for estimating the annual energy use of the proposed design and standard reference design based on estimates of energy use.

ENERGY COST. The total estimated annual cost for purchased energy for the building functions regulated by this code, including applicable demand charges.

ENERGY RECOVERY VENTILATION SYSTEM. Systems that employ air-to-air heat exchangers to recover energy from exhaust air for the purpose of preheating, precooling, humidifying or dehumidifying outdoor ventilation air prior to supplying the air to a space, either directly or as part of an HVAC system.

ENERGY SIMULATION TOOL. An approved software program or calculation-based methodology that projects the annual energy use of a building.

EXTERIOR WALL. Walls including both above grade walls and basement walls.

FACTORY-ASSEMBLED GLAZED FENESTRATION PRODUCT. Fenestration products that are shipped to the field as factory-assembled units comprised of specified frame and glazing components including operable and fixed windows, and skylights.

FENESTRATION. Skylights, roof windows, vertical windows (fixed or moveable), opaque doors, glazed doors, glazed block, and combination opaque/glazed doors. Fenestration includes products with glass and non-glass glazing materials.

HEAT TRAP. An arrangement of piping and fittings, such as elbows, or a commercially available heat trap that prevents thermosyphoning of hot water during standby periods.

HEATED SLAB. Slab-on-grade construction in which the heating elements, hydronic tubing, or hot air distribution system is in contact with, or placed within or under the slab.

HUMIDISTAT. A regulatory device, actuated by changes in humidity, used for automatic control of relative humidity.

INFILTRATION. The uncontrolled inward air leakage into a building caused by the pressure effects of wind or the effect of differences in the indoor and outdoor air density or both.

INSULATING SHEATHING. An insulating board with a core material having a minimum *R*-value of R-2.

LABELED. Devices, equipment, or materials to which have been affixed a label, seal, symbol or other identifying mark of a nationally recognized testing laboratory, inspection agency or other organization concerned with product evaluation that maintains periodic inspection of the production of the above-labeled items that attests to compliance with a specific standard.

LISTED. Equipment, appliances, assemblies or materials included in a list published by an approved testing laboratory, inspection agency or other organization concerned with product evaluation that maintains periodic inspection of production of listed equipment, appliances, assemblies or material, and whose listing states either that the equipment, appliances, assemblies, or material meets nationally recognized standards or has been tested and found suitable for use in a specified manner.

LOW-VOLTAGE LIGHTING. Lighting equipment powered through a transformer such as a cable conductor, a rail conductor and track lighting.

MANUAL. Capable of being operated by personal intervention (see "Automatic")

PROPOSED DESIGN. A description of the proposed building used to estimate annual energy use for determining compliance based on total building performance.

READILY ACCESSIBLE. Capable of being reached quickly for operation, renewal or inspection without requiring those to whom ready access is requisite to climb over or remove obstacles or to resort to portable ladders or access equipment (see "Accessible").

REPAIR. The reconstruction or renewal of any part of an existing building.

RESIDENTIAL BUILDING. For this code, includes R-3 buildings, as well as R-2 and R-4 buildings three stories or less in height above grade.

R-VALUE (THERMAL RESISTANCE). The inverse of the time rate of heat flow through a body from one of its bounding surfaces to the other surface for a unit temperature difference between the two surfaces, under steady state conditions, per unit area ($h \cdot ft^2 \cdot °F/Btu$) [($m^2 \cdot K$)/W].

ROOF ASSEMBLY. A system designed to provide weather protection and resistance to design loads. The system consists of a roof covering and roof deck or a single component serving as both the roof covering and the roof deck. A roof assembly includes the roof covering, underlayment, roof deck, insulation, vapor retarder and interior finish.

SCREW LAMP HOLDERS. A lamp base that requires a screw-in-type lamp, such as a compact-fluorescent, incandescent, or tungsten-halogen bulb.

SERVICE WATER HEATING. Supply of hot water for purposes other than comfort heating.

SITE-BUILT GLAZED PRODUCT. Fenestration products that are designed to be field glazed or field assembled units comprised of specified frame and glazing components including operable and fixed windows, curtain walls, window walls, storefronts, sloped glazing and skylights.

SKYLIGHT. Glass or other transparent or translucent glazing material installed at a slope of 15 degrees (0.26 rad) or more from vertical. Glazing material in skylights, including unit skylights, solariums, sunrooms, roofs and sloped walls is included in this definition.

SOLAR HEAT GAIN COEFFICIENT (SHGC). The ratio of the solar heat gain entering the space through the fenestration assembly to the incident solar radiation. Solar heat gain includes directly transmitted solar heat and absorbed solar radiation which is then reradiated, conducted or convected into the space.

STANDARD REFERENCE DESIGN. A version of the proposed design that meets the minimum requirements of this code and is used to determine the maximum annual energy use requirement for compliance based on total building performance.

SUNROOM. A one-story structure attached to a dwelling with a glazing area in excess of 40 percent of the gross area of the structure's exterior walls and roof.

THERMAL ISOLATION. Physical and space conditioning separation from conditioned space(s). The conditioned space(s) shall be controlled as separate zones for heating and cooling or conditioned by separate equipment.

THERMOSTAT. An automatic control device used to maintain temperature at a fixed or adjustable set point.

***U*-FACTOR (THERMAL TRANSMITTANCE).** The coefficient of heat transmission (air to air) through a building component or assembly, equal to the time rate of heat flow per unit area and unit temperature difference between the warm side and cold side air films (Btu/h · ft^2 · °F) [W/(m^2 · K)].

VAPOR RETARDER. A vapor resistant material, membrane or covering such as foil, plastic sheeting, or insulation facing having a permeance rating of 1 perm (5.7 x 10^{-11} kg/Pa · s · m^2) or less when tested in accordance with the dessicant method using Procedure A of ASTM E 96. Vapor retarders limit the amount of moisture vapor that passes through a material or wall assembly.

VENTILATION. The natural or mechanical process of supplying conditioned or unconditioned air to, or removing such air from, any space.

VENTILATION AIR. That portion of supply air that comes from outside (outdoors) plus any recirculated air that has been treated to maintain the desired quality of air within a designated space.

ZONE. A space or group of spaces within a building with heating or cooling requirements that are sufficiently similar so that desired conditions can be maintained throughout using a single controlling device.

Delete Chapter 3 in its entirety and replace as shown: (EC48-03/04; EC7-03/04)

CHAPTER 3
CLIMATE ZONES

SECTION 301
CLIMATE ZONES

301.1 General. Climate zones from Figure 301.1 or Table 301.1 shall be used in determining the applicable requirements from Chapters 4 and 8. Locations not in Table 301.1 (outside the US) shall be assigned a climate zone based on Section 301.3.

301.2 Warm humid counties. Warm humid counties are listed in Table 301.2.

301.3 International climate zones. The climate zone for any location outside the United States shall be determined by applying Table 301.3(1) and then Table 301.3(2).

301.3.1 Warm humid criteria. "Warm humid" locations shall be defined as locations where either of the following conditions occurs:

1. 67°F (19.4°C) or higher wet-bulb temperature for 3,000 or more hours during the warmest six consecutive months of the year;

2. 73°F (22.8°C) or higher wet-bulb temperature for 1,500 or more hours during the warmest six consecutive months of the year.

2004 SUPPLEMENT TO THE IECC

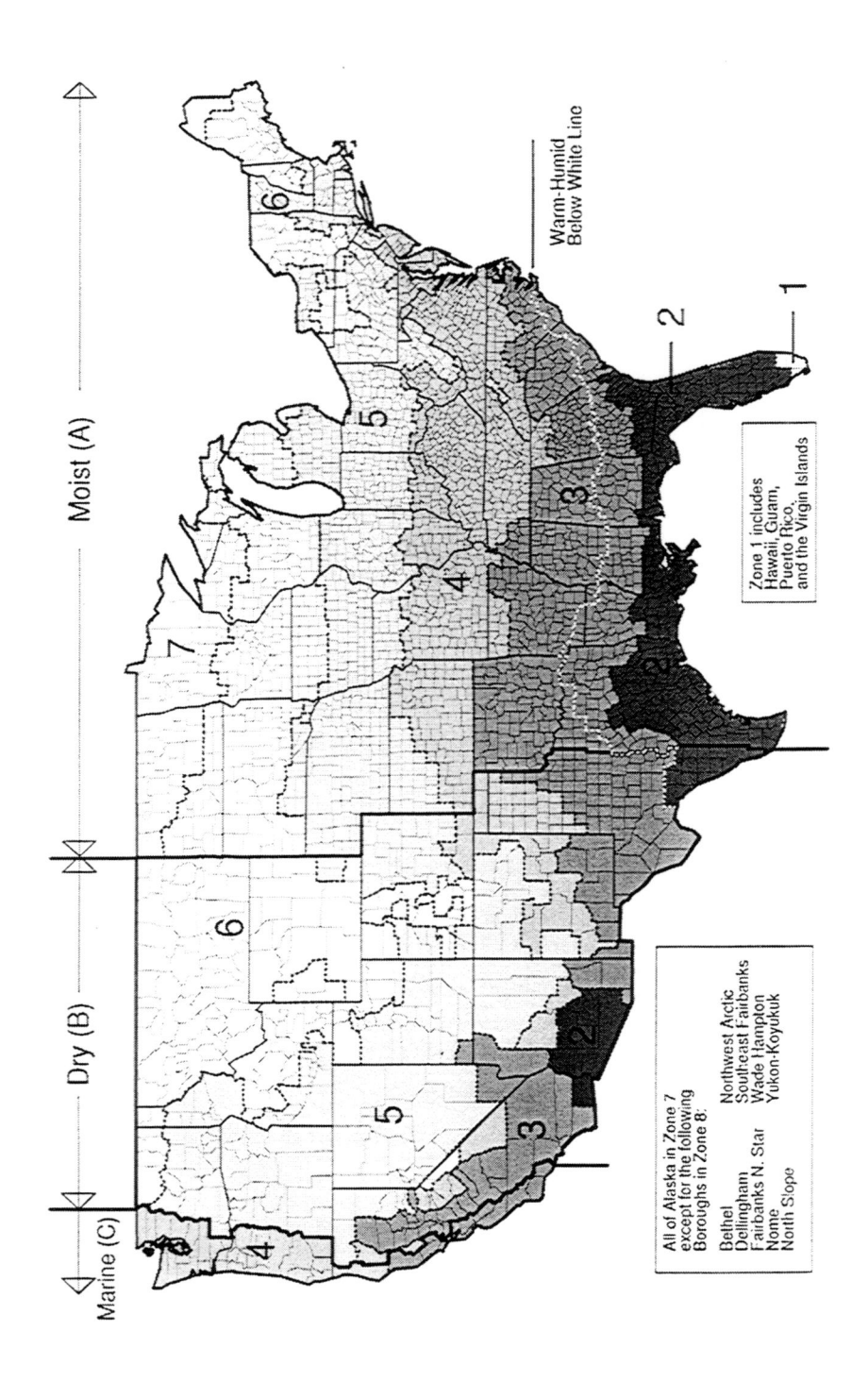

Figure 301.1 CLIMATE ZONES

TABLE 301.1
CLIMATE ZONES BY STATE AND COUNTY

Alabama
Zone 3 except
Zone 2
Baldwin
Mobile

Alaska
Zone 7 except
Zone 8
Bethel
Dellingham
Fairbanks North Star
Nome
North Slope
Northwest Arctic
Southeast Fairbanks
Wade Hampton
Yukon-Koyukuk

Arizona
Zone 3 except
Zone 2
La Paz
Maricopa
Pima
Pinal
Yuma
Zone 4
Gila
Yavapai
Zone 5
Apache
Coconino
Navajo

Arkansas
Zone 3 except
Zone 4
Baxter
Benton
Boone
Carroll
Fulton
Izard
Madison
Marion
Newton
Searcy
Stone
Washington

California
Zone 3 Dry except
Zone 2
Imperial
Zone 3 Marine
Alameda
Marin

Mendocino
Monterey
Napa
San Benito
San Francisco
San Luis Obispo
San Mateo
Santa Barbara
Santa Clara
Santa Cruz
Sonoma
Ventura
Zone 4 Dry
Amador
Calaveras
El Dorado
Inyo
Lake
Mariposa
Trinity
Tuolumne
Zone 4 Marine
Del Norte
Humboldt
Zone 5
Lassen
Modoc
Nevada
Plumas
Sierra
Siskiyou
Zone 6
Alpine
Mono

Colorado
Zone 5 except
Zone 4
Baca
Las Animas
Otero
Zone 6
Alamosa
Archuleta
Chaffee
Conejos
Costilla
Custer
Dolores
Eagle
Moffat
Ouray
Rio Blanco
Saguache
San Miguel
Zone 7
Clear Creek
Grand

Gunnison
Hinsdale
Jackson
Lake
Mineral
Park
Pitkin
Rio Grande
Routt
San Juan
Summit

Connecticut
Zone 5

Delaware
Zone 4

Dist Of Columbia
Zone 4

Florida
Zone 2 except
Zone 1
Broward
Dade
Monroe

Georgia
Zone 3 except
Zone 2
Appling
Atkinson
Bacon
Baker
Berrien
Brantley
Brooks
Bryan
Camden
Charlton
Chatham
Clinch
Colquitt
Cook
Decatur
Echols
Effingham
Evans
Glynn
Grady
Jeff Davis
Lanier
Liberty
Long
Lowndes
McIntosh
Miller

Mitchell
Pierce
Seminole
Tattnall
Thomas
Toombs
Ware
Wayne
Zone 4
Banks
Catoosa
Chattooga
Dade
Dawson
Fannin
Floyd
Franklin
Gilmer
Gordon
Habersham
Hall
Lumpkin
Murray
Pickens
Rabun
Stephens
Towns
Union
Walker
White
Whitfield

Hawaii
Zone 1

Idaho
Zone 6 except
Zone 5
Ada
Benewah
Canyon
Cassia
Clearwater
Elmore
Gem
Gooding
Idaho
Jerome
Kootenai
Latah
Lewis
Lincoln
Minidoka
Nez Perce
Owyhee
Payette
Power
Shoshone

Twin Falls
Washington

Illinois
Zone 5 except
Zone 4
Alexander
Bond
Christian
Clay
Clinton
Crawford
Edwards
Effingham
Fayette
Franklin
Gallatin
Hamilton
Hardin
Jackson
Jasper
Jefferson
Johnson
Lawrence
Macoupin
Madison
Marion
Massac
Monroe
Montgomery
Perry
Pope
Pulaski
Randolph
Richland
Saline
Shelby
St Clair
Union
Wabash
Washington
Wayne
White
Williamson

Indiana
Zone 5 except
Zone 4
Brown
Clark
Crawford
Daviess
Dearborn
Dubois
Floyd
Gibson
Greene
Harrison

2004 SUPPLEMENT TO THE IECC

Jackson
Jefferson
Jennings
Knox
Lawrence
Martin
Monroe
Ohio
Orange
Perry
Pike
Posey
Ripley
Scott
Spencer
Sullivan
Switzerland
Vanderburgh
Warrick
Washington

Iowa
Zone 5 except
Zone 6
Allamakee
Black Hawk
Bremer
Buchanan
Buena Vista
Butler
Calhoun
Cerro Gordo
Cherokee
Chickasaw
Clay
Clayton
Delaware
Dickinson
Emmet
Fayette
Floyd
Franklin
Grundy
Hamilton
Hancock
Hardin
Howard
Humboldt
Ida
Kossuth
Lyon
Mitchell
O'Brien
Osceola
Palo Alto
Plymouth
Pocahontas
Sac
Sioux
Webster

Winnebago
Winneshiek
Worth
Wright

Kansas
Zone 4 except
Zone 5
Cheyenne
Cloud
Decatur
Ellis
Gove
Graham
Greeley
Hamilton
Jewell
Lane
Logan
Mitchell
Ness
Norton
Osborne
Phillips
Rawlins
Republic
Rooks
Scott
Sheridan
Sherman
Smith
Thomas
Trego
Wallace
Wichita

Kentucky
Zone 4

Louisiana
Zone 2 except
Zone 3
Bienville
Bossier
Caddo
Caldwell
Catahoula
Claiborne
Concordia
De Soto
East Carroll
Franklin
Grant
Jackson
La Salle
Lincoln
Madison
Morehouse
Natchitoches

Ouachita
Red River
Richland
Sabine
Tensas
Union
Vernon
Webster
West Carroll
Winn

Maine
Zone 6 except
Zone 7
Aroostook

Maryland
Zone 4 except
Zone 5
Garrett

Massachusetts
Zone 5

Michigan
Zone 5 except
Zone 6
Alcona
Alger
Alpena
Antrim
Arenac
Benzie
Charlevoix
Cheboygan
Clare
Crawford
Delta
Dickinson
Emmet
Gladwin
Grand Traverse
Huron
Iosco
Isabella
Kalkaska
Lake
Leelanau
Manistee
Marquette
Mason
Mecosta
Menominee
Missaukee
Montmorency
Newaygo
Oceana
Ogemaw
Oscoda

Otsego
Presque Isle
Roscommon
Sanilac
Wexford
Zone 7
Baraga
Chippewa
Gogebic
Houghton
Iron
Keweenaw
Luce
Mackinac
Ontonagon
Schoolcraft

Minnesota
Zone 6 except
Zone 7
Aitkin
Becker
Beltrami
Carlton
Cass
Clay
Clearwater
Cook
Crow Wing
Grant
Hubbard
Itasca
Kanabec
Kittson
Koochiching
Lake Of The
Woods Mahnomen
Marshall
Mille Lacs
Norman
Otter Tail
Pennington
Pine
Polk
Red Lake
Roseau
St Louis
Wadena
Wilkin

Mississippi
Zone 3 except
Zone 2
Hancock
Harrison
Jackson
Pearl River
Stone

Missouri
Zone 4 except
Zone 5
Adair
Andrew
Atchison
Buchanan
Caldwell
Chariton
Clark
Clinton
Daviess
De Kalb
Gentry
Grundy
Harrison
Holt
Knox
Lewis
Linn
Livingston
Macon
Marion
Mercer
Nodaway
Pike
Putnam
Ralls
Schuyler
Scotland
Shelby
Sullivan
Worth

Montana
Zone 6

Nebraska
Zone 5

Nevada
Zone 5 except
Zone 3
Clark

New Hampshire
Zone 6 except
Zone 5
Cheshire
Hillsborough
Rockingham
Strafford

New Jersey
Zone 4 except
Zone 5
Bergen
Hunterdon
Mercer

IECC-10

Morris	Oneida	Wake	**Oregon**	Mellette
Passaic	Otsego	Warren	Zone 4 Marine	Todd
Somerset	Schoharie	Wilkes	except	Tripp
Sussex	Schuyler	Yadkin	Zone 5 Dry	Union
Warren	St Lawrence	Zone 5	Baker	Yankton
	Steuben	Alleghany	Crook	
New Mexico	Sullivan	Ashe	Deschutes	**Tennessee**
Zone 4 except	Tompkins	Avery	Gilliam	Zone 4 except
Zone 3	Ulster	Mitchell	Grant	Zone 3
Chaves	Warren	Watauga	Harney	Chester
Dona Ana	Wyoming	Yancey	Hood River	Crockett
Eddy			Jefferson	Dyer
Hidalgo	**North Carolina**	**North Dakota**	Klamath	Fayette
Lea	Zone 3 except	Zone 7 except	Lake	Hardeman
Luna	Zone 4	Zone 6	Malheur	Hardin
Otero	Alamance	Adams	Morrow	Haywood
Zone 5	Alexander	Billings	Sherman	Henderson
Catron	Bertie	Bowman	Umatilla	Lake
Cibola	Buncombe	Burleigh	Union	Lauderdale
Colfax	Burke	Dickey	Wallowa	Madison
Harding	Caldwell	Dunn	Wasco	McNairy
Los Alamos	Caswell	Emmons	Wheeler	Shelby
McKinley	Catawba	Golden Valley		Tipton
Mora	Chatham	Grant	**Pennsylvania**	
Rio Arriba	Cherokee	Hettinger	Zone 5 except	**Texas**
San Juan	Clay	La Moure	Zone 4	Zone 2 Moist
San Miguel	Cleveland	Logan	Bucks	except
Sandoval	Davie	McIntosh	Chester	Zone 2 Dry
Santa Fe	Durham	McKenzie	Delaware	Bandera
Taos	Forsyth	Mercer	Montgomery	Dimmit
Torrance	Franklin	Morton	Philadelphia	Edwards
	Gates	Oliver	York	Kinney
New York	Graham	Ransom	Zone 6	La Salle
Zone 5 except	Granville	Richland	Cameron	Maverick
Zone 4	Guilford	Sargent	Clearfield	Medina
Bronx	Halifax	Sioux	Elk	Real
Kings	Harnett	Slope	McKean	Uvalde
Nassau	Haywood	Stark	Potter	Val Verde
New York	Henderson		Susquehanna	Webb
Queens	Hertford	**Ohio**	Tioga	Zapata
Richmond	Iredell	Zone 5 except	Wayne	Zavala
Suffolk	Jackson	Zone 4		Zone 3 Dry
Westchester	Lee	Adams	**Rhode Island**	Andrews
Zone 6	Lincoln	Brown	Zone 5	Baylor
Allegany	Macon	Clermont		Borden
Broome	Madison	Gallia	**South Carolina**	Brewster
Cattaraugus	McDowell	Hamilton	Zone 3	Callahan
Chenango	Nash	Lawrence		Childress
Clinton	Northampton	Pike	**South Dakota**	Coke
Delaware	Orange	Scioto	Zone 6 except	Coleman
Essex	Person	Washington	Zone 5	Collingsworth
Franklin	Polk		Bennett	Concho
Fulton	Rockingham	**Oklahoma**	Bon Homme	Cottle
Hamilton	Rutherford	Zone 3 Moist except	Charles Mix	Crane
Herkimer	Stokes	Zone 4 Dry	Clay	Crockett
Jefferson	Surry	Beaver	Douglas	Crosby
Lewis	Swain	Cimarron	Gregory	Culberson
Madison	Transylvania	Texas	Hutchinson	Dawson
Montgomery	Vance		Jackson	Dickens

2004 SUPPLEMENT TO THE IECC

Ector	Zone 3 Moist	Rusk	Duchesne	Jackson
El Paso	Archer	Sabine	Morgan	Jefferson
Fisher	Blanco	San Augustine	Rich	Kanawha
Foard	Bowie	San Saba	Summit	Lincoln
Gaines	Brown	Shelby	Uintah	Logan
Garza	Burnet	Smith	Wasatch	Mason
Glasscock	Camp	Somervell		McDowell
Hall	Cass	Tarrant	**Vermont**	Mercer
Hardeman	Clay	Titus	Zone 6	Mingo
Haskell	Collin	Upshur		Monroe
Hemphill	Comanche	Van Zandt	**Virginia**	Morgan
Howard	Cooke	Wood	Zone 4	Pleasants
Hudspeth	Dallas	Zone 4		Putnam
Irion	Delta	Armstrong	**Washington**	Ritchie
Jeff Davis	Denton	Bailey	Zone 4 Marine	Roane
Jones	Eastland	Briscoe	except	Tyler
Kent	Ellis	Carson	Zone 5 Dry	Wayne
Kerr	Erath	Castro	Adams	Wirt
Kimble	Fannin	Cochran	Asotin	Wood
King	Franklin	Dallam	Benton	Wyoming
Knox	Gillespie	Deaf Smith	Chelan	
Loving	Grayson	Donley	Columbia	**Wisconsin**
Lubbock	Gregg	Floyd	Douglas	Zone 6 except
Lynn	Hamilton	Gray	Franklin	Zone 7
Martin	Harrison	Hale	Garfield	Ashland
Mason	Henderson	Hansford	Grant	Bayfield
Mcculloch	Hood	Hartley	Kittitas	Burnett
Menard	Hopkins	Hockley	Klickitat	Douglas
Midland	Hunt	Hutchinson	Lincoln	Florence
Mitchell	Jack	Lamb	San Juan	Forest
Motley	Johnson	Lipscomb	Skamania	Iron
Nolan	Kaufman	Moore	Spokane	Langlade
Pecos	Kendall	Ochiltree	Walla Walla	Lincoln
Presidio	Lamar	Oldham	Whitman	Oneida
Reagan	Lampasas	Parmer	Yakima	Price
Reeves	Llano	Potter	Zone 6 Dry	Sawyer
Runnels	Montague	Randall	Ferry	Taylor
Schleicher	Stephens	Roberts	Okanogan	Vilas
Scurry	Wichita	Sherman	Pend Oreille	Washburn
Shackelford	Wise	Swisher	Stevens	
Sterling	Young	Yoakum		**Wyoming**
Stonewall	Marion		**West Virginia**	Zone 6 except
Sutton	Mills	**Utah**	Zone 5 except	Zone 5
Taylor	Morris	Zone 5 except	Zone 4	Goshen
Terrell	Nacogdoches	Zone 3	Berkeley	Platte
Terry	Navarro	Washington	Boone	Zone 7
Throckmorton	Palo Pinto	Zone 6	Braxton	Lincoln
Tom Green	Panola	Box Elder	Cabell	Sublette
Ward	Parker	Cache	Calhoun	Teto
Wheeler	Rains	Carbon	Clay	
Wilbarger	Red River	Daggett	Gilmer	
Winkler	Rockwall			

IECC-12

TABLE 301.2
WARM HUMID COUNTIES

Alabama
Autauga
Baldwin
Barbour
Bullock
Butler
Choctaw
Clarke
Coffee
Conecuh
Covington
Crenshaw
Dale
Dallas
Elmore
Escambia
Geneva
Henry
Houston
Lowndes
Macon
Marengo
Mobile
Monroe
Montgomery
Perry
Pike
Russell
Washington
Wilcox

Arkansas
Columbia
Hempstead
Lafayette
Little River
Miller
Sevier
Union

Florida
All

Georgia
All in Zone 2 Plus
Ben Hill
Bleckley
Bulloch
Calhoun
Candler
Chattahoochee
Clay
Coffee
Crisp
Dodge
Dooly
Dougherty
Early
Emanuel
Houston
Irwin
Jenkins
Johnson
Laurens
Lee
Macon
Marion
Montgomery
Peach
Pulaski
Quitman
Randolph
Schley
Screven
Stewart
Sumter
Taylor
Telfair
Terrell
Tift
Treutlen
Turner
Twiggs
Webster
Wheeler
Wilcox
Worth

Louisiana
All in Zone 2 Plus
Bienville
Bossier
Caddo
Caldwell
Catahoula
Claiborne
Concordia
De Soto
Franklin
Grant
Jackson
La Salle
Lincoln
Madison
Natchitoches
Ouachita
Red River
Richland
Sabine
Tensas
Union
Vernon
Webster
Winn

Mississippi
All in Zone 2 Plus
Adams
Amite
Claiborne
Copiah
Covington
Forrest
Franklin
George
Greene
Hinds
Jefferson
Jefferson Davis
Jones
Lamar
Lawrence
Lincoln
Marion
Perry
Pike
Rankin
Simpson
Smith
Walthall
Warren
Wayne
Wilkinson

North Carolina
Brunswick
Carteret
Columbus
New Hanover
Onslow
Pender

South Carolina
Allendale
Bamberg
Barnwell
Beaufort
Berkeley
Charleston
Colleton
Dorchester
Georgetown
Hampton
Horry
Jasper

Texas
All in Zone 2 Plus
Blanco
Bowie
Brown
Burnet
Camp
Cass
Collin
Comanche
Dallas
Delta
Denton
Ellis
Erath
Franklin
Gillespie
Gregg
Hamilton
Harrison
Henderson
Hood
Hopkins
Hunt
Johnson
Kaufman
Kendall
Lamar
Lampasas
Llano
Marion
Mills
Morris
Nacogdoches
Navarro
Palo Pinto
Panola
Parker
Rains
Red River
Rockwall
Rusk
Sabine
San Augustine
San Saba
Shelby
Smith
Somervell
Tarrant
Titus
Upshur
Van Zandt
Wood

TABLE 301.3(1)
INTERNATIONAL CLIMATE ZONE DEFINITIONS

MAJOR CLIMATE TYPE DEFINITIONS

Marine (C) Definition - Locations meeting all four criteria:
1. Mean temperature of coldest month between −3°C (27°F) and 18°C (65°F)
2. Warmest month mean < 22°C (72°F)
3. At least four months with mean temperatures over 10°C (50°F)
4. Dry season in summer. The month with the heaviest precipitation in the cold season has at least three times as much precipitation as the month with the least precipitation in the rest of the year. The cold season is October through March in the Northern Hemisphere and April through September in the Southern Hemisphere.

Dry (B) Definition - Locations meeting the following criteria: Not Marine and
$P_{in} < 0.44 \times (TF - 19.5)$ $[P_{cm} < 2.0 \times (TC + 7)$ in SI units]
where:
P = Annual precipitation in inches (cm)
T = Annual mean temperature in °F (°C)

Moist (A) Definition - Locations that are not Marine and not Dry.

For SI: °C = [(°F)-32]/1.8; 1 inch = 2.54 cm

TABLE 301.3(2)
INTERNATIONAL CLIMATE ZONE DEFINITIONS

ZONE NUMBER	THERMAL CRITERIA IP Units	SI Units
1	9000 < CDD50°F	5000 < CDD10°C
2	6300 < CDD50°F ≤ 9000	3500 < CDD10°C ≤ 5000
3A and 3B	4500 < CDD50°F ≤ 6300 AND HDD65°F ≤ 5400	2500 < CDD10°C ≤ 3500 AND HDD18°C ≤ 3000
4A and 4B	CDD50°F ≤ 4500 AND HDD65°F ≤ 5400	CDD10°C ≤ 2500 AND HDD18°C ≤ 3000
3C	HDD65°F ≤ 3600	HDD18°C ≤ 2000
4C	3600 < HDD65°F ≤ 5400	2000 < HDD18°C ≤ 3000
5	5400 < HDD65°F ≤ 7200	3000 < HDD18°C ≤ 4000
6	7200 < HDD65°F ≤ 9000	4000 < HDD18°C ≤ 5000
7	9000 < HDD65°F ≤ 12600	5000 < HDD18°C ≤ 7000
8	12600 < HDD65°F	7000 < HDD18 °C

For SI: °C = [(°F)-32]/1.8

SECTION 302
DESIGN CONDITIONS

302.1 Interior design conditions. The interior design temperatures used for heating and cooling load calculations shall be a maximum of 72°F (22°C) for heating and minimum of 78°F (26°C) for cooling.

Delete Chapter 4 in its entirety and replace as shown (EL3-03/04; EC48-03/04; EC50-03/04; EC53-03/04)

CHAPTER 4
RESIDENTIAL ENERGY EFFICIENCY

SECTION 401
GENERAL

401.1 Scope. This chapter applies to residential buildings.

401.2 Compliance. Compliance shall be demonstrated by meeting each of the applicable provisions of this chapter.

401.3 Certificate. A permanent certificate shall be posted on or in the electrical distribution panel. The certificate shall be completed by the builder or registered design professional. The certificate shall list the predominant R-values of insulation installed in or on ceiling/roof, walls, foundation (slab, basement wall, crawlspace wall and/or floor) and ducts outside conditioned spaces; U-factors for fenestration; and, where requirements apply, the solar heat gain coefficient (SHGC) of fenestration. Where there is more than one value for each component, the certificate shall list the value covering the largest area. The certificate shall list the type and efficiency of heating, cooling and service water heating equipment.

SECTION 402
BUILDING THERMAL ENVELOPE

402.1 Insulation and fenestration criteria. The building thermal envelope shall meet the requirements of Table 402.1 based on the climate zone specified in Chapter 3.

402.1.1 R-value computation. Insulation material used in layers, such as framing cavity insulation and insulating sheathing, shall be summed to compute the component R-value. The manufacturer's settled R-value shall be used for blown insulation. Computed R-values shall not include an R-value for other building materials or air films.

402.1.2 U-factor alternative. An assembly with a U-factor equal to or less than that specified in Table 402.1.2 shall be permitted as an alternative to the R-value in Table 402.1.

Exception: For mass walls not meeting the criterion for insulation location in Section 402.2.3, the U-factor shall be permitted to be:

1. U-factor of 0.17 in Climate Zone 1

2. U-factor of 0.14 in Climate Zone 2

3. U-factor of 0.12 in Climate Zone 3

402.1.3 Total UA alternative. If the total building thermal envelope UA (sum of U-factor times assembly area) is less than or equal to the total UA resulting from using the U-factors in Table 402.1.2 (multiplied by the same assembly area as in the proposed building), the building shall be considered in compliance with Table 402.1. The UA calculation shall be done using a method consistent with the ASHRAE *Handbook of Fundamentals* and shall include the thermal bridging effects of framing materials. The SHGC requirements shall be met in addition to UA compliance.

402.2 Specific insulation requirements.

402.2.1 Ceilings with attic spaces. When Section 402.1 would require R-38 in the ceiling, R-30 shall be deemed to satisfy the requirement for R-38 wherever the full height of uncompressed R-30 insulation extends over the wall top plate at the eaves. Similarly R-38 shall be deemed to satisfy the requirement for R-49 wherever the full height of uncompressed R-38 insulation extends over the wall top plate at the eaves.

402.2.2 Ceilings without attic spaces. Where Section 402.1 would require insulation levels above R-30 and the design of the roof/ceiling assembly does not allow sufficient space for the required insulation, the minimum required insulation for such roof/ceiling assemblies shall be R-30. This reduction of insulation from the requirements of Section 402.1 shall be limited to 500 ft^2 of ceiling area.

402.2.3 Mass walls. Mass walls for the purposes of this Chapter shall be considered walls of concrete block, concrete, insulated concrete form (ICF), masonry cavity, brick (other than brick veneer), earth (adobe, compressed earth block, rammed earth) and solid timber/logs. The provisions of Section 402.1 for mass walls shall be applicable when at least 50 percent of the required insulation R-value is on the exterior of, or integral to, the wall. Walls that do not meet this criterion for insulation placement shall meet the wood frame wall insulation requirements of Section 402.1.

Exception: For walls that do not meet the criterion for insulation placement, the minimum added insulation R-value shall be permitted to be:

1. R-value of 4 in Climate Zone 1
2. R-value of 6 in Climate Zone 2
3. R-value of 8 in Climate Zone 3

402.2.4 Steel-frame ceilings, walls and floors. Steel-frame ceilings, walls and floors shall meet the insulation requirements of Table 402.2.4 or shall meet the *U*-factor requirements in Table 402.1.2. The calculation of the *U*-factor for a steel-frame envelope assembly shall use a series-parallel path calculation method.

402.2.5 Floors. Floor insulation shall be installed to maintain permanent contact with the underside of the subfloor decking.

402.2.6 Basement walls. Walls associated with conditioned basements shall be insulated from the top of the basement wall down to 10 feet below grade or to the basement floor, whichever is less. Walls associated with unconditioned basements shall meet this requirement unless the floor overhead is insulated in accordance with Sections 402.1 and 402.2.5.

402.2.7 Slab-on-grade floors. Slab-on-grade floors with a floor surface less than 12 inches (305 mm) below grade shall be insulated in accordance with Table 402.1. The insulation shall extend downward from the top of the slab on the outside or inside of the foundation wall. Insulation located below grade shall be extended the distance provided in Table 402.1 by any combination of vertical insulation, insulation extending under the slab or insulation extending out from the building. Insulation extending away from the building shall be protected by pavement or by a minimum of 10 inches (254 mm) of soil. The top edge of the insulation installed between the exterior wall and the edge of the interior slab shall be permitted to be cut at a 45-degree (0.79 rad) angle away from the exterior wall. Slab-edge insulation is not required in jurisdictions designated by the code official as having a very heavy termite infestation.

402.2.8 Crawl space walls. As an alternative to insulating floors over crawl spaces, crawl space walls shall be permitted to be insulated when the crawl space is not vented to the outside. Crawl space wall insulation shall be permanently fastened to the wall and extend downward from the floor to the finished grade level and then vertically and/or horizontally for at least an additional 24 inches (610 mm). Exposed earth in unvented crawl space foundations shall be covered with a continuous vapor retarder. All joints of the vapor retarder shall overlap by 6 inches (153 mm) and be sealed or taped. The edges of the vapor retarder shall extend at least 6 inches (153 mm) up the stem wall and shall be attached to the stem wall.

402.2.9 Masonry veneer. Insulation shall not be required on the horizontal portion of the foundation that supports a masonry veneer.

402.2.10 Thermally isolated sunroom insulation. The minimum ceiling insulation *R*-values shall be R-19 in zones 1 through 4 and R-24 in zones 5 though 8. The minimum wall *R*-value shall be R-13 in all zones. New wall(s) separating a sunroom from conditioned space shall meet the building thermal envelope requirements.

402.3 Fenestration.

402.3.1 *U*-factor. An area-weighted average of fenestration products shall be permitted to satisfy the *U*-factor requirements.

402.3.2 Glazed fenestration SHGC. An area-weighted average of fenestration products more than 50 percent glazed shall be permitted to satisfy the SHGC requirements.

402.3.3 Glazed fenestration exemption. Up to 15 ft^2 (1.4 m^2) of glazed fenestration per dwelling unit shall be permitted to be exempt from *U*-factor and SHGC requirements in Section 402.1.

402.3.4 Opaque door exemption. One opaque door assembly is exempted from the *U*-factor requirement in Section 402.1.

402.3.5 Thermally isolated sunroom *U*-factor. For Zones 4 through 8, the maximum fenestration *U*-factor shall be 0.50 and the maximum skylight *U*-factor shall be 0.75. New windows and doors separating the sunroom from conditioned space shall meet the building thermal envelope requirements.

402.3.6 Replacement fenestration. Where some or all of an existing fenestration unit is replaced with a new fenestration product, including frame, sash and glazing, the replacement fenestration unit shall meet the applicable requirements for *U*-factor and SHGC in Table 402.1.

402.4 Air leakage.

402.4.1 Building thermal envelope. The building thermal envelope shall be durably sealed to limit infiltration. The sealing methods between dissimilar materials shall allow for differential expansion and contraction. The following shall be caulked, gasketed, weatherstripped or otherwise sealed with an air barrier material, suitable film or solid material:

1. All joints, seams and penetrations.

2. Site-built windows, doors and skylights.

3. Openings between window and door assemblies and their respective jambs and framing.

4. Utility penetrations.

5. Dropped ceilings or chases adjacent to the thermal envelope.

6. Knee walls.

7. Walls and ceilings separating a garage from conditioned spaces.

8. Behind tubs and showers on exterior walls.

9. Common walls between dwelling units.

10. Other sources of infiltration.

402.4.2 Fenestration air leakage. Windows, skylights and sliding glass doors shall have an air infiltration rate of no more than 0.3 cfm/ft^2 (1.5 L/s/m^2), and swinging doors no more than 0.5 cfm/ft^2 (2.6 L/s/m^2), when tested according to NFRC 400, AAMA/WDMA 101/I.S.2, or AAMA/WDMA101/I.S.2/NAFS by an accredited, independent laboratory and listed and labeled by the manufacturer.

Exceptions: Site-built windows, skylights and doors.

402.4.3 Recessed lighting. Recessed luminaires installed in the building thermal envelope shall be sealed to limit air leakage between conditioned and unconditioned spaces by being:

1. IC-rated and labeled with enclosures that are sealed or gasketed to prevent air leakage to the ceiling cavity or unconditioned space; or

2. IC-rated and labeled as meeting ASTM E 283 when tested at 1.57 psi (75 Pa) pressure differential with no more than 2.0 cfm (0.944 L/s) of air movement from the conditioned space to the ceiling cavity; or

3. Located inside an airtight sealed box with clearances of at least 0.5 inch (12.7 mm) from combustible material and 3 inches (76 mm) from insulation.

402.5 Moisture control. The building design shall not create conditions of accelerated deterioration from moisture condensation. Above-grade frame walls, floors and ceilings not ventilated to allow moisture to escape shall be provided with an approved vapor retarder. The vapor retarder shall be installed on the warm-in-winter side of the thermal insulation.

Exceptions:

1. In construction where moisture or its freezing will not damage the materials.

2. Frame walls, floors and ceilings in jurisdictions in Zones 1 through 4. (Crawl space floor vapor retarders are not exempted.)

3. Where other approved means to avoid condensation are provided.

402.5.1 Maximum fenestration *U*-factor and SHGC. The area weighted average maximum fenestration *U*-factor permitted using trade offs from Section 402.1.3 or Section 404 in Zones 4 through 8 shall be 0.40. The area weighted average maximum fenestration SHGC permitted using trade-offs from Section 404 in Zones 1 through 3 shall be 0.50.

SECTION 403
SYSTEMS

403.1 Controls. At least one thermostat shall be provided for each separate heating and cooling system.

403.2 Ducts.

403.2.1 Insulation. Supply and return ducts shall be insulated to a minimum of R-8. Ducts in floor trusses shall be insulated to a minimum of R-6.

Exception: Ducts or portions thereof located completely inside the building thermal envelope.

403.2.2 Sealing. All ducts, air handlers, filter boxes, and building cavities used as ducts shall be sealed. Joints and seams shall comply with Section M1601.3.1 of the *International Residential Code*.

403.2.3 Building cavities. Building framing cavities shall not be used as supply ducts.

403.3 Mechanical system piping insulation. Mechanical system piping capable of carrying fluids above 105°F (41°C) or below 55°F (13°C) shall be insulated to a minimum of R-2.

403.4 Circulating hot water systems. All circulating service hot water piping shall be insulated to at least R-2. Circulating hot water systems shall include an automatic or readily accessible manual switch that can turn off the hot water circulating pump when the system is not in use.

403.5 Mechanical ventilation. Outdoor air intakes and exhausts shall have automatic or gravity dampers that close when the ventilation system is not operating.

403.6 Equipment sizing. Heating and cooling equipment shall be sized in accordance with Section M1401.3 of the *International Residential Code*.

TABLE 402.1
INSULATION AND FENESTRATION REQUIREMENTS BY COMPONENT[a]

CLIMATE ZONE	FENESTRATION U-FACTOR	SKYLIGHT[b] U-FACTOR	GLAZED FENESTRATION SHGC	CEILING R-VALUE	WOOD FRAME WALL R-VALUE	MASS WALL R-VALUE	FLOOR R-VALUE	BASEMENT[c] WALL R-VALUE	SLAB[d] R-VALUE & DEPTH	CRAWL SPACE[c] WALL R-VALUE
1	1.20	0.75	0.40	30	13	3	13	0	0	0
2	0.75	0.75	0.40	30	13	4	13	0	0	0
3	0.65	0.65	0.40[e]	30	15	5	19	0	0	5 / 13
4 except Marine	0.40	0.60	NR	38	15	5	19	10 / 13	10, 2 ft	10 / 13
5 and Marine 4	0.35	0.60	NR	38	21 or 15+5[g]	13	30[f]	10 / 13	10, 2 ft	10 / 13
6	0.35	0.60	NR	49	21 or 15+5[g]	15	30[f]	10 / 13	10, 4 ft	10 / 13
7 and 8	0.35	0.60	NR	49	21	19	30[f]	10 /13	10, 4 ft	10 / 13

For SI: 1 foot = 304.8 mm.

a. R-values are minimums. U-factors and SHGC are maximums. R-19 shall be permitted to be compressed into a 2 × 6 cavity.
b. The fenestration U-factor column excludes skylights. The SHGC column applies to all glazed fenestration.
c. The first R-value applies to continuous insulation, the second to framing cavity insulation; either insulation meets the requirement.
d. R-5 shall be added to the required slab edge R-values for heated slabs.
e. There are no SHGC requirements in the Marine zone.
f. Or insulation sufficient to fill the framing cavity, R-19 minimum.
g "15+5" means R-15 cavity insulation plus R-5 insulated sheathing. If structural sheathing covers 25 percent or less of the exterior, R-5 sheathing is not required where structural sheathing is used. If structural sheathing covers more than 25 percent of exterior, structural sheathing shall be supplemented with insulated sheathing of at least R-2.

TABLE 402.1.2
EQUIVALENT U-FACTORS[a]

CLIMATE ZONE	FENESTRATION U-FACTOR	SKYLIGHT U-FACTOR	CEILING U-FACTOR	FRAME WALL U-FACTOR	MASS WALL U-FACTOR	FLOOR U-FACTOR	BASEMENT WALL U-FACTOR	CRAWL SPACE WALL U-FACTOR
1	1.20	0.75	0.035	0.082	0.197	0.064	0.360	0.477
2	0.75	0.75	0.035	0.082	0.165	0.064	0.360	0.477
3	0.65	0.65	0.035	0.082	0.141	0.047	0.360	0.136
4 except Marine	0.40	0.60	0.030	0.082	0.141	0.047	0.059	0.065
5 and Marine 4	0.35	0.60	0.030	0.060	0.082	0.033	0.059	0.065
6	0.35	0.60	0.026	0.060	0.060	0.033	0.059	0.065
7 and 8	0.35	0.60	0.026	0.057	0.057	0.033	0.059	0.065

a. Non-fenestration U-factors shall be obtained from measurement, calculation or an approved source.

TABLE 402.2.4
STEEL-FRAME CEILING, WALL AND FLOOR INSULATION (R-VALUE)

WOOD FRAME R-VALUE REQUIREMENT	COLD-FORMED STEEL EQUIVALENT R-VALUE[1]
Steel Truss Ceilings[2]	
R-30	R-38 or R-30+3 or R-26+5
R-38	R-49 or R-38+3
R-49	R-38+5
Steel Joist Ceilings[2]	
R-30	R-38 in 2x4 or 2x6 or 2x8 R-49 in any framing
R-38	R-49 in 2x4 or 2x6 or 2x8 or 2x10
Steel Framed Wall	
R-13	R-13+5 or R-15+4 or R-21+3
R-19	R-13+9 or R-19+8 or R-25+7
R-21	R-13+10 or R-19+9 or R-25+8
Steel Joist Floor	
R-13	R-19 in 2x6 R-19+6 in 2x8 or 2x10
R-19	R-19+6 in 2x6 R-19+12 in 2x8 or 2x10

1. Cavity insulation R-value is listed first, followed by continuous insulation R-value.
2. Insulation exceeding the height of the framing shall cover the framing.

SECTION 404
SIMULATED PERFORMANCE ALTERNATIVE

404.1 Scope. This section establishes criteria for compliance using simulated energy performance analysis. Such analysis shall include heating, cooling, and service water heating energy only.

404.2 Mandatory requirements. Compliance with this Section requires that the criteria of Sections 401, 402.4, 402.5 and 403 be met.

404.3 Performance-based compliance. Compliance based on simulated energy performance requires that a proposed residence (proposed design) be shown to have an annual energy cost that is less than or equal to the annual energy cost of the standard reference design. Energy prices shall be taken from a source approved by the code official, such as the Department of Energy, Energy Information Administration's *State Energy Price and Expenditure Report*. Code officials shall be permitted to require time-of-use pricing in energy cost calculations.

Exception: Jurisdictions that require site energy (1kWh = 3,413 Btu) rather than energy cost as the metric of comparison.

404.4 Documentation

404.4.1 Compliance software tools. Documentation verifying that the methods and accuracy of the compliance software tools conform to the provisions of this section shall be provided to the code official.

404.4.2 Compliance report. Compliance software tools shall generate a report that documents that the proposed design has annual energy costs less than or equal to the annual energy costs of the standard reference design. The compliance documentation shall include the following information:

1. Address of the residence;

2. An inspection checklist documenting the building component characteristics of the proposed design as listed in Table 404.5.2(1). The inspection checklist shall show the estimated annual energy cost for both the standard reference design and the proposed design;

3. Name of individual completing the compliance report;

4. Name and version of the compliance software tool.

404.4.3 Additional documentation. The code official shall be permitted to require the following documents:

1. Documentation of the building component characteristics of the standard reference design.

2. A certification signed by the builder providing the building component characteristics of the proposed design as given in Table 404.5.2(1).

404.5 Calculation procedure.

404.5.1 General. Except as specified by this section, the standard reference design and proposed design shall be configured and analyzed using identical methods and techniques.

404.5.2 Residence specifications. The standard reference design and proposed design shall be configured and analyzed as specified by Table 404.5.2(1). Table 404.5.2(1) shall include by reference all notes contained in Table 402.1.

404.6 Calculation software tools.

404.6.1 Minimum capabilities. Calculation procedures used to comply with this section shall be software tools capable of calculating the annual energy consumption of all building elements that differ between the standard reference design and the proposed design and shall include the following capabilities:

1. Computer generation of the standard reference design using only the input for the proposed design. The calculation procedure shall not allow the user to directly modify the building component characteristics of the standard reference design.

2. Calculation of whole-building (as a single zone) sizing for the heating and cooling equipment in the standard reference design residence in accordance with Section M1401.3 of the *International Residential Code*.

3. Calculations that account for the effects of indoor and outdoor temperatures and part-load ratios on the performance of heating, ventilating and air conditioning equipment based on climate and equipment sizing.

4. Printed code official inspection checklist listing each of the proposed design component characteristics from Table 404.5.2(1) determined by the analysis to provide compliance, along with their respective performance ratings (e.g. *R*-Value, *U*-Factor, SHGC, HSPF, AFUE, SEER, EF, etc.).

404.6.2 Approved tools. Performance analysis tools shall be approved. Tools are permitted to be approved based on meeting a specified threshold for a jurisdiction, such as an accredited home energy rating system (HERS) tool. The code official shall be permitted to approve tools for a specified application or limited scope.

404.6.3 Input values. When calculations require input values not specified by Sections 402, 403 and 404, those input values shall be taken from an approved source.

TABLE 404.5.2(1)
SPECIFICATIONS FOR THE STANDARD REFERENCE AND PROPOSED DESIGNS

BUILDING COMPONENT	STANDARD REFERENCE DESIGN	PROPOSED DESIGN
Above-grade walls:	Type: mass wall if proposed wall is mass: otherwise wood frame Gross area: same as proposed U-Factor: from Table 402.1.2 Solar absorptance = 0.75 Emittance = 0.90	As proposed As proposed As proposed As proposed As proposed
Basement and crawlspace walls:	Type: same as proposed Gross area: same as proposed U-Factor: from Table 402.1.2 with insulation layer on interior side of walls	As proposed As proposed As proposed
Above-grade floors:	Type: wood frame Gross area: same as proposed U-Factor: from Table 402.1.2	As proposed As proposed As proposed
Ceilings:	Type: wood frame Gross area: same as proposed U-Factor: from Table 402.1.2	As proposed As proposed As proposed
Roofs:	Type: composition shingle on wood sheathing Gross area: same as proposed Solar absorptance = 0.75 Emittance = 0.90	As proposed As proposed As proposed As proposed
Attics:	Type: vented with aperture = 1 ft^2 per 300 ft^2 ceiling area	As proposed
Foundations:	Type: same as proposed	As proposed
Doors:	Area: 40 ft^2 Orientation: North U-factor: same as fenestration from Table 402.1.2	As proposed As proposed As proposed

Glazing:[a]	Total area[b] =18% of conditioned floor area
	  Orientation: equally distributed to four cardinal compass orientations (N, E, S, & W)
	U-factor: from Table 402.1.2
	SHGC: from Table 402.1 except that for climates with no requirement (NR) SHGC = 0.40 shall be used
	Interior shade fraction:
	  Summer (all hours when cooling is required) = 0.70
	    Winter (all hours when heating is required) = 0.85
	External shading: none
		As proposed
		
		As proposed
		
		As proposed
		As proposed
		
		Same as standard reference design[c]
		
		As proposed

Skylights	None	As proposed
Thermally isolated sunrooms	None	As proposed
Air exchange rate	Specific Leakage Area (SLA)[d] = 0.00048 assuming no energy recovery	For residences that are not tested, the same as the standard reference design For residences without mechanical ventilation that are tested in accordance with ASHRAE Standard 119, Section 5.1, the measured air exchange rate[e] but not less than 0.35 ACH. For residences with mechanical ventilation that are tested in accordance with ASHRAE Standard 119, Section 5.1, the measured air exchange rate[e] combined with the mechanical ventilation rate,[f] which shall not be less than 0.01 x CFA + 7.5 x (N_{br}+1). where: CFA = conditioned floor area N_{br} = number of bedrooms

TABLE 404.5.2(1) (continued)
SPECIFICATIONS FOR THE STANDARD REFERENCE AND PROPOSED DESIGNS

BUILDING COMPONENT	STANDARD REFERENCE DESIGN	PROPOSED DESIGN
Mechanical ventilation:	None, except where mechanical ventilation is specified by the proposed design, in which case: Annual vent fan energy use: kWh/yr = $0.03942*CFA + 29.565*(N_{br}+1)$ where: CFA = conditioned floor area N_{br} = number of bedrooms	As proposed
Internal gains:	IGain = $17,900 + 23.8*CFA + 4104*N_{br}$ (Btu/day per dwelling unit)	Same as standard reference design
Internal mass:	An internal mass for furniture and contents of 8 pounds per square foot of floor area.	Same as standard reference design, plus any additional mass specifically designed as a thermal storage element[g] but not integral to the building envelope or structure.
Structural mass:	For masonry floor slabs, 80% of floor area covered by R-2 carpet and pad, and 20% of floor directly exposed to room air; For masonry basement walls, as proposed, but with insulation required by Table 402.1.2 located on the interior side of the walls; For other walls, for ceilings, floors, and interior walls, wood frame construction.	As proposed As proposed As proposed
Heating systems[h,i]	Fuel type: same as proposed design Efficiencies: Electric: air-source heat pump with prevailing federal minimum efficiency Nonelectric furnaces: natural gas furnace with prevailing federal minimum efficiency Nonelectric boilers: natural gas boiler with prevailing federal minimum efficiency Capacity: sized in accordance with Section M1401.3 of the *International Residential Code*.	As proposed As proposed As proposed As proposed As proposed
Cooling systems[h,j]	Fuel type: Electric Efficiency: in accordance with prevailing federal minimum standards Capacity: sized in accordance with Section M1401.3 of the *International Residential Code*.	As proposed As proposed As proposed
Service Water Heating[h,k]	Fuel type: same as proposed design Efficiency: in accordance with prevailing Federal minimum standards Use (gal/day): $30 + 10*N_{br}$ Tank temperature: 120°F	As proposed As proposed Same as standard reference Same as standard reference
Thermal distribution systems:	A thermal distribution system efficiency (DSE) of 0.80 shall be applied to both the heating and cooling system efficiencies.	Same as standard reference design, except as specified by Table 404.5.2(2).

TABLE 404.5.2(1) (continued)
SPECIFICATIONS FOR THE STANDARD REFERENCE AND PROPOSED DESIGNS

BUILDING COMPONENT	STANDARD REFERENCE DESIGN	PROPOSED DESIGN
Thermostat	Type: manual, cooling temperature set point = 78°F; heating temperature set point = 68°F	Same as standard reference design

For SI: 1 square foot = 0.93 m^2; 1 British thermal unit = 1055J; 1 pound per square foot = 4.88 kg/m^2; 1 gallon (U.S.) = 3.785 L; °C = (°F-32)/1.8.

a. Glazing shall be defined as sunlight-transmitting fenestration, including the area of sash, curbing or other framing elements, that enclose conditioned space. Glazing includes the area of sunlight-transmitting fenestration assemblies in walls bounding conditioned basements. For doors where the sunlight-transmitting opening is less than 50% of the door area, the glazing area is the sunlight transmitting opening area. For all other doors, the glazing area is the rough frame opening area for the door including the door and the frame.
b. For residences with conditioned basements, R-2 and R-4 residences and townhouses, the following formula shall be used to determine glazing area:
   $AF = 0.18 \times AFL \times FA \times F$
   where:
   AF = Total glazing area.
   AFL = Total floor area of directly conditioned space.
   FA = (Above-grade thermal boundary gross wall area)/(above-grade boundary wall area + 0.5 x below-grade boundary wall area).
   F = (Above-grade thermal boundary wall area)/(above-grade thermal boundary wall area + common wall area) or 0.56, whichever is greater.
   and where:
   Thermal boundary wall is any wall that separates conditioned space from unconditioned space or ambient conditions.
   Above-grade thermal boundary wall is any thermal boundary wall component not in contact with soil.
   Below-grade boundary wall is any thermal boundary wall in soil contact.
   Common wall area is the area of walls shared with an adjoining dwelling unit.
c. For fenestrations facing within 15 degrees (0.26 rad) of true south that are directly coupled to thermal storage mass, the winter interior shade fraction shall be permitted to be increased to 0.95 in the proposed design.
d. Where Leakage Area (L) is defined in accordance with Section 5.1 of ASHRAE 119 and where:
   SLA = L/CFA
   where L and CFA are in the same units.
e. Tested envelope leakage shall be determined and documented by an independent party approved by the code official. Hourly calculations as specified in the 2001 ASHRAE *Handbook of Fundamentals*, Chapter 26, page 26.21, equation 40 (Sherman-Grimsrud model) or the equivalent shall be used to determine the energy loads resulting from infiltration.
f. The combined air exchange rate for infiltration and mechanical ventilation shall be determined in accordance with equation 43 of 2001 ASHRAE *Handbook of Fundamentals* page 26.24 and the " Whole-house Ventilation" provisions of 2001 ASHRAE *Handbook of Fundamentals*, page 26.19 for intermittent mechanical ventilation.
g. Thermal Storage Element shall mean a component not part of the floors, walls or ceilings that is part of a passive solar system, and that provides thermal storage such as enclosed water columns, rock beds, or phase-change containers. A thermal storage element must be in the same room as fenestration that faces within 15 degrees (0.26 rad) of true south, or must be connected to such a room with pipes or ducts that allow the element to be actively charged.
h. For a proposed design with multiple heating, cooling or water heating systems using different fuel types, the applicable standard reference design system capacities and fuel types shall be weighted in accordance with their respective loads as calculated by accepted engineering practice for each equipment and fuel type present.
i. For a proposed design without a proposed heating system, a heating system with the prevailing federal minimum efficiency shall be assumed for both the standard reference design and proposed design. For electric heating systems, the prevailing federal minimum efficiency air-source heat pump shall be used for the standard reference design.
j. For a proposed design home without a proposed cooling system, an electric air conditioner with the prevailing federal minimum efficiency shall be assumed for both the standard reference design and the proposed design.
k. For a proposed design with a nonstorage-type water heater, a 40-gallon storage-type water heater with the prevailing Federal minimum Energy Factor for the same fuel as the predominant heating fuel type shall be assumed. For the case of a proposed design without a proposed water heater, a 40-gallon storage-type water heater with the prevailing federal minimum efficiency for the same fuel as the predominant heating fuel type shall be assumed for both the proposed design and standard reference design.

TABLE 404.5.2(2)
DEFAULT DISTRIBUTION SYSTEM EFFICIENCIES FOR PROPOSED DESIGNS[a]

DISTRIBUTION SYSTEM CONFIGURATION AND CONDITION:	FORCED AIR SYSTEMS	HYDRONIC SYSTEMS[b]
Distribution system components located in unconditioned space	0.80	0.95
Distribution systems entirely located in conditioned space[c]	0.88	1.00
Proposed "reduced leakage" with entire air distribution system located in the conditioned space[d]	0.96	
Proposed "reduced leakage" air distribution system with components located in the unconditioned space	0.88	
"Ductless" systems[e]	1.00	

For SI: 1 cubic foot per minute = 0.47 L/s; 1 square foot = 0.093 m^2; 1 pound per square inch = 6895 Pa; 1 inch water gauge = 1250 Pa.

a. Default values given by this table are for untested distribution systems, which must still meet minimum requirements for duct system insulation.
b. Hydronic Systems shall mean those systems that distribute heating and cooling energy directly to individual spaces using liquids pumped through closed loop piping and that do not depend on ducted, forced air flows to maintain space temperatures.
c. Entire system in conditioned space shall mean that no component of the distribution system, including the air handler unit, is located outside of the conditioned space.
d. Proposed "reduced leakage" shall mean leakage to outdoors not greater than 3 cfm per 100 ft^2 of conditioned floor area and total leakage not greater than 9 cfm per 100 ft^2 of conditioned floor area at a pressure differential of 0.02 inches w.g. (25 Pa) across the entire system, including the manufacturer's air handler enclosure. Total leakage of not greater than 3 cfm per 100 ft^2 of conditioned floor area at a pressure difference of 25 Pascals across the entire system, including the manufacturer's air handler enclosure, shall be deemed to meet this requirement without measurement of leakage to outdoors. This performance shall be specified as required in the construction documents and confirmed through field-testing of installed systems as documented by an approved independent party.
e. Ductless systems may have forced airflow across a coil but shall not have any ducted airflows external to the manufacturer's air handler enclosure.

Delete Chapter 5 in its entirety: (EC48-03/04)

Delete Chapter 6 in its entirety: (EC48-03/04)

Delete Chapter 7 in its entirety: Existing provisions revised and combined with Chapter 8. (EC30-03/04)

Delete Chapter 8 in its entirety and replace as shown. (EC30-03/04; EC31-03/04; EC35-03/04; EC36-03/04; EC37-03/04; EC38-03/04; EC39-03/04; EC40-03/04; EC41-03/04; EC42-03/04; EC44-03/04; EC45-03/04; EC46-03/04 and EL3-03/04) The following Chapter 8 includes both sections which are revised by the listed code changes as well as all remaining portions of Chapter 8 which were not affected by any of the changes. A detailed index which lists the section numbers and what code change affected the section, if any, is posted on the ICC website at http://www.iccsafe.org.

CHAPTER 8
BUILDING DESIGN FOR COMMERCIAL BUILDINGS

SECTION 801
GENERAL

801.1 Scope. The requirements contained in this chapter are applicable to commercial buildings, or portions of commercial buildings. These commercial buildings shall meet either the requirements of ASHRAE/IESNA 90.1, *Energy Standard for Buildings Except for Low-Rise Residential Buildings*, or the requirements contained in this chapter.

801.2 Application. The requirements in Sections 802, 803, 804 and 805 shall each be satisfied on an individual basis. Where one or more of these sections is not satisfied, compliance for that section(s) shall be demonstrated in accordance with the applicable provisions of ASHRAE/IESNA 90.1.

> **Exception:** Buildings conforming to Section 806, provided Sections 802.1.2, 802.3, 803.2.1 or 803.3.1 as applicable, 803.2.2 or 803.3.2 as applicable, 803.2.3 or 803.3.3 as applicable, 803.2.8 or 803.3.6 as applicable, 803.2.9 or 803.3.7 as applicable, 804, 805.2, 805.3, 805.4, 805.6 and 805.7 are each satisfied.

SECTION 802
BUILDING ENVELOPE REQUIREMENTS

802.1 General. Walls, roof assemblies, floors, glazing and slabs on grade which are part of the building envelope for buildings where the window and glazed door area is not greater than 50 percent of the gross area of above-grade walls shall meet the requirements of Sections 802.2.1 through 802.2.9, as applicable. Buildings with more glazing shall meet the applicable provisions of ASHRAE/IESNA 90.1.

802.1.1 Classification of walls. Walls associated with the building envelope shall be classified in accordance with Section 802.1.1.1, 802.1.1.2 or 802.1.1.3.

802.1.1.1 Above-grade walls. Above-grade walls are those walls covered by Section 802.2.1 on the exterior of the building and completely above grade or walls that are more than 15 percent above grade.

802.1.1.2 Below-grade walls. Below-grade walls covered by Section 802.2.8 are basement or first-story walls associated with the exterior of the building that are at least 85 percent below grade.

802.1.1.3 Interior walls. Interior walls covered by Section 802.2.9 are those walls not on the exterior of the building and that separate conditioned and unconditioned space.

802.1.2 Moisture control. All framed walls, floors and ceilings not ventilated to allow moisture to escape shall be provided with an approved vapor retarder having a permeance rating of 1 perm (5.7×10^{-11} kg/Pa·s·m^2) or less, when tested in accordance with the dessicant method using Procedure A of ASTM E 96. The vapor retarder shall be installed on the warm-in-winter side of the insulation.

Exceptions:

1. Buildings located in Climate Zones 1 through 3 as indicated in Figure 301.1 and Table 301.1.

2. In construction where moisture or its freezing will not damage the materials.

3. Where other approved means to avoid condensation in unventilated framed wall, floor, roof and ceiling cavities are provided.

802.2 Criteria. The building envelope components shall meet each of the applicable requirements in Tables 802.2(1), 802.2(2), and 802.2(3) based on the percentage of wall that is glazed. The percentage of wall that is glazed shall be determined by dividing the aggregate area of rough openings for glazing (windows and glazed doors) in all above-grade walls associated with the building envelope by the total gross area of all above-grade exterior walls that are a part of the building envelope. In buildings with multiple types of building envelope construction, each building envelope construction type shall be evaluated separately. Where Table 802.2(1), 802.2(2), or 802.2(3) does not list a particular construction type, the applicable provisions of ASHRAE/IESNA 90.1 shall be used in lieu of Section 802.

802.2.1 Above-grade walls. The minimum thermal resistance (R-value) of the insulating material(s) installed in the wall cavity between the framing members and continuously on the walls shall be as specified in Table 802.2(1), based on framing type and construction materials used in the wall assembly. The R-value of integral insulation installed in concrete masonry units (CMU) shall not be used in determining compliance with Table 802.2(1). "Mass walls" shall include walls weighing at least (1) 35 pounds per square foot (170 kg/m^2) of wall surface area or (2) 25 pounds per square foot (120 kg/m^2) of wall surface area if the material weight is not more than 120 pounds per cubic foot (1,900 kg/m^3).

802.2.2 Opaque doors. Opaque doors (doors having less than 50 percent glass area) shall meet the applicable requirements for doors as specified in Table 802.2(1) and be considered as part of the gross area of above-grade walls that are part of the building envelope.

802.2.3 Windows and glass doors. The maximum solar heat gain coefficient (SHGC) and thermal transmittance (U-factor) of window assemblies and glass doors located in the building envelope shall be as specified in Table 802.2(2), based on the window projection factor.

The window projection factor shall be determined in accordance with Equation 8-1.

PF = A/B   (Equation 8-1)

where:

PF = Projection factor (decimal).
A = Distance measured horizontally from the furthest continuous extremity of any overhang, eave, or permanently attached shading device to the vertical surface of the glazing.
B = Distance measured vertically from the bottom of the glazing to the underside of the overhang, eave, or permanently attached shading device.

Where different windows or glass doors have different PF values, they shall each be evaluated separately, or an area-weighted PF value shall be calculated and used for all windows and glass doors.

802.2.4 Roof assembly. The minimum thermal resistance (R-value) of the insulating material installed either between the roof framing or continuously on the roof assembly shall be as specified in Table 802.2.(1), based on construction materials used in the roof assembly.

Exception: Continuously insulated roof assemblies where the thickness of insulation varies 1 inch (25.4 mm) or less and where the area weighted U-factor is equivalent to the same assembly with the R-value specified in Table 802.2(1).

802.2.5 Skylights. Skylights located in the building envelope shall be limited to 3 percent of the gross roof assembly area and shall have a maximum thermal transmittance (*U*-factor) and SHGC of the skylight assembly as specified in Table 802.2(2).

802.2.6 Floors over outdoor air or unconditioned space. The minimum thermal resistance (*R*-value) of the insulating material installed either between the floor framing or continuously on the floor assembly shall be as specified in Table 802.2(1), based on construction materials used in the floor assembly.

802.2.7 Slabs on grade. The minimum thermal resistance (*R*-value) of the insulation around the perimeter of unheated or heated slab-on-grade floors shall be as specified in Table 802.2(1). The insulation shall be placed on the outside of the foundation or on the inside of the foundation wall. The insulation shall extend downward from the top of the slab for a minimum distance as shown in the table or to the top of the footing, whichever is less, or downward to at least the bottom of the slab and then horizontally to the interior or exterior for the total distance shown in the table.

802.2.8 Below-grade walls. The minimum thermal resistance (*R*-value) of the insulating material installed in, or continuously on, the below-grade walls shall be as specified in Table 802.2(1), and shall extend to a depth of 10 feet (3048 mm) below the outside finish ground level, or to the level of the floor, whichever is less.

802.2.9 Interior walls. The minimum thermal resistance (*R*-value) of the insulating material installed in the wall cavity or continuously on the interior walls shall be the same as that specified for above-grade walls in Table 802.2(1).

802.3 Air leakage. The requirements for air leakage shall be as specified in Sections 802.3.1 through 802.3.7.

802.3.1 Window and door assemblies. The air leakage of window and sliding or swinging door assemblies that are part of the building envelope shall be determined in accordance with AAMA/WDMA 101/I.S.2 or 101/I.S.2/NAFS-02, or NFRC 400 by an accredited, independent laboratory, and labeled and certified by the manufacturer and shall not exceed the values in Section 402.4.2.

Exception: Site-constructed windows and doors that are weatherstripped or sealed in accordance with Section 802.3.3.

802.3.2 Curtain wall, storefront glazing and commercial entrance doors. Curtain wall, storefront glazing and commercial-glazed swinging entrance doors and revolving doors shall be tested for air leakage at 1.57 pounds per square foot (psf) (75 Pa) in accordance with ASTM E 283. For curtain walls and storefront glazing, the maximum air leakage rate shall be 0.3 cubic feet per minute per square foot (cfm/ft^2) (5.5 m^3/h × m^2) of fenestration area. For commercial glazed swinging entrance doors and revolving doors, the maximum air leakage rate shall be 1.00 cfm/ft^2 (18.3 m^3/h × m^2) of door area when tested in accordance with ASTM E 283.

802.3.3 Sealing of the building envelope. Openings and penetrations in the building envelope shall be sealed with caulking materials or closed with gasketing systems compatible with the construction materials and location. Joints and seams shall be sealed in the same manner or taped or covered with a moisture vapor-permeable wrapping material. Sealing materials spanning joints between construction materials shall allow for expansion and contraction of the construction materials.

802.3.4 Outdoor air intakes and exhaust openings. Stair and elevator shaft vents and other outdoor air intakes and exhaust openings integral to the building envelope shall be equipped with not less than a Class I motorized, leakage-rated damper with a maximum leakage rate of 4 cfm/ft^2 (6.8 L/s · m^2) at 1.0 inch water gauge (w.g.) (1250 Pa) when tested in accordance with AMCA 500D.

Exception: Gravity (nonmotorized) dampers are permitted to be used in buildings less than three stories in height above grade.

802.3.5 Loading dock weatherseals. Cargo doors and loading dock doors shall be equipped with weatherseals to restrict infiltration when vehicles are parked in the doorway.

802.3.6 Vestibules. A door that separates conditioned space from the exterior shall be protected with an enclosed vestibule, with all doors opening into and out of the vestibule equipped with self-closing devices. Vestibules shall be designed so that in passing through the vestibule it is not necessary for the interior and exterior doors to open at the same time.

Exceptions:

1. Buildings in Climate Zones 1 and 2 as indicated in Figure 301.1 and Table 301.1.

2. Doors not intended to be used as a building entrance door, such as doors to mechanical or electrical equipment rooms.

3. Doors opening directly from a guestroom or dwelling unit.

4. Doors that open directly from a space less than 3,000 square feet (298 m^2) in area.

5. Revolving doors.

6. Doors used primarily to facilitate vehicular movement or material handling and adjacent personnel doors.

802.3.7 Recessed luminaires. When installed in the building envelope, recessed luminaires shall meet one of the following requirements:

1. Type IC rated, manufactured with no penetrations between the inside of the recessed fixture and ceiling cavity and sealed or gasketed to prevent air leakage into the unconditioned space.

2. Type IC or non-IC rated, installed inside a sealed box constructed from a minimum 0.5-inch-thick (12.7 mm) gypsum wallboard or constructed from a preformed polymeric vapor barrier, or other airtight assembly manufactured for this purpose, while maintaining required clearances of not less than 0.5 inch (12.7 mm) from combustible material and not less than 3 inches (76 mm) from insulation material.

3. Type IC rated, in accordance with ASTM E 283 admitting no more than 2.0 cubic feet per minute (cfm) (0.944 L/s) of air movement from the conditioned space to the ceiling cavity. The luminaire shall be tested at 1.57 psf (75 Pa) pressure difference and shall be labeled.

TABLE 802.2(1)
BUILDING ENVELOPE REQUIREMENTS – OPAQUE ELEMENTS

CLIMATE ZONE	1	2	3	4 except Marine	5 and Marine 4	6	7	8
Roofs								
Insulation entirely above deck	R-15 ci	R-15 ci	R-15 ci	R-15 ci	R-20 ci	R-20 ci	R-25 ci	R-25 ci
Metal buildings (with R-5 thermal blocks[a])[b]	R-19 + R-10	R-19	R-19	R-19	R-19	R-19	R-19 + R-10	R-19 + R-10
Attic and other	R-30	R-30	R-30	R-30	R-30	R-30	R-38	R-38
Walls, Above Grade								
Mass	NR	NR	R-5.7 ci[c,e]	R-5.7 ci[c]	R-7.6 ci	R-9.5 ci	R-11.4 ci	R-13.3 ci
Metal building[b]	R-13	R-13	R-13	R-13	R-13 + R-13	R-13 + R-13	R-13 + R-13	R-13 + R-13
Metal framed	R-13	R-13	R-13	R-13	R-13 + R-3.8 ci	R-13 + R-3.8 ci	R-13 + R-7.5 ci	R-13 + R-7.5 ci
Wood framed and other	R-13	R-13	R-13	R-13	R-13	R-13	R-13	R-13 + R-7.5 ci
Walls, Below Grade								
Below grade wall[d]	NR	NR	NR	NR	NR	NR	R-7.5 ci	R-7.5 ci
Floors								
Mass	NR	R-5 ci	R-5 ci	R-10 ci	R-10 ci	R-10 ci	R-15 ci	R-15 ci
Joist/Framing	NR	R-19	R-19	R-19	R-19	R-30	R-30	R-30
Slab-on-Grade Floors								
Unheated Slabs	NR	NR	NR	NR	NR	NR	NR	R-10 for 24 in. below
Heated Slabs	R-7.5 for 12 in. below	R-7.5 for 12 in. below	R-7.5 for 12 in. below	R-7.5 for 12 in. below	R-7.5 for 24 in. below	R-10 for 36 in. below	R-10 for 36 in. below	R-10 for 48 in. below
Opaque Doors								
Swinging	U – 0.70	U – 0.70	U – 0.70	U – 0.70	U – 0.70	U – 0.70	U – 0.70	U – 0.50
Roll-up or sliding	U – 1.45	U – 1.45	U – 1.45	U – 1.45	U – 1.45	U – 0.50	U – 0.50	U – 0.50

For SI: 1 inch = 25.4 mm.
ci — Continuous Insulation
NR – No Requirement

a. Thermal blocks are a minimum R-5 of rigid insulation, which extends 1" beyond the width of the purlin on each side, perpendicular to the purlin.
b. Assembly descriptions can be found in Table 802.2(3)
c. R-5.7 ci may be substituted with concrete block walls complying with ASTM C90, ungrouted or partially grouted at 32 in. or less on center vertically and 48 in. or less on center horizontally, with ungrouted cores filled with material having a maximum thermal conductivity of 0.44 Btu-in./h-ft^2 F.
d. When heated slabs are placed below grade, below grade walls must meet the exterior insulation requirements for perimeter insulation according to the heated slab-on-grade construction.
e. Insulation is not required for mass walls in Climate Zone 3A located below the "Warm-Humid" line, and in Zone 3B.

TABLE 802.2(2)
BUILDING ENVELOPE REQUIREMENTS

Climate Zone	1	2	3	4 except Marine	5 and Marine 4	6	7	8
Windows (40% maximum)								
Factory-assembled glazed fenestration products								
U-Factor	1.20	0.75	0.65	0.40	0.35	0.35	0.35	0.35
SHGC	0.40	0.40	0.40	0.40	0.40	0.40	NR	NR
Site-built glazed products								
U-Factor	1.20	0.75	0.65	0.50	0.45	0.45	0.45	0.45
SHGC: PF < 0.25	0.25	0.25	0.25	0.40	0.40	0.40	NR	NR
SHGC: 0.25 < PF < 0.5	0.33	0.33	0.33	NR	NR	NR	NR	NR
SHGC: PF ≥ 0.5	0.40	0.40	0.40	NR	NR	NR	NR	NR
Skylights (3% maximum)								
Glass								
U-Factor	1.60	1.05	0.90	0.60	0.60	0.60	0.60	0.60
SHGC	0.40	0.40	0.40	0.40	0.40	0.40	NR	NR
Plastic								
U-Factor	1.90	1.90	1.30	1.30	1.30	0.90	0.90	0.60
SHGC	0.35	0.35	0.35	0.62	0.62	0.62	NR	NR

NR = No requirement
PF = Projection factor (see Section 802.2.3)

TABLE 802.2(3)
METAL BUILDING ASSEMBLY DESCRIPTIONS

Roofs	Description	Reference
R-19 + R-10	Filled cavity roof Thermal blocks are a minimum, R-5 of rigid insulation, which extends 1 in. beyond the width of the purlin on each side, perpendicular to the purlin This construction is R-10 insulation batts draped perpendicularly over the purlins, with enough looseness to allow R-19 batt to be laid above it, parallel to the purlins. Thermal blocks are then placed above the purlin/batt, and the roof deck is secured to the purlins. In the metal building industry, this is known as the "sag and bag" insulation system.	ASHRAE/IESNA 90.1 Table A-2
R-19	Standing seam with single insulation layer Thermal blocks are a minimum R-5 of rigid insulation, which extends 1 in. beyond the width of the purlin on each side, perpendicular to the purlin. This construction R-19 insulation batts draped perpendicularly over the purlins. Thermal blocks are then placed above the purlin/batt, and the roof deck is secured to the purlins.	ASHRAE/IESNA 90.1 Table A-2
Walls		
R-13	Single insulation layer The first layer of R-13 insulation batts is installed continuously perpendicular to the girts and is compressed as the metal skin is attached to the girts.	ASHRAE/IESNA 90.1 Table A-9
R-13 + R-13	Double insulation layer The first layer of R-13 insulation batts is installed continuously perpendicular to the girts, and is compressed as the metal skin is attached to the girts. The second layer of R-13 insulation batts is installed within the framing cavity.	ASHRAE/IESNA 90.1 Table A-9

For SI: 1 inch = 25.4 mm.

SECTION 803
BUILDING MECHANICAL SYSTEMS

803.1 General. This section covers the design and construction of mechanical systems and equipment serving the building heating, cooling or ventilating needs.

803.1.1 Compliance. Compliance with Section 803 shall be achieved by meeting either Section 803.2 or 803.3.

803.2 Simple HVAC systems and equipment. This section applies to buildings served by unitary or packaged HVAC equipment listed in Tables 803.2.2(1) through 803.2.2(5), each serving one zone and controlled by a single thermostat in the zone served. It also applies to two-pipe heating systems serving one or more zones, where no cooling system is installed.

This section does not apply to fan systems serving multiple zones, nonunitary or nonpackaged HVAC equipment and systems or hydronic or steam heating and hydronic cooling equipment and distribution systems that provide cooling or cooling and heating which are covered by Section 803.3.

803.2.1 Calculation of heating and cooling loads. Design loads shall be determined in accordance with the procedures described in the ASHRAE *Fundamentals Handbook*. Heating and cooling loads shall be adjusted to account for load reductions that are achieved when energy recovery systems are utilized in the HVAC system in accordance with the ASHRAE *HVAC Systems and Equipment Handbook*. Alternatively, design loads shall be determined by an approved equivalent computation procedure, using the design parameters specified in Chapter 3.

803.2.1.1 Equipment and system sizing. Heating and cooling equipment and systems capacity shall not exceed the loads calculated in accordance with Section 803.2.1. A single piece of equipment providing both heating and cooling must satisfy this provision for one function with the capacity for the other function as small as possible, within available equipment options.

803.2.2 HVAC equipment performance requirements. Equipment shall meet the minimum efficiency requirements of Tables 803.2.2(1), 803.2.2(2), 803.2.2(3), 803.2.2(4) and 803.2.2(5), when tested and rated in accordance with the applicable test procedure. The efficiency shall be verified through data furnished by the manufacturer or through certification under an approved certification program. Where multiple rating conditions or performance requirements are provided, the equipment shall satisfy all stated requirements

803.2.3 Temperature and humidity controls. Requirements for temperature and humidity controls shall be as specified in Sections 803.2.3.1 through 803.2.3.3.

803.2.3.1 Temperature controls. Each heating and cooling system shall have at least one solid-state programmable thermostat. The thermostat shall have the capability to set back or shut down the system based on day of the week and time of day, and provide a readily accessible manual override that will return to the presetback or shutdown schedule without reprogramming.

Exceptions:

1. HVAC systems serving hotel/motel guestrooms.

2. Packaged terminal air conditioners, packaged terminal heat pumps and room air conditioner systems.

803.2.3.2 Heat pump supplementary heat. Heat pumps having supplementary electric-resistance heat shall have controls that, except during defrost, prevent supplemental heat operation when the heat pump can meet the heating load.

803.2.3.3 Humidity controls. When humidistats are installed, they shall have the capability to prevent the use of fossil fuel or electric power to achieve a humidity below 60 percent when the system controlled is cooling, and above 30 percent when the system controlled is heating.

Exceptions:

1. Systems serving spaces where specific humidity levels are required to satisfy process needs, such as computer rooms, museums, surgical suites and buildings with refrigerating systems, such as supermarkets, refrigerated warehouses and ice arenas.

2. Systems where humidity is removed as the result of the use of a desiccant system with energy recovery.

3. Reheat systems utilizing site-recovered (including condenser heat) or site-solar energy sources.

TABLE 803.2.2(1)
UNITARY AIR CONDITIONERS AND CONDENSING UNITS, ELECTRICALLY OPERATED, MINIMUM EFFICIENCY REQUIREMENTS

EQUIPMENT TYPE	SIZE CATEGORY	SUBCATEGORY OR RATING CONDITION	MINIMUM EFFICIENCY [b]	TEST PROCEDURE [a]
Air conditioners, Air cooled	<65,000 Btu/h[d]	Split system	10.0 SEER	ARI 210/240
		Single package	9.7 SEER	
	≥ 65,000 Btuh/h and < 135,000 Btu/h	Split system and single package	10.3 EER[c]	
	≥ 135,000 Btu/h and < 240,000 Btu/h	Split system and single package	9.7 EER[c]	ARI 340/360
	≥ 240,000 Btu/h and < 760,000 Btu/h	Split system and single package	9.5 EER[c] 9.7 IPLV[c]	
	≥ 760,000 Btu/h	Split system and single package	9.2 EER[c] 9.4 IPLV[v]	
Air conditioners, Water and evaporatively cooled	< 65,000 Btu/h	Split system and single package	12.1 EER	ARI 210/240
	≥ 65,000 Btu/h and < 135,000 Btu/h	Split system and single package	11.5 EER[c]	
	≥135,000 Btu/h and < 240,000 Btu/h	Split system and single package	11.0 EER[c]	ARI 340/360
	≥ 240,000 Btu/h	Split system and single package	11.0 EER[c] 10.3 IPLV[c]	

For SI: 1 British thermal unit per hour = 0.2931 W.

a. Chapter 10 contains a complete specification of the referenced test procedure, including the referenced year version of the test procedure.
b. IPLVs are only applicable to equipment with capacity modulation.
c. Deduct 0.2 from the required EERs and IPLVs for units with a heating section other than electric resistance heat.
d. Single-phase air-cooled air conditioners < 65,000 Btu/h are regulated by the National Appliance Energy Conservation Act of 1987 (NAECA), SEER values are those set by NAECA.

2004 SUPPLEMENT TO THE IECC

TABLE 803.2.2(2)
UNITARY AND APPLIED HEAT PUMPS, ELECTRICALLY OPERATED,
MINIMUM EFFICIENCY REQUIREMENTS

EQUIPMENT TYPE	SIZE CATEGORY	SUBCATEGORY OR RATING CONDITION	MINIMUM EFFICIENCY [b]	TEST PROCEDURE [a]
Air cooled (Cooling mode)	< 65,000 Btu/h[d]	Split system	10.0 SEER	ARI 210/240
		Single package	9.7 SEER	
	≥ 65,000 Btu/h and < 135,000 Btu/h	Split system and single package	10.1 EER[c]	
	≥ 135,000 Btu/h and < 240,000 Btu/h	Split system and single package	9.3 EER[c]	ARI 340/360
	≥ 240,000 Btu/h	Split system and single package	9.0 EER[c] 9.2 IPLV[c]	
Water source (Cooling mode)	< 17,000 Btu/h	86°F entering water	11.2 EER	ARI/ASHRAE-13256-1
	≥ 17,000 Btu/h and < 135,000 Btu/h	86°F entering water	12.0 EER	ARI/ASHRAE-13256-1
Groundwater source (Cooling mode)	< 135,000 Btu/h	59°F entering water	16.2 EER	ARI/ASHRAE-13256-1
Ground source (Cooling mode)	< 135,000 Btu/h	77°F entering water	13.4 EER	ARI/ASHRAE 13256-1
Air cooled (Heating mode)	< 65,000 Btu/h[d] (Cooling capacity)	Split system	6.8 HSPF	ARI 210/240
		Single package	6.6 HSPF	
	≥ 65,000 Btu/h and < 135,000 Btu/h (Cooling capacity)	47°F db/43°F wb outdoor air	3.2 COP	
	≥ 135,000 Btu/h (Cooling capacity)	47°F db/43°F wb outdoor air	3.1 COP	ARI 340/360
Water source (Heating mode)	< 135,000 Btu/h (Cooling capacity)	68°F entering water	4.2 COP	ARI/ASHRAE-13256-1
Groundwater source (Heating mode)	< 135,000 Btu/h (Cooling capacity)	50°F entering water	3.6 COP	ARI/ASHRAE-13256-1
Ground Source (Heating mode)	< 135,000 Btu/h (Cooling capacity)	32°F entering water	3.1 COP	ARI/ASHRAE-13256-1

For SI: °C = [(°F) - 32] / 1.8, 1 British thermal unit per hour = 0.2931 W.
db = dry-bulb temperature, °F
wb = wet-bulb temperature, °F

a. Chapter 10 contains a complete specification of the referenced test procedure, including the referenced year version of the test procedure.
b. IPLVs and Part load rating conditions are only applicable to equipment with capacity modulation.
c. Deduct 0.2 from the required EERs and IPLVs for units with a heating section other than electric resistance heat.
d. Single-phase air-cooled heat pumps < 65,000 Btu/h are regulated by the National Appliance Energy Conservation Act of 1987 NAECA), SEER and HSPF values are those set by NAECA.

TABLE 803.2.2(3)
PACKAGED TERMINAL AIR CONDITIONERS AND PACKAGED TERMINAL HEAT PUMPS

EQUIPMENT TYPE	SIZE CATEGORY (INPUT)	SUBCATEGORY OR RATING CONDITION	MINIMUM EFFICIENCY [b]	TEST PROCEDURE [a]
PTAC (Cooling mode) New construction	All capacities	95°F db outdoor air	12.5 - (0.213 · Cap/1000) EER	ARI 310/380
PTAC (Cooling mode) Replacements[c]	All capacities	95°F db outdoor air	10.9 - (0.213 · Cap/1000) EER	
PTHP (Cooling mode) New construction	All capacities	95°F db outdoor air	12.3 - (0.213 · Cap/1000) EER	
PTHP (Cooling mode) Replacements[c]	All capacities	95°F db outdoor air	10.8 - (0.213 · Cap/1000) EER	
PTHP (Heating mode) New construction	All capacities	—	3.2 - (0.026 · Cap/1000) COP	
PTHP (Heating mode) Replacements[c]	All capacities	—	2.9 - (0.026 · Cap/1000) COP	

For SI: °C - [(°F) - 32] / 1.8, 1 British thermal unit per hour - 0.2931 W
db = dry-bulb temperature, °F
wb = wet-bulb temperature, °F

a. Chapter 10 contains a complete specification of the referenced test procedure, including the referenced year version of the test procedure.
b. Cap means the rated cooling capacity of the product in Btu/h. If the unit's capacity is less than 7,000 Btu/h, use 7,000 Btu/h in the calculation. If the unit's capacity is greater than 15,000 Btu/h, use 15,000 Btu/h in the calculation.
c. Replacement units must be factory labeled as follows: "MANUFACTURED FOR REPLACEMENT APPLICATIONS ONLY: NOT TO BE INSTALLED IN NEW CONSTRUCTION PROJECTS." Replacement efficiencies apply only to units with existing sleeves less than 16 inches (406 mm) high and less than 42 inches (1067 mm) wide.

TABLE 803.2.2(4)
WARM AIR FURNACES AND COMBINATION WARM AIR FURNACES/AIR-CONDITIONING UNITS, WARM AIR DUCT FURNACES AND UNIT HEATERS, MINIMUM EFFICIENCY REQUIREMENTS

EQUIPMENT TYPE	SIZE CATEGORY (INPUT)	SUBCATEGORY OR RATING CONDITION	MINIMUM EFFICIENCY [d, e]	TEST PROCEDURE [a]
Warm air furnaces, gas fired	< 225,000 Btu/h	—	78% AFUE or 80% E_t^c	DOE 10 CFR Part 430 or ANSI Z21.47
	≥ 225,000 Btu/h	Maximum capacity[c]	80% E_t^f	ANSI Z21.47
Warm furnaces, oil fired	< 225,000 Btu/h	—	78% AFUE or 80% E_t^c	DOE 10 CFR Part 430 or UL 727
	≥ 225,000 Btu/h	Maximum capacity[b]	81% E_t^g	UL 727
Warm air duct furnaces, gas fired	All capacities	Maximum capacity[b]	80% E_c	ANSI Z83.8
Warm air unit heaters, gas fired	All capacities	Maximum capacity[b]	80% E_c	ANSI Z83.8
Warm air unit heaters, oil fired	All capacities	Maximum capacity[b]	80% E_c	UL 731

For SI: 1 British thermal unit per hour=0.2931 W.

a. Chapter 10 contains a complete specification of the referenced test procedure, including the referenced year version of the test procedure.
b. Minimum and maximum ratings as provided for and allowed by the unit's controls.

(Footnotes to Table 803.2.2(4) [continued]

c. Combination units not covered by the National Appliance Energy Conservation Act of 1987(NAECA) (3-phase power or cooling capacity greater than or equal to 65,000 Btu/h [19 kW]) shall comply with either rating.
d. E_t = Thermal efficiency. See test procedure for detailed discussion.
e. E_c = Combustion efficiency (100% less flue losses). See test procedure for detailed discussion.
f. E_c = Combustion efficiency. Units must also include an IID, have jackets not exceeding 0.75 percent of the input rating, and have either power venting or a flue damper. A vent damper is an acceptable alternative to a flue damper for those furnaces where combustion air is drawn from the conditioned space.
g. E_t = Thermal efficiency. Units must also include an IID, have jacket losses not exceeding 0.75 percent of the input rating, and have either power venting or a flue damper. A vent damper is an acceptable alternative to a flue damper for those furnaces where combustion air is drawn from the conditioned space.

TABLE 803.2.2(5)
BOILERS, GAS- AND OIL-FIRED, MINIMUM EFFICIENCY REQUIREMENTS

EQUIPMENT TYPE[f]	SIZE CATEGORY (INPUT)	SUBCATEGORY OR RATING CONDITION	MINIMUM EFFICIENCY [c, d, e]	TEST PROCEDURE [a]
Boilers, Gas fired	< 300,000 Btu/h	Hot water	80% AFUE	DOE 10 CFR Part 430
		Steam	75% AFUE	
	≥ 300,000 Btu/h and ≤ 2,500,000 Btu/h	Minimum capacity[b]	75% E_t	H.I. HBS
	> 2,500,000 Btu/h[f]	Hot water	80% E_c	
		Steam	80% E_c	
Boilers, Oil fired	< 300,000 Btu/h	—	80% AFUE	DOE 10 CFR Part 430
	≥ 300,000 Btu/h and ≤ 2,500,000 Btu/h	Minimum capacity[b]	78% E_t	H.I. HBS
	> 2,500,000 Btu/h[f]	Hot water	83% E_c	
		Steam	83% E_c	
Boilers, Oil fired (Residual)	≥ 300,000 Btu/h and ≤ 2,500,000 Btu/h	Minimum capacity[b]	78% E_t	H.I. HBS
	> 2,500,000 Btu/h[f]	Hot water	83% E_c	
		Steam	83% E_c	

For SI: 1 British thermal unit per hour = 0.2931 W.
a. Chapter 10 contains a complete specification of the referenced test procedure, including the referenced year version of the test procedure.
b. Minimum ratings as provided for and allowed by the unit's controls.
c. E_c = Combustion efficiency (100 percent flue losses). See reference document for detailed information.
d. E_t = Thermal efficiency. See reference document for detailed information.
e. Alternative test procedures used at the manufacturer's option are ASME PTC-4.1 for units greater than 5,000,000 Btu/h input, or ANSI Z21.13 for units greater than or equal to 300,000 Btu/h and less than or equal to 2,500,000 Btu/h input.
f. These requirements apply to boilers with rated input of 8,000,000 Btu/h or less that are not packaged boilers, and to all packaged boilers. Minimum efficiency requirements for boilers cover all capacities of packaged boilers.

803.2.4 Hydronic system controls. Hydronic systems of at least 300,000 Btu/h (87,930 W) design output capacity supplying heated and chilled water to comfort conditioning systems shall include controls that meet the requirements of Section 803.3.3.7.

803.2.5 Ventilation. Ventilation, either natural or mechanical, shall be provided in accordance with Chapter 4 of the *International Mechanical Code*. Where mechanical ventilation is provided, the system shall provide the capability to reduce the outdoor air supply to the minimum required by Chapter 4 of the *International Mechanical Code*.

803.2.5.1 Energy recovery ventilation systems. Individual fan systems that have both a design supply air capacity of 5,000 cfm (2.36 m^3/s) or greater and a minimum outside air supply of 70 percent or greater of the design supply air quantity shall have an energy recovery system that provides a change in the enthalpy of the outdoor air supply of 50 percent or more of the difference between the outdoor air and return air at design conditions. Provision shall be made to bypass or control the energy recovery system to permit cooling with outdoor air where cooling with outdoor air is required.

Exceptions: An energy recovery ventilation system shall not be required in any of the following conditions:

1. Where energy recovery systems are prohibited by the *International Mechanical Code*.

2. Laboratory fume hood systems with a total exhaust rate of 15,000 cfm (7.08 m^3/s) or less.

3. Laboratory fume hood systems with a total exhaust rate greater than 15,000 cfm (7.08 m^3/s) that include at least one of the following features:

   3.1 Variable air volume hood exhaust and room supply systems capable of reducing exhaust and makeup air volume to 50 percent or less of design values.

   3.2 Direct makeup (auxiliary) air supply equal to at least 75 percent of the exhaust rate, heated no warmer than 2°F (1.1°C) below room set point, cooled to no cooler than 3°F (1.7°C) above room set point, no humidification added, and no simultaneous heating and cooling used for dehumidification control.

4. Systems serving spaces that are not cooled and are heated to less than 60 °F (15.5°C).

5. Where more than 60 percent of the outdoor heating energy is provided from site-recovered or site solar energy.

6. Heating systems in climates with less than 3600 HDD.

7. Cooling systems in climates with a 1 percent cooling design wet-bulb temperature less than 64 °F (17.7°C).

8. Systems requiring dehumidification that employ series-style energy recovery coils wrapped around the cooling coil.

803.2.6 Cooling with outdoor air. Supply air economizers shall be provided on each cooling system as shown in Table 803.2.6(1).

Economizers shall be capable of operating at 100-percent outside air, even if additional mechanical cooling is required to meet the cooling load of the building. Where a single room or space is supplied by multiple air systems, the aggregate capacity of those systems shall be used in applying this requirement.

Exceptions:

1. Where the cooling equipment is covered by the minimum efficiency requirements of Table 803.2.2(1) or 803.2.2(2) and meets or exceeds the minimum cooling efficiency requirement (EER) by the percentages shown in Table 803.2.6(2).

2. Systems with air or evaporatively cooled condensers and which serve spaces with open case refrigeration or that require filtration equipment in order to meet the minimum ventilation requirements of Chapter 4 of the *International Mechanical Code*.

**TABLE 803.2.6(1)
ECONOMIZER REQUIREMENTS**

CLIMATE ZONES	ECONOMIZER REQUIREMENT
1A, 1B, 2A, 3A, 4A, 7, 8	No Requirement
2B, 3B, 3C, 4B, 4C, 5B, 5C, 6B	Economizers on All Cooling Systems ≥ 65,000 Btu/h
5A, 6A	Economizers on All Cooling Systems ≥ 135,000 Btu/h

For SI: 1 British thermal unit per hour = 0.293 W

TABLE 803.2.6(2)
EQUIPMENT EFFICIENCY PERFORMANCE EXCEPTION FOR ECONOMIZERS

CLIMATE ZONES	COOLING EQUIPMENT PERFORMANCE IMPROVEMENT (EER OR IPLV)
2B	10% Efficiency Improvement
3B	15% Efficiency Improvement
4B	20% Efficiency Improvement

803.2.7 Shutoff dampers. Outdoor air supply and exhaust ducts shall be provided with automatic means to reduce and shut off airflow.

Exceptions:

1. Systems serving areas designed for continuous operation.

2. Individual systems with a maximum 3,000 cfm (1416 L/s) airflow rate.

3. Systems with readily accessible manual dampers.

4. Where restricted by health and life safety codes.

803.2.8 Duct and plenum insulation and sealing. All supply and return air ducts and plenums shall be insulated with a minimum of R-5 insulation when located in unconditioned spaces and with a minimum of R-8 insulation when located outside the building. When located within a building envelope assembly, the duct or plenum shall be separated from the building exterior or unconditioned or exempt spaces by a minimum of R-8 insulation.

Exceptions:

1. When located within equipment.

2. When the design temperature difference between the interior and exterior of the duct or plenum does not exceed 15°F (8°C).

All joints, longitudinal and transverse seams and connections in ductwork, shall be securely fastened and sealed with welds, gaskets, mastics (adhesives), mastic-plus-embedded-fabric systems or tapes. Tapes and mastics used to seal ductwork shall be listed and labeled in accordance with UL 181A and shall be marked "181A-P" for pressure-sensitive tape, "181A-M" for mastic or "181A-H" for heat-sensitive tape. Tapes and mastics used to seal flexible air ducts and flexible air connectors shall comply with UL 181B and shall be marked "181B-FX" for pressure-sensitive tape or "181B-M" for mastic. Duct connections to flanges of air distribution system equipment shall be sealed and mechanically fastened. Unlisted duct tape is not permitted as a sealant on any metal ducts.

803.2.8.1 Duct construction. Ductwork shall be constructed and erected in accordance with the *International Mechanical Code*.

803.2.8.1.1 High- and medium-pressure duct systems. All ducts and plenums operating at a static pressures greater than 2 inches w.g. (500 Pa) shall be insulated and sealed in accordance with Section 803.2.8. Ducts operating at a static pressures in excess of 3 inches w.g. (750 Pa) shall be leak tested in accordance with Section 803.3.6. Pressure classifications specific to the duct system shall be clearly indicated on the construction documents in accordance with the *International Mechanical Code*.

803.2.8.1.2 Low-pressure duct systems. All longitudinal and transverse joints, seams and connections of supply and return ducts operating at a static pressure less than or equal to 2 inches w.g. (500 Pa) shall be securely fastened and sealed with welds, gaskets, mastics (adhesives), mastic-plus-embedded-fabric systems or tapes installed in accordance with the manufacturer's installation instructions. Pressure classifications specific to the duct system shall be clearly indicated on the construction documents in accordance with the *International Mechanical Code*.

Exception: Continuously welded and locking-type longitudinal joints and seams on ducts operating at static pressures less than 2 inches w.g. (500 Pa) pressure classification.

803.2.9 Piping insulation. All piping serving as part of a heating or cooling system shall be thermally insulated in accordance with Section 803.3.7.

803.3 Complex HVAC systems and equipment. This section applies to buildings served by HVAC equipment and systems not covered in Section 803.2.

803.3.1 Calculation of heating and cooling loads. Design loads shall be determined in accordance with Section 803.2.1.

803.3.1.1 Equipment and system sizing. Heating and cooling equipment and system capacity shall not exceed the loads calculated in accordance with Section 803.2.1.

Exceptions:

1. Required standby equipment and systems provided with controls and devices that allow

such systems or equipment to operate automatically only when the primary equipment is not operating.

2. Multiple units of the same equipment type with combined capacities exceeding the design load and provided with controls that have the capability to sequence the operation of each unit based on load.

803.3.2 HVAC equipment performance requirements. Equipment shall meet the minimum efficiency requirements of Tables 803.3.2(1) through 803.3.2(6) and Table 803.2.2(5), when tested and rated in accordance with the applicable test procedure. The efficiency shall be verified through certification under an approved certification program or, if no certification program exists, the equipment efficiency ratings shall be supported by data furnished by the manufacturer. Where multiple rating conditions or performance requirements are provided, the equipment shall satisfy all stated requirements. Where components, such as indoor or outdoor coils, from different manufacturers are used, calculations and supporting data shall be furnished by the designer that demonstrate that the combined efficiency of the specified components meets the requirements herein.

Where unitary or prepackaged equipment is used in a complex HVAC system and is not covered by Section 803.3.2, the equipment shall meet the applicable requirements of Section 803.2.2.

Exception: Equipment listed in Table 803.3.2(2) not designed for operation at ARI Standard test conditions of 44°F (7°C) leaving chilled water temperature and 85°F (29°C) entering condenser water temperature shall have a minimum full load COP and IPLV rating as shown in Tables 803.3.2(3) through 803.3.2(5) as applicable. The table values are only applicable over the following full load design ranges:

Leaving Chilled
Water Temperature: 40 to 48°F (4 to 9°C)

Entering Condenser
Water Temperature: 75 to 85°F (24 to 29°C)

Condensing Water
Temperature Rise: 5 to 15°F ($\Delta 3$ to $\Delta 8$°C)

Chillers designed to operate outside of these ranges are not covered by this code.

803.3.3 HVAC system controls. Each heating and cooling system shall be provided with thermostatic controls as required in Sections 803.3.3.1 through 803.3.3.8.

803.3.3.1 Thermostatic controls. The supply of heating and cooling energy to each zone shall be controlled by individual thermostatic controls capable of responding to temperature within the zone. Where humidification or dehumidification or both is provided, at least one humidity control device shall be provided for each humidity control system

Exception: Independent perimeter systems that are designed to offset only building envelope heat losses or gains or both serving one or more perimeter zones also served by an interior system provided:

1. The perimeter system includes at least one thermostatic control zone for each building exposure having exterior walls facing only one orientation (within +/- 45 degrees) (0.8 rad) for more than 50 contiguous feet (15.2 m); and,

2. The perimeter system heating and cooling supply is controlled by a thermostat(s) located within the zone(s) served by the system.

803.3.3.1.1 Heat pump supplementary heat. Heat pumps having supplementary electric resistance heat shall have controls that, except during defrost, prevent supplementary heat operation when the heat pump can meet the heating load.

803.3.3.2 Set point overlap restriction. Where used to control both heating and cooling, zone thermostatic controls shall provide a temperature range or deadband of at least 5°F ($\Delta 2.8$°C) within which the supply of heating and cooling energy to the zone is capable of being shut off or reduced to a minimum.

Exception: Thermostats requiring manual changeover between heating and cooling modes.

803.3.3.3 Off-hour controls. Each zone shall be provided with thermostatic setback controls that are controlled by either an automatic time clock or programmable control system.

Exceptions:

1. Zones that will be operated continuously.

2. Zones with a full HVAC load demand not exceeding 6,800 Btu/h (2 kW) and having a readily accessible manual shutoff switch.

TABLE 803.3.2(1)
CONDENSING UNITS, ELECTRICALLY OPERATED, MINIMUM EFFICIENCY REQUIREMENTS

EQUIPMENT TYPE	SIZE CATEGORY	MINIMUM EFFICIENCY[b]	TEST PROCEDURE[a]
Condensing units, air cooled	≥ 135,000 Btu/h	10.1 EER 11.2 IPLV	ARI 365
Condensing units Water or evaporatively cooled	≥ 135,000 Btu/h	13.1 EER 13.1 IPLV	

For SI: 1 British thermal unit per hour = 0.2931 W.
a. Chapter 10 contains a complete specification of the referenced test procedure, including the referenced year version of the test procedure.
b. IPLVs are only applicable to equipment with capacity modulation.

TABLE 803.3.2(2)
WATER CHILLING PACKAGES, MINIMUM EFFICIENCY REQUIREMENTS

EQUIPMENT TYPE	SIZE CATEGORY	MINIMUM EFFICIENCY[b]	TEST PROCEDURE[a]
Air cooled, with condenser, Electrically operated	< 150 tons	2.80 COP 2.80 IPLV	ARI 550/590
	≥ 150 tons	2.50 COP 2.50 IPLV	
Air cooled, without condenser, Electrically operated	All capacities	3.10 COP 3.10 IPLV	ARI 550/590
Water cooled, Electrically operated, Positive displacement (reciprocating)	All capacities	4.20 COP 4.65 IPLV	
Water cooled, Electrically operated, Positive displacement (rotary screw and scroll)	< 150 tons	4.45 COP 4.50 IPLV	ARI 550/590
	≥ 150 tons and < 300 tons	4.90 COP 4.95 IPLV	
	≥ 300 tons	5.50 COP 5.60 IPLV	
Water cooled, Electrically operated, centrifugal	< 150 tons	5.00 COP 5.00 IPLV	ARI 550/590
	≥ 150 tons and < 300 tons	5.55 COP 5.55 IPLV	
	≥ 300 tons	6.10 COP 6.10 IPLV	
Air cooled, absorption single effect	All capacities	0.60 COP	ARI 560
Water cooled, absorption single effect	All capacities	0.70 COP	
Absorption double effect, indirect-fired	All capacities	1.00 COP 1.05 IPLV	
Absorption double effect, direct-fired	All capacities	1.00 COP 1.00 IPLV	

For SI: 1 ton = 3.517 kW. °C = [(°F) - 32]/1.8.
a. Chapter 10 contains a complete specification of the referenced test procedure, including the referenced year version of the test procedure
b. The chiller equipment requirements do not apply for chillers used in low temperature applications where the design leaving fluid temperature is less than or equal to 40°F.

TABLE 803.3.2(3)
COPs AND IPLVs FOR NONSTANDARD CENTRIFUGAL CHILLERS < 150 TONS

Leaving chilled water temperature (°F)	Entering condenser water temperature (°F)	Lift[a] (°F)	2 gpm/ton	2.5 gpm/ton	3 gpm/ton	4 gpm/ton	5 gpm/ton	6 gpm/ton
\multicolumn{9}{c}{CENTRIFUGAL CHILLERS < 150 TONS, COP_{std} = 5.4}
			\multicolumn{6}{c}{Condenser flow rate}
			\multicolumn{6}{c}{Required COP and IPLV}
46	75	29	6.00	6.27	6.48	6.80	7.03	7.20
45	75	30	5.92	6.17	6.37	6.66	6.87	7.02
44	75	31	5.84	6.08	6.26	6.53	6.71	6.86
43	75	32	5.75	5.99	6.16	6.40	6.58	6.71
42	75	33	5.67	5.90	6.06	6.29	6.45	6.57
41	75	34	5.59	5.82	5.98	6.19	6.34	6.44
46	80	34	5.59	5.82	5.98	6.19	6.34	6.44
40	75	35	5.50	5.74	5.89	6.10	6.23	6.33
45	80	35	5.50	5.74	5.89	6.10	6.23	6.33
44	80	36	5.41	5.66	5.81	6.01	6.13	6.22
43	80	37	5.31	5.57	5.73	5.92	6.04	6.13
42	80	38	5.21	5.48	5.64	5.84	5.95	6.04
41	80	39	5.09	5.39	5.56	5.76	5.87	5.95
46	85	39	5.09	5.39	5.56	5.76	5.87	5.95
40	80	40	4.96	5.29	5.47	5.67	5.79	5.86
45	85	40	4.96	5.29	5.47	5.67	5.79	5.86
44	85	41	4.83	5.18	5.40	5.59	5.71	5.78
43	85	42	4.68	5.07	5.28	5.50	5.62	5.70
42	85	43	4.51	4.94	5.17	5.41	5.54	5.62
41	85	44	4.33	4.80	5.05	5.31	5.45	5.53
40	85	45	4.13	4.65	4.92	5.21	5.35	5.44
\multicolumn{3}{c}{Condenser ΔT[b]}	14.04	11.23	9.36	7.02	5.62	4.68

For SI: °C = [(°F) - 32] / 1.8, 1 gallon per minute = 3.785 L/min., 1 ton = 12,000 British thermal unit per hour = 3.517 kW.
a. Lift = Entering condenser water temperature (°F) - Leaving chilled water temperature (°F).
b. Condenser ΔT = Leaving condenser water temperature (°F) - Entering condenser water temperature (°F).

K_{adj} = 6.1507 - 0.30244(X) + 0.0062692(X)^2 - 0.000045595(X)
where: X = Condenser ΔT + Lift
COP_{adj} = K_{adj} x COP_{std}

TABLE 803.3.2(4)
COPs AND IPLVs FOR NONSTANDARD CENTRIFUGAL CHILLERS
≥ 150 TONS, ≤ 300 TONS

CENTRIFUGAL CHILLERS ≥ 150 Tons, ≤ 300 Tons
$COP_{std} = 5.55$

Leaving chilled water temperature (°F)	Entering condenser water temperature (°F)	Lift[a] (°F)	Condenser flow rate					
			2 gpm/ton	2.5 gpm/ton	3 gpm/ton	4 gpm/ton	5 gpm/ton	6 gpm/ton
			Required COP and IPLV					
46	75	29	6.17	6.44	6.66	6.99	7.23	7.40
45	75	30	6.08	6.34	6.54	6.84	7.06	7.22
44	75	31	6.00	6.24	6.43	6.71	6.90	7.05
43	75	32	5.91	6.15	6.33	6.58	6.76	6.89
42	75	33	5.83	6.07	6.23	6.47	6.63	6.75
41	75	34	5.74	5.98	6.14	6.36	6.51	6.62
46	80	34	5.74	5.98	6.14	6.36	6.51	6.62
40	75	35	5.65	5.90	6.05	6.26	6.40	6.51
45	80	35	5.65	5.90	6.05	6.26	6.40	6.51
44	80	36	5.56	5.81	5.97	6.17	6.30	6.40
43	80	37	5.46	5.73	5.89	6.08	6.21	6.30
42	80	38	5.35	5.64	5.8	6.00	6.12	6.20
41	80	39	5.23	5.54	5.71	5.91	6.03	6.11
46	85	39	5.23	5.54	5.71	5.91	6.03	6.11
40	80	40	5.10	5.44	5.62	5.83	5.95	6.03
45	85	40	5.10	5.44	5.62	5.83	5.95	6.03
44	85	41	4.96	5.33	5.55	5.74	5.86	5.94
43	85	42	4.81	5.21	5.42	5.66	5.78	5.86
42	85	43	4.63	5.08	5.31	5.56	5.69	5.77
41	85	44	4.45	4.93	5.19	5.46	5.60	5.69
40	85	45	4.24	4.77	5.06	5.35	5.50	5.59
Condenser ΔT[b]			14.04	11.23	9.36	7.02	5.62	4.68

For SI: °C = [(°F) - 32] / 1.8, 1 gallon per minute = 3.785 L/min., 1 ton = 12,000 British thermal unit per hour = 3.517 kW.
a. Lift = Entering condenser water temperature (°F) - Leaving chilled water temperature (°F).
b. Condenser ΔT = Leaving condenser water temperature (°F) - Entering condenser water temperature (°F).

$K_{adj} = 6.1507 - 0.30244(X) + 0.0062692(X)^2 - 0.000045595(X)$
where: X = Condenser ΔT + Lift
$COP_{adj} = K_{adj} \times COP_{std}$

2004 SUPPLEMENT TO THE IECC

TABLE 803.3.2(5)
COPs AND IPLVs FOR NONSTANDARD CENTRIFUGAL CHILLERS > 300 TONS

Leaving chilled water temperature (°F)	Entering condenser water temperature (°F)	Lift[a] (°F)	2 gpm/ton	2.5 gpm/ton	3 gpm/ton	4 gpm/ton	5 gpm/ton	6 gpm/ton
CENTRIFUGAL CHILLERS > 300 Tons, COP_{std} = 6.1								
			Condenser flow rate					
			Required COP and IPLV					
46	75	29	6.80	7.11	7.35	7.71	7.97	8.16
45	75	30	6.71	6.99	7.21	7.55	7.78	7.96
44	75	31	6.61	6.89	7.09	7.40	7.61	7.77
43	75	32	6.52	6.79	6.98	7.26	7.45	7.60
42	75	33	6.43	6.69	6.87	7.13	7.31	7.44
41	75	34	6.33	6.60	6.77	7.02	7.18	7.30
46	80	34	6.33	6.60	6.77	7.02	7.18	7.30
40	75	35	6.23	6.50	6.68	6.91	7.06	7.17
45	80	35	6.23	6.50	6.68	6.91	7.06	7.17
44	80	36	6.13	6.41	6.58	6.81	6.95	7.05
43	80	37	6.02	6.31	6.49	6.71	6.85	6.94
42	80	38	5.90	6.21	6.40	6.61	6.75	6.84
41	80	39	5.77	6.11	6.30	6.52	6.65	6.74
46	85	39	5.77	6.11	6.30	6.52	6.65	6.74
40	80	40	5.63	6.00	6.20	6.43	6.56	6.65
45	85	40	5.63	6.00	6.20	6.43	6.56	6.65
44	85	41	5.47	5.87	6.10	6.33	6.47	6.55
43	85	42	5.30	5.74	5.98	6.24	6.37	6.46
42	85	43	5.11	5.60	5.86	6.13	6.28	6.37
41	85	44	4.90	5.44	5.72	6.02	6.17	6.27
40	85	45	4.68	5.26	5.58	5.90	6.07	6.17
Condenser ΔT[b]			14.04	11.23	9.36	7.02	5.62	4.68

For SI: °C = [(°F) - 32] / 1.8, 1 gallon per minute = 3.785 L/min., 1 ton = 12,000 British thermal unit per hour = 3.517 kW.
a. Lift = Entering condenser water temperature (°F) - Leaving chilled water temperature (°F).
b. Condenser ΔT = Leaving condenser water temperature (°F) - Entering condenser water temperature (°F).

K_{adj} = 6.1507 - 0.030244(X) + 0.0062692(X)^2 - 0.000045595(X)
where: X = Condenser ΔT + Lift
COP_{adj} = K_{adj} x COP_{std}

TABLE 803.3.2(6)
PERFORMANCE REQUIREMENTS FOR HEAT REJECTION EQUIPMENT

EQUIPMENT TYPE	TOTAL SYSTEM HEAT REJECTION CAPACITY AT RATED CONDITIONS	SUBCATEGORY OR RATING CONDITION	PERFORMANCE REQUIRED[a, b]	TEST PROCEDURE[c]
Propeller or axial fan cooling towers	All	95°F entering water 85°F leaving water 75°F wb outdoor air	≥ 38.2 gpm/hp	CTI ATC-105 and CTI STD-201
Centrifugal fan cooling towers	All	95°F entering water 85°F leaving water 75°F wb outdoor air	≥ 20.0 gpm/hp	CTI ATC-105 and CTI STD-201
Air cooled condensers	All	125°F condensing temperature R-22 test fluid 190°F entering gas temperature 15°F subcooling 95°F entering db	≥ 176,000 Btu/h · hp (69 COP)	ARI 460

For SI: °C = [(°F) - 32] / 1.8, 1 British thermal unit per hour = 0.2931 W, 1 gallon per minute per horsepower = 0.846 L/s · kW.
wb = wet-bulb temperature, °F
a. For purposes of this table, cooling tower performance is defined as the maximum flow rating of the tower units (gpm) divided by the fan nameplate rated motor power units (hp).
b. For purposes of this table, air-cooled condenser performance is defined as the heat rejected from the refrigerant units (Btu/h) divided by the fan nameplate rated motor power units (hp).
c. Chapter 10 contains a complete specification of the referenced test procedure, including the referenced year version of the test procedure.

803.3.3.3.1 Thermostatic setback capabilities. Thermostatic setback controls shall have the capability to set back or temporarily operate the system to maintain zone temperatures down to 55°F (13°C) or up to 85°F (29°C).

803.3.3.3.2 Automatic setback and shutdown capabilities. Automatic time clock or programmable controls shall be capable of starting and stopping the system for seven different daily schedules per week and retaining their programming and time setting during a loss of power for at least 10 hours. Additionally, the controls shall have: a manual override that allows temporary operation of the system for up to 2 hours; a manually operated timer capable of being adjusted to operate the system for up to 2 hours; or an occupancy sensor.

803.3.3.4 Shutoff damper controls. Both outdoor air supply and exhaust ducts shall be equipped with gravity or motorized dampers that will automatically shut when the systems or spaces served are not in use.

Exception: Individual supply systems with a design airflow rate of 3,000 cfm (1416 L/s) or less.

803.3.3.5 Economizers. Supply air economizers shall be provided on each cooling system according to Table 803.2.6(1). Economizers shall be capable of operating at 100 percent outside air, even if additional mechanical cooling is required to meet the cooling load of the building.

Exceptions:

1. Systems utilizing water economizers that are capable of cooling supply air by direct or indirect evaporation or both and providing 100 percent of the expected system cooling load at outside air temperatures of 50°F (10°C) dry bulb/45°F (7°C) wet bulb and below.

2. Where the cooling equipment is covered by the minimum efficiency requirements of Table 803.2.2(1), 803.2.2(2), or 803.3.2(1) and meets or exceeds the minimum EER by the percentages shown in Table 803.2.6(2)

3. Where the cooling equipment is covered by the minimum efficiency requirements of Table 803.3.2(2) and meets or exceeds the minimum integrated part load value (IPLV) by the percentages shown in Table 803.2.6(2).

803.3.3.6 Variable air volume (VAV) fan control. Individual VAV fans with motors of 25 horsepower (18.8 kW) or greater shall be:

1. Driven by a mechanical or electrical variable speed drive; or

2. The fan motor shall have controls or devices that will result in fan motor demand of no more than 30 percent of their design wattage at 50 percent of

design air flow when static pressure set point equals one-third of the total design static pressure, based on manufacturer's certified fan data.

803.3.3.7 Hydronic systems controls. The heating of fluids that have been previously mechanically cooled and the cooling of fluids that have been previously mechanically heated shall be limited in accordance with Sections 803.3.3.7.1 through 803.3.3.7.3. Hydronic heating systems comprised of multiple-packaged boilers and designed to deliver conditioned water or steam into a common distribution system shall include automatic controls capable of sequencing operation of the boilers. Hydronic heating systems comprised of a single boiler and greater than 500,000 Btu/h input design capacity shall include either a multistaged or modulating burner.

803.3.3.7.1 Three-pipe system. Hydronic systems that use a common return system for both hot water and chilled water are prohibited.

803.3.3.7.2 Two-pipe changeover system. Systems that use a common distribution system to supply both heated and chilled water shall be designed to allow a dead band between changeover from one mode to the other of at least 15°F (8.3°C) outside air temperatures; be designed to and provided with controls that will allow operation in one mode for at least 4 hours before changing over to the other mode; and be provided with controls that allow heating and cooling supply temperatures at the changeover point to be no more than 30°F (16.7°C) apart.

803.3.3.7.3 Hydronic (water loop) heat pump systems. Hydronic heat pumps connected to a common heat pump water loop with central devices for heat rejection and heat addition shall have controls that are capable of providing a heat pump water supply temperature dead band of at least 20°F (11.1°C) between initiation of heat rejection and heat addition by the central devices. For Climate Zones 3 through 8 as indicated in Figure 301.1 and Table 301.1, if a closed-circuit cooling tower is used, either an automatic valve shall be installed to bypass all but a minimal flow of water around the tower, or lower leakage positive closure dampers shall be provided. If an open-circuit tower is used directly in the heat pump loop, an automatic valve shall be installed to bypass all heat pump water flow around the tower. If an open-circuit cooling tower is used in conjunction with a separate heat exchanger to isolate the cooling tower from the heat pump loop, then heat loss shall be controlled by shutting down the circulation pump on the cooling tower loop. Each hydronic heat pump on the hydronic system having a total pump system power exceeding 10 horsepower (hp) (7.5 kW) shall have a two-position valve.

> **Exception:** Where a system loop temperature optimization controller is installed and can determine the most efficient operating temperature based on real time conditions of demand and capacity, dead bands of less than 20°F (11.1°C) shall be permitted.

803.3.3.7.4 Part load controls. Hydronic systems greater than or equal to 300,000 Btu/h (87,930 W) in design output capacity supplying heated or chilled water to comfort conditioning systems shall include controls that have the capability to:

1. Automatically reset the supply-water temperatures using zone-return water temperature, building-return water temperature, or outside air temperature as an indicator of building heating or cooling demand. The temperature shall be capable of being reset by at least 25 percent of the design supply-to-return water temperature difference; or

2. Reduce system pump flow by at least 50 percent of design flow rate utilizing adjustable speed drive(s) on pump(s), or multiple-staged pumps where at least one-half of the total pump horsepower is capable of being automatically turned off or control valves designed to modulate or step down, and close, as a function of load, or other approved means.

803.3.3.7.5 Pump isolation. Chilled water plants including more than one chiller shall have the capability to reduce flow automatically through the chiller plant when a chiller is shut down. Chillers piped in series for the purpose of increased temperature differential, shall be considered as one chiller.

Boiler plants including more than one boiler shall have the capability to reduce flow automatically through the boiler plant when a boiler is shut down.

803.3.3.8 Heat rejection equipment fan speed control. Each fan powered by a motor of 7.5 hp (5.6 kW) or larger shall have the capability to operate that fan at two-thirds of full speed or less, and shall have controls that automatically change the fan speed to control the leaving fluid temperature or condensing temperature/pressure of the heat rejection device.

> **Exception:** Factory-installed heat rejection devices within HVAC equipment tested and rated in accordance with Tables 803.3.2(1) through 803.3.2(6).

803.3.4 Requirements for complex mechanical systems serving multiple zones. Sections 803.3.4.1 through 803.3.4.3 shall apply to complex mechanical systems serving multiple zones. Supply air systems serving multiple zones shall be VAV systems which, during periods of occupancy, are designed and capable of being controlled to reduce primary air supply to each zone to one of the following before reheating, recooling or mixing takes place:

1. Thirty percent of the maximum supply air to each zone.

2. Three hundred cfm (142 L/s) or less where the maximum flow rate is less than 10 percent of the total fan system supply airflow rate.

3. The minimum ventilation requirements of Chapter 4 of the *International Mechanical Code*.

Exception: The following define when individual zones or when entire air distribution systems are exempted from the requirement for VAV control:

1. Zones where special pressurization relationships or cross-contamination requirements are such that VAV systems are impractical.

2. Zones or supply air systems where at least 75 percent of the energy for reheating or for providing warm air in mixing systems is provided from a site-recovered or site-solar energy source.

3. Zones where special humidity levels are required to satisfy process needs.

4. Zones with a peak supply air quantity of 300 cfm (142 L/s) or less and where the flow rate is less than 10 percent of the total fan system supply airflow rate.

5. Zones where the volume of air to be reheated, recooled or mixed is no greater than the volume of outside air required to meet the minimum ventilation requirements of Chapter 4 of the *International Mechanical Code*.

6. Zones or supply air systems with thermostatic and humidistatic controls capable of operating in sequence the supply of heating and cooling energy to the zone(s) and which are capable of preventing reheating, recooling, mixing or simultaneous supply of air that has been previously cooled, either mechanically or through the use of economizer systems, and air that has been previously mechanically heated.

803.3.4.1 Single duct variable air volume (VAV) systems, terminal devices. Single duct VAV systems shall use terminal devices capable of reducing the supply of primary supply air before reheating or recooling takes place.

803.3.4.2 Dual duct and mixing VAV systems, terminal devices. Systems that have one warm air duct and one cool air duct shall use terminal devices which are capable of reducing the flow from one duct to a minimum before mixing of air from the other duct takes place.

803.3.4.3 Single fan dual duct and mixing VAV systems, economizers. Individual dual duct or mixing heating and cooling systems with a single fan and with total capacities greater than 90,000 Btu/h [(26 375 W) 7.5 tons] shall not be equipped with air economizers.

803.3.5 Ventilation. Ventilation shall be in accordance with Section 803.2.5.

803.3.6 Duct and plenum insulation and sealing. All ducts and plenums shall be insulated and sealed in accordance with Section 803.2.8.

Ducts designed to operate at static pressures in excess of 3 inches w.g. (746 Pa) shall be leak-tested in accordance with the SMACNA *HVAC Air Duct Leakage Test Manual* with the rate of air leakage (*CL*) less than or equal to 6.0 as determined in accordance with Equation 8-2.

$$CL = F \times P^{0.65} \qquad \text{(Equation 8-2)}$$

where:

F = The measured leakage rate in cfm per 100 square feet of duct surface.

P = The static pressure of the test.

Documentation shall be furnished by the designer demonstrating that representative sections totaling at least 25 percent of the duct area have been tested and that all tested sections meet the requirements of this section.

803.3.7 Piping insulation. All piping serving as part of a heating or cooling system shall be thermally insulated in accordance with Table 803.3.7.

Exceptions:

1. Factory-installed piping within HVAC equipment tested and rated in accordance with a test procedure referenced by this code.

2. Piping that conveys fluids that have a design operating temperature range between 55°F (13°C) and 105°F (41°C).

3. Piping that conveys fluids that have not been heated or cooled through the use of fossil fuels or electric power.

4. Runout piping not exceeding 4 feet (1219 mm) in length and 1 inch (25 mm) in diameter between the control valve and HVAC coil.

803.3.8 HVAC system completion. Prior to the issuance of a certificate of occupancy, the design professional shall provide evidence of system completion in accordance with Sections 803.3.8.1 through 803.3.8.3.

TABLE 803.3.7
MINIMUM PIPE INSULATION[a]
(thickness in inches)

FLUID	NOMINAL PIPE DIAMETER	
	≤ 1.5"	> 1.5"
Steam	1 ½	3
Hot water	1	2
Chilled water, brine or refrigerant	1	1 1/2

For SI: 1 inch = 25.4 mm, British thermal unit per inch/h · ft² · °F = W per 25 mm/K · m²

a. Based on insulation having a conductivity (k) not exceeding 0.27 Btu per inch/h · ft² · °F.

803.3.8.1 Air system balancing. Each supply air outlet and zone terminal device shall be equipped with means for air balancing in accordance with the requirements of Chapter 6 of the *International Mechanical Code*. Discharge dampers are prohibited on constant volume fans and variable volume fans with motors 25 hp (18.6 kW) and larger.

803.3.8.2 Hydronic system balancing. Individual hydronic heating and cooling coils shall be equipped with means for balancing and pressure test connections.

803.3.8.3 Manuals. The construction documents shall require that an operating and maintenance manual be provided to the building owner by the mechanical contractor. The manual shall include, at least, the following:

1. Equipment capacity (input and output) and required maintenance actions.

2. Equipment operation and maintenance manuals.

3. HVAC system control maintenance and calibration information, including wiring diagrams, schematics, and control sequence descriptions. Desired or field-determined setpoints shall be permanently recorded on control drawings, at control devices or, for digital control systems, in programming comments.

4. A complete written narrative of how each system is intended to operate.

803.3.9 Heat recovery for service water heating. Condenser heat recovery shall be installed for heating or reheating of service hot water provided the facility operates 24 hours a day, the total installed heat capacity of water-cooled systems exceeds 6,000,000 Btu/hr of heat rejection, and the design service water heating load exceeds 1,000,000 Btu/h.

The required heat recovery system shall have the capacity to provide the smaller of:

1. Sixty percent of the peak heat rejection load at design conditions; or

2. The preheating required to raise the peak service hot water draw to 85°F (29°C).

Exceptions:

1. Facilities that employ condenser heat recovery for space heating or reheat purposes with a heat recovery design exceeding 30 percent of the peak water-cooled condenser load at design conditions.

2. Facilities that provide 60 percent of their service water heating from site solar or site recovered energy or from other sources.

803.3.10 Energy recovery ventilation systems. Individual fan systems that have both a design supply air capacity of 5,000 cfm (2.36 m³/s) or greater and a minimum outside air supply of 70 percent or greater of the design supply air quantity shall have an energy recovery system that provides a change in the enthalpy of the outdoor air supply of 50 percent or more of the difference between the outdoor air and return air at design conditions. Provision shall be made to bypass or control the energy recovery system to permit cooling with outdoor air where cooling with outdoor air is required.

Exception: An energy recovery ventilation system shall not be required in any of the following conditions:

1. Where energy recovery systems are prohibited by the *International Mechanical Code*.

2. Laboratory fume hood systems with a total exhaust rate of 15,000 cfm (7.08 m³/s) or less.

3. Laboratory fume hood systems with a total exhaust rate greater than 15,000 cfm (7.08 m³/s) that include at least one of the following features:

   3.1. Variable-air-volume hood exhaust and room supply systems capable of reducing exhaust and makeup air volume to 50 percent or less of design values.

   3.2 Direct makeup (auxiliary) air supply equal to at least 75 percent of the exhaust rate, heated no warmer than 2°F

(1.1°C) below room set point, cooled to no cooler than 3 °F (1.7°C) above room set point, no humidification added, and no simultaneous heating and cooling used for dehumidification control.

4. Systems serving spaces that are not cooled and are heated to less than 60°F (15.5°C).

5. Where more than 60 percent of the outdoor heating energy is provided from site-recovered or site solar energy.

6. Heating systems in climates with less than 3600 HDD.

7. Cooling systems in climates with a 1 percent cooling design wet-bulb temperature less than 64 °F (17.7°C).

8. Systems requiring dehumidification that employ series-style energy recovery coils wrapped around the cooling coil.

SECTION 804
SERVICE WATER HEATING

804.1 General. This section covers the minimum efficiency of, and controls for, service water-heating equipment and insulation of service hot water piping.

804.2 Service water-heating equipment performance efficiency. Water-heating equipment and hot water storage tanks shall meet the requirements of Table 804.2. The efficiency shall be verified through data furnished by the manufacturer or through certification under an approved certification program.

804.3 Temperature controls. Service water-heating equipment shall be provided with controls to allow a setpoint of 110°F (43°C) for equipment serving dwelling units and 90°F (32°C) for equipment serving other occupancies. The outlet temperature of lavatories in public facility rest rooms shall be limited to 110°F (43°C).

TABLE 804.2
MINIMUM PERFORMANCE OF WATER-HEATING EQUIPMENT

EQUIPMENT TYPE	SIZE CATEGORY (input)	SUBCATEGORY OR RATING CONDITION	PERFORMANCE REQUIRED[a,b]	TEST PROCEDURE
Water heaters, Electric	≤ 12 W	Resistance	$0.93 - 0.00132V$, EF	DOE 10 CFR Part 430
	> 12 W	Resistance	$1.73V + 155$, SL, Btu/h	ANSI Z21.10.3
	≤ 24 amps and ≤ 250 volts	Heat pump	$0.93 - 0.00132V$, EF	DOE 10 CFR Part 430
Storage water heaters, Gas	≤ 75,000 Btu/h	≥ 20 gal	$0.62 - 0.0019V$, EF	DOE 10 CFR Part 430
	> 75,000 Btu/h and ≤ 155,000 Btu/h	< 4,000 Btu/h/gal	80% E_t $(Q/800 + 110\sqrt{V})$, SL, Btu/h	ANSI Z21.10.3
	> 155,000 Btu/h	< 4,000 Btu/h/gal	80% E_t $(Q/800 + 110\sqrt{V})$, SL, Btu/h	
Instantaneous water heaters, Gas	> 50,000 Btu/h and < 200,000 Btu/h[c]	≥ 4,000 Btu/h/gal and < 2 gal	$0.62 - 0.0019V$, EF	DOE 10 CFR Part 430
	≥ 200,000 Btu/h	≥ 4,000 Btu/h/gal and < 10 gal	80% E_t	ANSI Z21.10.3
	≥ 200,000 Btu/h	≥ 4,000 Btu/h/gal and ≥ 10 gal	80% E_t $(Q/800 + 110\sqrt{V})$, SL, Btu/h	
Storage water heaters, Oil	≤ 105,000 Btu/h	≥ 20 gal	$0.59 - 0.0019V$, EF	DOE 10 CFR Part 430
	> 105,000 Btu/h	< 4,000 Btu/h/gal	78% E_t $(Q/800 + 110\sqrt{V})$, SL, Btu/h	ANSI Z21.10.3
Instantaneous water heaters, Oil	≤ 210,000 Btu/h	≥ 4,000 Btu/h/gal and < 2 gal	$0.59 - 0.0019V$, EF	DOE 10 CFR Part 430
	> 210,000 Btu/h	≥ 4,000 Btu/h/gal and ≥ 10 gal	80% E_t	ANSI Z21.10.3
	> 210,000 Btu/h	≥ 4,000 Btu/h/gal and < 10 gal	78% E_t $(Q/800 + 110\sqrt{V})$, SL, Btu/h	
Hot water supply boilers, Gas and Oil	≥ 300,000 Btu/h and < 12,500,000 Btu/h	≥ 4,000 Btu/h/gal and < 10 gal	80% E_t	ANSI Z21.10.3
Hot water supply boilers, Gas and Oil		≥ 4,000 Btu/h/gal and ≥ 10 gal	80% E_t $(Q/800 + 110\sqrt{V})$, SL, Btu/h	
Pool heaters, Gas and Oil	All	—	78% E_t	ASHRAE 146
Unfired storage tanks	All	—	≤ 6.5 Btu/h · ft²	(none)

For SI: °C = [(°F) - 32] / 1.8 British thermal unit per hour = 0.2931 W, 1 gallon = 3.785 L, 1 British thermal unit per hour per gallon = 0.078 W/L.

a. Energy factor (EF) and thermal efficiency (E_t) are minimum requirements. In the EF equation, V is the rated volume in gallons.
b. Standby loss (SL) is the maximum Btu/h based on a nominal 70°F temperature difference between stored water and ambient requirements. In the SL equation, Q is the nameplate input rate in Btu/h. In the SL equation for electric water heaters, V is the rated volume in gallons. In the SL equation for oil and gas water heaters and boilers, V is the rated volume in gallons.
c. Instantaneous water heaters with input rates below 200,000 Btu/h must comply with these requirements if the water heater is designed to heat water to temperatures 180°F or higher.

804.4 Heat traps. Water-heating equipment not supplied with integral heat traps and serving noncirculating systems shall be provided with heat traps on the supply and discharge piping associated with the equipment.

804.5 Pipe insulation. For automatic-circulating hot water systems, piping shall be insulated with 1 inch (25 mm) of insulation having a conductivity not exceeding 0.27 Btu per inch/h × ft^2 × °F (1.53 W per 25 mm/m^2 × K). The first 8 feet (2438 mm) of piping in noncirculating systems served by equipment without integral heat traps shall be insulated with 0.5 inch (12.7 mm) of material having a conductivity not exceeding 0.27 Btu per inch/h × ft^2 × °F (1.53 W per 25 mm/m^2 × K).

804.6 Hot water system controls. Automatic-circulating hot water system pumps or heat trace shall be arranged to be conveniently turned off automatically or manually when the hot water system is not in operation.

804.7 Pools. Pools shall be provided with energy conserving measures in accordance with Sections 804.7.1 through 804.7.3.

804.7.1 Pool heaters. All pool heaters shall be equipped with a readily accessible on-off switch to allow shutting off the heater without adjusting the thermostat setting. Pool heaters fired by natural gas shall not have continuously burning pilot lights.

804.7.2 Time switches. Time switches that can automatically turn off and on heaters and pumps according to a preset schedule shall be installed on swimming pool heaters and pumps.

Exceptions:

1. Where public health standards require 24-hour pump operation.

2. Where pumps are required to operate solar-and waste-heat-recovery pool heating systems.

804.7.3 Pool covers. Heated pools shall be equipped with a vapor retardant pool cover on or at the water surface. Pools heated to more than 90°F (32°C) shall have a pool cover with a minimum insulation value of R-12 (R -2.1).

Exception: Pools deriving over 60 percent of the energy for heating from site-recovered energy or solar energy source.

SECTION 805
ELECTRICAL POWER AND LIGHTING SYSTEMS

805.1 General. This section covers lighting system controls, the connection of ballasts, the maximum lighting power for interior applications, and minimum acceptable lighting equipment for exterior applications.

Exception: Lighting within dwelling units.

805.2 Lighting controls. Lighting systems shall be provided with controls as required in Sections 805.2.1, 805.2.2, 805.2.3 and 805.2.4.

805.2.1 Interior lighting controls. Each area enclosed by walls or floor-to-ceiling partitions shall have at least one manual control for the lighting serving that area. The required controls shall be located within the area served by the controls or be a remote switch that identifies the lights served and indicates their status.

Exceptions:

1. Areas designated as security or emergency areas that must be continuously lighted.

2. Lighting in stairways or corridors that are elements of the means of egress.

805.2.2 Additional controls. Each area that is required to have a manual control shall have additional controls that meet the requirements of Sections 805.2.2.1 and 805.2.2.2.

Exceptions:

1. Areas that have only one luminaire.

2. Areas that are controlled by an occupant-sensing device.

3. Corridors, storerooms, restrooms or public lobbies.

4. Guestrooms (See Section 805.2.3).

805.2.2.1 Light reduction controls. Each area that is required to have a manual control shall also allow the occupant to reduce the connected lighting load in a reasonably uniform illumination pattern by at least 50 percent. Lighting reduction shall be achieved by one of the following or other approved method:

1. Controlling all lamps or luminaires;

2. Dual switching of alternate rows of luminaires, alternate luminaires or alternate lamps;

3. Switching the middle lamp luminaires independently of the outer lamps; or

4. Switching each luminaire or each lamp.

Exceptions:

1. Areas that have only one luminaire.

2. Areas that are controlled by an occupant-sensing device.

3. Corridors, storerooms, restrooms or public lobbies.

4. Guestrooms (see Section 805.2.3).

5. Spaces that use less than 0.6 watts per square foot (6.5 W/m²).

805.2.2.2 Automatic lighting shutoff. Buildings larger than 5,000 square feet (465 m²) shall be equipped with an automatic control device to shut off lighting in those areas. This automatic control device shall function on either:

1. A scheduled basis, using time-of-day, with an independent program schedule that controls the interior lighting in areas that do not exceed 25,000 square feet (2323 m²) and are not more than one floor; or

2. An unscheduled basis by occupant intervention.

805.2.2.2.1 Occupant override. Where an automatic time switch control device is installed to comply with Section 805.2.2.2, Item 1, it shall incorporate an override switching device that:

1. Is readily accessible.

2. Is located so that a person using the device can see the lights or the area controlled by that switch, or so that the area being lit is annunciated.

3. Is manually operated.

4. Allows the lighting to remain on for no more than 2 hours when an override is initiated.

5. Controls an area not exceeding 5,000 square feet (465 m²).

Exceptions:

1. In malls and arcades, auditoriums, single-tenant retail spaces, industrial facilities and arenas, where captive-key override is utilized, override time may exceed 2 hours.

2. In malls and arcades, auditoriums, single-tenant retail spaces, industrial facilities and arenas, the area controlled may not exceed 20,000 square feet (1860 m²).

805.2.2.2.2 Holiday scheduling. If an automatic time switch control device is installed in accordance with Section 805.2.2.2, Item 1, it shall incorporate an automatic holiday scheduling feature that turns off all loads for at least 24 hours, then resumes the normally scheduled operation.

Exception: Retail stores and associated malls, restaurants, grocery stores, churches and theaters.

805.2.3 Guestrooms. Guestrooms in hotels, motels, boarding houses or similar buildings shall have at least one master switch at the main entry door that controls all permanently wired luminaires and switched receptacles, except those in the bathroom(s). Suites shall have a control meeting these requirements at the entry to each room or at the primary entry to the suite.

805.2.4 Exterior lighting controls. Automatic switching or photocell controls shall be provided for all exterior lighting not intended for 24-hour operation. Automatic time switches shall have a combination seven-day and seasonal daylight program schedule adjustment, and a minimum 4-hour power backup.

805.3 Tandem wiring. The following luminaires located within the same area shall be tandem wired:

1. Flourescent luminaires equipped with one, three or odd- numbered lamp configurations, that are recess-mounted within 10 feet (3048 mm) center-to-center of each other.

2. Flourescent luminaires equipped with one, three or any other odd-numbered lamp configuration, that are pendant- or surface-mounted within 1 foot (305 mm) edge-to-edge of each other.

Exceptions:

1. Where electronic high-frequency ballasts are used.

2. Luminaires on emergency circuits.

3. Luminaires with no available pair in the same area.

805.4 Exit signs. Internally illuminated exit signs shall not exceed 5 Watts per side.

805.5 Interior lighting power requirements. A building complies with this section if its total connected lighting power calculated under Section 805.5.1 is no greater than the interior lighting power calculated under Section 805.5.2.

805.5.1 Total connected interior lighting power. The total connected interior lighting power (Watts) shall be the sum of the watts of all interior lighting equipment as determined in accordance with Sections 805.5.1.1 through 805.5.1.4.

Exceptions: The connected power associated with the following lighting equipment is not included in calculating total connected lighting power.

1. Specialized medical, dental and research lighting.

2. Professional sports arena playing field lighting.

3. Display lighting for exhibits in galleries, museums and monuments.

4. Guestroom lighting in hotels, motels, boarding houses or similar buildings.

5. Emergency lighting automatically off during normal building operation.

805.5.1.1 Screw lamp holders. The wattage shall be the maximum labeled wattage of the luminaire.

805.5.1.2 Low-voltage lighting. The wattage shall be the specified wattage of the transformer supplying the system.

805.5.1.3 Other luminaires. The wattage of all other lighting equipment shall be the wattage of the lighting equipment verified through data furnished by the manufacturer or other approved sources.

805.5.1.4 Line-voltage lighting track and plug-in busway. The wattage shall be the greater of the wattage of the luminaires determined in accordance with Sections 805.5.1.1 through 805.5.1.3 or 30 W/linear foot (98W/lin m).

805.5.2 Interior lighting power. The interior lighting power shall be calculated using Section 805.5.2.1 or 805.5.2.2 as applicable.

805.5.2.1 Entire building method. Under this approach, the interior lighting power (Watts) is the value from Table 805.5.2 for the building type times the floor area of the entire building. The interior lighting power (Watts) shall not be increased by the allowances contained in the footnotes of Table 805.5.2 when using the entire building method.

805.5.2.2 Tenant area or portion of building method. The total interior lighting power (Watts) is the sum of all interior lighting powers for all areas in the building covered in this permit. The interior lighting power is the floor area for each area type listed in Table 805.5.2 times the value from Table 805.5.2 for that area. For the purposes of this method, an "area" shall be defined as all contiguous spaces that accommodate or are associated with a single area type as listed in Table 805.5.2. When this method is used to calculate the total interior lighting power for an entire building, each area type shall be treated as a separate area.

805.6 Exterior lighting. When the power for exterior lighting is supplied through the energy service to the building, all exterior lighting, other than low-voltage landscape lighting, shall have a source efficacy of at least 45 lumens per Watt.

Exception: Where approved because of historical, safety, signage or emergency considerations.

805.7 Electrical energy consumption. In buildings having individual dwelling units, provisions shall be made to determine the electrical energy consumed by each tenant by separately metering individual dwelling units.

SECTION 806
TOTAL BUILDING PERFORMANCE

806.1 General. The proposed design complies with this section where annual energy costs of the proposed design as determined in accordance with Section 806.3 do not exceed those of the standard design as determined in accordance with Section 806.4.

TABLE 805.5.2
INTERIOR LIGHTING POWER

BUILDING OR AREA TYPE	ENTIRE BUILDING (W/ft²)	TENANT AREA OR PORTION OF BUILDING (W/ft²)
Auditorium	Not Applicable	1.8
Automotive facility	0.9	Not Applicable
Bank/financial institution[a]	Not Applicable	1.5
Classroom/lecture hall[b]	Not Applicable	1.4
Convention, conference or meeting center[a]	1.2	1.3
Corridor, restroom, support area	Not Applicable	0.9
Courthouse/town hall	1.2	Not Applicable
Dining[a]	Not Applicable	0.9
Dormitory	1.0	Not Applicable
Exercise center[a]	1.0	0.9
Exhibition hall	Not Applicable	1.3
Grocery store[c]	1.5	1.6
Gymnasium playing surface	Not Applicable	1.4
Hotel function[a]	1.0	1.3
Industrial work, < 20-foot ceiling height	Not Applicable	1.2
Industrial work, ≥ 20-foot ceiling height	Not Applicable	1.7
Kitchen	Not Applicable	1.2
Library[a]	1.3	1.7
Lobby - hotel[a]	Not Applicable	1.1
Lobby - other[a]	Not Applicable	1.3
Mall, arcade or atrium	Not Applicable	0.6
Medical and clinical care[b, d]	1.2	1.2
Motel	1.0	Not Applicable
Multifamily	0.7	Not Applicable
Museum[b]	1.1	1.0
Office[b]	1.0	1.1
Parking garage	0.3	Not Applicable
Penitentiary	1.0	Not Applicable
Police/fire station	1.0	Not Applicable
Post office	1.1	Not Applicable

Table 805.5.2 (continued)

BUILDING OR AREA TYPE	ENTIRE BUILDING (W/ft²)	TENANT AREA OR PORTION OF BUILDING (W/ft²)
Religious worship[a]	1.3	2.4
Restaurant[a]	1.6	0.9
Retail sales, wholesale showroom[c]	1.5	1.7
School	1.2	Not Applicable
Storage, industrial and commercial	0.8	0.8
Theaters - motion picture	1.2	1.2
Theaters - performance[a]	1.6	2.6
Transportation	1.0	Not Applicable
Other	0.6	1.0

For SI: 1 foot = 304.8 mm, 1 Watts per square foot = W/0.0929 m²

a. Where lighting equipment is specified to be installed for decorative appearances in addition to lighting equipment specified for general lighting and is switched or dimmed on circuits different from the circuits for general lighting, the smaller of the actual wattage of the decorative lighting equipment or 1.0 W/ft² times the area of the space that the decorative lighting equipment is in shall be added to the interior lighting power determined in accordance with this line item.
b. Where lighting equipment is specified to be installed to meet requirements of visual display terminals as the primary viewing task, the smaller of the actual wattage of the lighting equipment or 0.35 W/ft² times the area of the space that the lighting equipment is in shall be added to the interior lighting power determined in accordance with this line item.
c. Where lighting equipment is specified to be installed to highlight specific merchandise in addition to lighting equipment specified for general lighting and is switched or dimmed on circuits different from the circuits for general lighting, the smaller of the actual wattage of the lighting equipment installed specifically for merchandise, or 1.6 W/ft² times the area of the specific display, or 3.9 W/ft² times the actual case or shelf area for displaying and selling fine merchandise such as jewelry, fine apparel and accessories, or china and silver, shall be added to the interior lighting power determined in accordance with this line item.
d. Where lighting equipment is specified to be installed, the smaller of the actual wattage of the lighting equipment, or 1.0 W/ft² times the area of the emergency, recovery, medical supply and pharmacy space shall be added to the interior lighting power determined in accordance with this line item.

806.2 Analysis procedures. Sections 806.2.1 through 806.2.8 shall be applied in determining total building performance.

806.2.1 Energy analysis. Annual (8,760 hours) energy costs for the standard design and the proposed design shall each be determined using the same approved energy analysis simulation tool.

806.2.2 Climate data. The climate data used in the energy analysis shall cover a full calendar year (8,760 hours) and shall reflect approved coincident hourly data for temperature, solar radiation, humidity and wind speed for the building location.

806.2.3 Energy rates. The annual energy costs shall be estimated using energy rates published by the serving energy supplier and which would apply to the actual building or *DOE State-Average Energy Prices* published by DOE's Energy Information Administration and which would apply to the actual building.

806.2.4 Nondepletable energy. Nondepletable energy collected off site shall be treated and priced the same as purchased energy. Energy from nondepletable energy sources collected on site shall be omitted from the annual energy cost of the proposed design. The analysis and performance of any nondepletable energy system shall be determined in accordance with accepted engineering practice using approved methods.

806.2.5 Building operation. Building operation shall be simulated for a full calendar year (8,760 hours). Operating schedules shall include hourly profiles for daily operation and shall account for variations between weekdays, weekends, holidays, and any seasonal operation. Schedules shall model the time-dependent variations of occupancy, illumination, receptacle loads, thermostat settings, mechanical ventilation, HVAC equipment availability, service hot water usage, and any process loads.

806.2.6 Simulated loads. The following systems and loads shall be modeled in determining total building performance: heating systems, cooling systems, fan systems, lighting power, receptacle loads, and process loads that exceed 1.0 W/ft^2 (W/0.0929 m^2) of floor area of the room or space in which the process loads are located.

> **Exception:** Systems and loads serving required emergency power only.

806.2.7 Service water-heating systems. Service water-heating systems that are other than combined service hot water/space-heating systems shall be be omitted from the energy analysis provided all requirements in Section 804 have been met.

806.2.8 Exterior lighting. Exterior lighting systems shall be the same as in the standard and proposed designs.

806.3 Determining energy costs for the proposed design. Building systems and loads shall be simulated in the Proposed design in accordance with Sections 806.3.1 and 806.3.2.

806.3.1 HVAC and service water-heating equipment. All HVAC and service water-heating equipment shall be simulated in the proposed design using capacities, rated efficiencies and part-load performance data for the proposed equipment as provided by the equipment manufacturer.

806.3.2 Features not documented at time of permit. If any feature of the proposed design is not included in the building permit application, the energy performance of that feature shall be assumed to be that of the corresponding feature used in the calculations required in Section 806.4.

806.4 Determining energy costs for the standard design. Sections 806.4.1 through 806.4.7 shall be used in determining the annual energy costs of the Standard design.

806.4.1 Equipment efficiency. The space-heating, space-cooling, service water-heating, and ventilation systems and equipment shall meet, but not exceed, the minimum efficiency requirements of Sections 803 and 804.

806.4.2 HVAC system capacities. HVAC system capacities in the standard design shall be established such that no smaller number of unmet heating and cooling load hours and no larger heating and cooling capacity safety factors are provided than in the proposed design.

806.4.3 Envelope. The performance of elements of the thermal envelope of the standard design shall be determined in accordance with the requirements of Section 802.2 as applicable.

806.4.4 Identical characteristics. The heating/cooling system zoning, the orientation of each building feature, the number of floors and the gross envelope areas of the standard design shall be the same as those of the proposed design except as modified by Section 806.4.5 or 806.4.6.

> **Exception:** Permanent fixed or movable external shading devices for windows and glazed doors shall be excluded from the standard design.

806.4.5 Window area. The window area of the standard design shall be the same as the proposed design, or 35 percent of the above-grade wall area, whichever is less, and shall be distributed in a uniform pattern equally over each building facade.

806.4.6 Skylight area. The skylight area of the standard design shall be the same as the proposed design, or 3 percent of the gross area of the roof assembly, whichever is less.

806.4.7 Interior lighting. The lighting power for the standard design shall be the maximum allowed in accordance with Section 805.5. Where the occupancy of the building is not known, the lighting power density shall be 1.5 Watts per square foot (16.1 W/m^2).

806.5 Documentation. The energy analysis and supporting documentation shall be prepared by a registered design professional where required by the statutes of the jurisdiction in which the project is to be constructed. The information documenting compliance shall be submitted in accordance with Sections 806.5.1 through 806.5.4

806.5.1 Annual energy use and associated costs. The annual energy use and costs by energy source of the standard design and the proposed design shall be clearly indicated.

806.5.2 Energy-related features. A list of the energy-related features that are included in the proposed design and on which compliance with the provisions of the code are claimed shall be provided to the code official. This list shall include and prominently indicate all features that differ from those set forth in Section 806.4 and used in the energy analysis between the standard design and the proposed design.

806.5.3 Input and output report(s). Input and output report(s) from the energy analysis simulation program containing the complete input and output files, as applicable. The output file shall include energy use totals and energy use by energy source and end-use served, total hours that space conditioning loads are not met and any errors or warning messages generated by the simulation tool as applicable.

806.5.4 Written explanation(s). An explanation of any error or warning messages appearing in the simulation tool output shall be provided in a written, narrative format.

Delete Chapter 9 in its entirety. Climate maps revised and relocated to Table 301.1 and Figure 301.1: (EC48-03/04)

2004 SUPPLEMENT TO THE IECC

CHAPTER 10
REFERENCED STANDARDS

Delete Chapter 10 in its entirety and replace as shown:

This chapter lists the standards that are referenced in various sections of this document. The standards are listed by the promulgating agency of the standard, the standard identification, the effective date and title, and the section or sections of this document that reference the standard. The application of the referenced standards shall be specified in Section 107.

AAMA

American Architectural Manufacturers Association
1827 Walden Office Square
Suite 104
Schaumburg, IL 60173-4268

Standard reference number	Title	Referenced in code section number
101/I.S.2-97	Voluntary Specifications for Aluminum, Vinyl (PVC) and Wood Windows and Glass Doors	402.4.2, 802.3.1
101/I.S.2/NAFS-02	Voluntary Performance Specifications for Windows, Skylights and Glass Doors	402.4.2, 802.3.1

AMCA

Air Movement and Control Association International
30 West University Drive
Arlington Heights, IL 60004-1806

Standard reference number	Title	Referenced in code section number
500-D-98	Laboratory Methods for Testing Dampers for Rating	802.3.4

ANSI

American National Standards Institute
25 West 43rd Street - 4th Floor
New York, NY 10036

Standard reference number	Title	Referenced in code section number
Z21.10.3-01	Gas Water Heaters - Volume III - Storage Water Heaters with Input Ratings Above 75,000 Btu Per Hour, Circulating and Instantaneous	Table 804.2
Z21.13-00	Gas-Fired Low Pressure Steam and Hot Water Boilers - with Addenda Z21.13a	Table 803.2.2(5)
Z21.47-01	Gas-Fired Central Furnaces - with Addenda Z21.47a- 2001 and Z21.47b	Table 803.2.2(4)
Z83.8-02	Gas Unit Heaters and Gas-Fired Duct Furnaces	Table 803.2.2(4)

ARI

Air Conditioning and Refrigeration Institute
4100 North Fairfax Drive, Suite 200
Arlington, VA 22203

Standard reference number	Title	Referenced in code section number
210/240-03	Unitary Air-Conditioning and Air-Source Heat Pump Equipment	Table 803.2.2(1), Table 803.2.2.(2)
310/380-93	Standard for Packaged Terminal Air-Conditioning and Heat Pumps	Table 803.2.2.(3)
340/360-2000	Commercial and Industrial Unitary Air-Conditioning and Heat Pump Equipment	Table 803.2.2(1), Table 803.2.2(2)
365-02	Commercial and Industrial Unitary Air-Conditioning Condensing Units	Table 803.3.2(1)
460-00	Remote Mechanical-Draft Air-Cooled Refrigerant Condensers	Table 803.3.2(6)
550/590-98	Water Chilling Packages Using the Vapor Compression Cycle w/ Addenda	Table 803.3.2(2)
560-00	Absorption Water Chilling Packages	Table 803.3.2(2)
13256-1(1998)	Water-source Heat Pumps - Testing and Rating for Performance - Part 1: Water-to-Air and Brine-to-Air Heat Pumps	Table 803.2.2(2)

ASHRAE

American Society of Heating, Refrigerating and Air-Conditioning Engineers, Inc.
1791 Tullie Circle, NE
Atlanta, GA 30329-2305

Standard reference number	Title	Referenced in code section number
119-88 (Reaffirmed 1994)	Air Leakage Performance for Detached Single-Family Residential Buildings	Table 404.5.2(1)
146-1998	Testing and Rating Pool Heaters	Table 804.2
13256-1(1998)	Water-Source Heat Pumps - Testing and Rating for Performance - Part 1: Water-to-Air and Brine-to-Air Heat Pumps	Table 803.2.2(2)
90.1-2001	Energy Standard for Buildings Except Low-Rise Residential Buildings	801.1, 801.2, 802.1, 802.2, Table 802.2(3)
ASHRAE-2001	ASHRAE Handbook of Fundamentals - 2001	402.1.3, Table 404.5.2(1), 803.2.1
ASHRAE-2000	ASHRAE HVAC Systems and Equipment Handbook-2000	803.2.1

ASME

American Society of Mechanical Engineers
Three Park Avenue
New York, NY 10016-5990

Standard reference number	Title	Referenced in code section number
PTC 4.1 - 1964	Steam Generating Units	Table 803.2.2(5)

ASTM

ASTM International
100 Barr Harbor Drive
West Conshohocken, PA   19428-2859

Standard reference number	Title	Referenced in code section number
C90-01A	Specification for Loadbearing Concrete Masonry Units	Table 802.2(1)
E 96-00e01	Standard Test Method for Water Vapor Transmission of Materials	202, 802.1.2
E283--99	Test Method for Determining the Rate of Air Leakage Through Exterior Windows, Curtain Walls and Doors Under Specified Pressure Differences Across the Specimen	402.4.3, 802.3.2, 802.3.7

CTI

Cooling Technology Institue
2611 FM 1960 West, Suite H-200
Houston, TX   77068-3730

Standard reference number	Title	Referenced in code section number
ATC-105 (2000)	Acceptance Test Code	Table 803.3.2(6)
STD-201 (2002)	Certification Standard for Commercial Water Cooling Towers	Table 803.3.2(6)

DOE

U.S. Department of Energy
C/o Superintendent of Documents
U.S. Government Printing Office
Washington, DC   20402-9325

Standard reference number	Title	Referenced in code section number
10 CFR Part 430 Subpart B, Appendix E (1998)	Uniform Test Method for Measuring the Energy Consumption of Water Heaters	Table 804.2
10 CFR Part 430 Subpart B, Appendix N (1998)	Uniform Test Method for Measuring the Energy Consumption of Furnaces and Boilers	Table 803.2.2(4), Table 803.2.2(5)
DOE/EIA –0376 (Current Edition)	State Energy Prices and Expenditure Report	806.2.3

HI

Hydronics Institute, Division of the Gas Appliance Manufacturers Association.
P.O. Box 218
Berkeley Heights, NJ 07054

Standard reference number	Title	Referenced in code section number
HBS	I=B=R - Testing and Rating Standard for Heating Boilers, 1989 Ed	Table 803.2.2(5)

ICC

International Code Council, Inc.
5203 Leesburg Pike, Suite 600
Falls Church, VA 22041-3401

Standard reference number	Title	Referenced in code section number
IBC–03	International Building Code®	102.2, 201.3
ICC EC–03	ICC Electrical Code™	201.3
IEBC–03	International Existing Building Code™	101.2
IFC–03	International Fire Code®	201.3
IFGC–03	International Fuel Gas Code®	201.3
IMC–03	International Mechanical Code®	201.3, 803.2.5, 803.2.5.1, 803.2.6, 803.2.8.1, 803.2.8.1.1, 803.2.8.1.2, 803.3.4, 803.3.8.1, 803.3.10
IPC–03	International Plumbing Code®	201.3
IRC-03	International Residential Code®	201.3, 403.2.2, 403.6, 404.6.1, Table 404.5.2(1)

IESNA

Illuminating Engineering Society of North America
120 Wall Street, 17th Floor
New York, NY 10005-4001

Standard reference number	Title	Referenced in code section number
90.1-2001	Energy Standard for Buildings Except Low-Rise Residential Buildings	801.1, 801.2, 802.1, 802.2, Table 802.2(3)

NFRC

National Fenestration Rating Council, Inc.
8484 Georgia Avenue, Suite 230
Silver Spring, MD 20910

Standard reference number	Title	Referenced in code section number
100-01	Procedure for Determining Fenestration Products U-Factors - Second Edition	102.1.3
200-01	Procedure for Determining Fenestration Product Solar Heat Gain Coefficients and Visible Transmittance at Normal Incidence - Second Edition	102.1.3
400-01	Procedure for Determining Fenestration Product Air Leakage - Second Edition	402.4.2, 802.3.1

SMACNA

Sheet Metal and Air Conditioning Contractors National Association, Inc.
4021 Lafayette Center Drive
Chantilly, VA 20151-1209

Standard reference number	Title	Referenced in code section number
SMACNA-85	HVAC Air Duct Leakage Test Manual	803.3.6

UL

Underwriters Laboratories, Inc.
333 Pfingsten Road
Northbrook, IL 60062-2096

Standard reference number	Title	Referenced in code section number
181A-98	Closure Systems for Use with Rigid Air Ducts and Air Connectors With Revisions through December 1998	803.2.8
181B-95	Closure Systems for Use with Flexible Air Ducts and Air Connectors With Revisions through May 2000	803.2.8
727-98	Oil-Fired Central Furnaces - with Revisions through January 1999	Table 803.2.2(4)
731-95	Oil-Fired Unit Heaters - with Revisions through January 1999	Table 803.2.2(4)

WDMA

Window and Door Manufacturers Association
1400 East Touhy Avenue, Suite 470
Des Plaines, IL 60018

Standard reference number	Title	Referenced in code section number
101/I.S.2-97	Voluntary Specifications for Aluminum, Vinyl (PVC) and Wood Windows And Glass Doors	402.4.2, 802.3.1
101/I.S.2/NAFS-02	Voluntary Specifications for Windows, Skylights and Glass Doors	402.4.2, 802.3.1

Delete Appendix in its entirety (EC48-03/04)

International Existing Building Code

2004 Supplement

INTERNATIONAL EXISTING BUILDING CODE 2004 SUPPLEMENT

CHAPTER 1
ADMINISTRATION

Section 104.10.1 Add new section to read as shown: (EB4-03/04)

104.10.1 Flood hazard areas. For existing buildings located in flood hazard areas for which repairs, alterations, and additions constitute substantial improvement, the code official shall not grant modifications to provisions related to flood resistance unless a determination is made that:

1. The applicant has presented good and sufficient cause that the unique characteristics of the size, configuration or topography of the site render compliance with the flood-resistant construction provisions inappropriate.

2. Failure to grant the modification would result in exceptional hardship.

3. The granting of the modification will not result in increased flood heights, additional threats to public safety, extraordinary public expense nor create nuisances, cause fraud on or victimization of the public or conflict with existing laws or ordinances.

4. The modification is the minimum necessary to afford relief, considering the flood hazard.

5. A written notice will be provided to the applicant specifying, if applicable, the difference between the design flood elevation and the elevation to which the building is to be built, stating that the cost of flood insurance will be commensurate with the increased risk resulting from the reduced floor elevation, and stating that construction below the design flood elevation increases risks to life and property.

Section 106.3.2 Change to read as shown: (EB6-03/04)

106.3.2 Previous approval. This code shall not require changes in the construction documents, construction or designated occupancy of a structure for which a lawful permit has been issued, and the construction of which has been pursued in good faith within 180 days after the effective date of this code and has not been abandoned.

CHAPTER 4
REPAIRS

Sections 407.1.1.2 and 407.1.1.3 Change to read as shown: (EB11-03/04)

407.1.1.2 IBC level seismic forces. When seismic forces are required to meet the *International Building Code* level, they shall be based on 100 percent of the values in the *International Building Code* or FEMA 356. Where the *International Building Code* is used, the R factor used for analysis in accordance with Chapter 16 of the *International Building Code* shall be the R factor specified for structural systems classified as "Ordinary" in accordance with Table 1617.6.2 unless it can be demonstrated that the structural system satisfies the proportioning and detailing requirements for systems classified as "Intermediate" or "Special". Where FEMA 356 is used, the FEMA 356 Basic Safety Objective (BSO) shall be used for buildings in Seismic Use Group I. For buildings in other Seismic Use Groups the applicable FEMA 356 performance levels shown in Table 407.1.1.2 for BSE-1 and BSE-2 Earthquake Hazard Levels shall be used.

407.1.1.3 Reduced IBC level seismic forces. When seismic forces are permitted to meet reduced *International Building Code* levels, they shall be based on 75 percent of the assumed forces prescribed in the *International Building Code*, applicable chapters in Appendix A of this code (GSREB), the applicable performance level of ASCE 31 as shown in Table 407.1.1.2, or the applicable performance level for the BSE-1 Earthquake Hazard Level of FEMA 356 shown in Table 407.1.1.2. Where the *International Building Code* is used, the R factor used for analysis in accordance with Chapter 16 of the *International Building Code* shall be the R factor as specified in Section 407.1.1.2 of this code.

CHAPTER 6
ALTERATIONS—LEVEL 2

Section 604.2.1 Change to read as shown: (EB13-03/04)

604.2.1 High-rise buildings. In high-rise buildings, work areas that include exits or corridors shared by more than one tenant or that serve an occupant load greater than 30 shall be provided with automatic sprinkler protection in the entire work area where the work area is located on a floor that has a sufficient sprinkler water supply system from an existing standpipe or a sprinkler riser serving that floor.

Section 605.3.1.1 Change to read as shown: (EB20-03/04)

605.3.1.1 Single-exit buildings. Only one exit is required from buildings and spaces of the following occupancies:

1. In Group A, B, E, F, M, U and S occupancies, a single exit is permitted in the story at the level of exit discharge when the occupant load of the story does not exceed 50 and the exit access travel distance does not exceed 75 feet (22 860 mm).

IEBC-1

2. Group B, F-2 and S-2 occupancies not more than two stories in height that are not greater than 3,000 square feet per floor (279 m^2), when the exit access travel distance does not exceed 75 feet (22 860 mm). The minimum fire-resistance rating of the exit enclosure and of the opening protection shall be 1 hour.

3. Open parking structures where vehicles are mechanically parked.

4. In community residences for the developmentally disabled, the maximum occupant load excluding staff is 12.

5. Groups R-1 and R-2 not more than two stories in height, when there are not more than four dwelling units per floor and the exit access travel distance does not exceed 50 feet (15 240 mm). The minimum fire-resistance rating of the exit enclosure and of the opening protection shall be 1 hour.

6. In multilevel dwelling units in buildings of occupancy Group R-1 or R-2, an exit shall not be required from every level of the dwelling unit provided that one of the following conditions is met:

    6.1. The travel distance within the dwelling unit does not exceed 75 feet (22 860mm); or

    6.2. The building is not more than three stories in height and all third-floor space is part of one or more dwelling units located in part on the second floor; and no habitable room within any such dwelling unit shall have a travel distance that exceeds 50 feet (15 240 mm) from the outside of the habitable room entrance door to the inside of the entrance door to the dwelling unit.

7. In Group R-2, H-4, H-5 and I occupancies and in rooming houses and childcare centers, a single exit is permitted in a one-story building with a maximum occupant load of 10 and the exit access travel distance does not exceed 75 feet (22 860 mm).

8. In buildings of Group R-2 occupancy that are equipped throughout with an automatic fire sprinkler system, a single exit shall be permitted from a basement or story below grade if every dwelling unit on that floor is equipped with an approved window providing a clear opening of at least 5 square feet (0.47 m^2) in area, a minimum net clear opening of 24 inches (610 mm) in height and 20 inches (508 mm) in width, and a sill height of not more than 44 inches (1118 mm) above the finished floor.

9. In buildings of Group R-2 occupancy of any height with not more than four dwelling units per floor; with a smokeproof enclosure or outside stair as an exit; and with such exit located within 20 feet (6096 mm) of travel to the entrance doors to all dwelling units served thereby.

10 In buildings of Group R-3 occupancy equipped throughout with an automatic fire sprinkler system, only one exit shall be required from basements or stories below grade.

CHAPTER 9
ADDITIONS

Section 901.1 Change Section 901.1 to read as shown: (EB4-03/04)

901.1 Scope. An addition to a building or structure shall comply with the building, plumbing, electrical and mechanical codes without requiring the existing building or structure to comply with any requirements of those codes or of these provisions.

Section 901.1.1 Add new section as shown: (EB4-03/04)

901.1.1 Flood hazard areas. In flood hazard areas, the existing building is subject to the requirements of Section 903.5.

Section 905.1 Change to read as shown: (EB30-03/04)

905.1 Minimum requirements. Accessibility provisions for new construction shall apply to additions. An addition that affects the accessibility to, or contains an area of, primary function shall comply with the requirements of Section 506.2.

2004 SUPPLEMENT TO THE IEBC

Change, delete or add the following referenced standards to read as shown: (EB23-03/04)
(STANDARDS NOT SHOWN REMAIN UNCHANGED)

CHAPTER 14
REFERENCED STANDARDS

ASME
American Society of Mechanical Engineers
Three Park Avenue
New York, NY 10016-5990

Standard reference number	Title	Referenced in code section
A 17.1-00	Safety Code for Elevators and Escalators - with A17.1a-2002 Addenda	506.1.2
A18.1-99	Safety Standard for Platform Lifts and Stairway Chairlifts - with A18.1a-2001 Addenda and A18.1b-2001 Addenda	506.1.3
A112.19.2M-98	Vitreous China Plumbing Fixtures (Reaffirmed 2002)	410.1

FEMA
Federal Emergency Management Agency
500 C Street, SW
Washington, DC 20472

Standard reference number	Title	Referenced in code section
Pub 368 (2000)	NEHRP Recommended Provisions for Seismic Regulations for New Buildings and Other Structures	A202

NFPA
National Fire Protection Association
1 Batterymarch Park
Quincy, MA 02269-9101

Standard reference number	Title	Referenced in code section
13R-02	Installation of Sprinkler Systems in Residential Occupancies up to and Including Four Stories in Height	604.2.5
72-02	National Fire Alarm Code	604.2.5, 604.4
99-02	Health Care Facilities	408.1

International Fire Code

2004 Supplement

INTERNATIONAL FIRE CODE 2004 SUPPLEMENT

CHAPTER 1
ADMINISTRATION

Section 105.6.15 Change to read as shown: (F5-03/04)

105.6.15 Explosives. An operational permit is required for the manufacture, storage, handling, sale or use of any quantity of explosive, explosive material, fireworks or pyrotechnic special effects within the scope of Chapter 33.

> **Exception:** Storage in Group R-3 occupancies of smokeless propellant, black powder and small arms primers for personal use, not for resale and in accordance with Section 3306.

Table 105.6.21 Add entries to read as shown: (F6-03/04)

TABLE 105.6.21
PERMIT AMOUNTS FOR HAZARDOUS MATERIALS

Type of Material	Amount
Oxidizing materials	
Gases	See Section 105.6.9
Liquids	
Class 4	Any Amount
Class 3	1 gallon [a]
Class 2	10 gallons
Class 1	55 gallons
Solids	
Class 4	Any Amount
Class 3	10 pounds [b]
Class 2	100 pounds
Class 1	500 pounds

(Portions of table not shown do not change)

a. 20 gallons when Table 2703.1.1(1) Note k applies and hazard identification signs in accordance with Section 2703.5 are provided for quantities of 20 gallons or less.
b. 200 pounds when Table 2703.1.1(1) Note k applies and hazard identification signs in accordance with Section 2703.5 are provided for quantities of 200 pounds or less.

CHAPTER 2
DEFINITIONS

SECTION 202
DEFINITIONS

Add new definitions to read as shown: (F8-03/04; F163-03/04)

DAY BOX. See Section 2702.1

EMERGENCY SHUTOFF VALVE. A valve designed to shut off the flow of gases or liquids.

EMERGENCY SHUTOFF VALVE, MANUAL. A manually operated valve designed to shut off the flow of gases or liquids.

EMERGENCY SHUTOFF VALVE, AUTOMATIC. A fail-safe automatic-closing valve designed to shut off the flow of gases or liquids, initiated by a control system that is activated by automatic means.

Change definition of Occupancy Classifications High Hazard Group H to read as shown: (F193-03/04; G27-03/04)

High-Hazard Group H. High-Hazard Group H occupancy includes, among others, the use of a building or structure, or a portion thereof, that involves the manufacturing, processing, generation or storage of materials that constitute a physical or health hazard in quantities in excess of quantities allowed in control areas constructed and located as required in Section 2703.8.3. Hazardous uses are classified in Groups H-1, H-2, H-3, H-4 and H-5 and shall be in accordance with this code and the requirements of Section 415 of the *International Building Code*.

> **Exceptions:** The following shall not be classified in Group H, but shall be classified in the occupancy which they most nearly resemble.
>
> 1. Buildings and structures that contain not more than the maximum allowable quantities per control area of hazardous materials as shown in Tables 307.7(1) and 307.7(2) *of the International Building Code* provided that such buildings are maintained in accordance with this code.
>
> 2. Buildings utilizing control areas in accordance with Section 414.2 of the *International Building Code* that contain not more than the maximum allowable quantities per control area of hazardous materials as shown in Tables 307.7(1) and 307.7(2) of the *International Building Code*.
>
> 3. Buildings and structures occupied for the application of flammable finishes, provided that such buildings or areas conform to the requirements of Section 416 of the *International Building Code* and Chapter 15 of this code.
>
> 4. Wholesale and retail sales and storage of flammable and combustible liquids in mercantile occupancies conforming to Chapter 34 of this code.
>
> 5. Closed systems housing flammable or combustible liquids or gases utilized for the operation of machinery or equipment.
>
> 6. Cleaning establishments that utilize combustible liquid solvents having a flash point

of 140°F (60°C) or higher in closed systems employing equipment listed by an approved testing agency, provided that this occupancy is separated from all other areas of the building by 1-hour fire-resistance-rated fire barriers.

7. Cleaning establishments which utilize a liquid solvent having a flash point at or above 200°F (93°C).

8. Liquor stores and distributors without bulk storage.

9. Refrigeration systems.

10. The storage or utilization of materials for agricultural purposes on the premises.

11. Stationary batteries utilized for facility emergency power, uninterrupted power supply or telecommunication facilities provided that the batteries are provided with safety venting caps and ventilation is provided in accordance with the *International Mechanical Code*.

12. Corrosives shall not include personal or household products in their original packaging used in retail display or commonly used building materials.

13. Buildings and structures occupied for aerosol storage shall be classified as Group S-1, provided that such buildings conform to the requirements of Chapter 28 of this code.

14. Display and storage of nonflammable solid and nonflammable or noncombustible liquid hazardous materials in quantities not exceeding the maximum allowable quantity per control area in Group M or S occupancies complying with Section 414.2.4 of the *International Building Code*.

15. The storage of black powder, smokeless propellant and small arms primers in Groups M and R-3 and special industrial explosive devices in Groups B, F, M and S, provided such storage conforms to the quantity limits and requirements of this code.

Change definition of Occupancy Classification Moderate hazard storage Group S-1 to read as shown: (G46-03/04)

Moderate-hazard storage, Group S-1. Buildings occupied for storage uses which are not classified as Group S-2 including, but not limited to, storage of the following:

Aerosols, Levels 2 and 3
Aircraft repair hangar
Bags; cloth, burlap and paper
Bamboos and rattan
Baskets
Belting; canvas and leather
Books and paper in rolls or packs
Boots and shoes
Buttons, including cloth covered, pearl or bone
Cardboard and cardboard boxes
Clothing, woolen wearing apparel
Cordage
Dry boat storage (indoor)
Furniture
Furs
Glues, mucilage, pastes and size
Grains
Horns and combs, other than celluloid
Leather
Linoleum
Lumber
Motor vehicle repair garages (complying with the *International Building Code* and containing less than the maximum allowable quantities of hazardous materials)
Photo engravings
Resilient flooring
Silks
Soaps
Sugar
Tires, bulk storage of
Tobacco, cigars, cigarettes and snuff
Upholstery and mattresses
Wax candles

Add new definitions as shown: (F8-03/04; F170-03/04; F183-03/04)

OPERATING LINE. See Section 3302.1.

PHYSIOLOGICAL WARNING THRESHOLD. See Section 3702.1.

PUBLIC TRAFFIC ROUTE (PTR). See Section 3302.1.

QUANTITY-DISTANCE (Q-D). See Section 3302.1.

Minimum Separation Distance (D_0). See Section 3302.1.
Intraline Distance (ILD) or Intraplant Distance (IPD). See Section 3302.1.
Inhabited Building Distance (IBD). See Section 3302.1.
Intermagazine Distance (IMD). See Section 3302.1.

REMOTELY LOCATED, MANUALLY ACTIVATED SHUTDOWN CONTROL. A control system that is designed to initiate shutdown of the flow of gases or liquids that is manually activated from a point located some distance from the delivery system.

CHAPTER 3
GENERAL PRECAUTIONS AGAINST FIRE

Section 305.1 Change to read as shown: (EL3-03/04)

305.1 Clearance from ignition sources. Clearance between ignition sources, such as luminaires, heaters and flame-producing devices and combustible materials shall be maintained in an approved manner.

Section 308.3.8 Add new section to read as shown: (F13-03/04)

308.3.8 Group R-2 dormitories: Candles, incense and similar open-flame-producing items shall not be allowed in sleeping units in Group R-2 dormitory occupancies.

Section 309.5 Change to read as shown: (F14-03/04)

309.5 Refueling. Powered industrial trucks using liquid fuel or LP-gas shall be refueled outside of buildings or in areas specifically approved for that purpose. Fixed fuel-dispensing equipment and associated fueling operations shall be in accordance with Chapter 22. Other fuel-dispensing equipment and operations, including cylinder exchange for LP-gas-fueled vehicles, shall be in accordance with Chapter 34 for flammable and combustible liquids or Chapter 38 for LP-gas.

CHAPTER 5
FIRE SERVICE FEATURES

Section 502.1 Change definition of "Key Box" to read as shown: (F26-03/04)

502.1 Definitions. The following words and terms shall, for the purposes of this chapter and as used elsewhere in this code, have the meanings shown herein.

KEY BOX. A secure device with a lock operable only by a fire department master key, and containing building entry keys and other keys that may be required for access in an emergency.

CHAPTER 6
BUILDING SERVICES AND SYSTEMS

SECTION 602
DEFINITIONS

602.1 Definitions. The following words and terms shall, for the purposes of this chapter and as used elsewhere in this code, have the meanings shown herein.

Add new definitions to read as shown: (F39-03/04)

NICKEL CADMIUM (Ni-Cd) BATTERY. An alkaline storage battery in which the positive active material is nickel oxide, the negative contains cadmium, and the electrolyte is potassium hydroxide.

NON-RECOMBINANT BATTERY. A storage battery in which, under conditions of normal use, hydrogen and oxygen gasses created by electrolysis are vented into the air outside of the battery

RECOMBINANT BATTERY. A storage battery in which, under conditions of normal use, hydrogen and oxygen gases created by electrolysis are converted back into water inside the battery instead of venting into the air outside of the battery.

Revise definition of "Stationary Storage Battery" to read as shown: (F39-03/04)

STORAGE BATTERY, STATIONARY. A group of electrochemical cells interconnected to supply a nominal voltage of DC power to a suitably connected electrical load, designed for service in a permanent location. The number of cells connected in series determines the nominal voltage rating of the battery. The size of the cells determines the discharge capacity of the entire battery. After discharge, it may be restored to a fully charged condition by an electric current flowing in a direction opposite to the flow of current when the battery is discharged.

Section 604.2.14.2 Change to read as shown: (EL3-03/04)

604.2.14.2 Separate circuits and luminaires. Separate lighting circuits and luminaires shall be required to provide sufficient light with an intensity of not less than 1 foot-candle (11 lux) measured at floor level in all means of egress corridors, stairways, smoke-proof enclosures, elevator cars and lobbies, and other areas which are clearly a part of the escape route.

Section 604.3 Change to read as shown: (F34-03/04)

604.3 Maintenance. Emergency and standby power systems shall be maintained in accordance with NFPA 110 and NFPA 111 such that the system is capable of supplying service within the time specified for the type and duration required.

Section 605.4.1 Change to read as shown: (F35-03/04)

605.4.1 Power tap design. Relocatable power taps shall be of the polarized or grounded type, equipped with overcurrent protection, and shall be listed in accordance with UL 1363.

Section 606.9.3 Emergency control box. Delete section without substitution: (F36-03/04)

Section 606.9.3.1 Location. Delete section without substitution: (F36-03/04)

Section 606.9.3.2 Construction. Delete section without substitution: (F36-03/04)

Section 606.9.3.3 Operational procedure. Delete section without substitution: (F36-03/04)

Section 606.9.3.4 Identification. Delete section without substitution: (F36-03/04)

Section 606.9.3.5 Instructions. Delete section without substitution: (F36-03/04)

Section 606.10 Add new section and subsections as shown: (F36-03/04)

606.10 Emergency pressure control system. Refrigeration systems containing more than 6.6 pounds (3 kg) of flammable, toxic or highly toxic refrigerant or ammonia shall be provided with an emergency pressure control system in accordance with Sections 606.10.1 and 606.10.2.

606.10.1 Automatic crossover valves. Each high-and intermediate-pressure zone in a refrigeration system shall be provided with a single automatic valve providing a crossover connection to a lower pressure zone. Automatic crossover valves shall comply with Sections 606.10.1.1 through 606.10.1.3.

606.10.1.1 Over-pressure limit set point. Automatic crossover valves shall be arranged to automatically relieve excess system pressure to a lower pressure zone if the pressure in a high-or intermediate-pressure zone rises to within 15 psi (108.4 kPa) of the set point for emergency pressure-relief devices.

606.10.1.2 Manual operation. When required by the fire code official, automatic crossover valves shall be capable of manual operation.

606.10.1.3 System design pressure. Refrigeration system zones that are connected to a higher pressure zone by an automatic crossover valve shall be designed to safely contain the maximum pressure that can be achieved by interconnection of the two zones.

606.10.2 Automatic emergency stop.

606.10.2.1 Operation of an automatic crossover valve. Operation of an automatic crossover valve shall cause all compressors on the affected system to immediately stop. Dedicated pressure-sensing devices located immediately adjacent to crossover valves shall be permitted as a means for determining operation of a valve. To ensure that the automatic crossover valve system provides a redundant means of stopping compressors in an over-pressure condition, high-pressure cutout sensors associated with compressors shall not be used as a basis for determining operation of a crossover valve.

606.10.2.2 Over-pressure in low pressure zone. The lowest pressure zone in a refrigeration system shall be provided with a dedicated means of determining a rise in system pressure to within 15 psi (103.4 kPa) of the set point for emergency pressure-relief devices. Activation of the over-pressure sensing device shall cause all compressors on the affected system to immediately stop.

Section 608 Stationary Lead-Acid Battery Systems. Delete entire section and substitute as shown: (F39-03/04)

SECTION 608
STATIONARY STORAGE BATTERY SYSTEMS

608.1 Scope. Stationary storage battery systems having an electrolyte capacity of more than 50 gallons (189L) used for facility standby power, emergency power, or uninterrupted power supplies shall comply with this section and with Table 608.1.

TABLE 608.1
BATTERY REQUIREMENTS

Requirement	Non-Recombinant — Flooded Lead Acid Batteries	Non-Recombinant — Flooded Nickel Cadmium (Ni-Cd) Batteries	Recombinant — Valve Regulated Lead Acid (VRLA) Batteries
Safety Caps (608.2)	Venting caps (608.2.1)	Venting caps (608.2.1)	Self-resealing flame-arresting caps (608.2.2)
Thermal Runaway Management	Not required	Not required	Required (608.3)
Spill Control	Required (608.5)	Required (608.5)	Not required
Neutralization	Required (608.5.1)	Required (608.5.1)	Required (608.5.2)
Ventilation	Required (608.6.1; 608.6.2)	Required (608.6.1; 608.6.2)	Required (608.6.1; 608.6.2)
Signage	Required (608.7)	Required (608.7)	Required (608.7)
Seismic Control	Required (608.8)	Required (608.8)	Required (608.8)
Fire Detection	Required (608.9)	Required (608.9)	Required (608.9)

608.2 Safety caps. Safety caps for stationary storage battery systems shall comply with Sections 608.2.1 and 608.2.2.

608.2.1 Non-recombinant batteries. Vented lead-acid, nickel-cadmium, or other types of non-recombinantbatteries shall be provided with safety venting caps

608.2.2 Recombinant batteries. Valve-regulated lead-acid (VRLA) or other types of sealed, recombinant batteries shall be equipped with self-resealing flame-arresting safety vents.

608.3 Thermal runaway. VRLA battery systems shall be provided with a listed device or other approved method to preclude, detect, and control thermal runaway.

608.4 Room design and construction. Enclosure of stationary battery systems shall comply with the *International Building Code*. Battery systems shall be allowed to be in the same room with the equipment they support.

608.4.1 Separate rooms. When stationary batteries are installed in a separate equipment room accessible only to authorized personnel, they shall be permitted to be installed on an open rack for ease of maintenance.

608.4.2 Occupied work centers. When a system of VRLA or other type of sealed, non-venting batteries is situated in an occupied work center, it shall be allowed to be housed in a noncombustible cabinet or other enclosure to prevent access by unauthorized personnel.

608.4.3 Cabinets. When stationary batteries are contained in cabinets in occupied work centers, the cabinet enclosures shall be located within 10 feet (3048 mm) of the equipment that they support.

608.5 Spill control and neutralization. An approved method and materials for the control and neutralization of a spill of electrolyte shall be provided in areas containing lead-acid, nickel-cadmium, or other types of batteries with free-flowing liquid electrolyte. For purposes of this paragraph, a "spill" is defined as any unintentional release of electrolyte.

Exception: (VRLA) or other types of sealed batteries with immobilized electrolyte shall not require spill control.

608.5.1 Non-recombinant battery neutralization. For battery systems containing lead-acid, nickel-cadmium, or other types of batteries with free-flowing electrolyte, the method and materials shall be capable of neutralizing a spill from the largest lead-acid battery to a pH between 7.0 and 9.0.

608.5.2 Recombinant battery neutralization: For VRLA or other types of sealed batteries with immobilized electrolyte, the method and material shall be capable of neutralizing a spill of 3.0 percent of the capacity of the largest VRLA cell or block in the room to a pH between 7.0 and 9.0.

608.6 Ventilation. Ventilation of stationary storage battery systems shall comply with Sections 608.6.1 and 608.6.2.

608.6.1 Room ventilation. Ventilation shall be provided in accordance with the *International Mechanical Code* and the following:

1. The ventilation system shall be designed to limit the maximum concentration of hydrogen to 1.0 percent of the total volume of the room; or

2. Continuous ventilation shall be provided at a rate of not less than 1 cubic foot per minute per square foot (1 ft³/min/ft²) [0.0051m³/s m²] of floor area of the room.

608.6.2 Cabinet ventilation. When VRLA batteries are installed inside a cabinet, the cabinet shall be approved for use in occupied spaces and shall be mechanically or naturally vented by one of the following methods:

1. The cabinet ventilation shall limit the maximum concentration of hydrogen to 1% of the total volume of the cabinet during the worst-case event of simultaneous "boost" charging of all the batteries in the cabinet; or

2. When calculations are not available to substantiate the ventilation rate, continuous ventilation shall be provided at a rate of not less than 1 cubic foot per minute per square foot (1 ft³/min/ft² or 0.0051m³/s m²) of floor area covered by the cabinet. The room in which the cabinet is installed shall also be ventilated as required in 608.6.1.

608.7 Signs. Signs shall comply with Sections 608.7.1 and 608.7.2.

608.7.1 Equipment room and building signage. Doors into electrical equipment rooms or buildings containing stationary battery systems shall be provided with approved signs. The signs shall state that:

1. The room contains energized battery systems.

2. The room contains energized electrical circuits.

3. The battery electrolyte solutions are corrosive liquids.

608.7.2 Cabinet signage. Cabinets shall have exterior labels that identify the manufacturer and model number of the system and electrical rating (voltage and current) of the contained battery system. There shall be signs within the cabinet that indicate the relevant electrical, chemical, and fire hazards.

608.8 Seismic protection. The battery systems shall be seismically braced in accordance with the *International Building Code*.

608.9 Smoke detection. An approved automatic smoke detection system shall be installed in accordance with Section 907.2 in rooms containing stationary battery systems.

Section 609 Valve-Regulated Lead Acid (VRLA) Battery Systems. Delete entire section without substitution: (F39-03/04)

CHAPTER 7
FIRE-RESISTANCE-RATED CONSTRUCTION

Section 703.1 Revise to read as shown: (F41-03/04)

703.1 Maintenance. The required fire-resistance rating of fire-resistance-rated construction (including walls, fire stops, shaft enclosures, partitions, smoke barriers, floors, fire resistive coatings and sprayed fire-resistant materials applied to structural members, and fire-resistive joint systems) shall be maintained. Such elements shall be properly repaired, restored or replaced when damaged, altered, breached or penetrated. Openings made therein for the passage of pipes, electrical conduit, wires, ducts, air transfer openings, and holes made for any reason shall be protected with approved methods capable of resisting the passage of smoke and fire. Openings through fire-resistance-rated assemblies shall be protected by self-closing or automatic-closing doors of approved construction meeting the fire protection requirements for the assembly.

CHAPTER 8
INTERIOR FINISH, DECORATIVE MATERIALS AND FURNISHINGS

Section 803.2.1 Change to read as shown: (F42-03/04)

803.2.1 Foam plastics. Exposed foam plastic materials and unprotected materials containing foam plastic used for decorative purposes or stage scenery or exhibit booths shall have a maximum heat release rate of 100 kilowatts (kW) when tested in accordance with UL 1975.

Exceptions:

1. Individual foam plastic items or items containing foam plastic where the foam plastic does not exceed 1 pound (0.45 kg) in weight.

2. Cellular or foam plastic shall be allowed for trim not in excess of 10 percent of the wall or ceiling area, provided it is not less than 20 pounds per cubic foot (320 kg/m³) in density, is limited to 0.5 inch (12.7 mm) in thickness and 8 inches (203 mm) in width, and complies with the requirements for Class B interior wall and ceiling finish, except that the smoke-developed index shall not be limited.

Section 803.2.2 Change to read as shown: (F43-03/04)

803.2.2 Motion picture screens. The screens upon which motion pictures are projected shall either meet the flame propagation performance criteria of NFPA 701, or shall comply with the requirements for a Class B interior finish in accordance with Section 803 of the *International Building Code*.

Section 803.5.2 Change to read as shown: (F44-03/04)

803.5.2 Upholstered furniture heat release rate. Newly introduced upholstered furniture shall have limited rates of heat release when tested in accordance with ASTM E 1537, as follows.

1. The peak rate of heat release for the single upholstered furniture item shall not exceed 250 kW.

   Exception: Upholstered furniture in rooms or spaces protected by an approved automatic sprinkler system installed in accordance with Section 903.3.1.1.

2. The total energy released by the single upholstered furniture item during the first 5 minutes of the test shall not exceed 40 megajoules (MJ).

   Exception: Upholstered furniture in rooms or spaces protected by an approved automatic sprinkler system installed in accordance with Section 903.3.1.1.

Section 803.5.3 Change to read as shown: (F45-03/04)

803.5.3 Mattresses, heat-release rate. Newly introduced mattresses in Group I-2 occupancies shall have limited rates of heat release when tested in accordance with ASTM E 1590.

1. The peak rate of heat release for the mattress shall not exceed 250 kW.

   Exception: Mattresses in rooms or spaces protected by an approved automatic sprinkler system installed in accordance with Section 903.3.1.1.

2. The total energy released by the mattress during the first 5 minutes of the test shall not exceed 40 MJ.

   Exception: Mattresses in rooms or spaces protected by an approved automatic sprinkler system installed in accordance with Section 903.3.1.1.

Section 803.6.3 Change to read as shown: (F45-03/04)

803.6.3 Mattresses, heat-release rate. Newly introduced mattresses in Group I-1 occupancies shall have limited rates of heat release when tested in accordance with ASTM E 1590.

1. The peak rate of heat release for the mattress shall not exceed 250 kW.

   Exception: Mattresses in rooms or spaces protected by an approved automatic sprinkler system.

2. The total energy released by the mattress during the first 5 minutes of the test shall not exceed 40 MJ.

   Exception: Mattresses in rooms or spaces protected by an approved automatic sprinkler system.

Section 803.7.2 Change to read as shown: (F44-03/04)

803.7.2 Upholstered furniture heat release rate. Newly introduced upholstered furniture shall have limited rates of heat release when tested in accordance with ASTM E 1537, as follows:

1. The peak rate of heat release for the single upholstered furniture item shall not exceed 250 kW.

   Exceptions:

   1. In Use Condition I, II and III occupancies, as defined in the *International Building Code*, upholstered furniture in rooms or spaces protected by approved smoke detectors that initiate, without delay, an alarm that is audible in that room or space.

   2. Upholstered furniture in rooms or spaces protected by an approved automatic sprinkler system installed in accordance with Section 903.3.1.1.

2. The total energy released by the single upholstered furniture item during the first 5 minutes of the test shall not exceed 40 MJ.

   Exception: Upholstered furniture in rooms or spaces protected by an approved automatic sprinkler system installed in accordance with Section 903.3.1.1.

Section 803.7.4 Change to read as shown: (F45-03/04)

803.7.4 Mattresses, heat release rate. Newly introduced mattresses in detention and correctional occupancies shall have limited rates of heat release when tested in accordance with ASTM E 1590 as follows:

1. The peak rate of heat release for the mattress shall not exceed 250 kW.

   Exception: Mattresses in rooms or spaces protected by an approved automatic sprinkler system installed in accordance with Section 903.3.1.1.

2. The total energy released by the mattress during the first 5 minutes of the test shall not exceed 40 MJ.

> **Exception:** Mattresses in rooms or spaces protected by an approved automatic sprinkler system installed in accordance with Section 903.3.1.1.

Section 804.4 Change to read as shown: (F43-03/04)

804.4 Artificial vegetation. Artificial decorative vegetation shall meet the flame propagation performance criteria of NFPA 701. Meeting the flame propagation performance criteria of NFPA 701 shall be documented and certified by the manufacturer in an approved manner.

Section 805.1 Change to read as shown: (F43-03/04; F47-03/04)

805.1 General. In occupancies of Groups A, E, I and R-1 and dormitories in Group R-2, curtains, draperies, hangings and other decorative materials suspended from walls or ceilings shall meet the flame propagation performance criteria of NFPA 701 in accordance with Section 805.2 or be noncombustible.

In Groups I-1 and I-2, combustible decorations shall meet the flame propagation performance criteria of NFPA 701 unless the decorations, such as photographs and paintings, are of such limited quantities that a hazard of fire development or spread is not present. In Group I-3, combustible decorations are prohibited.

Fixed or movable walls and partitions, paneling, wall pads and crash pads, applied structurally or for decoration, acoustical correction, surface insulation or other purposes, shall be considered interior finish if they cover 10 percent or more of the wall or of the ceiling area, and shall not be considered decorations or furnishings.

Section 805.1.2 Change to read as shown: (F43-03/04)

805.1.2 Combustible decorative materials. The permissible amount of decorative materials meeting the flame propagation performance criteria of NFPA 701 shall not exceed 10 percent of the aggregate area of walls and ceilings.

> **Exception:** In auditoriums of Group A, the permissible amount of decorative material meeting the flame propagation performance criteria of NFPA 701 shall not exceed 50 percent of the aggregate area of walls and ceiling where the building is equipped throughout with an automatic sprinkler system in accordance with Section 903.3.1.1, and where the material is installed in accordance with Section 803.4 of the *International Building Code*.

Section 805.2 Change to read as shown: (F43-03/04; F48-03/04)

805.2 Acceptance criteria and reports. Where required by Section 805.1, decorative materials shall be tested by an approved agency and meet the flame propagation performance criteria of NFPA 701, or such materials shall be noncombustible. Reports of test results shall be prepared in accordance with NFPA 701 and furnished to the fire code official upon request.

Section 806.1.1 Change to read as shown: (F49-03/04)

806.1.1 Requirements based on occupancy. Interior finish shall be restricted by combustibility, flame spread and smoke developed and according to occupancy group in accordance with Table 806.3. Motion picture screens in Group A occupancies shall be restricted according to occupancy group in accordance with Table 806.3 and Section 803.2.2.

CHAPTER 9
FIRE PROTECTION SYSTEMS

Section 903.2 Change to read as shown: (FS40-03/04)

903.2 Where required. Approved automatic sprinkler systems in new buildings and structures shall be provided in the locations described in this section.

> **Exception:** Spaces or areas in telecommunications buildings used exclusively for telecommunications equipment, associated electrical power distribution equipment, batteries and standby engines, provided those spaces or areas are equipped throughout with an automatic fire alarm system and are separated from the remainder of the building by fire barriers consisting of not less than 1 hour fire-resistance-rated walls and 2-hour fire-resistance-rated floor/ceiling assemblies.

Section 903.2.1.2 Change to read as shown: (F58-03/04)

903.2.1.2 Group A-2. An automatic sprinkler system shall be provided for Group A-2 occupancies where one of the following conditions exists:

1. The fire area exceeds 5,000 square feet (464.5 m^2);

2. The fire area has an occupant load of 100 or more; or

3. The fire area is located on a floor other than the level of exit discharge.

TABLE 903.2.13 Change the table by deleting the last row and replacing it with new rows to read as shown: (F140-03/04)

**TABLE 903.2.13
ADDITIONAL REQUIRED FIRE-EXTINGUISHING SYSTEMS**

SECTION	SUBJECT
914.1.2	Covered malls
914.2.1	High rise buildings
914.3.1	Atriums
914.4.1	Underground structures
914.5.1	Stages
914.6.2	Special amusement buildings
914.7.2, 914.7.2.3	Aircraft hangars
914.9.1	Flammable finishes

(Portions of table not shown do not change)

Section 904.11.1 Change to read as shown: (F88-03/04)

904.11.1 Manual system operation. A manual actuation device shall be located at or near a means of egress from the cooking area, a minimum of 10 feet (3048 mm) and a maximum of 20 feet (6096 mm) from the kitchen exhaust system. The manual actuation device shall be installed not more than 48 inches (1200 mm), nor less than 42 inches (1067 mm) above the floor and shall clearly identify the hazard protected. The manual actuation shall require a maximum force of 40 pounds (178 N) and a maximum movement of 14 inches (356 mm) to actuate the fire suppression system.

Exception: Automatic sprinkler systems shall not be required to be equipped with manual actuation means.

Section 905.3.3 Change to read as shown: (F97-03/04)

905.3.3 Covered mall buildings. A covered mall building shall be equipped throughout with a standpipe system where required by Section 905.3.1. Covered mall buildings not required to be equipped with a standpipe system by Section 905.3.1 shall be equipped with Class I hose connections connected to a system sized to deliver 250 gallons per minute (946.4 L/min) at the most hydraulically remote outlet. Hose connections shall be provided at each of the following locations:

1. Within the mall at the entrance to each exit passageway or corridor.

2. At each floor-level landing within enclosed stairways opening directly on the mall.

3. At exterior public entrances to the mall.

Section 905.3.4 Change to read as shown: (F98-03/04)

905.3.4 Stages. Stages greater than 1,000 square feet in area (93 m^2) shall be equipped with a Class III wet standpipe system with 1.5-inch and 2.5-inch (38 mm and 64 mm) hose connections on each side of the stage.

Exception: Where the building or area is equipped throughout with an automatic sprinkler system, a 1½ inch hose connection shall be installed and shall be allowed to be supplied from the automatic sprinkler system and shall have a flow rate of not less than 100 gpm (378.5 L/m) at a minimum residual pressure of 65 psi (448 kPa), added to the sprinkler flow demand, at the most hydraulically remote hose connection.

Section 905.3.7 Add new section to read as shown: (F99-03/04)

905.3.7 Marinas and boatyards. Marinas and boatyards shall be equipped throughout with standpipe systems in accordance with NFPA 303.

Section 905.4 Change to read as shown: (F100-03/04)

905.4 Location of Class I standpipe hose connections. Class I standpipe hose connections shall be provided in all of the following locations:

1. In every required stairway, a hose connection shall be provided for each floor level above or below grade. Hose connections shall be located at an intermediate floor level landing between floors, unless otherwise approved by the fire code official.

2. On each side of the wall adjacent to the exit opening of a horizontal exit.

    Exception: Where floor areas adjacent to a horizontal exit are reachable from exit stairway hose connections by a 30-foot (9144 mm) hose stream from a nozzle attached to 100 feet (30480 mm) of hose, a hose connection shall not be required at the horizontal exit.

3. In every exit passageway at the entrance from the exit passageway to other areas of a building.

4. In covered mall buildings, adjacent to each exterior public entrance to the mall and adjacent to each

entrance from an exit passageway or exit corridor to the mall.

5. Where the roof has a slope less than four units vertical in 12 units horizontal (33.3-percent slope), each standpipe shall be provided with a hose connection located either on the roof or at the highest landing of stairways with stair access to the roof. An additional hose connection shall be provided at the top of the most hydraulically remote standpipe for testing purposes.

6. Where the most remote portion of a nonsprinklered floor or story is more than 150 feet (45 720 mm) from a hose connection or the most remote portion of a sprinklered floor or story is more than 200 feet (60 960 mm) from a hose connection, the fire code official is authorized to require that additional hose connections be provided in approved locations.

Section 906.2 Change to read as shown: (F104-03//04)

906.2 General requirements. Fire extinguishers shall be selected, installed and maintained in accordance with this section and NFPA 10.

Exceptions:

1. The travel distance to reach an extinguisher shall not apply to the spectator seating portions of Group A-5 occupancies.
2. The use of a supervised, listed electronic monitoring device shall be allowed in lieu of 30-day interval inspections, when approved.

Section 907.2.6 Change to read as shown: (F111-03/04)

907.2.6 Group I. A manual fire alarm system shall be installed in Group I occupancies. An electrically supervised, automatic smoke detection system shall be provided in accordance with Sections 907.2.6.1 and 907.2.6.2.

Exception: Manual fire alarm boxes in patient sleeping areas of Group I-1 and I-2 occupancies shall not be required at exits if located at all nurses' control stations or other constantly attended staff locations, provided such stations are visible and continuously accessible and that travel distances required in Section 907.4.1 are not exceeded.

Section 907.2.6.1 Add new section as shown: (F111-03/04)

907.2.6.1 Group I-1. Corridors, habitable spaces other than sleeping rooms and kitchens and waiting areas that are open to corridors shall be equipped with an automatic smoke detection system.

Exceptions:

1. Smoke detection in habitable spaces is not required where the facility is equipped throughout with an automatic sprinkler system.
2. Smoke detection is not required for exterior balconies.

(Renumber subsequent sections)

Section 907.2.6.1 Change to read as shown: (F111-03/04)

907.2.6.1 Group I-2. Corridors in nursing homes (both intermediate care and skilled nursing facilities), detoxification facilities and spaces permitted to be open to the corridors by IBC Section 407.2 shall be equipped with an automatic fire detection system. Hospitals shall be equipped with smoke detection as required in Section 407.2.

Exceptions:

1. Corridor smoke detection is not required in smoke compartments that contain patient sleeping rooms where patient sleeping units are provided with smoke detectors that comply with UL 268. Such detectors shall provide a visual display on the corridor side of each patient sleeping unit and shall provide an audible and visual alarm at the nursing station attending each unit.
2. Corridor smoke detection is not required in smoke compartments that contain patient sleeping rooms where patient sleeping unit doors are equipped with automatic door-closing devices with integral smoke detectors on the unit sides installed in accordance with their listing, provided that the integral detectors perform the required alerting function.

Section 907.2.9 Change to read as shown: (F114-03/04)

907.2.9 Group R-2. A manual fire alarm system shall be installed in Group R-2 occupancies where:

1. Any dwelling unit or sleeping unit is located three or more stories above the lowest level of exit discharge;
2. Any dwelling unit or sleeping unit is located more than one story below the highest level of exit discharge of exits serving the dwelling unit or sleeping unit; or
3. The building contains more than 16 dwelling units or sleeping units.

Exceptions:

1. A fire alarm system is not required in buildings not more than two stories in height where all dwelling units or sleeping units and contiguous attic and crawl spaces are separated from each other and public or common areas by at least 1-hour fire partitions and each dwelling unit or sleeping unit has an exit directly to a public way, exit court or yard.

2. Manual fire alarm boxes are not required throughout the building when the following conditions are met:

    2.1. The building is equipped throughout with an automatic sprinkler system in accordance with Section 903.3.1.1 or Section 903.3.1.2; and

    2.2. The notification appliances will activate upon sprinkler flow.

3. A fire alarm system is not required in buildings that do not have interior corridors serving dwelling units and are protected by an approved automatic sprinkler system installed in accordance with Section 903.3.1.1 or 903.3.1.2, provided that dwelling units either have a means of egress door opening directly to an exterior exit access that leads directly to the exits or are served by open-ended corridors designed in accordance with Section 1022.6, Exception 4.

Sections 907.2.12.2, 907.2.12.2.1 and 907.2.12.2.2 Change to read as shown: (F120-03/04)

907.2.12.2 Emergency voice/alarm communication system. The operation of any automatic fire detector, sprinkler water-flow device or manual fire alarm box shall automatically sound an alert tone followed by voice instructions giving approved information and directions for a general or staged evacuation on a minimum of the alarming floor, the floor above, and the floor below in accordance with the building's fire safety and evacuation plans required by Section 404. Speakers shall be provided throughout the building by paging zones. As a minimum, paging zones shall be provided as follows:

1. Elevator groups.
2. Exit stairways.
3. Each floor.
4. Areas of refuge as defined in Section 1002.

Exception: In Group I-1 and I-2 occupancies, the alarm shall sound in a constantly attended area and a general occupant notification shall be broadcast over the overhead page.

907.2.12.2.1 Manual override. A manual override for emergency voice communication shall be provided on a selective and all-call basis for all paging zones.

907.2.12.2.2 Live voice messages. The emergency voice/alarm communication system shall also have the capability to broadcast live voice messages through paging zones on a selective and all-call basis.

Section 909.4.6 Change to read as shown: (F117-03/04)

909.4.6 Duration of operation. All portions of active or passive smoke control systems shall be capable of continued operation after detection of the fire event for a period of not less than either 20 minutes or 1.5 times the calculated egress time, whichever is less.

Section 909.5.2 Change to read as shown: (FS53-03/04)

909.5.2 Opening protection. Openings in smoke barriers shall be protected by automatic-closing devices actuated by the required controls for the mechanical smoke control system. Door openings shall be protected by door assemblies complying with Section 715.3.3.

Exceptions:

1. Passive smoke control systems with automatic-closing devices actuated by spot-type smoke detectors listed for releasing service installed in accordance with Section 907.10.

2. Fixed openings between smoke zones that are protected utilizing the airflow method.

3. In Group I-2, where such doors are installed across corridors, a pair of opposite-swinging doors without a center mullion shall be installed having vision panels with fire-protection-rated glazing materials in fire-protection-rated frames, the area of which shall not exceed that tested. The doors shall be close fitting within operational tolerances and shall not have undercuts, louvers or grilles. The doors shall have head and jamb stops, astragals or rabbets at meeting edges, and automatic-closing devices. Positive-latching devices are not required.

4. Group I-3.

5. Openings between smoke zones with clear ceiling heights of 14 feet (4267 mm) or greater and bank-down capacity of greater than 20 minutes as determined by the design fire size.

Section 909.8.1 Change to read as shown: (FS118-03/04; FS119-03/04)

909.8.1 Exhaust rate. The height of the lowest horizontal surface of the accumulating smoke layer shall be maintained at least 6 feet (1829 mm) above any walking surface which forms a portion of a required egress system within the smoke zone. The required exhaust rate for the zone shall be the largest of the calculated plume mass-flow rates for the possible plume configurations. Provisions shall be made for natural or mechanical supply of air from outside or adjacent smoke zones to make up for the air exhausted. Makeup airflow rates, when measured at the potential fire location, shall not increase the smoke production rate beyond the capabilities of the smoke control system. The temperature of the makeup air shall be such that it does not expose temperature-sensitive fire protection systems beyond their limits.

Section 909.8.3 Balcony spill plumes. Delete without substitution: (FS120-03/04)

Section 909.8.4 Window plumes. Delete without substitution: (FS121-03/04)

Sections 909.9 Change to read as shown: (FS122-03/04)

909.9 Design fire. The design fire shall be based on a rational analysis performed by the registered design professional and approved by the building official. The design fire shall be based on the analysis in accordance with Section 909.4 and this section.

Section 909.9.2 Change to read as shown: (F122-03/04

909.9.2 Separation distance. Determination of the design fire shall include consideration of the type of fuel, fuel spacing and configuration.

Section 909.16 Change to read as shown: (F125-03/04)

909.16 Fire-fighter's smoke control panel. A fire-fighter's smoke control panel for fire department emergency response purposes only shall be provided and shall include manual control or override of automatic control for mechanical smoke control systems. The panel shall be located in a fire command center complying with Section 509 in high rise buildings or buildings with smoke-protected assembly seating. In all other buildings, the fire-fighter's smoke control panel shall be installed in an approved location adjacent to the fire alarm control panel. The fire-fighter's smoke control panel shall comply with Sections 909.16.1 through 909.16.3.

Section 910.1 Add new section as shown. (F126-03/04)

910.1 General. Where required by this code or otherwise installed, smoke and heat vents, or mechanical smoke exhaust systems, and draft curtains shall conform to the requirements of this section.

Exceptions:

1. Frozen food warehouses used solely for storage of Class I and Class II commodities where protected by an approved automatic sprinkler system.

2. Where areas of buildings are equipped with early suppression fast-response (ESFR) sprinklers, automatic smoke and heat vents shall not be required within these areas.

Section 910.2 Change to read as shown: (F127-03/04)

910.2 Where required. Smoke and heat vents shall be installed in the roofs of one-story buildings or portions thereof occupied for the uses set forth in Sections 910.2.1 through 910.2.4.

Section 910.2.2 Change to read as shown: (F132-03/04)

910.2.2 Group H-2 or H-3. Buildings and portions thereof used as a Group H-2 or H-3 occupancy having more than 15,000 square feet (1394 m^2) in single floor area.

Exception: Buildings of noncombustible construction containing only noncombustible materials.

Table 910.3 Change table to read as shown: (F134-03/04; F135-03/04)

TABLE 910.3
REQUIREMENTS FOR DRAFT CURTAINS AND SMOKE AND HEAT VENTS[a]

OCCUPANCY GROUP AND COMMODITY CLASSIFICATION	VENT AREA TO FLOOR AREA RATIO[c]
Groups F-1 and S-1	(No change)

Table 910.3 (continued)

High-piled Storage (see Section 910.2.3 I-IV (Option 1)	(No change)
High-piled Storage (see Section 910.2.3 I-IV (Option 2)	
High-piled Storage (see Section 910.2.3 High Hazard (Option 1)	
High-piled Storage (see Section 910.2.3) High Hazard (Option 2)	

(Portions of the table not shown do not change)
a. and b. (No change to current text)
c. Where draft curtains are not required, the vent area to floor area ratio shall be calculated based on a minimum draft curtain depth of 6 feet (Option 1).

Section 910.3.1 Add new section to read as shown: (F127-03/04)

910.3.1 Design. Smoke and heat vents shall be listed and labeled.

(Renumber subsequent sections)

Section 910.3.1 Change to read as shown: (F127-03/04)

910.3.1 Vent operation. Smoke and heat vents shall be capable of being operated by approved automatic and manual means. Automatic operation of smoke and heat vents shall conform to the provisions of Sections 910.3.2.1 through 910.3.2.3.

Section 914 Add new section to read as shown: (F140-03/04)

SECTION 914
FIRE PROTECTION BASED ON SPECIAL DETAILED REQUIREMENTS OF USE AND OCCUPANCY

914.1 Scope. This section shall specify where fire protection systems are required based on detailed requirements of use and occupancy.

914.1.1 Covered mall buildings.

914.1.2 Automatic sprinkler system. The covered mall building and buildings connected shall be equipped throughout with an automatic sprinkler system in accordance with Section 903.1.1 which shall comply with the following:

1. The automatic sprinkler system shall be complete and operative throughout occupied space in the covered mall building prior to occupancy of any of the tenant spaces. Unoccupied tenant spaces shall be similarity protected unless provided with an approved alternate protection.

2. Sprinkler protection for mall shall be independent from that provided for tenant spaces or anchors. Where tenant spaces are supplied by the same system, they shall be independently controlled.

Exception: An automatic sprinkler system shall not be required in space or areas of open parking garages connected in accordance with Section 406.2 of the *International Building Code*.

914.1.3 Standpipe systems. The covered mall building shall be equipped throughout with a standpipe system in accordance with Section 905.

914.1.4 Emergency voice/alarm communication system. Covered mall buildings exceeding 50,000 square feet (4645m²) in total floor area shall be provided with an emergency voice/alarm communication system. Emergency voice/alarm communication systems serving a mall, required or otherwise, shall be accessible to the fire department. The system shall be provided in accordance with Section 907.2.12.2.

914.1.5 Fire department access to equipment. Rooms or areas containing controls for air-conditioning systems, automatic fire-extinguishing systems or other detection, suppression or control elements shall be identified for use by the fire department.

914.2 High-rise buildings.

914.2.1 Automatic sprinkler systems. Buildings and structures shall be equipped throughout with an automatic sprinkler system in accordance with Section 903.3.1.1 and a secondary water supply where required by Section 903.3.5.2

Exception: An automatic sprinkler system shall not be required in spaces or areas of:

1. Open parking garages in accordance with Section 406.3 of the *International Building Code*.

2. Telecommunication equipment buildings used exclusively for telecommunications equipment, associated electrical power distribution equipment, batteries and standby engines, provided that those spaces or areas are equipped throughout with an automatic fire detection system in accordance with Section 907.2 and are separated from the remainder of

the building with fire barriers consisting of 1-hour fire-resistance-rated walls and 2 hour fire resistance-rated floor/ceiling assemblies.

914.2.2 Automatic fire detection. Smoke detection shall be provided in accordance with Section 907.2.12.

914.2.3 Emergency voice/alarm communication systems. An emergency voice/alarm communication system shall be provided in accordance with Section 907.2.12.2.

914.2.4 Fire department communication system. A two-way fire department communication system shall be provided for the fire department use in accordance with Section 907.2.12.3

914.2.5 Fire command. A fire command center complying with Section 509 shall be provided in a location approved by the fire chief.

914.3 Atriums.

914.3.1 Automatic sprinkler protection. An approved automatic sprinkler system shall be installed throughout the entire building.

Exceptions:

1. That area of a building adjacent to or above the atrium need not be sprinklered provided that portion of the building is separated from the atrium portion by a 2-hour fire barrier wall or horizontal assembly or both.

2. Where the ceilings of the atrium is more than 55 feet (16764mm) above the floor, sprinkler protection at the ceiling of the atrium is not required.

914.3.2 Fire alarm system. A fire alarm system shall be provided where required by Section 907.2.13.

914.4 Underground buildings.

914.4.1 Automatic sprinkler system. The highest level of exit discharge serving the underground portions of the building and all levels below shall be equipped with an automatic sprinkler system installed in accordance with Section 903.3.1.1. Water-flow switches and control valves shall be supervised in accordance with Section 903.4.

914.4.2 Smoke control system.

914.4.2.1 Control system. A smoke control system is required to control the migration of products of combustion in accordance with Section 909 and provisions of this section. Smoke control shall restrict movement of smoke to the general area of fire origin and maintain means of egress in a usable condition.

914.4.2.2 Smoke exhaust system. Where compartmentation is required by Section 405.4 of the *International Building Code*, each compartment shall have an independent smoke-control system. The system shall be automatically activated and capable of manual operation in accordance with Section 907.2.18.

914.4.3 Fire alarm systems. A fire alarm system shall be provided where required by Section 907.2.19.

914.4.4 Public address. A public address system shall be provided where required by Section 907.2.19.1.

914.4.5 Standpipe system. The underground building shall be provided throughout with a standpipe system in accordance with Section 905.

914.5 Stages.

914.5.1 Automatic sprinklers system. Stages shall be equipped with an automatic fire-extinguishing system in accordance with Chapter 9. The system shall be installed under the roof and gridiron, in the tie and fly galleries, and in places behind the proscenium wall of the stage and in dressing rooms, lounges, workshops, and storerooms accessory to such stages.

Exceptions:

1. Sprinklers are not required under stage areas less than 4 feet (1219mm) in clear height utilized exclusively for storage of tables and chairs, provided the concealed space is separated from the adjacent spaces by not less than ⅝ inch (15.9mm) Type X gypsum board.

2. Sprinklers are not required for stages 1,000 square feet (93m²) or less in area and 50 feet (15240 mm) or less in height where curtains, scenery, or other combustible hangings are not retractable vertically. Combustible hangings shall be limited to a single curtain, borders, legs, and a single backdrop.

914.5.2 Standpipes. Standpipe systems shall be provided in accordance with Section 905.

914.6 Special amusement buildings.

914.6.1 Automatic fire detection. Special amusement buildings shall be equipped with an automatic fire detection system in accordance with Section 907.

914.6.2 Automatic sprinkler system. Special amusement buildings shall be equipped throughout with an automatic sprinkler system in accordance with Section 903.3.1.1. Where the special amusement building is temporary, the sprinkler water supply shall be of an approved temporary means.

Exception: Automatic sprinklers are not required where the total floor area of a temporary special amusement building is less than 1,000 square feet (93 m²) and the travel distance from any point to an exit is less than 50 feet (15240 mm).

914.7 Aircraft related occupancies.

914.7.1 Automatic fire detection systems. Airport traffic control towers shall be provided with an automatic fire detection system installed in accordance with Section 907.2.

914.7.2 Fire suppression. Aircraft hangers shall be provided with fire suppression as required by NFPA 409.

Exception: Group II hangers as defined in NFPA 409 storing private aircraft without major maintenance or overhaul are exempt from foam suppression requirements.

914.7.2.3 Finishing. The process of "doping" involving the use of a volatile flammable solvent, or of painting, shall be carried on in a separate detached building equipped with automatic fire-extinguishing equipment in accordance with Section 903.

914.8.3 Residential aircraft hangers.

914.8.3.1 Smoke detection. Smoke alarms shall be provided within the hanger in accordance with Section 907.2.21.

914.8.4 Aircraft paint hangers.

914.8.4.1 Fire suppression. Aircraft paint hangers shall be provided with fire suppression as required by NFPA 409.

914.9 Application of flammable finishes.

914.9.1 Fire protection. An automatic fire-extinguishing system shall be provided in all spray, dip and immersing spaces and storage rooms, and shall be installed in accordance with Chapter 9.

914.10 Drying rooms.

914.10.1 Fire protection. Drying rooms designed for high-hazard materials and processes, including special occupancies as provided for in Chapter 4 of the *International Building Code*, shall be protected by an approved automatic fire-extinguishing system complying with the provisions of Chapter 9.

CHAPTER 10
MEANS OF EGRESS

Section 1002.1 Change definition of ACCESSIBLE MEANS OF EGRESS to read as shown: (E6-03/04)

ACCESSIBLE MEANS OF EGRESS. A continuous and unobstructed way of egress travel from any accessible point in a building or facility to a public way.

Section 1003.2 Change to read as shown: (E8-03/04)

1003.2 Ceiling height. The means of egress shall have a ceiling height of not less than 7 feet 6 inches (2286 mm).

Exceptions:

1. Sloped ceilings in accordance with Section 1208.2 of the *International Building Code*.

2. Ceilings of dwelling units and sleeping units within residential occupancies in accordance with Section 1208.2 of the *International Building Code*.

3. Allowable projections in accordance with Section 1003.3.

4. Stair headroom in accordance with Section 1009.2

5. Door height in accordance with Section 1008.1.1.

Section 1004.1 Change to read as shown: (E9-03/04)

1004.1 Design occupant load. In determining means of egress requirements, the number of occupants for whom means of egress facilities shall be provided shall be determined in accordance with this section. Where occupants from accessory areas egress through a primary space, the calculated occupant load for the primary space shall include the total occupant load of the primary space plus the number of occupants egressing through it from the accessory area.

Section 1004.1.1 Actual number. Delete section without substitution: (E9-03/04)

(Renumber subsequent sections and table)

Section 1004.1.2 Change to read as shown: (E9-03/04)

1004.1.2 Areas without fixed seating. The number of occupants computed at the rate of one occupant per unit of area as prescribed in Table 1004.1.1. For areas without fixed seating, the occupant load shall not be less than that number determined by dividing the floor area under consideration by the occupant-per-unit-of-area factor assigned to the occupancy as set forth in Table 1004.1.1. Where an intended use is not listed in Table 1004.1.1, the building official shall establish a use based on a listed use that most nearly resembles the intended use.

Exception: Where approved by the building official, the actual number of occupants for whom each occupied space, floor or building is designed, although less than those determined by calculation, shall be permitted to be used in the determination of the design occupant load.

Section 1004.1.3 Number by combination. Delete section without substitution: (E9-03/04)

Section 1004.2 Change to read as shown: (E9-03/04)

1004.2 Increased occupant load. The occupant load permitted in any building or portion thereof is permitted to be increased from that number established for the occupancies in Table 1004.1.1 provided that all other requirements of the code are also met based on such modified number and the occupant load shall not exceed one occupant per 5 square feet (0.47 m^2) of occupiable floor space. Where required by the building official, an approved aisle, seating or fixed equipment diagram substantiating any increase in occupant load shall be submitted. Where required by the building official, such diagram shall be posted.

Section 1004.7 Change to read as shown: (E12-03/04)

1004.7 Fixed seating. For areas having fixed seats and aisles, the occupant load shall be determined by the number of fixed seats installed therein. The occupant load for area in which fixed seating is not installed, such as waiting spaces and wheelchair spaces, shall be determined in accordance with Section 1004.1.2 and added to the number of fixed seats.

For areas having fixed seating without dividing arms, the occupant load shall not be less than the number of seats based on one person for each 18 inches (457 mm) of seating length.

The occupant load of seating booths shall be based on one person for each 24 inches (610 mm) of booth seat length measured at the backrest of the seating booth.

Section 1006.3 Change to read as shown: (E3-03/04; E14-03/04)

1006.3 Illumination emergency power. The power supply for means of egress illumination shall normally be provided by the premise's electrical supply.

In the event of power supply failure, an emergency electrical system shall automatically illuminate the following areas:

1. Aisles and unenclosed egress stairways in rooms and spaces which require two or more means of egress.

2. Corridors, exit enclosures and exit passageways in buildings required to have two or more exits.

3. Exterior egress components at other than the level of exit discharge until exit discharge is accomplished for buildings required to have two or more exits.

4. Interior exit discharge elements, as permitted in Section 1023.1, in buildings required to have two or more exits.

5. The portion of the exterior exit discharge immediately adjacent to exit discharge doorways in buildings required to have two or more exits.

The emergency power system shall provide power for a duration of not less than 90 minutes and shall consist of storage batteries, unit equipment or an on-site generator. The installation of the emergency power system shall be in accordance with Section 604.

Section 1007.1 Change to read as shown: (E16-03/04)

1007.1 Accessible means of egress required. Accessible means of egress shall comply with this section. Accessible spaces shall be provided with not less than one accessible means of egress. Where more than one means of egress is required by Section 1014.1 or 1018.1 from any accessible space, each accessible portion of the space shall be served by not less than two accessible means of egress.

Exceptions:

1. Accessible means of egress are not required in alterations to existing buildings.

2. One accessible means of egress is required from an accessible mezzanine level in accordance with Section 1007.3, 1007.4 or 1007.5.

3. In assembly spaces with sloped floors, one accessible means of egress is required from a space where the common path of travel of the accessible route for access to the wheelchair spaces meets the requirements in Section 1024.8.

Section 1007.2 Change to read as shown: (E17-03/04)

1007.2 Continuity and components. Each required accessible means of egress shall be continuous to a public way and shall consist of one or more of the following components:

1. Accessible routes complying with Section 1104.

2. Stairways within exit enclosures complying with Sections 1007.3 and 1019.1.

3. Elevators complying with Section 1007.4.
4. Platform lifts complying with Section 1007.5.
5. Horizontal exits.
6. Smoke barriers.
7. Ramps complying with Section 1010.

Exceptions:

1. Where the exit discharge is not accessible, an exterior area for assisted rescue must be provided in accordance with Section 1007.8.
2. Where the exit stairway is open to the exterior, the accessible means of egress shall include either an area of refuge in accordance with Section 1007.6 or an exterior area for assisted rescue in accordance with Section 1007.8.

Section 1008.1.1 Change to read as shown: (E20-03/04)

1008.1.1 Size of doors. The minimum width of each door opening shall be sufficient for the occupant load thereof and shall provide a clear width of not less than 32 inches (813 mm). Clear openings of doorways with swinging doors shall be measured between the face of the door and the stop, with the door open 90 degrees (1.57 rad). Where this section requires a minimum clear width of 32 inches (813 mm) and a door opening includes two door leaves without a mullion, one leaf shall provide a clear opening width of 32 inches (813 mm). The maximum width of a swinging door leaf shall be 48 inches (1219 mm) nominal. Means of egress doors in an occupancy in Group I-2 used for the movement of beds shall provide a clear width not less than 41½ inches (1054 mm). The height of doors shall not be less than 80 inches (2032 mm).

Exceptions:

1. The minimum and maximum width shall not apply to door openings that are not part of the required means of egress in occupancies in Groups R-2 and R-3 as applicable in Section 1001.1.
2. Door openings to resident sleeping units in occupancies in Group I-3 shall have a clear width of not less than 28 inches (711 mm).
3. Door openings to storage closets less than 10 square feet (0.93 m²) in area shall not be limited by the minimum width.
4. Width of door leafs in revolving doors that comply with Section 1008.1.3.1 shall not be limited.
5. Door openings within a dwelling unit or sleeping unit shall not be less than 78 inches (1981 mm) in height.
6. Exterior door openings in dwelling units and sleeping units, other than the required exit door, shall not be less than 76 inches (1930 mm) in height.
7. Interior egress doors within a dwelling unit or sleeping unit which is not required to be an Accessible unit, Type A unit or Type B unit.
8. Door openings required to be accessible within Type B dwelling units shall have a minimum clear width of 31³/₄ inches (806 mm).

Section 1008.1.2 Change to read as shown: (E22-03/04)

1008.1.2 Door swing. Egress doors shall be side-hinged swinging.

Exceptions:

1. Private garages, office areas, factory and storage areas with an occupant load of 10 or less.
2. Group I-3 occupancies used as a place of detention.
3. Doors within or serving a single dwelling unit in Groups R-2 and R-3 as applicable in Section 1001.1.
4. In other than Group H occupancies, revolving doors complying with Section 1008.1.3.1.
5. In other than Group H occupancies, horizontal sliding doors complying with Section 1008.1.3.3 are permitted in a means of egress.
6. Power-operated doors in accordance with Section 1008.1.3.2.
7. Doors serving a bathroom within an individual sleeping unit in Group R-1.

Doors shall swing in the direction of egress travel where serving an occupant load of 50 or more persons or a Group H occupancy.

The opening force for interior side-swinging doors without closers shall not exceed a 5-pound (22 N) force. For other side-swinging, sliding and folding doors, the door latch shall release when subjected to a 15-pound (67 N) force. The door shall be set in motion when subjected to a 30-pound (133 N) force. The door shall swing to a full-open position when subjected to a 15-

IFC-17

pound (67 N) force. Forces shall be applied to the latch side.

Section 1008.1.6 Change to read as shown: (E25-03/04; E26-03/04)

1008.1.6 Thresholds. Thresholds at doorways shall not exceed 0.75 inch (19.1 mm) in height for sliding doors serving dwelling units or 0.5 inch (12.7 mm) for other doors. Raised thresholds and floor level changes greater than 0.25 inch (6.4 mm) at doorways shall be beveled with a slope not greater than one unit vertical in two units horizontal (50-percent slope).

> **Exception:** The threshold height shall be limited to 7¾ inches (197 mm) where the occupancy is Group R-2 or R-3 as applicable in Section 1001.1, the door is an exterior door that is not a component of the required means of egress, the door, other than an exterior storm or screen door, does not swing over the landing or step and the doorway is not on an accessible route as required by Chapter 11 of the *International Building Code*.

Section 1008.1.8.5 Change to read as shown: (E28-03/04)

1008.1.8.5 Unlatching. The unlatching of any door or leaf shall not require more than one operation.

> **Exceptions:**
> 1. Places of detention or restraint.
> 2. Where manually operated bolt locks are permitted by Section 1008.1.8.4.
> 3. Doors with automatic flush bolts as permitted by Section 1008.1.8.3, Exception 3.
> 4. Doors from individual dwelling units and guestrooms of Group R occupancies as permitted by Section 1008.1.8.3, Exception 4.

Section 1008.1.8.7 Change to read as shown: (E29-03/04)

1008.1.8.7 Stairway doors. Interior stairway means of egress doors shall be openable from both sides without the use of a key or special knowledge or effort.

> **Exceptions:**
> 1. Stairway discharge doors shall be openable from the egress side and shall only be locked from the opposite side.
> 2. This section shall not apply to doors arranged in accordance with Section 403.12.
> 3. In stairways serving not more than four stories, doors are permitted to be locked from the side opposite the egress side, provided they are openable from the egress side and capable of being unlocked simultaneously without unlatching upon a signal from the fire command station or a single location inside the main entrance to the building.

Section 1008.1.9 Change to read as shown: (E31-03/04; E33-03/04; E34-03/04)

1008.1.9 Panic and fire exit hardware. Where panic and fire exit hardware is installed, it shall comply with the following:

1. The actuating portion of the releasing device shall extend at least one-half of the door leaf width.
2. The maximum unlatching force shall not exceed 15 pounds (67 N).

Each door in a means of egress from a Group A or E occupancy having an occupant load of more than 50 and any Group H-1, H-2, H-3 or H-5 occupancy, shall not be provided with a latch or lock unless it is panic hardware or fire exit hardware.

> **Exception:** A main exit of a Group A use in compliance with Section 1008.1.8.3, Item 2.

Electrical rooms with equipment rated 1200 amperes or more and over 6 feet (1.9 m) wide that contain overcurrent devices, switching devices, or control devices, with exit access doors must be equipped with panic hardware and doors must swing in the direction of egress.

If balanced doors are used and panic hardware is required, the panic hardware shall be the push-pad type and the pad shall not extend more then one-half the width of the door measured from the latch side.

Section 1008.2.1 Change to read as shown: (E35-03/04)

1008.2.1 Stadiums. Panic hardware is not required on gates surrounding stadiums where such gates are under constant immediate supervision while the public is present, and further provided that safe dispersal areas based on 3 square feet (0.28 m²) per occupant are located between the fence and enclosed space. Such required safe dispersal areas shall not be located less the 50 feet (15 240 mm) from the enclosed space. See Section 1023.6 for means of egress from safe dispersal areas.

Section 1009.1 Change to read as shown: (E37-03/04)

1009.1 Stairway width. The width of stairways shall be determined as specified in Section 1005.1, but such width shall not be less than 44 inches (1118 mm). See Section 1007.3 for accessible means of egress stairways.

Exceptions:

1. Stairways serving an occupant load of 50 or less shall have a width of not less than 36 inches (914 mm).

2. Spiral stairways as provided for in Section 1009.9.

3. Aisle stairs complying with Section 1024.

4. Where an incline platform lift or stairway chairlift is installed on stairways serving occupancies in Group R-3, or within dwelling units in occupancies in Group R-2, both as applicable in Section 1001.1, a clear passage width not less than 20 inches (508 mm) shall be provided. If the seat and platform can be folded when not in use, the distance shall be measured from the folded position.

Section 1009.3 Change to read as shown: (E40-03/04)

1009.3 Stair treads and risers. Stair riser heights shall be 7 inches (178 mm) maximum and 4 inches (102 mm) minimum. Stair tread depths shall be 11 inches (279 mm) minimum. The riser height shall be measured vertically between the leading edges of adjacent treads. The tread depth shall be measured horizontally between the vertical planes of the foremost projection of adjacent treads and at right angle to the tread's leading edge. Winder treads shall have a minimum tread depth of 11 inches (279 mm) measured at a right angle to the tread's leading edge at a point 12 inches (305 mm) from the side where the treads are narrower and a minimum tread depth of 10 inches (254 mm).

Exceptions:

1. Alternating tread devices in accordance with Section 1009.10.

2. Spiral stairways in accordance with Section 1009.9.

3. Aisle stairs in assembly seating areas where the stair pitch or slope is set, for sightline reasons, by the slope of the adjacent seating area in accordance with Section 1024.11.2.

4. In occupancies in Group R-3, as applicable in Section 101.2, within dwelling units in occupancies in Group R-2, as applicable in Section 101.2, and in occupancies in Group U, which are accessory to an occupancy in Group R-3, as applicable in Section 101.2, the maximum riser height shall be 7.75 inches (197 mm) and the minimum tread depth shall be 10 inches (254mm), the minimum winder tread depth at the walk line shall be 10 inches (254 mm), and the minimum winder tread depth shall be 6 inches (152 mm). A nosing not less than 0.75 inch (19.1 mm) but not more than 1.25 inches (32 mm) shall be provided on stairways with solid risers where the tread depth is less than 11 inches (279 mm).

5. See the *International Existing Building Code* for the replacement of existing stairways.

Section 1009.3.1 Add new section to read as shown: (E40-03/04)

1009.3.1 Winder treads. Winder treads are not permitted in means of egress stairways except within a dwelling unit.

Exceptions:

1. Circular stairways in accordance with Section 1009.7

2. Spiral stairways in accordance with Section 1009.9

(Renumber subsequent sections)

Section 1009.3.1 Change to read as shown: (E40-03/04)

1009.3.1 Dimensional uniformity. Stair treads and risers shall be of uniform size and shape. The tolerance between the largest and smallest riser height or between the largest and smallest tread depth shall not exceed 0.375 inch (9.5 mm) in any flight of stairs. The greatest winder tread depth at the 12-inch (305 mm) walk line within any flight of stairs shall not exceed the smallest by more than 0.375 inch (9.5 mm) measured at a right angle to the treads leading edge.

Exceptions:

1. Nonuniform riser dimensions of aisle stairs complying with Section 1024.11.2.

2. Consistently shaped winders, complying with Section 1009.3, differing from rectangular treads in the same stairway flight.

Where the bottom or top riser adjoins a sloping public way, walkway or driveway having an established grade and serving as a landing, the bottom or top riser is permitted to be reduced along the slope to less than 4 inches (102 mm) in height with the variation in height of the bottom or top riser not to exceed one unit vertical in 12 units horizontal (8-per-cent slope) of stairway width. The nosings or leading edges of treads at such nonuniform height risers shall have a distinctive marking stripe, different from any other nosing marking provided on the stair flight. The distinctive marking stripe shall be visible in descent of the stair and shall have a slip-resistant surface. Marking stripes shall have a width of at least 1 inch (25 mm) but not more than 2 inches (51 mm).

Section 1009.5.2 Change to read as shown: (E46-03/04)

1009.5.2 Outdoor conditions. Outdoor stairways and outdoor approaches to stairways shall be designed so that water will not accumulate on walking surfaces.

Section 1009.5.3 (relocated from Section1019.1.5) Change to read as shown: (E96-03/04)

1009.5.3 Enclosures under stairways. The walls and soffits within enclosed usable spaces under enclosed and unenclosed stairways shall be protected by 1-hour fire-resistance-rated construction, or the fire-resistance rating of the stairway enclosure, whichever is greater. Access to the enclosed space shall not be directly from within the stair enclosure.

> **Exception:** Spaces under stairways serving and contained within a single residential dwelling unit in Group R-2 or R-3 as applicable in Section 1001.1 shall be permitted to be protected on the enclosed side with ½ inch (12.7 mm) gypsum board.

There shall be no enclosed usable space under exterior exit stairways unless the space is completely enclosed in 1-hour fire-resistance-rated construction. The open space under exterior stairways shall not be used for any purpose.

Section 1009.8 Delete section without substitution: (E40-03/04)

Section 1009.12.2 Add new section to read as shown: (E51-03/04)

1009.12.2 Protection at roof hatch openings. Where the roof hatch opening providing the required access is located within 10 feet (3048 mm) of the roof edge, such roof access or roof edge shall be protected by guards installed in accordance with the provisions of Section 1012.

Section 1010.2 Change to read as shown: (E52-03/04)

1010.2 Slope. Ramps used as part of a means of egress shall have a running slope not steeper than one unit vertical in 12 units horizontal (8-percent slope). The slope of other pedestrian ramps shall not be steeper than one unit vertical in eight units horizontal (12.5-percent slope).

> **Exception:** Aisle ramp slope in occupancies of Group A shall comply with Section 1024.11.

Section 1010.6.3 Change to read as shown: (E53-03/04)

1010.6.3 Length. The landing length shall be 60 inches (1525 mm) minimum.

> **Exceptions:**
>
> 1. Landings in nonaccessible Group R-2 and R-3 individual dwelling units, as applicable in Section 1001.1, are permitted to be 36 inches (914 mm) minimum.
>
> 2. Where the ramp is not a part of an accessible route, the length of the landing shall not be required to be more than 48 inches (1220 mm) in the direction of travel.

Section 1010.7 Change to read as shown: (E1-03/04)

1010.7 Ramp construction. All ramps shall be built of materials consistent with the types permitted for the type of construction of the building; except that wood handrails shall be permitted for all types of construction. Ramps used as an exit shall conform to the applicable requirements of Sections 1019.1 and 1019.1.1 through 1019.1.3 for exit enclosures.

Section 1010.7.2 Change to read as shown: (E46-03/04)

1010.7.2 Outdoor conditions. Outdoor ramps and outdoor approaches to ramps shall be designed so that water will not accumulate on walking surfaces.

Section 1010.9 Change to read as shown: (E54-03/04)

1010.9 Edge protection. Edge protection complying with Section 1010.9.1, 1010.9.2 or 1010.9.3 shall be provided on each side of ramp runs and at each side of ramp landings.

> **Exceptions:**
>
> 1. Edge protection is not required on ramps not required to have handrails, provided they have flared sides that comply with the ICC A117.1 curb ramp provisions.
>
> 2. Edge protection is not required on the sides of ramp landings serving an adjoining ramp run or stairway.

3. Edge protection is not required on the sides of ramp landings having a vertical dropoff of not more than 0.5 inch (13 mm) within 10 inches (254 mm) horizontally of the required landing area.

Section 1010.9.3 Add new section to read as shown: (E54-03/04)

1010.9.3 Extended floor or ground surface. The floor or ground surface of the ramp run or landing shall extend 12 inches (305 mm) minimum beyond the inside face of a handrail complying with Section 1009.11.

Section 1011.1 Change to read as shown: (E3-03/04)

1011.1 Where required. Exits and exit access doors shall be marked by an approved exit sign readily visible from any direction of egress travel. Access to exits shall be marked by readily visible exit signs in cases where the exit or the path of egress travel is not immediately visible to the occupants. Exit sign placement shall be such that no point in a corridor is more than 100 feet (30 480 mm) or the listed viewing distance for the sign, whichever is less, from the nearest visible exit sign.

Exceptions:

1. Exit signs are not required in rooms or areas which require only one exit or exit access.

2. Main exterior exit doors or gates which obviously and clearly are identifiable as exits need not have exit signs where approved by the fire code official.

3. Exit signs are not required in occupancies in Group U and individual sleeping units or dwelling units in Group R-1, R-2 or R-3.

4. Exit signs are not required in sleeping areas in occupancies in Group I-3.

5. In occupancies in Groups A-4 and A-5, exit signs are not required on the seating side of vomitories or openings into seating areas where exit signs are provided in the concourse that are readily apparent from the vomitories. Egress lighting is provided to identify each vomitory or opening within the seating area in an emergency.

Section 1011.5.1 Change to read as shown: (E59-03/04)

1011.5.1 Graphics. Every exit sign and directional exit sign shall have plainly legible letters not less than 6 inches (152 mm) high with the principal strokes of the letters not less than 0.75 inch (19.1 mm) wide. The word "EXIT" shall have letters having a width not less than 2 inches (51 mm) wide except the letter "I," and the minimum spacing between letters shall not be less than 0.375 inch (9.5 mm). Signs larger than the minimum established in this section shall have letter widths, strokes and spacing in proportion to their height.

The word "EXIT" shall be in high contrast with the background and shall be clearly discernible when the exit sign illumination means is or is not energized. If a chevron directional indicator is provided as part of the exit sign, the construction shall be such that the direction of the chevron directional indicator cannot be readily changed.

Section 1012.5 Change to read as shown: (E51-03/04; E67-03/04)

1012.5 Mechanical equipment. Guards shall be provided where appliances, equipment, fans, roof hatch openings or other components that require service are located within 10 feet (3048 mm) of a roof edge or open side of a walking surface and such edge or open side is located more than 30 inches (762 mm) above the floor, roof or grade below. The guard shall be constructed so as to prevent the passage of a 21-inch (533 mm) diameter sphere. The guard shall extend not less than 30 inches beyond each end of such appliance, equipment, fan or component.

Section 1012.6 Add new section to read as shown: (E51-03/04)

1012.6 Roof access. Guards shall be provided where the roof hatch opening is located within 10 feet (3048 mm) of a roof edge or open side of a walking surface and such edge or open side is located more than 30 inches (762 mm) above the floor, roof or grade below. The guard shall be constructed so as to prevent the passage of a 21-inch (533 mm) diameter sphere.

Section 1013.2 Change to read as shown: (E68-03/04)

1013.2 Egress through intervening spaces. Egress through intervening spaces shall be controlled in accordance with this section.

1. Egress from a room or space shall not pass through adjoining or intervening rooms or areas, except where such adjoining rooms or areas are accessory to the area served; are not a high-hazard occupancy and provide a discernible path of egress travel to an exit.

    Exception: Means of egress are not prohibited through adjoining or intervening rooms or spaces in a Group H occupancy when the adjoining or intervening rooms or spaces are the same or a lesser hazard occupancy group.

2. Egress shall not pass through kitchens, storage rooms, closets or spaces used for similar purposes.

Exception: Means of egress are not prohibited through a kitchen area serving adjoining rooms constituting part of the same dwelling unit or sleeping unit.

3. An exit access shall not pass through a room that can be locked to prevent egress.

4. Means of egress from dwelling units or sleeping areas shall not lead through other sleeping areas, toilet rooms or bathrooms.

Section 1013.2.2 Change to read as shown: (E3-03/04)

1013.2.2 Group I-2. Habitable rooms or suites in Group I-2 occupancies shall have an exit access door leading directly to a corridor.

Exceptions:

1. Rooms with exit doors opening directly to the outside at ground level.

2. Patient sleeping rooms are permitted to have one intervening room if the intervening room is not used as an exit access for more than eight patient beds.

3. Special nursing suites are permitted to have one intervening room where the arrangement allows for direct and constant visual supervision by nursing personnel.

4. For rooms other than patient sleeping rooms, suites of rooms are permitted to have one intervening room if the travel distance within the suite to the exit access door is not greater than 100 feet (30 480 mm) and are permitted to have two intervening rooms where the travel distance within the suite to the exit access door is not greater than 50 feet (15 240 mm).

Suites of sleeping rooms shall not exceed 5,000 square feet (465 m^2). Suites of rooms, other than patient sleeping rooms, shall not exceed 10,000 square feet (929 m^2). Any patient sleeping room, or any suite that includes patient sleeping rooms, of more than 1,000 square feet (93 m^2) shall have at least two exit access doors remotely located from each other. Any room or suite of rooms, other than patient sleeping rooms, of more than 2,500 square feet (232 m^2) shall have at least two access doors remotely located from each other. The travel distance between any point in a Group I-2 occupancy and an exit access door in the room shall not exceed 50 feet (15 240 mm). The travel distance between any point in a suite of sleeping rooms and an exit access door of that suite shall not exceed 100 feet (30 480 mm).

Section 1013.5 Change to read as shown: (E46-03/04)

1013.5 Egress balconies. Balconies used for egress purposes shall conform to the same requirements as corridors for width, headroom, dead ends and projections.

Section 1016.4 Change to read as shown: (E3-03/04)

1016.4 Air movement in corridors. Corridors shall not serve as supply, return, exhaust, relief or ventilation air ducts or plenums.

Exceptions:

1. Use of a corridor as a source of makeup air for exhaust systems in rooms that open directly onto such corridors, including toilet rooms, bathrooms, dressing rooms, smoking lounges and janitor closets, shall be permitted provided that each such corridor is directly supplied with outdoor air at a rate greater than the rate of makeup air taken from the corridor.

2. Where located within a dwelling unit, the use of corridors for conveying return air shall not be prohibited.

3. Where located within tenant spaces of 1,000 square feet (93 m^2) or less in area, utilization of corridors for conveying return air is permitted.

Section 1018.1 Change to read as shown: (E89-03/04)

1018.1 Minimum number of exits. All rooms and spaces within each story shall be provided with and have access to the minimum number of approved independent exits required by Table 1018.1 based on the occupant load of the story, except as modified in Section 1014.1 or 1018.2. For the purposes of this chapter, occupied roofs shall be provided with exits as required for stories. The required number of exits from any story, basement or individual space shall be maintained until arrival at grade or the public way.

Table 1018.1 Change entire table to read as shown: (E89-03/04)

TABLE 1018.1
MINIMUM NUMBER OF EXITS
FOR OCCUPANT LOAD

OCCUPANT LOAD (persons per story)	MINIMUM NUMBER OF EXITS (per story)
1-500	2
501-1,000	3
More than 1,000	4

2004 SUPPLEMENT TO THE IFC

Section 1019.1 Change to read as shown: (E90-03/04; E91-03/04; E93-03/04)

1019.1 Enclosures required. Interior exit stairways and interior exit ramps shall be enclosed with fire barriers constructed in accordance with Section 706 of the *International Building Code*. Exit enclosures shall have a fire-resistance rating of not less than 2 hours where connecting four stories or more and not less than 1 hour where connecting less than four stories. The number of stories connected by the shaft enclosure shall include any basements but not any mezzanines. An exit enclosure shall not be used for any purpose other than means of egress.

Exceptions:

1. In other than Group H and I occupancies, a stairway is not required to be enclosed when serving an occupant load of less than 10 either not more than one story above or one story below the level of exit discharge, but not both.

2. Exits in buildings of Group A-5 where all portions of the means of egress are essentially open to the outside need not be enclosed.

3. Stairways serving and contained within a single residential dwelling unit or sleeping unit in occupancies in Group R-1. R-2 or R-3 are not required to be enclosed.

4. Stairways that are not a required means of egress element are not required to be enclosed where such stairways comply with Section 707.2 of the *International Building Code*.

5. Stairways in open parking structures which serve only the parking structure are not required to be enclosed.

6. Stairways in occupancies in Group I-3 as provided for in Section 408.3.6 of the *International Building Code* are not required to be enclosed.

7. Means of egress stairways as required by Section 410.5.4 of the *International Building Code* are not required to be enclosed.

8. In other than occupancy Groups H and I, a maximum of 50 percent of egress stairways serving one adjacent floor are not required to be enclosed, provided at least two means of egress are provided from both floors served by the unenclosed stairways. Any two such interconnected floors shall not be open to other floors.

9. In other than occupancy Groups H and I, interior egress stairways serving only the first and second stories of a building equipped throughout with an automatic sprinkler system in accordance with Section 903.3.1.1 are not required to be enclosed, provided at least two means of egress are provided from both floors served by the unenclosed stairways. Such interconnected stories shall not be open to other stories.

Section 1019.1.3 Change to read as shown: (FS53-03/04)

1019.1.3 Ventilation. Equipment and ductwork for exit enclosure ventilation shall comply with one of the following items:

1. Such equipment and ductwork shall be located exterior to the building and shall be directly connected to the exit enclosure by ductwork enclosed in construction as required for shafts.

2. Where such equipment and ductwork is located within the exit enclosure, the intake air shall be taken directly from the outdoors and the exhaust air shall be discharged directly to the outdoors, or such air shall be conveyed through ducts enclosed in construction as required for shafts.

3. Where located within the building, such equipment and ductwork shall be separated from the remainder of the building, including other mechanical equipment, with construction as required for shafts.

In each case, openings into the fire-resistance-rated construction shall be limited to those needed for maintenance and operation and shall be protected by opening protectives in accordance with Section 715 for shaft enclosures.

Exit enclosure ventilation systems shall be independent of other building ventilation systems.

Section 1019.1.4 Change to read as shown: (E1-03/04)

1019.1.4 Exit enclosure exterior walls. Exterior walls of an exit enclosure shall comply with the requirements of Section 704 of the *International Building Code* for exterior walls. Where nonrated walls or unprotected openings enclose the exterior of the stairway and the walls or openings are exposed by other parts of the building at an angle of less than 180 degrees (3.14 rad), the building exterior walls within 10 feet (3048 mm) horizontally of a nonrated wall or unprotected opening shall be constructed as required for a minimum 1-hour fire-resistance rating with ¾-hour opening protectives. This construction shall extend vertically from the ground to a point 10 feet (3048 mm) above the topmost landing of the stairway or to the roof line, whichever is lower.

Section 1019.1.5 Relocate to be Section 1009.5.3: (E96-03/04)

Section 1019.1.7 Change to read as shown: (E1-03/04)

1019.1.7 Stairway floor number signs. A sign shall be provided at each floor landing in interior exit enclosures connecting more than three stories designating the floor level, the terminus of the top and bottom of the stair enclosure and the identification of the stair. The signage shall also state the story of, and the direction to the exit discharge and the availability of roof access from the stairway for the fire department. The sign shall be located 5 feet (1524 mm) above the floor landing in a position that is readily visible when the doors are in the open and closed positions.

Section 1024.3 Change to read as shown: (E105-03/04)

1024.3 Assembly other exits. In addition to having access to a main exit, each level of an occupancy in Group A having an occupant load greater than 300 shall be provided with additional means of egress that shall provide an egress capacity for at least one-half of the total occupant load served by that level and comply with Section 1014.2.

> **Exception:** In assembly occupancies where there is no well-defined main exit or where multiple main exits are provided, exits shall be permitted to be distributed around the perimeter of the building provided that the total width of egress is not less than 100 percent of the required width.

Section 1024.5.1 Change to read as shown: (E1-03/04)

1024.5.1 Enclosure of balcony openings. Interior stairways and other vertical openings shall be enclosed in an exit enclosure as provided in Section 1019.1, except that stairways are permitted to be open between the balcony and the main assembly floor in occupancies such as theaters, churches and auditoriums. At least one accessible means of egress is required from a balcony or gallery level containing accessible seating locations in accordance with Section 1007.3 or 1007.4.

CHAPTER 13
COMBUSTIBLE DUST-PRODUCING OPERATIONS

Table 1304.1 Change entire table to read as shown: (F188-03/04)

TABLE 1304.1
EXPLOSION PROTECTION STANDARDS

Standard	Subject
NFPA 61	Agricultural and Food Products
NFPA 69	Explosion Prevention
NFPA 85	Boiler and Combustion Systems Hazards
NFPA 120	Coal Preparation Plants
NFPA 484	Combustible Metals, Metal Powders and Metal Dusts
NFPA 654	Combustible Particulate Solids
NFPA 655	Sulfur
NFPA 664	Prevention of Fires and Explosions in Wood Processing and Woodworking Facilities
ICC *Electrical Code*	Electrical Installations

CHAPTER 14
FIRE SAFETY DURING CONSTRUCTION AND DEMOLITION

Section 1411.3 Add new section to read as shown: (E145-03/04)

1411.3 Stairway floor number signs. Temporary stairway floor number signs shall be provided in accordance with the requirements of Section 1019.1.7.

CHAPTER 15
FLAMMABLE FINISHES

Section 1503.2.1.1 Change to read as shown: (F142-03/04)

1503.2.1.1 Spray and vapor areas. Electrical wiring and equipment in spray and vapor areas shall be of an explosion-proof type approved for use in such hazardous locations. Such areas shall be considered to be Class I, Division 1 or Class II, Division 1 hazardous locations in accordance with the *ICC Electrical Code*.

Section 1503.2.2 Change to read as shown: (F142-03/04)

1503.2.2 Open flames and sparks. Open flames and spark-producing devices shall not be located in spray or vapor areas and shall not be located within 20 feet (6096 mm) of such areas unless separated by a permanent partition.

> **Exception:** Drying and baking apparatus complying with Section 1504.7.2.

Section 1503.2.4 Change to read as shown: (F142-03/04)

1503.2.4 Equipment enclosures. Equipment or apparatus that is capable of producing sparks or particles of hot metal that would fall into a spray or vapor area shall be totally enclosed.

Section 1503.2.6 Change to read as shown: (F142-03/04)

1503.2.6 Smoking prohibited. Smoking shall be prohibited in spray or vapor areas. "No Smoking" signs complying with Section 310 shall be conspicuously posted in such areas.

Sections 1504.1.3 and 1504.1.3.1 Change to read as shown: (F142-03/04)

1504.1.3 Spraying areas. Spraying areas shall be designed and constructed in accordance with the *International Building Code* and Sections 1504.1.3.1, 1504.2, 1504.3, 1504.4, 1504.5 and 1504.6 of this code.

1504.1.3.1 Floor. Combustible floor construction in spraying areas shall be covered by approved, noncombustible, nonsparking material, except where combustible coverings, such as thin paper or plastic and strippable coatings are utilized over noncombustible materials to facilitate cleaning operations in spraying areas.

Section 1504.2.2 Change to read as shown: (F142-03/04)

1504.2.2 Recirculation. Air exhausted from spraying operations shall not be recirculated.

Exceptions:

1. Air exhausted from spraying operations is allowed to be recirculated as makeup air for unmanned spray operations provided that:

    1.1. The solid particulate has been removed.

    1.2. The vapor concentration is less than 25 percent of the LFL.

    1.3. Approved equipment is used to monitor the vapor concentration.

    1.4. When the vapor concentration exceeds 25 percent of the LFL, the following shall occur:
        a. An alarm shall sound; and
        b. Spray operations shall automatically shut down.

    1.5. In the event of shutdown of the vapor concentration monitor, 100 percent of the air volume specified in Section 1504.2 is automatically exhausted.

2. Air exhausted from spraying operations is allowed to be recirculated as makeup air to manned spraying operations where all of the conditions provided in Exception 1 are included in the installation and documents have been prepared to show that the installation does not pose a life safety hazard to personnel inside the spray booth, spray area or spray room.

Section 1504.2.6 Change to read as shown: (F143-03/04)

1504.2.6 Termination point. The termination point for exhaust ducts discharging to the atmosphere shall not be less than the following distances:

1. Ducts conveying explosive or flammable vapors, fumes or dusts: 30 feet (9144 mm) from the property line; 10 feet (3048 mm) from openings into the building; 6 feet (1829 mm) from exterior walls and roofs; 30 feet (9144 mm) from combustible walls or openings into the building which are in the direction of the exhaust discharge; 10 feet (3048 mm) above adjoining grade.

2. Other product-conveying outlets: 10 feet (3048 mm) from the property line; 3 feet (914 mm) from exterior walls and roofs; 10 feet (3048 mm) from openings into the building; 10 feet (3048 mm) above adjoining grade.

Sections 1504.4 and 1504.5 Change to read as shown: (F142-03/04)

1504.4 Different coatings. Spray booths, spray rooms and spray areas shall not be alternately utilized for different types of coating materials where the combination of materials is conducive to spontaneous ignition, unless all deposits of one material are removed from the booth, room or area and exhaust ducts prior to spraying with a different material.

1504.5 Illumination. Where spraying areas, spray rooms or spray booths are illuminated through glass panels or other transparent materials, only fixed lighting units shall be utilized as a source of illumination.

Sections 1504.5.1, 1504.5.2 and 1504.5.3 Change to read as shown: (EL3-03/04)

1504.5.1 Glass panels. Panels for luminaires or for observation shall be of heat-treated glass, wired glass or hammered-wire glass and shall be sealed to confine vapors, mists, residues, dusts and deposits to the spraying area. Panels for luminaires shall be separated from the fixture to prevent the surface temperature of the panel from exceeding 200°F (93°C).

1504.5.2 Exterior luminairies. Luminaires attached to the walls or ceilings of a spraying area, but which are outside of any classified area and are separated from the spraying area by vapor-tight glass panels, shall be suitable for use in ordinary hazard locations. Such fixtures shall be serviced from outside the spraying area.

1504.5.3 Integral luminaires. Luminaires that are an integral part of the walls or ceiling of a spraying area are allowed to be separated from the spraying area by glass panels that are an integral part of the luminaire. Such luminaires shall be listed for use in Class I, Division 2 or Class II, Division 2 locations, whichever is applicable, and also shall be suitable for accumulations of deposits of combustible residues. Such luminaires are allowed to be serviced from inside the spraying area.

CHAPTER 16
FRUIT AND CROP RIPENING

Section 1604.4 Change to read as shown: (EL3-03/04)

1604.4 Lighting. Lighting shall be by approved electric lamps or luminaires only.

CHAPTER 17
FUMIGATION AND THERMAL INSECTICIDAL FOGGING

Section 1703.5 Change to read as shown: (F43-03/04)

1703.5 Sealing of buildings. Paper and other similar materials that do not meet the flame propagation performance criteria of NFPA 701 shall not be used to wrap or cover a building in excess of that required for the sealing of cracks, casements and similar openings.

CHAPTER 18
SEMICONDUCTOR FABRICATION FACILITIES

Section 1803.1 Change to read as shown: (F146-03/04)

1803.1 Emergency control station. An emergency control station shall be provided on the premises at an approved location outside of the fabrication area, and shall be continuously staffed by trained personnel. The emergency control station shall receive signals from emergency equipment and alarm and detection systems. Such emergency equipment and alarm and detection systems shall include, but not be limited to, the following where such equipment or systems are required to be provided either in this chapter or elsewhere in this code:

1. Automatic sprinkler system alarm and monitoring systems.
2. Manual fire alarm systems.
3. Emergency alarm systems.
4. Continuous gas detection systems.
5. Smoke detection systems.
6. Emergency power system.
7. Automatic detection and alarm systems for pyrophoric liquids and Class 3 water reactive liquids required in Section 1805.2.2.4.
8. Exhaust ventilation flow alarm devices for pyrophoric liquids and Class 3 water reactive liquids cabinet exhaust ventilation systems required in Section 1805.2.2.4.

Section 1803.3.9 Add new section to read as shown: (F146-03/04)

1803.3.9 Cabinets containing pyrophoric liquids or water-reactive Class 3 liquids. Cabinets in fabrication areas containing pyrophoric liquids or Class 3 water reactive liquids in containers or in amounts greater than 0.5 gallon (2 L) shall comply with Section 1805.2.2.4.

Sections 1803.10.4 and 1803.10.4.1 Change to read as shown: (F144-03/04)

1803.10.4 Exhaust ducts for HPM. An approved automatic sprinkler system shall be provided in exhaust ducts conveying gases, vapors, fumes, mists or dusts generated from HPM in accordance with this section and the *International Mechanical Code*.

1803.10.4.1 Metallic and noncombustible nonmetallic exhaust ducts. An approved automatic sprinkler system shall be provided in metallic and noncombustible nonmetallic exhaust ducts when all of the following conditions apply:

1. When the largest cross-sectional diameter is equal to or greater than 10 inches (254 mm).
2. The ducts are within the building.
3. The ducts are conveying flammable gases, vapors or fumes.

Section 1803.13 Change to read as shown: (F183-03/04)

1803.13 Continuous gas detection systems. A continuous gas detection system shall be provided for HPM gases when the physiological warning threshold level of the gas is at a higher level than the accepted permissible exposure limit (PEL) for the gas and for flammable gases in accordance with this section.

Section 1803.14.1 Change to read as shown: (F144-03/04; F146-03/04)

1803.14.1 Where required. Exhaust ventilation systems shall be provided in the following locations in accordance with the requirements of this section and the *International Building Code*:

1. Fabrication areas: Exhaust ventilation for fabrication areas shall comply with the *International Building Code*. The fire code official

is authorized to require additional manual control switches.

2. Workstations: A ventilation system shall be provided to capture and exhaust gases, fumes and vapors at workstations.

3. Liquid storage rooms: Exhaust ventilation for liquid storage rooms shall comply with Section 2704.3.1 and the *International Building Code*.

4. HPM rooms: Exhaust ventilation for HPM rooms shall comply with Section 2704.3.1 and the *International Building Code.*

5. Gas cabinets: Exhaust ventilation for gas cabinets shall comply with Section 2703.8.6.2. The gas cabinet ventilation system is allowed to connect to a workstation ventilation system. Exhaust ventilation for gas cabinets containing highly toxic or toxic gases shall also comply with Chapter 37.

6. Exhausted enclosures: Exhaust ventilation for exhausted enclosures shall comply with Section 2703.8.5.2. Exhaust ventilation for exhausted enclosures containing highly toxic or toxic gases shall also comply with Chapter 37.

7. Gas rooms: Exhaust ventilation for gas rooms shall comply with Section 2703.8.4.2. Exhaust ventilation for gas cabinets containing highly toxic or toxic gases shall also comply with Chapter 37.

8. Cabinets containing pyrophoric liquids or Class 3 water-reactive liquids. Exhaust ventilation for cabinets in fabrication areas containing pyrophoric liquids or Class 3 water-reactive liquids as required in Section 1805.2.2.4.

Section 1803.15.1 Change to read as shown: (F146-03/04)

1803.15.1 Required electrical systems. Emergency power shall be provided for electrically operated equipment and connected control circuits for the following systems:

1. HPM exhaust ventilation systems.

2. HPM gas cabinet ventilation systems.

3. HPM exhausted enclosure ventilation systems.

4. HPM gas room ventilation systems.

5. HPM gas detection systems.

6. Emergency alarm systems.

7. Manual fire alarm systems.

8. Automatic sprinkler system monitoring and alarm systems.

9. Automatic alarm and detection systems for pyrophoric liquids and Class 3 water-reactive liquids required in Section 1805.2.2.4.

10. Flow alarm switches for pyrophoric liquids and Class 3 water-reactive liquids cabinet exhaust ventilation systems required in Section 1805.2.2.4.

11. Electrically operated systems required elsewhere in this code or in the *International Building Code* applicable to the use, storage or handling of HPM.

Section 1804.2 Change to read as shown: (F145-03/04)

1804.2 Fabrication areas. Hazardous materials storage and the maximum quantities of hazardous materials in use and storage allowed in fabrication areas shall be in accordance with this section.

1804.2.1 Location of HPM storage in fabrication areas. Storage of HPM in fabrication areas shall be within approved or listed storage cabinets, gas cabinets, exhausted enclosures or within a workstation.

Flammable and combustible liquid storage cabinets shall comply with Section 3404.3.2.

Hazardous materials storage cabinets shall comply with Section 2703.8.7.

Gas cabinets shall comply with Section 2703.8.6. Gas cabinets for highly toxic or toxic gases shall also comply with Section 3704.1.2.

Exhausted enclosures shall comply with Section 2703.8.5. Exhausted enclosures for highly toxic or toxic gases shall also comply with Section 3704.1.3.

Workstations shall comply with Section 1805.2.2.

1804.2.2 Maximum aggregate quantities in fabrication areas. The aggregate quantities of hazardous materials stored or used in a single fabrication area shall be limited as specified in this section.

Exception: Fabrication areas containing quantities of hazardous materials not exceeding the maximum allowable quantities per control area established by Sections 2703.1.1, 3404.3.4 and 3404.3.5.

1804.2.2.1 Storage and use in fabrication areas. The maximum quantities of hazardous materials stored or used in a single fabrication area shall not exceed the quantities set forth in Table 1804.2.2.1.

(Renumber current Table 1804.2.1 to Table 1804.2.2.1; no change to table.)

2004 SUPPLEMENT TO THE IFC

1804.2.2.2 HPM storage in fabrication areas. The maximum quantities of HPM stored in a single fabrication area shall not exceed the maximum allowable quantities per control area established by Sections 2703.1.1 and 3404.3.4.

Section 1804.3 Change to read as shown: (F145-03/04)

1804.3 Indoor storage outside of fabrication areas. The indoor storage of hazardous materials outside of fabrication areas shall be in accordance with this section.

1804.3.1 HPM storage. The indoor storage of HPM in quantities greater than those listed in Sections 2703.1.1 and 3404.3.4 shall be in a room complying with the requirements of the *International Building Code* and this code for a liquid storage room, HPM room or gas room as appropriate for the materials stored.

1804.3.2 Other hazardous materials storage. The indoor storage of other hazardous materials shall comply with Sections 2701, 2703 and 2704 and other applicable provisions of this code.

1804.3.3 Separation of incompatible hazardous materials. Incompatible hazardous materials in storage shall be separated from each other in accordance with Section 2703.9.8.

Section 1805.2 Change to read as shown: (F145-03/04; F146-03/04)

1805.2 Fabrication areas. The use of hazardous materials in fabrication areas shall be in accordance with Sections 1805.2.1 through 1805.2.3.4.

1805.2.1 Location of HPM in use in fabrication areas. Hazardous production materials in use in fabrication areas shall be within approved or listed gas cabinets, exhausted enclosures or a workstation.

1805.2.2 Maximum aggregate quantities in fabrication areas. The aggregate quantities of hazardous materials in a single fabrication area shall comply with Section 1804.2.2, and Table 1804.2.2.1. The quantity of HPM in use at a workstation shall not exceed the quantities listed in Table 1805.2.2.

TABLE 1805.2.2
MAXIMUM QUANTITIES OF HPM AT A WORKSTATION[e]

HPM CLASSIFICATION	STATE	MAXIMUM QUANTITY
Flammable, highly toxic pyrophoric and toxic combined	Gas	3 cylinders
Flammable	Liquid Solid	15 gallons[a,b,c] 5 pounds[b,c]
Corrosive	Gas Liquid Solid	3 cylinders Use-Open System: 25 gallons[a,c] Use-Closed System: 150 gallons[a,c,f] 20 pounds[b,c]
Highly toxic	Liquid Solid	15 gallons[a,b] 5 pounds[b]
Oxidizer	Gas Liquid Solid	3 cylinders 12 gallons[a,b,c] 20 pounds[b,c]
Pyrophoric	Liquid Solid	0.5 gallon[d,g] See Table 1804.2.2.1
Toxic	Liquid Solid	15 gallons[a,b,c] 5 pounds[b,c]
Unstable reactive Class 3	Liquid Solid	0.5 gallon[b,c] 5 pounds[b,c]
Water reactive Class 3	Liquid Solid	0.5 gallon[d,g] See Table 1804.2.2.1

For SI 1 pound = 0.454 kg, 1 gallon = 3.785 L

a. DOT shipping containers with capacities of greater than 5.3 gallons shall not be located within a workstation.
b. Maximum allowable quantities shall be increased 100 percent for closed systems operations. When Note c also applies, the increase for both notes shall be allowed.
c. Quantities shall be allowed to be increased 100 percent when workstations are internally protected with an approved automatic fire-extinguishing or suppression system complying with Chapter 9. When Note b also applies, the increase for both notes shall be allowed. When Note f also applies, the maximum increase allowed for both Notes c and f shall not exceed 100 percent.
d. Allowed only in workstations that are internally protected with an approved automatic fire-extinguishing or fire protection system complying with Chapter 9 and compatible with the reactivity of materials in use at the workstation.
e. The quantity limits apply only to materials classified as HPM.
f. Quantities shall be allowed to be increase 100 percent for nonflammable, noncombustible corrosive liquids when the materials of construction for workstations are listed or approved for use without internal fire extinguishing or suppression system protection. When Note c also applies, the maximum increase allowed for both Notes c and f shall not exceed 100 percent.
g. A maximum quantity of 5.3 gallons shall be allowed at a workstation when conditions are in accordance with Section 1805.2.3.5.

1805.2.3 Workstations. Workstations in fabrication areas shall be in accordance with Sections 1805.2.3.1 through 1805.2.3.4.

1805.2.3.1 Construction. Workstations in fabrication areas shall be constructed of materials compatible with the materials used and stored at the workstation. The portion of the workstation that serves as a cabinet for HPM gases and HPM flammable liquids shall be noncombustible and, if of metal, shall be not less than 0.0478-inch (18 gage) (1.2 mm) steel.

1805.2.3.2 Protection of vessels. Vessels containing hazardous materials located in or connected to a workstation shall be protected as follows:

1. HPM: Vessels containing HPM shall be protected from physical damage and shall not project from the workstation.
2. Hazardous cryogenic fluids, gases and liquids: Hazardous cryogenic fluid, gas and liquid vessels located within a workstation shall be protected from seismic forces in an approved manner in accordance with the *International Building Code*.
3. Compressed gases: Protection for compressed gas vessels shall also comply with Section 3003.3.
9. Cryogenic fluids: Protection for cryogenic fluid vessels shall also comply with Section 3203.3.

1805.2.3.3 Drainage and containment for HPM liquids. Each workstation utilizing HPM liquids shall have all of the following:

1. Drainage piping systems connected to a compatible system for disposition of such liquids;
2. The work surface provided with a slope or other means for directing spilled materials to the containment or drainage system; and
3. An approved means of containing or directing spilled or leaked liquids to the drainage system.

1805.2.3.4 Clearances. Workstations where HPM is used shall be provided with horizontal servicing clearances of not less than 3 feet (914 mm) for electrical equipment, gas-cylinder connections and similar hazardous conditions. These clearances shall apply only to normal operational procedures and not to repair-or maintenance-related work.

1805.2.3.5 Pyrophoric liquids and Class 3 water-reactive liquids. Pyrophoric liquids and Class 3 water-reactive liquids in containers greater than 0.5-gallon (2 L) but not exceeding 5.3-gallon (20 L) capacity shall be allowed at workstations when located inside cabinets and the following conditions are met:

1. Maximum amount per cabinet: The maximum amount per cabinet shall be limited to 5.3 gallons (20 L).
2. Cabinet construction: Cabinets shall be constructed in accordance with the following:
   2.1. Cabinets shall be constructed of not less than 0.097-inch (2.5 mm) (12 gauge) steel.
   2.2. Cabinets shall be permitted to have self-closing limited access ports or noncombustible windows that provide access to equipment controls.
   2.3. Cabinets shall be provided with self-closing or manual closing doors. Manual closing doors shall be equipped with a door switch that will initiate local audible and visual alarms when the door is in the open position.
3. Cabinet exhaust ventilation system: An exhaust ventilation system shall be provided for cabinets and shall comply with the following:
   3.1. The system shall be designed to operate at a negative pressure in relation to the surrounding area.
   3.2. The system shall be equipped with a pressure monitor and a flow switch alarm monitored at the on-site emergency control station.
4. Cabinet spill containment: Spill containment shall be provided in each cabinet, with the spill containment capable of holding the contents of the aggregate amount of liquids in containers in each cabinet.
5. Valves: Valves in supply piping between the product containers in the cabinet and the workstation served by the containers shall fail in the closed position upon power failure, loss of exhaust ventilation and upon actuation of the fire control system.
6. Fire detection system: Each cabinet shall be equipped with an automatic fire detection system complying with the following conditions:
   6.1. Automatic detection system: UV/IR, High Sensitivity Smoke Detection (HSSD) or other approved detection systems shall be provided inside each cabinet.
   6.2. Automatic shut-Off: Activation of the detection system shall automatically close the shutoff valve, or shutoff valves, at the source on the liquid supply.
   6.3 Alarms and signals: Activation of the detection system shall initiate a local alarm within the fabrication area and transmit a signal to the emergency control station. The alarms and signals shall be both visual and audible.

CHAPTER 22
MOTOR FUEL-DISPENSING FACILITIES AND REPAIR GARAGES

Section 2205.2.2 Change to read as shown: (F8-03/04)

2205.2.2 Emergency shutoff valves. Automatic emergency shutoff valves required by Section 2206.7.4 shall be checked not less than once per year by manually tripping the hold-open linkage.

Table 2206.2.3 Change entire table to read as shown: (F150-03/04)

TABLE 2206.2.3
MINIMUM SEPARATION REQUIREMENTS FOR ABOVE-GROUND TANKS

CLASS OF LIQUID AND TANK TYPE	INDIVIDUAL TANK CAPACITY (gallons)	MINIMUM DISTANCE FROM NEAREST IMPORTANT BUILDING ON SAME PROPERTY (feet)	MINIMUM DISTANCE FROM NEAREST FUEL DISPENSER (feet)	MINIMUM DISTANCE FROM LOT LINE WHICH IS OR CAN BE BUILT UPON, INCLUDING THE OPPOSITE SIDE OF A PUBLIC WAY (feet)	MINIMUM DISTANCE FROM NEAREST SIDE OF ANY PUBLIC WAY (feet)	MINIMUM DISTANCE BETWEEN TANKS (feet)
Class I protected aboveground tanks	Less than or equal to 6,000	5	25 [a]	15	5	3
	Greater than 6,000	15	25 [a]	25	15	3
Class II and III protected above-ground tanks	Same as Class I	Same as Class I	Same as Class I	Same as Class I	Same as Class I	Same as Class I
Tanks in vaults	0-20,000	0	0	0	0	Separate compartment required for each tank
Other tanks	All	50	50	100	50	3

For SI: 1 foot = 304.8 mm, 1 gallon = 3.785 L.

a. At fleet vehicle motor fuel-dispensing facilities, no minimum separation distance is required.

Section 2206.7.4 Change to read as shown: (F8-03/04)

2206.7.4 Dispenser emergency valve. An approved automatic emergency shutoff valve designed to close in the event of a fire or impact shall be properly installed in the liquid supply line at the base of each dispenser supplied by a remote pump. The valve shall be installed so that the shear groove is flush with or within 0.5 inch (12.7 mm) of the top of the concrete dispenser island and there is clearance provided for maintenance purposes around the valve body and operating parts. The valve shall be installed at the liquid supply line inlet of each overhead-type dispenser. Where installed, a vapor return line located inside the dispenser housing shall have a shear section or approved flexible connector for the liquid supply line emergency shutoff valve to function. Emergency shutoff valves shall be installed and maintained in accordance with the manufacturer's instructions, tested at the time of initial installation and tested at least yearly thereafter in accordance with Section 2205.2.2.

Section 2206.7.7.1 Change to read as shown: (F151-03/04)

2206.7.7.1 Leak detection. Where remote pumps are used to supply fuel dispensers, each pump shall have installed on the discharge side a listed leak detection device that will detect a leak in the piping and dispensers and provide an indication. A leak detection device is not required if the piping from the pump discharge to under the dispenser is above ground and visible.

Sections 2208.3 and 2208.3.1 Change to read as shown: (F152-03/04)

2208.3 Location of dispensing operations and equipment. Compression, storage and dispensing equipment shall be located above ground, outside.

Exceptions:

1. Compression, storage or dispensing equipment shall be allowed in buildings of noncombustible construction, as set forth in the *International Building Code*, which are unenclosed for three quarters or more of the perimeter.

2. Compression, storage and dispensing equipment shall be allowed indoors or in vaults in accordance with Chapter 30.

2208.3.1 Location on property. In addition to the requirements of Section 2203.1, compression, storage and dispensing equipment not located in vaults complying with Chapter 30 shall be installed as follows:

1. Not beneath power lines.

2. Ten feet (3048 mm) or more from the nearest building or lot line which could be built on, public street, sidewalk or source of ignition.

   Exception: Dispensing equipment need not be separated from canopies that are constructed in accordance with the *International Building Code* and which provide weather protection for the dispensing equipment.

3. Twenty-five feet (7620 mm) or more from the nearest rail of any railroad track and 50 feet (15 240 mm) or more from the nearest rail of any railroad main track or any railroad or transit line where power for train propulsion is provided by an outside electrical source such as third rail or overhead catenary.

4. Fifty feet (15 240 mm) or more from the vertical plane below the nearest overhead wire of a trolley bus line.

Section 2208.7 Change to read as shown: (F8-03/04)

2208.7 Emergency shutdown control system. An emergency shutdown control system shall be located within 75 feet (22 860 mm) of, but not less than 25 feet (7620 mm) from, dispensers, and shall also be provided in the compressor area. Upon activation, the emergency shutdown shall automatically shut off the power supply to the compressor and close valves between the main gas supply and the compressor and between the storage containers and dispensers.

Section 2209.1 Change to read as shown: (F157-03/04)

2209.1 General. Hydrogen motor fuel-dispensing and generation facilities shall be in accordance with this section and Chapter 35. Where a fuel-dispensing facility also includes a repair garage, the repair operation shall comply with Section 2211.

Section 2209.2.1 Change to read as shown: (F157-03/04)

2209.2.1 Approved equipment. Cylinders, containers and tanks, pressure relief devices, including pressure valves, hydrogen vaporizers, pressure regulators and piping used for gaseous hydrogen systems shall be designed and constructed in accordance with Section 3003, 3203, NFPA 50A or NFPA 50B.

Section 2209.3.1 Change to read as shown: (F157-03/04)

2209.3.1 Separation from outdoor exposure hazards. Generation, compression and dispensing equipment shall be separated from other fuels or equivalent risks to life safety and buildings or public areas in accordance with Table 2209.3.1.

Section 2209.3.2 Change to read as shown: (F154-03/04; F155-03/04; F157-03/04)

2209.3.2 Location of dispensing operations and equipment. Generation, compression, storage and dispensing equipment shall be located in accordance with Sections 2209.3.2.1 through 2209.3.2.6.

2209.3.2.1 Outdoors. Generation, compression, storage or dispensing equipment shall be allowed outdoors in accordance with Section 2209.3.1.

2209.3.2.2 Weather protection. Generation, compression, storage or dispensing equipment shall be allowed under weather protection in accordance with the requirements of Section 2704.13 and constructed in a manner that prevents the accumulation of hydrogen gas.

2209.3.2.3 Indoors. Generation, compression, storage and dispensing equipment shall be allowed in indoor rooms or areas constructed in accordance with the requirements of the *International Building Code*, the *International Fuel Gas Code* and the *International Mechanical Code*.

2209.3.2.4 Gaseous hydrogen storage. Storage of gaseous hydrogen shall be in accordance with Chapters 30 and 35.

2209.3.2.5 Liquefied hydrogen storage. Storage of liquefied hydrogen shall be in accordance with Chapter 32.

2209.3.2.6 Canopy-tops. Gaseous hydrogen compression and storage equipment located on top of motor fuel-dispensing facility canopies shall be in accordance with Sections 2209.3.2.6.1 through 2209.3.2.6.3, Chapters 30 and 35, and the *International Fuel Gas Code*.

2209.3.2.6.1 Construction. Canopies shall be constructed in accordance with the motor fuel-dispensing facility canopy requirements of Section 406 of the *International Building Code*.

2209.3.2.6.2 Fire-extinguishing systems. Fuel-dispensing areas under canopies shall be equipped throughout with an approved automatic sprinkler system in accordance with Section 903.3.1.1. The design of the sprinkler system shall not be less than that required for Extra Hazard Group 2 occupancies. Operation of the sprinkler system shall activate the emergency functions of Sections 2209.3.2.6.2.1 and 2209.3.2.6.2.2.

2209.3.2.6.2.1 Emergency discharge. Operation of the automatic sprinkler system shall activate an automatic emergency discharge system, which will discharge the hydrogen gas from the equipment on the canopy-top through the vent pipe system.

2209.3.2.6.2.2 Emergency shutdown control. Operation of the automatic sprinkler system shall activate the emergency shutdown control required by Section 2209.5.3.

2209.3.2.6.3 Signage. Approved signage having 2-inch (51 mm) block letters shall be affixed at a conspicuous location on the exterior of the canopy structure stating: CANOPY TOP HYDROGEN STORAGE.

Section 2209.3.3 Change to read as shown: (F154-03/04; F155-03/04)

2209.3.3 Canopies. Dispensing equipment need not be separated from canopies of Types I or II construction that are constructed in accordance with Section 406.5 of the *International Building Code*, in a manner that would prevent the accumulation of hydrogen gas.

Section 2209.3.4 Change to read as shown: (F157-03/04)

2209.3.4 Overhead lines. Generation, compression, storage and dispensing equipment shall be separated from overhead electrical lines as follows:

1. Not less than 50 feet (15 240 mm) from the vertical plane below the nearest overhead wire of an electric trolley, train or bus line; and
2. Not less than 5 feet (1524 mm) from the vertical plane below the nearest overhead electrical wire.

Sections 2209.5.1 and 2209.5.2 Change to read as shown: (F8-03/04; F157-03/04)

2209.5.1 Protection from vehicles. Guard posts or other approved means shall be provided to protect hydrogen storage systems and use areas subject to vehicular damage in accordance with Section 312.

2209.5.2 Emergency shutoff valves. A manual emergency shutoff valve shall be provided to shut down the flow of gas from the hydrogen supply to the piping system.

Section 2209.5.2.1 Add new section to read as shown: (F8-03/04; F157-03/04)

2209.5.2.1 Identification. Emergency shutoff valves shall be identified and the location shall be clearly visible, accessible and indicated by means of a sign.

Section 2209.5.3 Change to read as shown: (F8-03/04; F157-03/04)

2209.5.3 Emergency shutdown controls. In addition to the manual emergency shutoff valve required by Section 2209.5.2, a remotely located, manually activated emergency shutdown control shall be provided. An emergency shutdown control shall be located within 75 feet (22 860 mm) of, but not less than 25 feet (7620 mm) from, dispensers and hydrogen generators.

Section 2209.5.3.1 Add new section to read as shown: (F8-03/04; F157-03/04)

2209.5.3.1 System requirements. Activation of the emergency shutdown control shall automatically shut off the power supply to all hydrogen storage, compression and dispensing equipment, shut off natural gas or other fuel supply to the hydrogen generator, and close valves between the main supply and the compressor and between the storage containers and dispensing equipment.

Section 2209.5.4 Change to read as shown: (F157-03/04)

2209.5.4 Venting of hydrogen systems. Hydrogen systems shall be equipped with pressure relief devices that will relieve excessive internal pressure.

2209.5.4.1 Location of discharge. Hydrogen vented from vent pipe systems serving pressure relief devices or purging systems shall not be discharged inside buildings or under canopies used for weather protection.

2209.5.4.2 Pressure relief devices. Portions of the system s ubject t o o verpressure s hall b e p rotected b y pressure relief devices designed and installed in accordance with the requirements of CGA S-1.1, S-1.2, S-1.3 or the ASME Boiler and Pressure Vessel Code, as applicable. Containers used for the storage of liquefied hydrogen shall be provided with pressure relief devices in accordance with Section 3203.2.

2209.5.4.2.1 Minimum rate of discharge. The minimum flow capacity of pressure relief devices on hydrogen storage containers shall be not less than that required by Section 2209.5.4.2 nor less than that required to accommodate a hydrogen compressor that fails to shut down or unload.

2209.5.4.3 Vent pipe. Stationary containers and tanks shall be provided with a vent pipe system that will divert gas discharged from pressure relief devices to the atmosphere. Vent pipe systems serving pressure relief devices and purging systems used for operational control shall be designed and constructed in accordance with Sections 2209.5.4.3.1 through 2209.5.4.3.6 and 2209.5.4.3.8.

2209.5.4.3.1 Vent pipe flow rates. Where above-ground storage of flammable or combustible liquids occurs and the storage is diked, or where no above-ground storage of flammable or combustible liquids exists, the sizing of the maximum flow capacity for the vent pipe need not include the pressure relief device flow capacity as a result of an "engulfing fire" of the hydrogen storage tanks (see Table 2209.3.1).

2209.5.4.3.2 Materials of construction. The vent pipe system shall be constructed of materials approved for hydrogen service in accordance with ASME B31.3 for the rated pressure, volume and temperature of gas to be transported. The vent piping shall be designed for the maximum back pressure within the pipe, but not less than 335 pounds per square inch gauge (psig) (2310 kPa).

2209.5.4.3.3 Structural support. The vent pipe system shall be supported to prevent structural collapse and shall be provided with a rain cap or other feature which would not limit or obstruct the gas flow from venting vertically upward.

2209.5.4.3.4 Obstructions. A means shall be provided to prevent water, ice and other debris from accumulating inside the vent pipe or obstructing the vent pipe.

2209.5.4.3.5 Height of vent and separation. The height (H) and separation distance (D) of the vent pipe shall meet the criteria set forth in Table 2209.5.4.3.5 for the combinations of maximum hydrogen flow rates and vent stack opening diameters listed. Alternative venting systems shall be allowed when in accordance with Section 2209.5.4.3.7.

TABLE 2209.5.4.3.5
VENT PIPE HEIGHT AND SEPARATION DISTANCE
VERSUS HYDROGEN FLOW RATE AND VENT PIPE DIAMETER [a,b,c,d,e,f]

Hydrogen Flow Rate	≤ 500 CFM at NTP[g]	>500 to ≤ 1000 CFM at NTP[g]	>1,000 to ≤ 2,000 CFM at NTP[g]	> 2,000 to ≤ 5,000 CFM at NTP[h]	> 5,000 to ≤ 10,000 CFM at NTP[h]	> 10,000 to ≤ 20,000 CFM at NTP[h]
Height (ft)	8	8	12	17	25	36
Distance (ft)	13	17	26	40	53	81

For SI: 1 inch = 25.4 mm, 1 foot = 304.8 mm, 1 Btu/ft^2 = 3.153W/m^2, 1 foot/second = 304.8 mm/sec.
a. Minimum distance to property line is 1.25D.
b. Designs seeking to achieve greater heights with commensurate reductions in separation distances shall be designed in accordance with accepted engineering practice.
c. With this table personnel on the ground or on the building/equipment are exposed to a maximum of 1,500 BTU/hr. ft^2, and are assumed to be provided with a means to escape to a shielded area within three minutes, including the case of a 30 ft./sec. wind.
d. Designs seeking to achieve greater radiant exposures to noncombustible equipment shall be designed in accordance with accepted engineering practice.
e. The analysis reflected in this table does not permit hydrogen air mixtures would to exceed one-half of the lower flammable limit (LFL) for hydrogen (2 percent by volume) at the building or equipment, including the case of a 30 ft./sec. wind.
f. See Figure 2209.5.4.3.5.
g. For vent pipe diameters up to and including 2 inches.
h. For vent pipe diameters up to and including 3 inches.

(Renumber Figure 2209.5.4.1 as Figure 2209.5.4.3.5; no change to figure.)

2209.5.4.3.6 Maximum flow rate. The vent pipe system shall be sized based on the maximum flow rate for the system served and be specified on the construction documents. The maximum flow rate shall be determined in accordance with the requirements of CGA S-1.3 using the aggregate gas flow rate from all connected vent, purge and relief devices that operate simultaneously during a venting operation, purging operation or emergency relief event.

2209.5.4.3.7 Alternative venting systems. Where alternative venting systems are used as an alternate to the requirements of Section 2209.5.4.3.5, an analysis of radiant heat exposures and hydrogen concentrations shall be provided. The analysis of exposure to radiant heat shall assume a wind speed of 30 feet/second (9.14 m/sec) and provide a design that limits radiant heat exposure to the maximum values shown in Table 2209.5.4.3.7(1). The analysis of exposure to hydrogen concentration shall provide a design that limits the maximum hydrogen concentration to the values shown in Table 2209.5.4.3.7(2).

TABLE 2209.5.4.3.7(1)
MAXIMUM RADIANT HEAT EXPOSURE

Exposed Object	Maximum Radiant Heat	Time Duration (minutes)
Personnel	1,500 Btu/hr·ft² (4732 W/m²)	3
Noncombustible equipment	8,000 Btu/hr·ft² (25 237 W/m²)	Any
Lot line	500 Btu/hr·ft² (1577 W/m²)	Any

TABLE 2209.5.4.3.7(2)
MAXIMUM HYDROGEN CONCENTRATION EXPOSURE

Exposed Object	Maximum Hydrogen Concentration
Personnel, buildings or equipment	50% LFL within a distance of D and H of Table 2209.5.4.3.5
Lot line	50% LFL within 1.25 times the distance of D and H of Table 2209.5.4.3.5

Sections 2209.5.4.1 and 2209.5.4.2 Relocated to Section 2209.5.4.3. (F157-03/04)

CHAPTER 23
HIGH-PILED COMBUSTIBLE STORAGE

Section 2301.1 Change to read as shown: (F158-03/04)

2301.1 Scope. High-piled combustible storage shall be in accordance with this chapter. In addition to the requirements of this chapter, the following material-specific requirements shall apply:

1. Aerosols shall be in accordance with Chapter 28.

2. Flammable and combustible liquids shall be in accordance with Chapter 34.

3. Hazardous materials shall be in accordance with Chapter 27.

4. Storage of combustible paper records shall be in accordance with NFPA 13 and NFPA 230.

5. Storage of combustible fibers shall be in accordance with Chapter 29.

6. Storage of miscellaneous combustible material shall be in accordance with Chapter 3.

Section 2304.2 Change to read as shown: (F158-03/04)

2304.2 Designation based on engineering analysis. The designation of a high-piled combustible storage area, or portion thereof, is allowed to be based on a lower hazard class than that of the highest class of commodity stored when a limited quantity of the higher hazard commodity has been demonstrated by engineering analysis to be adequately protected by the automatic sprinkler system provided. The engineering analysis shall consider the ability of the sprinkler system to deliver the higher density required by the higher hazard commodity. The higher density shall be based on the actual storage height of the pile or rack and the minimum allowable design area for sprinkler operation as set forth in the density/area figures provided in NFPA 13. The contiguous area occupied by the higher hazard commodity shall not exceed 120 square feet (111 m²), and additional areas of higher hazard commodity shall be separated from other such areas by 25 feet (7620 mm) or more. The sprinkler system shall be capable of delivering the higher density over a minimum area of 900 square feet (84 m²) for wet pipe systems and 1,200 square feet (11 m²) for dry pipe systems. The shape of the design area shall be in accordance with Section 903.

Section 2306.9.1 Change to read as shown: (F159-03/04)

2306.9.1 Width. Aisle width shall be in accordance with Sections 2306.9.1.1 and 2306.9.1.2.

Exceptions:

1. Aisles crossing rack structures or storage piles, which are used only for employee access, shall be a minimum of 24 inches (610 mm) wide.

2. Aisles separating shelves classified as shelf storage shall be a minimum of 30 inches (762 mm) wide.

Sections 2307.2 and 2307.2.1 Change to read as shown: (F158-03/04)

2307.2 Fire protection. Where automatic sprinklers are required by Table 2306.2, an approved automatic sprinkler system shall be installed throughout the building or to 1-hour fire-resistance-rated fire barrier walls constructed in accordance with the *International Building Code*. Openings in such walls shall be protected by opening protective assemblies having 1-hour fire protection ratings. The design and installation of the automatic sprinkler system and other applicable fire protection shall be in accordance with the *International Building Code* and NFPA 13.

2307.2.1 Shelf storage. Shelf storage greater than 12 feet (3658 mm) but less than 15 feet (4572 mm) in height shall be in accordance with the fire protection requirements set forth in NFPA 13. Shelf storage 15 feet (4572 mm) or more in height shall be protected in an approved manner with special fire protection, such as in-rack sprinklers.

Section 2308.2 Change to read as shown: (F158-03/04)

2308.2 Fire protection. Where automatic sprinklers are required by Table 2306.2, an approved automatic sprinkler system shall be installed throughout the building or to 1-hour fire barrier walls constructed in accordance with the *International Building Code*. Openings in such walls shall be protected by opening protective assemblies having 1-hour fire protection ratings. The design and installation of the automatic sprinkler system and other applicable fire protection shall be in accordance with Section 903.3.1.1 and the *International Building Code*.

Sections 2308.2.2 and 2308.2.2.1 Change to read as shown: (F158-0/04)

2308.2.2 Racks with solid shelving. Racks with solid shelving having an area greater than 32 square feet (3 m^2), measured between approved flue spaces at all four edges of the shelf, shall be in accordance with this section.

Exceptions:

1. Racks with mesh, grated, slatted or similar shelves having uniform openings not more than 6 inches (152 mm) apart, comprising at least 50 percent of overall shelf area, and with approved flue spaces, are allowed to be treated as racks without solid shelves.

2. Racks used for the storage of combustible paper records, with solid shelving, shall be in accordance with NFPA 13.

2308.2.2.1 Fire protection. Fire protection for racks with solid shelving shall be in accordance with NFPA 13.

Section 2308.4 Change to read as shown: (F158-03/04)

2308.4 Column protection. Steel building columns shall be protected in accordance with NFPA 230.

Section 2310.1 Change to read as shown: (F158-03/04)

2310.1 General. Records storage facilities used for the rack or shelf storage of combustible paper records greater than 12 feet (3658 mm) in height shall be in accordance with Sections 2306 and 2308 and NFPA 13 and NFPA 230. Palletized storage of records shall be in accordance with Section 2307.

CHAPTER 24
TENTS, CANOPIES AND OTHER MEMBRANE STRUCTURES

Section 2403.12.6.1 Change to read as shown: (EL3-03/04)

2403.12.6.1 Exit sign illumination. Exit signs shall be of an approved self-luminous type or shall be internally or externally illuminated by luminaires supplied in the following manner:

Section 2404.2 Change to read as shown: (F43-03/04)

2404.2 F lame p ropagation p erformance t reatment. Before a permit is granted, the owner or agent shall file with the fire code official a certificate executed by an approved testing laboratory, certifying that the tents, canopies and membrane structures and their appurtenances, sidewalls, drops and tarpaulins, floor coverings, bunting, combustible decorative materials and effects, including sawdust when used on floors or passageways, shall be composed of material meeting the flame propagation performance criteria of NFPA 701 or shall be treated with a flame retardant in an approved manner and meet the flame propagation performance criteria of NFPA 701, and that such flame propagation performance criteria are effective for the period specified by the permit.

Section 2404.4 Change to read as shown: (F43-03/04)

2404.4 Certification. An affidavit or affirmation shall be submitted to the fire code official and a copy retained on the premises on which the tent or air-supported structure is located. The affidavit shall attest to the following information relative to the flame propagation performance criteria of the fabric:

1. Names and a ddress of the owners of the tent, canopy or air-supported structure.

2. Date the fabric was last treated with flame-retardant solution.

3. Trade name or kind of chemical used in treatment.

4. Name of person or firm treating the material.

5. Name of testing agency and test standard by which the fabric was tested.

CHAPTER 25
TIRE REBUILDING AND TIRE STORAGE

Section 2501.1 Change to read as shown: (F158-03/04)

2501.1 Scope. Tire rebuilding plants, tire storage and tire byproduct facilities shall comply with this chapter, other applicable requirements of this code and NFPA 13 and NFPA 230. Tire storage in buildings shall also comply with Chapter 23.

CHAPTER 27
HAZARDOUS MATERIALS — GENERAL PROVISIONS

Table 2703.1.1(1) Change table to read as shown: (F163-03/04)

TABLE 2703.1.1(1)
MAXIMUM ALLOWABLE QUANTITY PER CONTROL AREA OF HAZARDOUS MATERIALS POSING A PHYSICAL HAZARD

(No change to current table body)

a. through d. (No change to current text)
e. Maximum allowable quantities shall be increased 100 percent when stored in approved storage cabinets, day boxes, gas cabinets, exhausted enclosures or safety cans. Where Note d also applies, the increase for both notes shall be applied accumulatively.
f. through n. (No change to current text)

Section 2702.1 Add new definition to read as shown: (F163-03/04)

2702.1 Definitions. The following words and terms shall, for the purposes of this chapter, Chapters 28 through 44, and as used elsewhere in this code, have the meanings shown herein.

DAY BOX. A portable magazine designed to hold explosive materials constructed in accordance with the requirements for a Type 3 magazine as defined and classified in Chapter 33.

Section 2703.2.2.1 Change to read as shown: (F8-03/04)

2703.2.2.1 Design and construction. Piping, tubing, valves, fittings and related components used for hazardous materials shall be in accordance with the following:

1. Piping, tubing, valves, fittings and related components shall be designed and fabricated from materials compatible with the material to be contained and shall be of adequate strength and durability to withstand the pressure, structural and seismic stress, and exposure to which they are subject.

2. Piping and tubing shall be identified in accordance with ANSI A13.1 to indicate the material conveyed.

3. Readily accessible manual valves, or automatic remotely activated fail-safe emergency shutoff valves shall be installed on supply piping and tubing at the following locations:

    3.1. The point of use.

    3.2. The tank, cylinder or bulk source.

4. Emergency shutoff valves shall be identified and the location shall be clearly visible, accessible and indicated by means of a sign.

5. Backflow prevention or check valves shall be provided when the backflow of hazardous materials could create a hazardous condition or cause the unauthorized discharge of hazardous materials.

6. Where gases or liquids having a hazard ranking of:
   Health hazard Class 3 or 4
   Flammability Class 4
   Reactivity Class 3 or 4
   in accordance with NFPA 704 are carried in pressurized piping above 15 pounds per square inch gauge (psig) (103 kPa), an approved means of leak detection and emergency shutoff or excess flow control shall be provided. Where the piping originates from within a hazardous material storage room or area, the excess flow control shall be located within the storage room or area. Where the piping originates from a bulk source, the excess flow control shall be located as close to the bulk source as practical.

Exceptions:

1. Piping for inlet connections designed to prevent backflow.

2. Piping for pressure relief devices.

Section 2703.2.4.2 Change to read as shown: (F169-03/04)

2703.2.4.2 Above-ground tanks. Above-ground stationary tanks used for the storage of hazardous materials shall be located and protected in accordance with the requirements for outdoor storage of the particular material involved.

Exception: Above-ground tanks that are installed in vaults complying with Section 3003.14 or 3404.2.8 shall not be required to comply with location and protection requirements for outdoor storage.

Sections 2703.8.3.1 and 2703.8.3.3 Change to read as shown: (G 94-03/04)

2703.8.3.1 Construction requirements. Control areas shall be separated from each other by fire barriers constructed in accordance with Section 706 of the *International Building Code*.

2703.8.3.3 Fire-resistance rating requirements. The required fire-resistance rating for fire barrier assemblies shall be in accordance with Table 2703.8.3.2. The floor construction of the control area and construction supporting the floor of the control area shall have a minimum 2-hour fire-resistance rating.

Table 2703.8.3.2 Change table to read as shown: (F165-03/04)

TABLE 2703.8.3.2
DESIGN AND NUMBER OF CONTROL AREAS

FLOOR LEVEL	PERCENTAGE OF THE MAXIMUM ALLOWABLE QUANTITY PER CONTROL AREA[a]	NUMBER OF CONTROL AREAS PER FLOOR	FIRE-RESISTANCE RATING FOR FIRE BARRIERS IN HOURS[b]

(Portions of table not shown do not change)

a. (No change to current text)
b. Fire barriers shall include walls and floors as necessary to provide separation from other portions of the building.

Section 2704.3.1 Change to read as shown: (F166-03/04)

2704.3.1 System requirements. Exhaust ventilation systems shall comply with all of the following:

1. Installation shall be in accordance with the *International Mechanical Code*.

2. Mechanical ventilation shall be at a rate of not less than 1 cubic foot per minute per square foot [0.00508 $m^3/(s \times m^2)$] of floor area over the storage area.

3. Systems shall operate continuously unless alternative designs are approved.

4. A manual shutoff control shall be provided outside of the room in a position adjacent to the access door to the room or in an approved location. The switch shall be a break-glass or other approved type and shall be labeled: VENTILATION SYSTEM EMERGENCY SHUTOFF.

5. Exhaust ventilation shall be designed to consider the density of the potential fumes or vapors released. For fumes or vapors that are heavier than air, exhaust shall be taken from a point within 12 inches (305 mm) of the floor. For fumes or vapors that are lighter than air, exhaust shall be taken from a point within 12 inches (305 mm) of the highest point of the room.

6. The location of both the exhaust and inlet air openings shall be designed to provide air movement across all portions of the floor or room to prevent the accumulation of vapors.

7. Exhaust air shall not be recirculated to occupied areas if the materials stored are capable of emitting hazardous vapors and contaminants have not been removed. Air-contaminated with explosive or flammable vapors, fumes or dusts; flammable, highly toxic or toxic gases; or radioactive materials shall not be recirculated.

Section 2705.2.1.1 Change to read as shown: (F144-03/04)

2705.2.1.1 Ventilation. Where gases, liquids or solids having a hazard ranking of 3 or 4 in accordance with NFPA 704 are dispensed or used, mechanical exhaust ventilation shall be provided to capture gases, fumes, mists or vapors at the point of generation.

Exception: Gases, liquids or solids which can be demonstrated not to create harmful gases, fumes, mists or vapors.

Table 2705.2.1.4 Change table to read as shown: (F167-03/04)

TABLE 2705.2.1.4
REQUIRED SECONDARY CONTAINMENT —
HAZARDOUS MATERIAL LIQUIDS USE

MATERIAL	INDOOR USE	OUTDOOR USE
	LIQUIDS	LIQUIDS

(Delete both "Solids" columns and all text therein)

(Portions of table not shown do not change)

CHAPTER 28
AEROSOLS

Section 2804.3.2.1 Change to read as shown: (F168-03/04)

2804.3.2.1 Chain-link fence enclosures. Chain-link fence enclosures required by Table 2804.3.2 shall comply with the following:

1. The fence shall not be less than No. 9 gage steel wire, woven into a maximum 2-inch (51 mm) diamond mesh.

2. The fence shall be installed from the floor to the underside of the roof or ceiling above.

3. Class IV and high-hazard commodities shall be stored outside of the aerosol storage area and a minimum of 8 feet (2438 mm) from the fence.

4. Access openings in the fence shall be provided with either self-closing or automatic-closing devices or a labyrinth opening arrangement preventing aerosol containers from rocketing through the access openings.

5. Not less than two means of egress shall be provided from the fenced enclosure.

Section 2806.2 Add new section to read as shown: (F168-03/04)

2806.2 Aerosol display and normal merchandising not exceeding 8 feet (2438 mm) high. Aerosol display and normal merchandising not exceeding 8 feet (2438 mm) in height shall be in accordance with Section 2806.2.1 through 2806.2.4.

2806.2.1 Maximum quantities in retail display areas. Aerosol products in retail display areas shall not exceed quantities needed for display and normal merchandising and shall not exceed the quantities in Table 2806.2.1.

TABLE 2806.2.1
MAXIMUM QUANTITIES OF LEVEL 2 AND 3 AEROSOL PRODUCTS IN RETAIL DISPLAY AREAS

MAXIMUM NET WEIGHT PER FLOOR (POUNDS)[a, B]			
Floor	Unprotected[a]	Protected in accordance with Section 2806.2[a, c]	Protected in accordance with Section 2806.3[c]
Basement	Not allowed	500	500
Ground	2500	10000	10000
Upper	500	2000	Not allowed

For SI: 1 pound = 0.454 kg, 1 square foot = 0.0929 m²
a. The total quantity shall not exceed 1,000 pounds net weight in any one 100-square foot retail display area.
b. Per 25,000-square-foot retail display area.
c. Minimum ordinary hazard Group 2 wet-pipe automatic sprinkler system throughout the retail sales occupancy.

2806.2.2 Display of containers. Level 2 and 3 aerosol containers shall not be stacked more than 6 feet (1829 mm) high from the base of the aerosol array to the top of the aerosol array unless the containers are placed on fixed shelving or otherwise secured in an approved manner. When storage or retail display is on shelves, the height of such storage or retail display to the top of aerosol containers shall not exceed 8 feet (2438 mm).

2806.2.3 Combustible cartons. Aerosol products located in retail display areas shall be removed from combustible cartons.

Exceptions:
1. Display areas that use a portion of combustible cartons which consist of only the bottom panel and not more than 2 inches (51 mm) of side panel are allowed.
2. When the display area is protected in accordance with Tables 6.3.2.7(a) through 6.3.2.7(l) of NFPA 30B, storage of aerosol products in combustible cartons is allowed.

2806.2.4 Retail display automatic sprinkler system. When an automatic sprinkler system is required for the protected retail display of aerosol products, the wet-pipe automatic sprinkler system shall be in accordance with Section 903.3.1.1. The minimum system design shall be for an Ordinary Hazard Group 2 occupancy. The system shall be provided throughout the retail display area.

Section 2806.3 Add new section and subsections to read as shown: (F168-03/04)

2806.3 Aerosol display and normal merchandising exceeding 8 feet (2438 mm) high. Aerosol display and merchandising exceeding 8 feet in height shall be in accordance with Sections 2806.3.1 through 2806.3.3.

2806.3.1 Maximum quantities in retail display areas. Aerosol products in retail display areas shall not exceed quantities needed for display and normal merchandising and shall not exceed the quantities in Table 2806.2.1, with fire protection in accordance with Section 2806.3.2.

2806.3.2 Automatic sprinkler protection. Aerosol display and merchandising areas shall be protected by an automatic sprinkler system based on the requirements set forth in NFPA 30B, Tables 6.3.2.7(a) through 6.3.2.7(l) and the following:

1. Protection shall be based on the highest level of aerosol product in the array and the packaging

method of the storage located more than 6 feet (1829 mm) above the finished floor.

2. When using the cartoned aerosol tables in NFPA 30B, uncartoned or display cut Level 2 and 3 aerosols shall be permitted not more than 6 feet (1829 mm) above the finished floor.

3. The design area for Level 2 and Level 3 aerosols shall extend not less than 20 feet (6096 mm) beyond the Level 2 and Level 3 aerosol display and merchandising areas.

4. Where ordinary and high temperature ceiling sprinkler systems are adjacent to each other, noncombustible draft curtains shall be installed at the interface.

2806.3.3 Separation of Level 2 and Level 3 aerosols areas.

1. Level 2 and Level 3 aerosol display and merchandising areas shall be separated from each other by not less than 25 feet (7620 mm). Also see Table 2806.2.1.

2. Level 2 and Level 3 aerosol display and merchandising areas shall be separated from flammable and combustible liquids storage and display areas by one or a combination of the following:

    a. Segregating areas from each other by horizontal distance of not less than 25 feet (7620 mm).

    b. Isolating areas from each other by a noncombustible partition extending not less than 18 inches (457 mm) above the merchandise.

    c. In accordance with Section 2806.5.

3. When option 2.b. above is used to separate Level 2 or Level 3 aerosols from flammable or combustible liquids, and the aerosol products are located within 25 feet (7620 mm) of flammable or combustible liquids, the area below the noncombustible partition shall be liquid-tight at the floor to prevent spilled liquids from flowing beneath the aerosol products.

Section 2806.4 Renumber only, from Section 2806.3 to read as shown: (F168-03/04)

2806.4 Maximum quantities in storage areas. Aerosol products in storage areas adjacent to retail display areas shall not exceed the quantities in Table 2806.4.

Renumber current Table 2806.3 to read as shown: (F168-03/04)

TABLE 2806.4
MAXIMUM STORAGE QUANTITIES FOR STORAGE AREAS ADJACENT TO RETAIL DISPLAY OF LEVEL 2 AND LEVEL 3 AEROSOLS

MAXIMUM NET WEIGHT PER FLOOR (POUNDS)			
		Separated	
Floor	Unseparated[a,b]	Storage Cabinets[b]	1-hour Occupancy Separation
Basement	Not allowed	Not allowed	Not allowed
Ground	2500	5000	In accordance with NFPA 30B, Sections 6.3.4.3 and 6.3.4.4
Upper	500	1000	In accordance with NFPA 30B, Sections 6.3.4.3 and 6.3.4.4

For SI: 1 pound = 0.454 kg, 1 square foot = 0.0929m²
a. and b. (No change to current text)

Sections 2806.4 and 2806.5 Relocate to Sections 2806.2.2 and 2806.2.3, respectively: (F168-03/04)

Section 2806.5 Add new section to read as shown: (F168-03/04)

2806.5 Special protection design for Level 2 and Level 3 aerosols adjacent to flammable and combustible liquids in double row racks. The display and merchandising of Level 2 and Level 3 aerosols adjacent to flammable and combustible liquids in double row racks shall be in accordance with Sections 2806.5.1 through 2806.5.8 or Section 2806.3.3.

2806.5.1 Fire protection. Fire protection for the display and merchandising of Level 2 and Level 3 aerosols in double-racks shall be in accordance with NFPA 30B, Table 7.4.1 and Figure 7.4.1.

2806.5.2 Cartoned products. Level 2 and Level 3 aerosols displayed or merchandised more than 8 feet (2438 mm) above the finished floor shall be in cartons.

2806.5.3 Shelving. Shelving in racks shall be limited to wire mesh shelving having uniform openings not more than 6 inches (152 mm) apart, with the openings comprising at least 50 percent of the overall shelf area.

2806.5.4 Aisles. Racks shall be arranged so that aisles not less than 7.5 feet (2286 mm) wide are maintained between rows of racks and adjacent solid-piled or palletized merchandise.

2806.5.5 Flue spaces. Flue spaces shall be as follows:

1. Transverse Flue Spaces. Nominal 3 inch (76 mm) transverse flue spaces shall be maintained between merchandise and rack uprights.

2. Longitudinal Flue Spaces. Nominal 6 inch (152 mm) longitudinal flue spaces shall be maintained.

2806.5.6 Horizontal barriers. Horizontal barriers constructed of minimum ⅜ inch (10 mm) thick plywood or minimum 0.034-inch (0.086 mm) (No. 22 gage) sheet metal shall be provided and located in accordance NFPA 30B, Table 7.4.1 and Figure 7.4.1 when in-rack sprinklers are installed.

2806.5.7 Class I, II, III, IV and plastic commodities. Class I, II, III, IV and plastic commodities located adjacent to Level 2 and Level 3 aerosols shall be protected in accordance with NFPA 13.

2806.5.8 Flammable and combustible liquids. Class I, II, III-A and III-B Liquids shall be allowed to be located adjacent to Level 2 and Level 3 aerosol products when the following conditions are met:

1. Class I, II, III-A and III-B Liquid Containers: Containers for Class I, II, III-A and III-B Liquids shall be limited to 1.06-gallon (4-L) metal relieving and non-relieving style containers and 5.3-gallon (20-L) metal relieving style containers.

2. Fire Protection for Class I, II, III-A and III-B Liquids: Fire sprinkler protection for Class I, II, III-A and III-B Liquids shall be in accordance with Chapter 34.

Section 2806.6 Aisles. Delete section without substitution: (F168-03/04)

Section 2806.7 relocate to Section 2806.2.4: (F168-03/04)

Section 2806.8 Storage automatic fire-extinguishing system. Delete section without substitution: (F168-03/04)

CHAPTER 30
COMPRESSED GASES

Section 3003.14 Add new section to read as shown: (F169-03/04)

3003.14 Vaults. Generation, compression, storage and dispensing equipment for compressed gases shall be permitted to be located in either above or below grade vaults complying with Sections 3003.14.1 through 3003.14.14.

3003.14.1 Listing required. Vaults shall be listed in by a nationally recognized testing laboratory.

Exception: Where approved by the fire code official, below-grade vaults are allowed to be constructed on site, provided that the design is in accordance with the *International Building Code* and that special inspections are conducted to verify structural strength and compliance of the installation with the approved design in accordance with the *International Building Code*, Section 1707. Installation plans for belowgrade vaults that are constructed on site shall be prepared by, and the design shall bear the stamp of, a professional engineer. Consideration shall be given to soil and hydrostatic loading on the floors, walls and lid; anticipated seismic forces; uplifting by ground water or flooding; and to loads imposed from above such as traffic and equipment loading on the vault lid.

3003.14.2 Design and construction. The vault shall completely enclose generation, compression, storage or dispensing equipment located in the vault. There shall be no openings in the vault enclosure except those necessary for vault ventilation and access, inspection, filling, emptying or venting of equipment in the vault. The walls and floor of the vault shall be constructed of reinforced concrete at least 6 inches (152mm) thick. The top of an above-grade vault shall be constructed of noncombustible material and shall be designed to be weaker than the walls of the vault, to ensure that the thrust of any explosion occurring inside the vault is directed upward.

The top of an at-grade or below-grade vault shall be designed to relieve safely or contain the force of an explosion occurring inside the vault. The top and floor of the vault and the tank foundation shall be designed to withstand the anticipated loading, including loading from vehicular traffic, where applicable. The walls and floor of a vault installed below grade shall be designed to withstand anticipated soil and hydrostatic loading. Vaults shall be designed to be wind and earthquake resistant, in accordance with the *International Building Code*.

3003.14.3 Secondary containment. Vaults shall be substantially liquid tight and there shall be no backfill within the vault. The vault floor shall drain to a sump. For premanufactured vaults, liquid tightness shall be certified as part of the listing provided by a nationally recognized testing laboratory. For field-erected vaults, liquid tightness shall be certified in an approved manner.

3003.14.4 Internal clearance. There shall be sufficient clearance within the vault to allow for visual inspection and maintenance of equipment in the vault.

3003.14.5 Anchoring. Vaults and equipment contained therein shall be suitably anchored to withstand uplifting by groundwater or flooding. The design shall verify that uplifting is prevented even when equipment within the vault is empty.

3003.14.6 Vehicle impact protection. Vaults shall be resistant to damage from the impact of a motor vehicle, or vehicle impact protection shall be provided in accordance with Section 312.

3003.14.7 Arrangement. Equipment in vaults shall be listed or approved for above-ground use. Where multiple vaults are provided, adjacent vaults shall be permitted to share a common wall. The common wall shall be liquid and vapor tight and shall be designed to withstand the load imposed when the vault on either side of the wall is filled with water.

3003.14.8 Connections. Connections shall be provided to permit venting of each vault to dilute, disperse and remove vapors prior to personnel entering the vault.

3003.14.9 Ventilation. Vaults shall be provided with an exhaust ventilation system installed in accordance with Section 2704.3. The ventilation system shall operate continuously or be designed to operate upon activation of the vapor or liquid detection system. The system shall provide ventilation at a rate of not less than 1 cubic foot per minute (cfm) per square foot of floor area [0.00508 m^3/s per m^2)], but not less than 150 cfm [0.071m^3/s per m2)]. The exhaust system shall be designed to provide air movement across all parts of the vault floor for gases having a density greater than air and across all parts of the vault ceiling for gases having a density less than air. Supply ducts shall extend to within 3 inches (76 mm), but not more than 12 inches (305 mm), of the floor. Exhaust ducts shall extend to within 3 inches (76 mm), but not more than 12 inches (305 mm) of the floor or ceiling, for heavier-than-air or lighter-than-air gases, respectively. The exhaust system shall be installed in accordance with the *International Mechanical Code*.

3003.14.10 Monitoring and detection. Vaults shall be provided with approved vapor and liquid detection systems and equipped with on-site audible and visual warning devices with battery backup. Vapor detection systems shall sound an alarm when the system detects vapors that reach or exceed 25 percent of the lower explosive limit (LEL) or ½ IDLH for the gas in the vault. Vapor detectors shall be located no higher than 12 inches (305 mm) above the lowest point in the vault for heavier-than-air gases and no lower than 12 inches (305 mm) below the highest point in the vault for lighter-than-air gases. Liquid detection systems shall sound an alarm upon detection of any liquid, including water. Liquid detectors shall be located in accordance with the manufacturer's instructions. Activation of either vapor or liquid detection systems shall cause a signal to be sounded at an approved, constantly attended location within the facility served by the tanks or at an approved location. Activation of vapor detection systems shall also shut off gas handling equipment in the vault and dispensers.

3003.14.11 Liquid removal. Means shall be provided to recover liquid from the vault. Where a pump is used to meet this requirement, the pump shall not be permanently installed in the vault. Electric-powered portable pumps shall be suitable for use in Class I, Division 1 locations, as defined in the *ICC Electrical Code*.

3003.14.12 Relief vents. Vent pipes for equipment in the vault shall terminate at least 12 feet (3658 mm) above ground level.

3003.14.13 Accessway. Vaults shall be provided with an approved personnel accessway with a minimum dimension of 30 inches (762 mm) and with a permanently affixed, nonferrous ladder. Accessways shall be designed to be nonsparking. Travel distance from any point inside a vault to an accessway shall not exceed 20 feet (6096 mm). At each entry point, a warning sign indicating the need for procedures for safe entry into confined spaces shall be posted. Entry points shall be secured against unauthorized entry and vandalism.

3003.14.14 Classified area. The interior of a vault containing a flammable gas shall be designated a Class I, Division 1 location, as defined in the *ICC Electrical Code*.

CHAPTER 32
CRYOGENIC FLUIDS

Sections 3204.4 Add new section to read as shown: (F155-03/04)

3204.4 Underground tanks. Underground tanks for the storage of liquid hydrogen shall be in accordance with this section.

3204.4.1 Construction. Storage tanks for liquid hydrogen shall be designed and constructed in accordance with *ASME Boiler and Pressure Vessel Code* (Section VIII, Division 1) and shall be vacuum jacketed in accordance with Section 3204.6.

3204.4.2 Location. Storage tanks shall be located outside in accordance with the following:

1. Tanks and associated equipment shall be located with respect to foundations and supports of other structures such that the loads carried by the latter cannot be transmitted to the tank.

2. The distance from any part of the tank to the nearest wall of a basement, pit, cellar, or lot line shall not be less than 3 feet (914 mm).

3. A minimum distance of 1 feet (1525 mm), shell to shell, shall be maintained between underground tanks.

3204.4.3 Depth, cover and fill. The tank shall be buried such that the top of the vacuum jacket is covered with a minimum of 1 foot of earth and with concrete a minimum of 4 inches (101mm) thick placed over the earthen cover. The concrete shall extend a minimum of one foot (0.3m) horizontally beyond the footprint of the tank in all directions. Underground tanks shall be set on firm foundations constructed in accordance with the *International Building Code* and surrounded with at least 6 inches (152 mm) of noncorrosive inert material, such as sand.

> **Exception:** The vertical extension of the vacuum jacket as required for service connections.

3204.4.4 Anchorage and security. Tanks and systems shall be secured against accidental dislodgement in accordance with this chapter.

3204.4.5 Venting of underground tanks. Vent pipes for underground storage tanks shall be in accordance with Sections 2209.5.4 and 3203.3.

3204.4.6 Underground liquid hydrogen piping. Underground liquid hydrogen piping shall be vacuum jacketed or protected by approved means and designed in accordance with this chapter.

3204.4.7 Overfill protection and prevention systems. An approved means or method shall be provided to prevent the overfill of all storage tanks.

Section 3204.6 Add new section to read as shown: (F155-03/04)

3204.6 Vacuum jacket construction. The vacuum jacket shall be designed and constructed in accordance with Section VIII of *ASME Boiler and Pressure Vessel Code* and shall be designed to withstand the anticipated loading, including loading from vehicular traffic, where applicable. Portions of the vacuum jacket installed below grade shall be designed to withstand anticipated soil and hydrostatic loading.

3204.6.1 Material. The vacuum jacket shall be constructed of stainless steel or other approved corrosion-resistant material.

3204.6.2 Corrosion protection. The vacuum jacket shall be protected by an engineered cathodic protection system. A cathodic protection system maintenance schedule shall be provided and reconciled by the owner/operator. Exposed components shall be inspected at least twice a year.

3204.6.3 Vacuum level monitoring. An approved method shall be provided to indicate loss of vacuum within the vacuum jacket(s).

Section 3205.1.2.4 Change to read as shown: (F155-03/04)

3205.1.2.4 Physical protection and support. Piping systems shall be supported and protected from physical damage. Piping passing through walls shall be protected from mechanical damage.

Section 3205.3.2 Change to read as shown (F8-03/04)

3205.3.2 Emergency shutoff valves. Accessible manual or automatic emergency shutoff valves shall be provided to shut off the cryogenic fluid supply in case of emergency. An emergency shutoff valve shall be located at the source of supply and at the point where the system enters the building.

CHAPTER 33
EXPLOSIVES AND FIREWORKS

Section 3301.8.1 Change to read as shown: (F170-03/04)

3301.8.1 Quantity of explosives. The quantity-distance (Q-D) tables in Sections 3304.5 and 3305.3 shall be used to provide the minimum separation distances from potential explosion sites as set forth in Tables 3301.8.1(1) through 3301.8.1(3). The classification of the explosives and the weight of the explosives are primary characteristics governing the use of these tables. The net explosive weight shall be determined in accordance with Sections 3301.8.1.1 through 3301.8.1.4.

Tables 3301.8.1(1), 3301.8.1(2) and 3301.8.1(3) Add new tables to read as shown: (F170-03/04)

TABLE 3301.8.1(1)
APPLICATION OF SEPARATION DISTANCE (Q-D) TABLES
DIVISION 1.1, 1.2 AND 1.5 EXPLOSIVES[a,b,c]

ITEM	MAGAZINE	Q-D	OPERATING BUILDING	Q-D	INHABITED BUILDING	Q-D	PUBLIC TRAFFIC ROUTE	Q-D
Magazine	Table 3304.5.2(2)	IMD	Table 3305.3	ILD or IPD	Table 3304.5.2(2)	IBD	Table 3304.5.2(2)	PTR
Operating Building	Table 3304.5.2(2)	ILD or IPD	Table 3305.3	ILD or IPD	Table 3304.5.2(2)	IBD	Table 3304.5.2(2)	PTR
Inhabited Building	Table 3304.5.2(2)	IBD	Table 3304.5.2(2)	IBD	NA	NA	NA	NA
Public traffic route	Table 3304.5.2(2)	PTR	Table 3304.5.2(2)	PTR	NA	NA	NA	NA

a. The minimum separation distance (D_0) shall be a minimum of 60 feet. Where a building or magazine containing explosives is barricaded, the minimum distance shall be 30 feet.
b. Linear interpolation between tabular values in the referenced Q-D tables shall not be allowed. Nonlinear interpolation of the values shall be allowed subject to an approved technical opinion and report prepared in accordance with Section 104.7.2.
c. For definitions of IBD, ILD, IMD, IPD and PTR, see Section 3302.1.

TABLE 3301.8.1(2)
APPLICATION OF SEPARATION DISTANCE (Q-D) TABLES
DIVISION 1.3 EXPLOSIVES[a,b,c]

ITEM	MAGAZINE	Q-D	OPERATING BUILDING	Q-D	INHABITED BUILDING	Q-D	PUBLIC TRAFFIC ROUTE	Q-D
Magazine	Table 3304.5.2(3)	IMD	Table 3304.5.2(3)	ILD or IPD	Table 3304.5.2(3)	IBD	Table 3304.5.2(3)	PTR
Operating Building	Table 3304.5.2(3)	ILD or IPD	Table 3304.5.2(3)	ILD or IPD	Table 3304.5.2(3)	IBD	Table 3304.5.2(3)	PTR
Inhabited Building	Table 3304.5.2(3)	IBD	Table 3304.5.2(3)	IBD	NA	NA	NA	NA
Public traffic route	Table 3304.5.2(3)	PTR	Table 3304.5.2(3)	PTR	NA	NA	NA	NA

a. The minimum separation distance (D_0) shall be a minimum of 50 feet.
b. Linear interpolation between tabular values in the referenced Q-D table shall be allowed.
c. For definitions of IBD, ILD, IMD, IPD and PTR, see Section 3302.1.

TABLE 3301.8.1(3)
APPLICATION OF SEPARATION DISTANCE (Q-D) TABLES
DIVISION 1.4 EXPLOSIVES[a,b,c]

ITEM	MAGAZINE	Q-D	OPERATING BUILDING	Q-D	INHABITED BUILDING	Q-D	PUBLIC TRAFFIC ROUTE	Q-D
Magazine	Table 3304.5.2(4)	IMD	Table 3304.5.2(4)	ILD or IPD	Table 3304.5.2(4)	IBD	Table 3304.5.2(4)	PTR
Operating Building	Table 3304.5.2(4)	ILD or IPD	Table 3304.5.2(4)	ILD or IPD	Table 3304.5.2(4)	IBD	Table 3304.5.2(4)	PTR
Inhabited Building	Table 3304.5.2(4)	IBD	Table 3304.5.2(4)	IBD	NA	NA	NA	NA
Public traffic route	Table 3304.5.2(4)	PTR	Table 3304.5.2(4)	PTR	NA	NA	NA	NA

a. The minimum separation distance (D_0) shall be a minimum of 50 feet.
b. Linear interpolation between tabular values in the referenced Q-D table shall not be allowed.
c. For definitions of IBD, ILD, IMD, IPD and PTR, see Section 3302.1.

Section 3302.1 Add new definitions to read as shown: (F170-03/04)

OPERATING LINE. A group of buildings, facilities, or workstations so arranged as to permit performance of the steps in the manufacture of an explosive or in the loading, assembly, modification, and maintenance of ammunition or devices containing explosive materials.

QUANTITY-DISTANCE (Q-D). The quantity of explosive material and separation distance relationships providing protection. These relationships are based on levels of risk considered acceptable for the stipulated exposures and are tabulated in the appropriate Q-D Tables. The separation distances specified afford less than absolute safety.

Minimum separation distance (D_0). The minimum separation distance between adjacent buildings occupied in conjunction with the manufacture, transportation, storage or use of explosive materials where one of the buildings contains explosive materials and the other building does not.

Intraline distance (ILD) or Intraplant distance (IPD). The distance to be maintained between any two operating buildings on an explosives manufacturing site when at least one contains or is designed to contain explosives, or the distance between a magazine and an operating building.

Inhabited building distance (IBD). The minimum separation distance between an operating building or magazine containing explosive materials and an inhabited building or site boundary.

Intermagazine distance (IMD). The minimum separation distance between magazines.

PUBLIC TRAFFIC ROUTE (PTR). Any public street, road, highway, navigable stream, or passenger railroad that is used for through traffic by the general public.

Section 3304.5.1.3 Change to read as shown: (F171-03/04)

3304.5.1.3 Quantity limit. Not more than 50 pounds (23 kg) of explosives or explosive materials shall be stored within an indoor magazine.

> **Exception:** Day boxes used for the storage of in-process material in accordance with Section 3305.6.4.1.

Section 3304.5.2 Change to read as shown: (F170-03/04)

3304.5.2 Outdoor magazines. All outdoor magazines other than Type 3 shall be located so as to comply with Table 3304.5.2(2), Table 3304.5.2(3) or Table 3304.5.2.4(4) as set forth in Tables 3301.8.1(1) through 3301.8.1(3). Where a magazine or group of magazines, as described in Section 3304.5.2.2, contains different classes of explosive materials, and Division 1.1 materials are present, the required separations for the magazine or magazine group as a whole shall comply with Table 3304.5.2(2).

Sections 3304.5.2.1 and 3304.5.2.2 Add new sections to read as shown: (F170-03/04)

3304.5.2.1 Separation. Where two or more storage magazines are located on the same property, each magazine shall comply with the minimum distances specified from inhabited buildings, public transportation routes and operating buildings. Magazines shall be separated from each other by not less than the intermagazine distances (IMD) shown for separation of magazines.

3304.5.2.2 Grouped magazines. Where two or more magazines are separated from each other by less than the intermagazine distances (IMD), such magazines as a group shall be considered as one magazine, and the total quantity of explosive materials stored in the group shall be treated as if stored in a single magazine. The location of the group of magazines shall comply with the intermagazine distances (IMD) specified from other magazines or magazine groups, inhabited buildings (IBD), public transportation routes (PTR) and operating buildings (ILD) or (IPD) as required.

Table 3304.5.2(1) APPLICATION OF SEPARATION DISTANCE TABLE. Delete table without substitution: (F170-03/04)

Table 3304.5.2(3) Change entire table to read as shown: (F170-03/04)

TABLE 3304.5.2(3)
TABLE OF DISTANCES (Q-D) FOR BUILDINGS CONTAINING EXPLOSIVES — DIVISION 1.3 MASS-FIRE HAZARD[a, b, c]

QUANTITY OF DIVISION 1.3 EXPLOSIVES (NET EXPLOSIVES WEIGHT)		DISTANCES IN FEET			
Pounds over	Pounds not over	Inhabited Building Distance (IBD)	Distance to Public Traffic Route (PTR)	Intermagazine Distance (IMD)	Intraline Distance (ILD) or Intraplant Distance (IPD)
0	1000	75	75	50	50
1000	5000	115	115	75	75
5000	10000	150	150	100	100
10000	20000	190	190	125	125
20000	30000	215	215	145	145
30000	40000	235	235	155	155
40000	50000	250	250	165	165
50000	60000	260	260	175	175
60000	70000	270	270	185	185
70000	80000	280	280	190	190
80000	90000	295	295	195	195
90000	100000	300	300	200	200
100000	200000	375	375	250	250
200000	300000	450	450	300	300

(Current notes and footnotes do not change)

Table 3304.5.2(4) Change entire table to read as shown: (F170-03/04)

TABLE 3304.5.2(4)
TABLE OF DISTANCES (Q-D) FOR BUILDINGS CONTAINING EXPLOSIVES —
DIVISION 1.4 [c]

QUANTITY OF DIVISION 1.4 EXPLOSIVES (NET EXPLOSIVES WEIGHT)		DISTANCES IN FEET			
Pounds over	Pounds not over	Inhabited Building Distance IBD	Distance to Public Traffic Route (PTR)	Intermagazine Distance[a, b] (IMD)	Intraline Distance (ILD) or Intraplant Distance[a] (IPD)
50	Not limited	100	100	50	50

(Current notes and footnotes do not change)

Table 3305.3 Change table heading to read shown: (F170-03/04)

TABLE 3305.3
MINIMUM INTRALINE (INTRAPLANT) SEPARATION DISTANCES (ILD OR IPD) BETWEEN
BARRICADED OPERATING BUILDINGS CONTAINING EXPLOSIVES —
DIVISION 1.1, 1.2 OR 1.5 — MASS EXPLOSION HAZARD[a]

NET EXPLOSIVE WEIGHT			NET EXPLOSIVE WEIGHT		
Pounds over	Pounds not over	Intraline Distance (ILD) or Intraplant Distance (IPD) (feet)	Pounds over	Pounds not over	Intraline Distance (ILD) or Intraplant Distance (IPD) (feet)

(Portions of table not shown do not change)

Section 3305.4 Change to read as shown: (F171-03/04)

3305.4 Separation of manufacturing operating buildings from inhabited buildings, public traffic routes, and magazines. When an operating building on an explosive materials plant site is designed to contain explosive materials, such building shall be located away from inhabited buildings, public traffic routes and magazines in accordance with Table 3304.5.2(2), 3304.5.2(3) or 3304.5.2(4) as appropriate, based on the maximum quantity of explosive materials permitted to be in the building at one time (see Section 3301.8).

Exception: Fireworks-manufacturing buildings constructed and operated in accordance with NFPA 1124.

Sections 3305.4.1 Add new section to read as shown: (F171-03/04)

3305.4.1 Determination of net explosive weight for operating buildings. In addition to the requirements of Section 3301.8 to determine the net explosive weight for materials stored or used in operating buildings, quantities of explosives materials stored in magazines located at distances less than intraline distances from the operating building shall be added to the contents of the operating building to determine the net explosive weight for the operating building.

3305.4.1.1 Indoor magazines. The storage of explosive materials located in indoor magazines in operating buildings shall be limited to a net explosive weight not to exceed 50 pounds.

3305.4.1.2 Outdoor magazines with a net explosive weight less than 50 pounds. The storage of explosive materials in outdoor magazines located at less than intraline distances from operating buildings shall be limited to a net explosive weight not to exceed 50 pounds.

3305.4.1.3 Outdoor magazines with a net explosive weight greater than 50 pounds. The storage of explosive materials in outdoor magazines in quantities exceeding 50 pounds net explosive weight shall be limited to storage in outdoor magazines located not less than intraline distances from the operating building in accordance with Section 3304.5.2.

3305.4.1.4 Net explosive weight of materials stored in combination indoor and outdoor magazines. The aggregate quantity of explosive materials stored in any combination of indoor magazines or outdoor magazines located at less than the intraline distances from an operating building shall not exceed 50 pounds.

Section 3308.2 Change to read as shown: (F174-03/04)

3308.2 Permit application. Prior to issuing permits for fireworks display, plans for the display, inspections of the display site, and demonstrations of the display operations shall be approved. A plan establishing procedures to follow and actions to be taken in the event that a shell fails to ignite in, or discharge from, a mortar or fails to function over the fallout area or other malfunctions shall be provided to the fire code official.

Section 3308.9 Change to read as shown: (F175-03/04)

3308.9 Post-display inspection. After the display, the firing crew shall conduct an inspection of the fallout area for the purpose of locating unexploded aerial shells or live components. This inspection shall be conducted before public access to the site shall be allowed. Where fireworks are displayed at night and it is not possible to inspect the site thoroughly, the operator or designated assistant shall inspect the entire site at first light. A report identifying any shells that fail to ignite in, or discharge from, a mortar or fail to function over the fallout area or otherwise malfunction shall be filed with the fire code official.

Section 3310.3 Add new section to read as shown: (E145-03/04)

3310.3 Stairway floor number signs. Temporary stairway floor number signs shall be provided in accordance with the requirements of Section 1019.1.7.

CHAPTER 34
FLAMMABLE AND COMBUSTIBLE LIQUIDS

Section 3404.2.13.1.4 Change to read as shown: (F177-03/04)

3404.2.13.1.4 Tanks abandoned in place. Tanks abandoned in place shall be abandoned as follows:

1. Flammable and combustible liquids shall be removed from the tank and connected piping.

2. The suction, inlet, gauge, vapor return and vapor lines shall be disconnected.

3. The tank shall be filled completely with an approved inert solid material.

   Exception: Residential heating oil tanks of 1,100 gallons (4164 L) or less, provided the fill line is permanently removed to a point below grade to prevent refilling of the tank.

4. Remaining underground piping shall be capped or plugged.

5. A record of tank size, location and date of abandonment shall be retained.

6. All exterior above-grade fill piping shall be permanently removed when tanks are abandoned or removed.

Section 3404.2.14.1 Change to read as shown (F178-03/04)

3404.2.14.1 Removal. Removal of above-ground and underground tanks shall be in accordance with all of the following:

1. Flammable and combustible liquids shall be removed from the tank and connecting piping.

2. Piping at tank openings which is not to be used further shall be disconnected.

3. Piping shall be removed from the ground.

   Exception: Piping is allowed to be abandoned in place where the fire code official determines that removal is not practical. Abandoned piping shall be capped and safeguarded as required by the fire code official.

4. Tank openings shall be capped or plugged, leaving a 0.125-inch to 0.25-inch-diameter (3.2 mm to 6.4 mm) opening for pressure equalization.

5. Tanks shall be purged of vapor and inserted prior to removal.

6. All exterior above grade fill and vent piping shall be permanently removed.

   Exception: Piping associated with bulk plants, terminal facilities and refineries.

Section 3404.3.3.9 Change to read as shown: (F158-03/04)

3404.3.3.9 Idle combustible pallets. Storage of empty or idle combustible pallets inside an unprotected liquid storage area shall be limited to a maximum pile size of 2,500 square feet (232 m^2) and to a maximum storage height of 6 feet (1829 mm). Storage of empty or idle combustible pallets inside a protected liquid storage area shall comply with NFPA 13 and NFPA 230. Pallet storage shall be separated from liquid storage by aisles that are at least 8 feet (2438 mm) wide.

Table 3404.3.6.3(2) Change second column of table in specified entries and add footnote "d" to read as shown: (F179-03/04)

TABLE 3404.3.6.3(2)
STORAGE ARRANGEMENTS FOR PALLETIZED OR SOLID-PILE STORAGE IN LIQUID STORAGE ROOMS AND LIQUID WAREHOUSES

CLASS	STORAGE LEVEL
IC	Ground floor[d] Upper floors Basements
II	Ground floor[d] Upper floor Basements

(Portions of table not shown do not change)

a. through c. (No change to current text)
d. For palletized storage of unsaturated polyester resins (UPR) in relieving-style metal containers with 50 percent or less by weight Class IC or Class II liquid and no Class IA or Class IB liquid, height and pile quantity limits shall be permitted to be 10 feet and 15,000 gallons, respectively, provided that such storage is protected by sprinklers in accordance with NFPA 30 and that the UPR storage area is not located in the same containment area or drainage path for other Class I or Class II liquids.

Table 3404.3.6.3(7) Change table title and add footnote "b" to read as shown: (F158-03/04)

TABLE 3404.3.6.3(7)
AUTOMATIC AFFF WATER PROTECTION REQUIREMENTS FOR RACK STORAGE OF LIQUIDS IN CONTAINERS GREATER THAN 5-GALLON CAPACITY[a,b]

(No change to contents of table)

a. (No change to current text)
b. Except as modified herein, in-rack sprinklers shall be installed in accordance with NFPA 13
c. and d. (No change to current text)

Section 3404.3.7.5.1 Change to read as shown: (F158-03/04)

3404.3.7.5.1 Fire-extinguishing systems. Liquid storage rooms shall be protected by automatic sprinkler systems installed in accordance with Chapter 9 and Tables 3404.3.6.3(4) through 3404.3.6.3(7) and Table 3404.3.7.5.1. In-rack sprinklers shall also comply with NFPA 13.

Automatic foam-water systems and automatic aqueous film-forming foam (AFFF) water sprinkler systems shall not be used except when approved.

Protection criteria developed from fire modeling or full-scale fire testing conducted at an approved testing laboratory are allowed in lieu of the protection as shown in Tables 3404.3.6.3(2) through 3404.3.6.3(7) and Table 3404.3.7.5.1 when approved.

Section 3404.3.8.4 Change to read as shown: (F158-03/04)

3404.3.8.4 Fire-extinguishing systems. Liquid storage warehouses shall be protected by automatic sprinkler systems installed in accordance with Chapter 9 and Tables 3404.3.6.3(4) through 3404.3.6.3(7) and Table 3404.3.7.5.1, or Section 4.8.2 and Tables 4.8.2(a) through (f) of NFPA 30. In-rack sprinklers shall also comply with NFPA 13.

Automatic foam water systems and automatic aqueous film-forming foam water sprinkler systems shall not be used except when approved.

Protection criteria developed from fire modeling or full-scale fire testing conducted at an approved testing laboratory are allowed in lieu of the protection as shown in Tables 3404.3.6.3(2) through 3404.3.6.3(7) and Table 3404.3.7.5.1 when approved.

Section 3405.3.8.4 Add new section to read as shown: (F180-03/04)

3405.3.8.4 Weather protection. Weather protection for outdoor use shall be in accordance with Section 2705.3.9.

SECTION 3502
DEFINITIONS

3502.1 Definitions: The following words and terms shall, for the purposes of this chapter and as used elsewhere in this code, have the meanings shown herein.

Section 3502.1 Add new definition to read as shown: (F181-03/04)

METAL HYDRIDE STORAGE SYSTEM. A system for the storage of hydrogen gas absorbed in solid material.

CHAPTER 35
FLAMMABLE GASES

Section 3503.1.3 Change section to read as shown: (F8-03/04)

3503.1.3 Emergency shutoff. Compressed gas systems conveying flammable gases shall be provided with approved manual or automatic emergency shutoff valves that can be activated at each point of use and each source.

Sections 3503.1.3.1 and 3503.1.3.2 Add new sections to read as shown: (F8-03/04)

3503.1.3.1 Shutoff at source. A manual or automatic fail-safe emergency shutoff valve shall be installed on supply piping at the cylinder or bulk source. Manual or automatic cylinder valves are allowed to be used as the required emergency shutoff valve when the source of supply is limited to unmanifolded cylinder sources.

3503.1.3.2 Shutoff at point of use. A manual or automatic emergency shutoff valve shall be installed on the supply piping at the point of use or at a point where the equipment using the gas is connected to the supply system.

Sections 3503.1.5 Add new section to read as shown: (F182-03/04)

3503.1.5 Electrical. Electrical wiring and equipment shall be installed and maintained in accordance with the ICC *Electrical Code*.

3503.1.5.1 Bonding of electrically conductive materials and equipment. Exposed noncurrent-carrying metal parts, including metal gas piping systems, that are part of flammable gas supply systems located in a hazardous (electrically classified) location shall be bonded to a grounded conductor in accordance with the provisions of the ICC *Electrical Code*.

3503.1.5.2 Static-producing equipment. Static-producing equipment located in flammable gas storage or use areas shall be grounded.

Sections 3503.1.6 Add new section to read as shown: (F181-03/04)

3503.1.6 Hydrogen gas absorbed in solids. The hazard classification of the metal hydride storage system, as required by Section 2701.2.2, shall be based on the hydrogen stored without regard to the metal hydride content.

3503.1.6.1 Listed system. Metal hydride storage systems shall be listed for the application and designed in a manner that prevents the removal of the metal hydride.

CHAPTER 37
HIGHLY TOXIC AND TOXIC MATERIALS

Section 3702.1 Add new definition to read as shown: (F183-03/04)

PHYSIOLOGICAL WARNING THRESHOLD LEVEL. A concentration of airborne contaminants, normally expressed in parts per million (ppm) or milligrams per cubic meter (mg/m³), that represents the concentration at which persons can sense the presence of the contaminant due to odor, irritation or other quick-acting physiological response. When used in conjunction with the Permissible Exposure Limit (PEL), the physiological warning threshold levels are those consistent with the classification system used to establish the PEL. See the definition of Permissible Exposure Limit (PEL).

Section 3704.2.2.7 Change to read as shown: (F184-03/04)

3704.2.2.7 Treatment systems. The exhaust ventilation from gas cabinets, exhausted enclosures and gas rooms and local exhaust systems required in Sections 3704.2.2.4 and 3704.2.2.5 shall be directed to a treatment system. The treatment system shall be utilized to handle the accidental release of gas and to process exhaust ventilation. The treatment system shall be designed in accordance with Sections 3704.2.2.7.1 through 3704.2.2.7.5 and Section 510 of the *International Mechanical Code*.

1. Highly toxic and toxic gases—storage. A treatment system is not required for cylinders, containers and tanks in storage when all of the following controls are provided:

   1.1. Valve outlets are equipped with gas-tight outlet plugs or caps.

   1.2. Handwheel-operated valves have handles secured to prevent movement.

   1.3. Approved containment vessels or containment systems are provided in accordance with Section 3704.2.2.3.

2. Toxic gases—use. Treatment systems are not required for toxic gases supplied by cylinders or portable tanks not exceeding 1,700 pounds water capacity when the following are provided:

   2.1. A gas detection system with a sensing interval not exceeding 5 minutes.

   2.2. An approved automatic-closing fail-safe valve located immediately adjacent to cylinder or portable tank valves. The fail-safe valve shall close when gas is detected at the permissible exposure limit (PEL) by a gas detection system monitoring the exhaust system at the point of discharge from the gas cabinet, exhausted enclosure, ventilated enclosure or gas room. The gas detection system shall comply with Section 3704.2.2.10.

Section 3704.2.2.10 Change to read as shown: (F183-03/04)

3704.2.2.10 Gas detection system. A gas detection system shall be provided to detect the presence of gas at or below the permissible exposure limit (PEL) or ceiling limit of the gas for which detection is provided. The system shall be capable of monitoring the discharge from the treatment system at or below one-half the IDLH limit.

Exception: A gas detection system is not required for toxic gases when the physiological warning threshold level for the gas is at a level below the accepted PEL for the gas.

CHAPTER 40
OXIDIZERS

Section 4003.1.2 Change to read as shown: (F8-03/04)

4003.1.2 Emergency shutoff. Compressed gas systems conveying oxidizing gases shall be provided with approved manual or automatic emergency shutoff valves that can be activated at each point of use and each source.

Sections 4003.1.2.1 and 4003.1.2.2 Add new sections to read as shown: (F8-03/04)

4003.1.2.1 Shutoff at source. A manual or automatic fail-safe emergency shutoff valve shall be installed on supply piping at the cylinder or bulk source. Manual or automatic cylinder valves are allowed to be used as the required emergency shutoff valve when the source of supply is limited to unmanifolded cylinder sources.

4003.1.2.2 Shutoff at point of use. A manual or automatic emergency shutoff valve shall be installed on the supply piping at the point of use or at a point where the equipment using the gas is connected to the supply system.

CHAPTER 41
PYROPHORIC MATERIALS

Section 4103.1.1 Change to read as shown: (F8-03/04)

4103.1.1 Emergency shutoff. Compressed gas systems conveying pyrophoric gases shall be provided with approved manual or automatic emergency shutoff valves that can be activated at each point of use and each source.

Sections 4103.1.1.1 and 4103.1.1.2 Add new sections as shown: (F8-03/04)

4103.1.1.1 Shutoff at source. An automatic emergency shutoff valve shall be installed on supply piping at the cylinder or bulk source. The shutoff valve shall be operated by a remotely located manually activated shutdown control located not less than 15 feet (4572 mm) from the source of supply. Manual or automatic cylinder valves are allowed to be used as the required emergency shutoff valve when the source of supply is limited to unmanifolded cylinder sources.

4103.1.1.2 Shutoff at point of use. A manual or automatic emergency shutoff valve shall be installed on the supply piping at the point of use or at a point where the equipment using the gas is connected to the supply system.

Section 4106.4.2 Change to read as shown: (F8-03/04)

4106.4.2 Remote manual shutdown. A remotely located manually activated shutdown control shall be provided outside each gas cabinet.

CHAPTER 44
WATER-REACTIVE SOLIDS AND LIQUIDS

Section 4402.1 Change definition to read as shown: (F187-03/04)

4402.1 Definition. The following word and term shall, for the purposes of this chapter and as used elsewhere in this code, have the meaning shown herein.

WATER-REACTIVE MATERIAL. A material that explodes; violently reacts; produces flammable, toxic or other hazardous gases; or evolves enough heat to cause auto-ignition or ignition of combustibles upon exposure to water or moisture. Water-reactive materials are subdivided as follows:

Class 3. Materials that react explosively with water without requiring heat or confinement.

Class 2. Materials that react violently with water or have the ability to boil water. Materials that produce flammable, toxic, or other hazardous gases, or evolve enough heat to cause autoignition or ignition of combustibles upon exposure to water or moisture.

Class 1. Materials that react with water with some release of energy, but not violently.

2004 SUPPLEMENT TO THE IFC

CHAPTER 45
REFERENCED STANDARDS

Change, delete or add the following referenced standards to read as shown: (E146-03/04; E147-03/04; F35-03/04; F44-03/04; F45-03/04; F99-03/04; F140-03/04; F158-03/04; F188-03/04) (STANDARDS NOT SHOWN DO NOT CHANGE)

AASHTO
American Association of State Highways and
444 North Capitol Street, NW – Suite 249
Washington, DC 20001

Standard reference number	Title	Referenced In code section number
HB-17-2002	Specification for Highway Bridges, 17th Edition 2002	503.2.6

API
American Petroleum Institute
1220 L Street, NW, Suite 900
Washington, DC 20005

Standard reference number	Title	Referenced In code section number
Std 653 (2001)	Tank Inspection, Repair, Alteration and Reconstruction	3406.7
Publ 2009 (2002)	Safe Welding and Cutting Practices in Refineries, Gas Plants and Petrochemical Plants	3406.7
Std 2015 (2001)	Requirements for Safe Entry and Cleaning of Petroleum Storage Tanks	3406.7, 3406.7.2
Publ 2023 (2001)	Guide for Safe Storage and Handling of Heated Petroleum-Derived Asphalt Products and Crude Oil Residue	3406.7, 3406.7.3
Publ 2028 (2002)	Flame Arrestors in Piping Systems	3404.2.7.3.2

ASME
American Society of Mechanical Engineers
Three Park Avenue
New York, NY 10016-5990

Standard reference number	Title	Referenced In code section number
A 17.1-2000	Safety Code for Elevators and Escalators - with A17.1a-2002 Addenda	607.1, 1007.4
A 17.3-2002	Safety Code for Existing Elevators and Escalators	607.1
A18.1-2003	Safety Standard for Platform Lifts and Stairway Chairlifts	1007.5
B 31.3-2002	Process Piping	2209.5.4.1, 2703.2.2.2
BPVC-2001	ASME Boiler & Pressure Vessel Code (2001 Edition) (Sections I, II, IV, V & VI)	2209.5.4.2.1, 3203.4.3, 3203.8, 3404.2.13.1.5

IFC-51

ASTM

ASTM International
100 Barr Harbor Drive
West Conshohocken, PA 19428-2859

Standard reference number	Title	Referenced In code section number
D 56-02a	Test Method for Flash Point by Tag Closed Tester	3402.1
D 86-02	Method for Distillation of Petroleum Products at Atmospheric Pressure	2702.1
D 92-02b	Test Method for Flash and Fire Points by Cleveland Open Cup	3401.2, 3402.1
D 93-02	Test Method for Flash Point by Pensky-Martens Closed Cup Tester	3402.1
E 84-03	Test Method for Surface Burning Characteristics of Building Materials	805.3.4, 806.2.3, 806.3
E 1537-02a	Test Method for Fire Testing of Upholstered Furniture	803.5.2
E 1590-02	Test Method for Fire Testing of Mattresses	803.5.3, 803.6.3, 803.7.4

BHMA

Builders Hardware Manufacturers' Association
355 Lexington Avenue, 17th Floor
New York, NY 10017-6603

Standard reference number	Title	Referenced In code section number
A 156.19-2002	American National Standard for Power Assist and Low Energy Power Operated Doors	1008.1.3.2

CGA

Compressed Gas Association
4221 Walney Road - 5th Floor
Chantilly, VA 20151-2923

Standard reference number	Title	Referenced In code section number
S-1.1 (2002)	Relief Device Standards - Part 1 - Cylinders for Compressed Gases	3203.2
V-1 (2002)	Gas Cylinder Valve Outlet and Inlet Connections	4106.1.3

ICC

International Code Council Inc.
5203 Leesburg Pike, Suite 600
Falls Church, VA 22041

Standard reference number	Title	Referenced In code section number
ICC/ANSI A117.1-03	Accessible and Usable Buildings and Facilities	907.10.14, 1007.6.5, 1010.1, 1010.6.5, 1010.9, 1011.3

NEMA

National Electrical Manufacturer's Association
1300 N. 17th Street, Suite 1874
Rosslyn, VA 22209

Standard reference number	Title	Referenced In code section number
250-2003	Enclosures for Electrical Equipment (1000 Volts Maximum)	3705.2

2004 SUPPLEMENT TO THE IFC

NFPA

National Fire Protection Association
1 Batterymarch Park
Quincy, MA 02269-9101

Standard reference number	Title	Referenced In code section number
10-02	Portable Fire Extinguishers	Table 901.6.1, 906.2, 906.3, Table 906.3(1), Table 906.3(2), 2106.3
11-02	Low-, Medium-, and High-Expansion Foam	904.7, 3404.2.9.1.2
13-02	Installation of Sprinkler Systems	Table 704.1, 903.3.1.1, 903.3.2, 905.3.5.1.1, 903.3.5.2, 904.11, 907.9, Table 2306.2, 2306.9, 2804.1, 3404.3.7.5.1, 3404.3.8.4
13D-02	Installation of Sprinkler Systems in One- and Two-Family Dwellings and Manufactured Homes	903.3.1.3, 903.3.5.1.1
13R-02	Installation of Sprinkler Systems in Residential Occupancies up to and Including Four Stories in Height	903.1.2, 903.3.1.2, 903.3.5.1.1, 903.3.5.1.2, 903.4
15-01	Water Spray Fixed Systems for Fire Protection	3404.2.9.1.3
17-02	Dry Chemical Extinguishing Systems	Table 901.6.1, 904.6, 904.11
17A-02	Wet Chemical Extinguishing Systems	Table 901.6.1, 904.5, 904.11
22-03	Water Tanks for Private Fire Protection	508.2.2
24-02	Installation of Private Fire Service Mains and Their Appurtenances	508.2.1, 1909.5
25-02	Inspection, Testing and Maintenance of Water-Based Fire Protection Systems	508.5.3, Table 901.6.1, 904.7.1, 912.6, 913.5,
30B-02	Manufacture and Storage of Aerosol Products	3403.6.2, 3403.6.2.1, 3404.2.7, 3404.2.7.1, 3404.2.7.2, 3404.2.7.3.6, 3404.2.7.4, 3404.2.7.6, 3404.2.7.7, 3404.2.7.8, 3404.2.7.9, 3404.2.9.2, 3404.2.9.3, 3404.2.9.5.1.1, 3404.2.9.5.1.2, 3404.2.9.5.1.3, 3404.2.9.5.1.4, 3404.2.9.5.1.5, 3404.2.9.5.2, 3404.2.9.6.4, 3404.2.10.2, 3404.2.11.4, 3404.2.11.5.2, 3404.2.12.1, 3404.3.1, 3404.3.6, 3404.3.7.2.3, 3404.3.8.4, 3406.8.3
40-01	Storage and Handling of Cellulose Nitrate Film	306.2
51-02	Design and Installation of Oxygen-Fuel Gas Systems for Welding, Cutting, and Allied Processes	2601.5, 2607.1, 2609.1
52-02	Compressed Natural Gas (CNG) Vehicular Fuel System Code	3001.1
57-02	Liquefied Natural Gas (LNG) Vehicular Fuel Systems Code	3001.1
61-02	Prevention of Fires and Dust Explosions in Agricultural and Food Processing Facilities	Table 1304.1
69-02	Explosion Prevention Systems	911.1, 911.3, Table 1304.1
72-02	National Fire Alarm Code	509.1, Table 901.6.1, 903.4.1, 904.3.5, 907.2, 907.2.1, 907.2.1.1, 907.2.10, 907.2.10.4, 907.2.11.2, 907.2.11.3, 907.2.12.2.3, 907.2.12.3, 907.3, 907.5, 907.6, 907.10.2, 907.11, 907.15, 907.17, 907.18, 907.20, 907.20.2, 907.20.5
99-02	Health Care Facilities	3006.4
110-02	Emergency and Standby Power Systems	604.1, 604.4, 913.5.2, 913.5.3
230-99	Standard for the Fire Protection of Storage	2301.1, 2308.4, 2310.1, 2501.1, 3404.3.3.9
231-98	DELETED	
231D-98	DELETED	
231C-98	DELETED	
265-02	Methods of Fire Tests for Evaluating Room Fire Growth Contribution of Textile Coverings on Full Height Panels and Walls	806.2.3, 806.2.3.1, 806.2.3.2
266-98	DELETED	
267-98	DELETED	
303-00	Fire Protection Standard for Marinas and Boatyards	905.3.7
407-01	Aircraft Fuel Servicing	1106.2, 1106.3
409-01	Aircraft Hangars	914.7.2, 914.8.4.1
430-00	Storage of Liquid and Solid Oxidizers	4004.1.4
480-98	DELETED (combined with 484-02)	
481-00	DELETED (combined with 484-02)	
482-96	DELETED (combined with 484-02)	
484-02	Combustible Metals, Metal Powders, and Metal Dusts	Table 1304.1
490-02	Storage of Ammonium Nitrate	3301.1.5

NFPA (continued)

495-01	Explosive Materials Code	911.1, 911.4, 3301.1.1, 3301.1.5, 3302.1, 3304.2, 3304.6.2, 3304.6.3, 3304.7.1, 3305.1, 3306.1, 3306.5.2.1, 3306.5.2.3, 3307.1, 3307.9, 3307.11, 3307.15
498-01	Safe Havens and Interchange Lots for Vehicles Transporting Explosives	3301.1.2
505-02	Powered Industrial Trucks Including Type Designations, Areas of Use, Conversions, Maintenance, and Operations	2703.7.3
650-98	DELETED (Combined into NFPA 654)	
651-98	DELETED (Combined into NFPA 484)	
655-01	Prevention of Sulfur Fires and Explosions	Table 1304.1
664-02	Prevention of Fires and Explosions in Wood Processing and Woodworking Facilities	Table 1304.1, 1905.3
704-01	System for the Identification of the Hazards of Materials for Emergency Response	606.7, 606.9.3.4, 1802.1, 2703.2.2.1, 2703.2.2.2, 2703.5, 2703.10.2, 2705.1.10, 2705.2.1.1, 2705.4.4, 3203.4.1, 3404.2.3.2
1122-02	Model Rocketry	3301.1.4
1124-03	Manufacture, Transportation, Storage, and Retail Sales of Fireworks and Pyrotechnic Articles	3302.1, 3304.2, 3305.1, 3305.3, 3305.4, 3305.5
1125-01	Manufacture of Model Rocket and High Power Rocket Motors	3301.1.4
1127-02	High Power Rocketry	3301.1.4

UL

Underwriters Laboratories, Inc.
333 Pfingsten Road
Northbrook, IL 60062-2096

Standard reference number	Title	Referenced In code section number
197-03	Commercial Electric Cooking Appliances	904.11
1363-96	Relocatable Power Taps - with Revisions through October 2001	605.4.1
2208-96	Solvent Distillation Units - with Revisions through August 2001	3405.4.1

APPENDIX D
FIRE APPARATUS ACCESS ROADS

Appendix Figure D103.1 Delete figures illustrating 70' Diameter Cul-De-Sac and 60' Hammerhead without substitution: (F190-03/04)

Appendix D103.5 Change to read as shown: (F191-03/04)

D103.5 Fire apparatus access road gates. Gates securing the fire apparatus access roads shall comply with all of the following criteria:

1. The minimum gate width shall be 20 feet (6096 mm).

2. Gates shall be of the swinging or sliding type.

3. Construction of gates shall be of materials that allow manual operation by one person.

4. Gate components shall be maintained in an operative condition at all times and replaced or repaired when defective.

5. Electric gates shall be equipped with a means of opening the gate by fire department personnel for emergency access. Emergency opening devices shall be approved by the fire code official.

6. Manual opening gates shall not be locked with a padlock or chain and padlock unless they are capable of being opened by means of forcible entry tools or a key box containing the key(s) to the lock is installed at the gate location.

7. Locking device specifications shall be submitted for approval by the fire code official.

Appendix D107.1 Change to read as shown: (F192-03/04)

D107.1 One- or two-family dwelling residential developments. Developments of one- or two-family dwellings where the number of dwelling units exceeds 30 shall be provided with separate and approved fire apparatus access roads, and shall meet the requirements of Section D104.3.

Exceptions:

1. Where there are more than 30 dwelling units on a single public or private fire apparatus access road and all dwelling units are equipped throughout with an approved automatic sprinkler system in accordance with Section 903.3.1.1, 903.3.1.2 or 903.3.1.3 of the *International Fire Code*, access from two directions shall not be required.

2. The number of dwelling units on a single fire apparatus access road shall not be increased unless fire apparatus access roads will connect with future development, as determined by the fire code official.

International Fuel Gas Code

2004 Supplement

INTERNATIONAL FUEL GAS CODE 2004 SUPPLEMENT

Chapter 1
ADMINISTRATION

Section 101.2 Change to read as shown: (FG4-03/04)

101.2 Scope. This code shall apply to the installation of fuel-gas piping systems, fuel-gas utilization equipment, gaseous hydrogen systems and related accessories in accordance with Sections 101.2.1 through 101.2.5.

Exceptions: (Unchanged)

CHAPTER 2
DEFINITIONS

SECTION 202
DEFINITIONS

Change the definition of "Approved" to read as shown: (FG2-03/04)

APPROVED. Acceptable to the code official or other authority having jurisdiction.

Change the definition of "Connector" to read as shown: (FG3-03/04)

CONNECTOR, CHIMNEY OR VENT. The pipe that connects an appliance to a chimney or vent.

Add new definition of "Connector, Appliance (Fuel)" as shown: (FG3-03/04)

CONNECTOR, APPLIANCE (fuel). Rigid metallic pipe and fittings, semi-rigid metallic tubing and fittings or a listed and labeled device that connects an appliance to the gas piping system.

Change the definition of "Fuel Gas" to read as shown: (FG4-03/04)

FUEL GAS. A natural gas, manufactured gas, liquefied petroleum gas or mixtures of these gases.

Delete the definition of "Mechanical Exhaust System" (FG5-03/04)

Change the definition of "Point of Delivery" to read as shown: (FG6-03/04)

POINT OF DELIVERY. For natural gas systems, the point of delivery is the outlet of the service meter assembly, or the outlet of the service regulator or service shutoff valve where a meter is not provided. Where a valve is provided at the outlet of the service meter assembly, such valve shall be considered to be downstream of the point of delivery. For undiluted liquefied petroleum gas systems, the point of delivery shall be considered the outlet of the first regulator that reduces pressure to 2 psig (13.8 kPag) or less.

Delete definition of "Regulator, Medium-Pressure (MP Regulator) and substitute as shown: (FG7-03/04)

REGULATOR, MEDIUM-PRESSURE (MP Regulator). A line pressure regulator that reduces gas pressure from the range of greater than 0.5 psig (3.4 kPa) and less than or equal to 5 psig (34.5 kPa) to a lower pressure.

Add new definitions of "Vent Piping, Breather, Relief" as shown: (FG30-03/04)

VENT PIPING

Breather. Piping run from a pressure regulating device to the outdoors, designed to provide a reference to atmospheric pressure. If the device incorporates an integral pressure relief mechanism, a breather vent can also serve as a relief vent.

Relief. Piping run from a pressure-regulating or pressure-limiting device to the outdoors, designed to provide for the safe venting of gas in the event of excessive pressure in the gas piping system.

CHAPTER 3
GENERAL REGULATIONS

Section 303.3 Change to read as shown: (FG8-03/04)

303.3 Prohibited locations. Appliances shall not be located in sleeping rooms, bathrooms, toilet rooms, storage closets or surgical rooms, or in a space that opens only into such rooms or spaces, except where the installation complies with one of the following:

1. The appliance is a direct-vent appliance installed in accordance with the conditions of the listing and the manufacturer's instructions.

2. Vented room heaters, wall furnaces, vented decorative appliances, vented gas fireplaces, vented gas fireplace heaters and decorative appliances for installation in vented solid-fuel-burning fireplaces are installed in rooms that meet the required volume criteria of Section 304.5.

3. A single wall-mounted unvented room heater is installed in a bathroom and such unvented room heater is equipped as specified in Section 621.6 and has an input rating not greater than 6,000 Btu/h (1.76 kW). The bathroom shall meet the required volume criteria of Section 304.5.

4. A single wall-mounted unvented room heater is installed in a bedroom and such unvented room heater is equipped as specified in Section 621.6 and has an input rating not greater than 10,000 Btu/h (2.93 kW). The bedroom shall meet the required volume criteria of Section 304.5.

5. The appliance is installed in a room or space that

opens only into a bedroom or bathroom, such room or space is used for no other purpose, and is provided with a solid weather-stripped door equipped with an approved self-closing device. All combustion air shall be taken directly from the outdoors in accordance with Section 304.6.

Section 305.3.1 Add new section as shown: (FG10-03/04)

305.3.1 Parking garages. Connection of a parking garage with any room in which there is a fuel-fired appliance shall be by means of a vestibule providing a two-doorway separation, except that a single door is permitted where the sources of ignition in the appliance are elevated in accordance with Section 305.3.

Section 306.3.1 Change to read as shown: (EL3-03/04)

306.3.1 Electrical requirements. A luminaire controlled by a switch located at the required passageway opening and a receptacle outlet shall be provided at or near the equipment location in accordance with the ICC *Electrical Code*.

Section 306.4.1 Change to read as shown: (EL3-03/04)

306.4.1 Electrical requirements. A luminaire controlled by a switch located at the required passageway opening and a receptacle outlet shall be provided at or near the equipment location in accordance with the ICC *Electrical Code*.

Section 306.5.1 Change to read as shown: (M14-03/04)

306.5.1 Sloped roofs. Where appliances are installed on a roof having a slope of three units vertical in 12 units horizontal (25-percent slope) or greater and having an edge more than 30 inches (762 mm) above grade at such edge, a level platform shall be provided on each side of the appliance to which access is required for service, repair or maintenance. The platform shall not be less than 30 inches (762 mm) in any dimension and shall be provided with guards. The guards shall extend not less than 42 inches above the platform, shall be constructed so as to prevent the passage of a 21-inch-diameter sphere and shall comply with the loading requirements for guards specified in the *International Building Code*.

Section 306.6 Change to read as shown: (FG13-03/04; M9-03/04)

306.6 Guards. Guards shall be provided where appliances or other components that require service and roof hatch openings are located within 10 feet (3048 mm) of a roof edge or open side of a walking surface and such edge or open side is located more than 30 inches (762 mm) above the floor, roof or grade below. The guard shall extend not less than 30 inches (762 mm) beyond each end of such appliances, components and roof hatch openings and the top of the guard shall be located not less than 42 inches (1067 mm) above the elevated surface adjacent to the guard. The guard shall be constructed so as to prevent the passage of a 21-inch-diameter (533 mm) sphere and shall comply with the loading requirements for guards specified in the *International Building Code*.

(Renumber subsequent sections)

Section 307.1 Add new section as shown: (FG15-03/04)

307.1 Evaporators and cooling coils. Condensate drainage systems shall be provided for equipment and appliances containing evaporators and cooling coils in accordance with the *International Mechanical Code*.

Section 307.4 Add new section as shown: (FG16-03/04)

307.4 Auxiliary drain pan. Category IV condensing appliances shall be provided with an auxiliary drain pan where damage to any building component will occur as a result of stoppage in the condensate drainage system. Such pan shall be installed in accordance with the applicable provisions of Section 307 of the *International Mechanical Code*.

> **Exception:** An auxiliary drain pan shall not be required for appliances that automatically shut down operation in the event of a stoppage in the condensate drainage system.

CHAPTER 4
GAS PIPING INSTALLATIONS

Section 401.5 Change to read as shown: (FG17-03/04)

401.5 Identification. For other than steel pipe, exposed piping shall be identified by a yellow label marked "Gas" in black letters. The marking shall be spaced at intervals not exceeding 5 feet (1524 mm). The marking shall not be required on pipe located in the same room as the equipment served.

Section 404.6 Change to read as shown: (FG21-03/04)

404.6 Piping in solid floors. Piping in solid floors shall be laid in channels in the floor and covered in a manner that will allow access to the piping with a minimum amount of damage to the building. Where such piping is subject to exposure to excessive moisture or corrosive substances, the piping shall be protected in an approved manner. As an alternative to installation in channels, the piping shall be installed in a conduit of Schedule 40 steel, wrought iron, PVC or ABS pipe with tightly sealed ends and joints. Both ends of such conduit shall extend not less than 2 inches (51 mm) beyond the point where the pipe emerges from the floor. The conduit shall be vented above grade to the outdoors and shall be installed so as to prevent the entry of water and insects.

Section 409.1.1 Change to read as shown: (FG25-03/04)

409.1.1 Valve approval. Shutoff valves shall be of an approved type. Shutoff valves shall be constructed of materials compatible with the piping. Shutoff valves shall comply with the standard that is applicable for the pressure and application, in accordance with Table 409.1.1.

Table 409.1.1 Add new table as shown: (FG25-03/04)

TABLE 409.1.1
MANUAL GAS VALVE STANDARDS

VALVE STANDARDS	APPLIANCE SHUTOFF VALVE APPLICATION UP TO ½ psig PRESSURE	OTHER VALVE APPLICATIONS			
		UP TO ½ psig PRESSURE	UP TO 2 psig PRESSURE	UP TO 5 psig PRESSURE	UP TO 125 psig PRESSURE
ANSI Z21.15	X				
CSA Requirement 3-88	X	X	If labeled 2G	If labeled 5G	
ASME B16.44	X	X	If labeled 2G	If labeled 5G	
ASME B16.33	X	X	X	X	X

Section 410.3 Change to read as shown: (FG30-03/04)

410.3 Venting of regulators. Pressure regulators that require a vent shall be vented directly to the outdoors. The vent shall be designed to prevent the entry of insects, water and foreign objects.

> **Exception:** A vent to the outdoors is not required for regulators equipped with and labeled for utilization with an approved vent-limiting device installed in accordance with the manufacturer's instructions.

Section 410.3.1 Add new section as shown: (FG30-03/04)

410.3.1 Vent piping. Vent piping shall be not smaller than the vent connection on the pressure regulating device. Vent piping serving relief vents and combination relief and breather vents shall be run independently to the outdoors and shall serve only a single device vent. Vent piping serving only breather vents is permitted to be connected in a manifold arrangement where sized in accordance with an approved design that minimizes back pressure in the event of diaphragm rupture.

Section 411.1 Change to read as shown: (FG31-03/04; FG32-03/04; FG33-03/04)

411.1 Connecting appliances. Appliances shall be connected to the piping system by one of the following:

1. Rigid metallic pipe and fittings.

2. Corrugated stainless steel tubing (CSST) where installed in accordance with the manufacturer's instructions.

3. Semirigid metallic tubing and metallic fittings. Lengths shall not exceed 6 feet (1829 mm) and shall be located entirely in the same room as the appliance. Semirigid metallic tubing shall not enter a motor-operated appliance through an unprotected knockout opening.

4. Listed and labeled appliance connectors in compliance with ANSI Z21.24 and installed in accordance with the manufacturer's installation instructions and located entirely in the same room as the appliance.

5. Listed and labeled quick-disconnect devices used in conjunction with listed and labeled appliance connectors.

6. Listed and labeled convenience outlets used in conjunction with listed and labeled appliance connectors.

7. Listed and labeled appliance connectors complying with ANSI Z21.69 and listed for use with food service equipment having casters, or that is otherwise subject to movement for cleaning, and other large movable equipment.

8. Listed and labeled outdoor appliance connectors in compliance with ANSI Z21.75/CSA 6.27 and installed in accordance with the manufacturer's installation instructions.

Section 411.1.2 Delete and substitute new Sections 411.1.2 through 411.1.2.4 as shown: (FG26-03/04; FG34-03/04)

411.1.2 Connector installation. Appliance fuel connectors shall be installed in accordance with the manufacturer's instructions and Sections 411.1.2.1 through 411.1.2.4.

411.1.2.1 Maximum length. Connectors shall have an overall length not to exceed 3 feet (914 mm), except for range and domestic clothes dryer connectors, which shall not exceed 6 feet (1829 mm) in overall length. Measurement shall be made along the centerline of the connector. Only one connector shall be used for each appliance.

Exception: Rigid metallic piping used to connect an appliance to the piping system shall be permitted to have a total length greater than 3 feet (914 mm), provided that the connecting pipe is sized as part of the piping system in accordance with Section 402, and the location of the equipment shutoff valve complies with Section 409.5.

411.1.2.2 Minimum size. Connectors shall have the capacity for the total demand of the connected appliance.

411.1.2.3 Prohibited locations and penetrations. Connectors shall not be concealed within, or extended through, walls, floors, partitions, ceilings or appliance housings.

Exception: Fireplace inserts factory-equipped with grommets, sleeves or other means of protection in accordance with the listing of the appliance.

411.1.2.4 Shutoff valve. A shutoff valve not less than the nominal size of the connector shall be installed ahead of the connector in accordance with Section 409.5.

Section 411.2 Add new section as shown: (FG33-03/04)

411.2 Manufactured home connections. Manufactured homes shall be connected to the distribution piping system by one of the following materials:

1. Metallic pipe in accordance with Section 403.4.

2. Metallic tubing in accordance with Section 403.5

3. Listed and labeled connectors in compliance with ANSI Z21.75/CSA 6.27 and installed in accordance with the manufacturer's installation instructions.

Section 413.2.3 Add new section as shown: (FG35-03/04)

413.2.3 General. Residential fueling appliances shall be listed. The capacity of a residential fueling appliance shall not exceed 5 standard cubic feet per minute (0.14 standard cubic meter/min) of natural gas.

Section 413.3 Change to read as shown: (FG35-03/04)

413.3 Location of dispensing operations and equipment. Compression, storage and dispensing equipment shall be located above ground outside.

Exceptions:

1. Compression, storage or dispensing equipment is allowed in buildings of noncombustible construction, as set forth in the *International Building Code*, which are unenclosed for three-quarters or more of the perimeter.

2. Compression, storage and dispensing equipment is allowed to be located indoors in accordance with the *International Fire Code*.

3. Residential fueling appliances and equipment shall be allowed to be installed indoors in accordance with the equipment manufacturer's instructions and Section 413.4.3.

Section 413.3.1 Change to read as shown: (FG35-03/04)

413.3.1 Location on property. In addition to the fuel-dispensing requirements of the *International Fire Code*, compression, storage and dispensing equipment other than residential fueling appliances shall not be installed:

1. Beneath power lines,

2. Less than 10 feet (3048 mm) from the nearest building or property line which could be built on, public street, sidewalk or source of ignition.

Exception: Dispensing equipment need not be separated from canopies providing weather protection for the dispensing equipment constructed in accordance with the *International Building Code*.

3. Less than 25 feet (7620 mm) from the nearest rail of any railroad track.

4. Less than 50 feet (15 240 mm) from the nearest rail of any railroad main track or any railroad or transit line where power for train propulsion is provided by an outside electrical source such as third rail or overhead catenary.

5. Less than 50 feet (15 240 mm) from the vertical plane below the nearest overhead wire of a trolley bus line.

Sections 413.4 through 413.4.3 Add new sections as shown: (FG35-03/04)

413.4 Residential fueling appliance installation. Residential fueling appliances shall be installed in accordance with Sections 413.4.1 through 413.4.3.

413.4.1 Gas connections. Residential fueling appliances shall be connected to the premises gas piping system without causing damage to the piping system or the connection to the internal appliance apparatus.

413.4.2 Outdoor installation. Residential fueling appliances located outdoors shall be installed on a firm, noncombustible base.

413.4.3 Indoor installation. Where located indoors, residential fueling appliances shall be vented to the outdoors. A gas detector set to operate at one-fifth of the lower limit of flammability of natural gas shall be installed in the room or space containing the appliance. The detector shall be located within 6 inches (152 mm) of the highest point in the room or space. The detector shall stop the operation of the appliance and activate an audible or a visual alarm.

(Renumber subsequent sections)

CHAPTER 5
CHIMNEYS AND VENTS

Section 505.1.1 Change to read as shown: (FG38-03/04; FG39-03/04)

505.1.1 Commercial cooking appliances vented by exhaust hoods. Where commercial cooking appliances are vented by means of the Type I or Type II kitchen exhaust hood system that serves such appliances, the exhaust system shall be fan powered and the appliances shall be interlocked with the exhaust hood system to prevent appliance operation when the exhaust hood system is not operating. Where a solenoid valve is installed in the gas piping as part of an interlock system, gas piping shall not be installed to bypass such valve. Dampers shall not be installed in the exhaust system.

Exception: An interlock between the cooking appliance(s) and the exhaust hood system shall not be required where heat sensors or other approved methods automatically activate the exhaust hood system when cooking operations occur.

CHAPTER 6
SPECIFIC APPLIANCES

Section 608.1 Change to read as shown: (FG40-03/04)

608.1 General. Vented wall furnaces shall be tested in accordance with ANSI Z21.86/CSA 2.32 and shall be installed in accordance with the manufacturer's installation instructions.

Section 609.1 Change to read as shown: (FG41-03/04)

609.1 General. Floor furnaces shall be tested in accordance with ANSI Z21.86/CSA 2.32 and shall be installed in accordance with the manufacturer's installation instructions.

Section 610.1 Change to read as shown: (FG42-03/04)

610.1 General. Duct furnaces shall be tested in accordance with ANSI Z83.8 or UL 795 and shall be installed in accordance with the manufacturer's installation instructions.

Section 622.1 Change to read as shown: (FG44-03/04)

622.1 General. Vented room heaters shall be tested in accordance with ANSI Z21.86/CSA 2.32, shall be designed and equipped as specified in Section 602.2 and shall be installed in accordance with the manufacturer's installation instructions.

Section 633 Change section title and Section 633.1 to read as shown: (FG45-03/04)

SECTION 633 (IFGC)
STATIONARY FUEL CELL
POWER SYSTEMS

633.1 General. Stationary fuel-cell power systems having a power output not exceeding 10 MW shall be tested in accordance with ANSI CSA America FC 1 and shall be installed in accordance with the manufacturer's installation instructions and NFPA 853.

CHAPTER 7
GASEOUS HYDROGEN SYSTEMS

Section 704.1.2.3 Change to read as shown: (FG47-03/04)

704.1.2.3 Piping design and construction. Piping and tubing materials shall be Type 304, Type 304L, Type 316 or Type 316L stainless steel or materials listed or

2004 SUPPLEMENT TO THE IFGC

approved for hydrogen service and the use intended through the full range of operating conditions to which they will be subjected. Piping systems shall be designed and constructed to provide allowance for expansion, contraction, vibration, settlement and fire exposure.

Section 705.2 Change to read as shown: (FG48-03/04)

705.2 Inspections. Inspections shall consist of a visual examination of the entire piping system installation and a pressure test. Hydrogen piping systems shall be inspected in accordance with this code. Inspection methods such as outlined in ASME B31.3 shall be permitted when specified by the design engineer and approved by the code official. Inspections shall be conducted or verified by the code official prior to system operation.

Section 705.3 Change to read as shown: (FG48-03/04)

705.3 Pressure tests. A hydrostatic or pneumatic leak test shall be performed. Testing of hydrogen piping systems shall utilize testing procedures identified in ASME B31.3 or other approved methods, provided the testing is performed in accordance with the minimum provisions specified in Sections 705.3.1 through 705.4.1.

Sections 705.3.1, 705.3.2, 705.3.3 and 705.3.4 Add new sections as shown: (FG48-03/04)

705.3.1 Hydrostatic leak tests. The hydrostatic test pressure shall be not less than 1½ times the maximum working pressure, and not less than 100 psig (689.5 kPa gauge).

705.3.2 Pneumatic leak tests. The pneumatic test pressure shall be not less than 1½ times the maximum working pressure for systems less than 125 psig (862 kPa gauge), and not less than 5 psig (34.5 kPa gauge) whichever is greater. For working pressures at or above 125 psig (862 kPa gauge), the pneumatic test pressure shall be not less than 110% of the maximum working pressure.

705.3.3 Test limits. Where the test pressure exceeds 125 psig (862 kPa gauge), the test pressure shall not exceed a value that produces hoop stress in the piping greater than 50 percent of the specified minimum yield strength of the pipe.

705.3.4 Test medium. Deionized water shall be utilized to perform hydrostatic pressure testing and shall be obtained from a potable source. The medium utilized to perform pneumatic pressure testing shall be air, nitrogen, carbon dioxide, or an inert gas. Oxygen shall not be used.

Renumber current Sections 705.3.1 and 705.3.2 to become Sections 705.3.5 and 705.3.6, respectively, and change to read as shown: (FG48-03/04)

705.3.5 Test duration. The minimum test duration shall be ½ hour. The test duration shall be not less than ½ hour for each 500 cubic feet (14.2 m^3) of pipe volume or fraction thereof. For piping systems having a volume of more than 24,000 cubic feet (680 m^3), the duration of the test shall not be required to exceed 24 hours. The test pressure required in Sections 705.3.1 and 705.3.2 shall be maintained for the entire duration of the test.

705.3.6 Test gauges. Gauges used for testing shall be as follows:

1. Tests requiring a pressure of 10 psig (68.95 kPa gauge) or less shall utilize a testing gauge having increments of 0.10 psi (0.6895 kPa) or less.

2. Tests requiring a pressure greater than 10 psig (68.98 kPa gauge) but less than or equal to 100 psig (689.5 kPa gauge) shall utilize a testing gauge having increments of 1 psi (6.895 kPa) or less.

3. Tests requiring a pressure greater than 100 psig (689.5 kPa gauge) shall utilize a testing gauge having increments of 2 psi (13.79 kPa) or less.

   Exception: Measuring devices having an equivalent level of accuracy and resolution shall be permitted where specified by the design engineer and approved by the code official.

Sections 705.3.7 through 705.3.7.3 Add new sections as shown: (FG48-03/04)

705.3.7 Test preparation. Pipe joints, including welds, shall be left exposed for examination during the test.

705.3.7.1 Expansion joints. Expansion joints shall be provided with temporary restraints, if required, for the additional thrust load under test.

705.3.7.2 Equipment disconnection. Where the piping system is connected to appliances, equipment or components designed for operating pressures of less than the test pressure, such appliances, equipment, and components shall be isolated from the piping system by disconnecting them and capping the outlet(s).

705.3.7.3 Equipment isolation. Where the piping system is connected to appliances, equipment or components designed for operating pressures equal to or greater than the test pressure, such appliances, equipment and components shall be isolated from the piping system by closing the individual appliance, equipment or component shutoff valve(s).

Sections 705.4 and 705.4.1 Change to read as shown: (FG48-03/04)

705.4 Detection of leaks and defects. The piping system shall withstand the test pressure specified for the test duration specified without showing any evidence of leakage or other defects. Any reduction of test pressures as indicated by pressure gauges shall indicate a leak within the system. Piping systems shall not be approved except where this reduction in pressure is attributed to some other cause.

705.4.1 Corrections. Where leakage or other defects are identified, the affected portions of the piping system shall be repaired and retested.

Sections 705.5 through 705.5.4 Add new sections as shown: (FG48-03/04)

705.5 Purging of gaseous hydrogen piping systems. Purging shall comply with Sections 705.5.1 through 705.5.4.

705.5.1 Removal from service. Where piping is to be opened for servicing, addition or modification, the section to be worked on shall be isolated from the supply at the nearest convenient point, and the line pressure vented to the outdoors. The remaining gas in this section of pipe shall be displaced with an inert gas.

705.5.2 Placing in operation. Prior to placing the system into operation, the air in the piping system shall be displaced with inert gas. The inert gas flow shall be continued without interruption until the vented gas is free of air. The inert gas shall then be displaced with hydrogen until the vented gas is free of inert gas. The point of discharge shall not be left unattended during purging. After purging, the vent opening shall be closed.

705.5.3 Discharge of purged gases. The open end of piping systems being purged shall not discharge into confined spaces or areas where there are sources of ignition except where precautions are taken to perform this operation in a safe manner by ventilation of the space, control of purging rate, and elimination of all hazardous conditions.

705.5.3.1 Vent pipe outlets for purging. Vent pipe outlets for purging shall be located such that the inert gas and fuel gas is released outdoors and not less than 8 feet (2438 mm) above the adjacent ground level. Gases shall be discharged upward or horizontally away from adjacent walls to assist in dispersion. Vent outlets shall be located such that the gas will not be trapped by eaves or other obstructions and shall be at least 5 feet (1524 mm) from building openings and lot lines of properties that can be built upon.

705.5.4 Placing equipment in operation. After the piping has been placed in operation, all equipment shall be purged in accordance with Section 707.2 and then placed in operation, as necessary.

Section 707.2 Change to read as shown: (FG48-03/04)

707.2 Purging. Purging of gaseous hydrogen systems, other than piping systems purged in accordance with Section 705.5, shall be in accordance with Section 2211.8 of the *International Fire Code* or in accordance with the system manufacturer's instructions.

2004 SUPPLEMENT TO THE IFGC

CHAPTER 8
REFERENCED STANDARDS

Change, delete or add the following referenced standards to read as shown: (FG25-03/04; FG31-03/04; FG33-03/04; FG40-03/04; FG41-03/04; FG42-03/04; FG44-03/04; FG45-03/04; FG48-03/04; FG49-03/04)
(STANDARDS NOT SHOWN REMAIN UNCHANGED)

ANSI

American National Standards Institute
25 West 43rd Street - 4th Floor
New York, NY 10036

Standard reference number	Title	Referenced In code section number
LC 1-97	Interior Fuel Gas Piping Systems Using Corrugated Stainless Steel Tubing - with Addenda LC 1a-1999 and LC 1b-2001	403.5.4
ANSI CSA-America FC 1-03	Stationary Fuel Cell Power Systems	633.1
Z21.1-00	Household Cooking Gas Appliances - with Addenda Z21.1a-2003	623.1
Z21.5.1-99	Gas Clothes Dryers - Volume I - Type 1 Clothes Dryers - with Addenda Z21.5.1a-2003	613.1
Z21.5.2-01	Gas Clothes Dryers - Volume II - Type 2 Clothes Dryers - with Addenda Z21.5.2a-2003	613.1, 614.3
Z 21.8-94 (R2002)	Installation of Domestic Gas Conversion Burners	619.1
Z21.10.1-01	Gas Water Heaters - Volume I Storage, Water Heaters with Input Ratings of 75,000 Btu per Hour or Less - with Addenda Z21.10.1a-2002	624.1
Z21.10.3-01	Gas Water Heaters - Volume III - Storage Water Heaters with Input Ratings Above 75,000 Btu Per Hour, Circulating and Instantaneous	624.1
Z21.11.1-91	DELETED	
Z21.11.2-02	Gas-Fired Room Heaters - Volume II - Unvented Room Heaters	621.1
Z21.13-00	Gas-Fired Low Pressure Steam and Hot Water Boilers - with Addenda Z21.13a-2002	631.1
Z21.15-97 (R2003)	Manually Operated Gas Valves for Appliances, Appliance Connector Valves and Hose End Valves - with Addenda Z21.15a-2001 (R2003)	409.1.1
Z21.19-02	Refrigerators Using Gas (R1999) Fuel	625.1
Z21.24-97	Connectors for Gas Appliances	411.1
Z21.40.1-96 (R2002)	Gas-Fired, Heat Activated Air Conditioning and Heat Pump Appliances - with Addenda Z21.40.1a- 1997 (R2002)	627.1
Z21.40.2-96 (R2002)	Gas-Fired Work Activated Air Conditioning and Heat Pump Appliances (Internal Combustion) - with Addenda Z21.40.2a-1997 (R2002)	627.1
Z21.42-93 (R2002)	Gas-Fired Illuminating Appliances	628.1
Z21.47-01	Gas-Fired Central Furnaces - with Addenda Z21.47a- 2001 and Z21.47b-2002	618.1
Z21.48-92	DELETED	
Z21.49-92	DELETED	
Z21.50-00	Vented Gas Fireplaces - with Addenda Z21.50a-2001 and Z21.50b-2002	604.1
Z21.56-01	Gas-Fired Pool Heaters	616.1
Z21.58-95 (R2002)	Outdoor Cooking Gas Appliances - with Addenda Z21.58a-1998 (R2002) and Z21.58b-2002	623.1
Z21.60-03	Decorative Gas Appliances for Installation in Solid-Fuel Burning Fireplaces	602.1
Z21.69-02	Connectors for Movable Gas Appliances	411.1
Z21.75/CSA 6.27-01	Connectors for Outdoor Gas Appliances and Manufactured Homes	411.1, 411.2
Z21.83-98	DELETED	
Z83.9-96	DELETED	
Z21.84-02	Manually Lighted, Natural Gas Decorative Gas Appliances for Installation in Solid-Fuel Burning Fireplaces	602.1, 602.2
Z21.86-00	Gas-Fired Vented Space Heating Appliances - with Addenda Z21.86a-2002 and Z21.86b-2002	608.1, 609.1, 622.1
Z21.88-02	Vented Gas Fireplace Heaters	605.1

2004 SUPPLEMENT TO THE IFGC

ANSI (continued)

Z83.4-99	Non-Recirculating Direct-Gas-Fired Industrial Air Heaters - with Addenda Z83.4a-2001 and Z83.4b-2002	611.1
Z83.8-02	Gas Unit Heaters and Gas-Fired Duct Furnaces	620.1
Z83.9-96	DELETED	
Z83.11-02	Gas Food Service Equipment	623.1
Z83.18-00	Recirculating Direct Gas-Fired Industrial Air Heaters - with Addenda Z83.18a-2001	612.1

ASME

American Society of Mechanical Engineers
Three Park Avenue
New York, NY 10016-5990

Standard reference number	Title	Referenced In code section number
B1.20.1-83 (Reaffirmed 2001)	Pipe Threads, General Purpose (inch)	403.9
B16.33-02	Manually Operated Metallic Gas Valves for Use in Gas Piping Systems up to 125 psig (Sizes 1/2 through 2)	409.1.1
B16.44-01	Manually Operated Metallic Gas Valves For Use in House Piping Systems	409.1.1
B31.3-02	Process Piping	704.12, 705.2, 705.3
BPVC-01	ASME Boiler & Pressure Vessel Code (2001 Edition) (Sections I, II, IV, V & IX)	631.1, 703.2.2, 703.3.3, 703.3.4
CSD-1-02	Controls and Safety Devices for Automatically Fired Boilers	631.1

ASTM

ASTM International
100 Barr Harbor Drive
West Conshohocken, PA 19428-2859

Standard reference number	Title	Referenced In code section number
A 53/A 53M-02	Specification for Pipe, Steel, Black and Hot Dipped, Zinc-Coated Welded and Seamless	403.4.2
A 106-02a	Specification for Seamless Carbon Steel Pipe for High-Temperature Service	403.4.2
A 254-97 (2002)	Specification for Copper Brazed Steel Tubing	403.5.1
B 88-02	Specification for Seamless Copper Water Tube	403.5.2
B 210-02	Specification for Aluminum and Aluminum-Alloy Drawn Seamless Tubes	403.5.3
B 241/B 241M-02	Specification for Aluminum and Aluminum-Alloy, Seamless Pipe and Seamless Extruded Tube	403.4.4, 403.5.3
B 280-02	Specification for Seamless Copper Tube for Air Conditioning and Refrigeration Field Service	403.5.2
C 315-02	Specification for Clay Flue Linings	501.12
D 2513-03	Specification for Thermoplastic Gas Pressure Pipe, Tubing, and Fittings	403.6, 403.6.1, 403.11, 404.14.2

CGA

Compressed Gas Association
4221 Wainey Road - 5th Floor
Chantilly, VA 20151-2923

Standard reference number	Title	Referenced In code section number
S-1.1 (2002)	Pressure Relief Device Standards - Part 1 - Cylinders for Compressed Gases	703.3

CSA

CSA America, Inc.
8501 E. Pleasant Valley Road
Cleveland, OH 44131-5575

Standard reference number	Title	Referenced In code section number
ANSI CSA America FC1-03	Stationary Fuel Cell Power Systems	633.1
CSA Requirement 3-88	Manually Operated Gas Valves For Use In House Piping Systems	409.1.1

MSS

Manufacturers Standardization Society of the Valve & Fittings Industry
127 Park Street, NE
Vienna, VA 22180-4602

Standard reference number	Title	Referenced In code section number
SP-6-01	Standard Finishes for Contact Faces of Pipe Flanges and Connecting-End Flanges of Valves and Fittings	403.12

NFPA

National Fire Protection Association
1 Batterymarch Park
Quincy, MA 02269-9101

Standard reference number	Title	Referenced In code section number
37-02	Installation and Use of Stationary Combustion Engines and Gas Turbines	616.1
51-02	Design and Installation of Oxygen-Fuel Gas Systems for Welding, Cutting, and Allied Processes	414.1

UL

Underwriters Laboratories, Inc.
333 Pfingsten Road
Northbrook, IL 60062-2096

Standard reference number	Title	Referenced In code section number
103-2001	Factory-Built Chimneys, for Residential Type and Building Heating Appliances	506.1
441-96	Gas Vents - with Revisions through December 1999	502.1

INTERNATIONAL MECHANICAL CODE®

2004 SUPPLEMENT

INTERNATIONAL MECHANICAL CODE 2004 SUPPLEMENT

CHAPTER 1
ADMINISTRATION

Section 106.3.1 Change to read as shown: (M1-03/04)

106.3.1 Construction documents. Construction documents, engineering calculations, diagrams and other data shall be submitted in two or more sets with each application for a permit. The code official shall require construction documents, computations and specifications to be prepared and designed by a registered design professional when required by state law. Where special conditions exist, the code official is authorized to require additional construction documents to be prepared by a registered design professional. Construction documents shall be drawn to scale and shall be of sufficient clarity to indicate the location, nature and extent of the work proposed and show in detail that the work conforms to the provisions of this code. Construction documents for buildings more than two stories in height shall indicate where penetrations will be made for mechanical systems, and the materials and methods for maintaining required structural safety, fire-resistance rating and fire blocking.

> **Exception:** The code official shall have the authority to waive the submission of construction documents, calculations or other data if the nature of the work applied for is such that reviewing of construction documents is not necessary to determine compliance with this code.

CHAPTER 2
DEFINITIONS

Section 202 Change definition of "Approved" to read as shown: (M3-03/04)

APPROVED. Acceptable to the code official or other authority having jurisdiction.

Section 202 Change definition of "Extra-Heavy-Duty Cooking Appliance" to read as shown: (M4-03/04)

EXTRA-HEAVY-DUTY COOKING APPLIANCE. Extra-heavy-duty cooking appliances include appliances utilizing solid fuel such as wood, charcoal, briquettes, and mesquite to provide all or part of the heat source for cooking.

CHAPTER 3
GENERAL REGULATIONS

Section 304.3.1 Add new section as shown: (M7-03/04)

304.3.1 Parking garages. Connection of a parking garage with any room in which there is a fuel-fired appliance shall be by means of a vestibule providing a two-doorway separation, except that a single door is permitted where the sources of ignition in the appliance are elevated in accordance with Section 304.3.

> **Exception:** This section shall not apply to appliance installations complying with Section 304.5.

Section 304.10 Change to read as shown: (M9-03/04)

304.10 Guards. Guards shall be provided where appliances, equipment, fans or other components that require service and roof hatch openings are located within 10 feet (3048 mm) of a roof edge or open side of a walking surface and such edge or open side is located more than 30 inches (762 mm) above the floor, roof or grade below. The guard shall extend not less than 30 inches (762 mm) beyond each end of such appliances, equipment, fans, components and roof hatch openings and the top of the guard shall be located not less than 42 inches (1067 mm) above the elevated surface adjacent to the guard. The guard shall be constructed so as to prevent the passage of a 21-inch-diameter (533 mm) sphere and shall comply with the loading requirements for guards specified in the *International Building Code*.

Section 306.3.1 Change to read as shown: (EL3-03/04)

306.3.1 Electrical requirements. A luminaire controlled by a switch located at the required passageway opening and a receptacle outlet shall be provided at or near the appliance location in accordance with the ICC *Electrical Code*.

Section 306.4.1 Change to read as shown: (EL3-03/04)

306.4.1 Electrical requirements. A luminaire controlled by a switch located at the required passageway opening and a receptacle outlet shall be provided at or near the appliance location in accordance with the ICC *Electrical Code*.

Section 306.6 Change to read as shown: (M14-03/04)

306.6 Sloped roofs. Where appliances, equipment, fans or other components that require service are installed on a roof having a slope of three units vertical in 12 units horizontal (25-percent slope) or greater and having an edge more than 30 inches (762 mm) above grade at such edge, a level platform shall be provided on each side of the appliance to which access is required for service, repair or maintenance. The platform shall be not less than 30 inches (762 mm) in any dimension and shall be provided with guards. The guards shall extend not less

2004 SUPPLEMENT TO THE IMC

than 42 inches above the platform, shall be constructed so as to prevent the passage of a 21-inch-diameter sphere and shall comply with the loading requirements for guards specified in the *International Building Code*.

Section 307.2.3 Change to read as shown: (M12-03/04 & M15-03/04)

307.2.3 Auxiliary and secondary drain systems. In addition to the requirements of Section 307.2.1, a secondary drain or auxiliary drain pan shall be required for each cooling or evaporator coil or fuel-fired appliance that produces condensate, where damage to any building components will occur as a result of overflow from the equipment drain pan or stoppage in the condensate drain piping. One of the following methods shall be used:

1. An auxiliary drain pan with a separate drain shall be provided under the coils on which condensation will occur. The auxiliary pan drain shall discharge to a conspicuous point of disposal to alert occupants in the event of a stoppage of the primary drain. The pan shall have a minimum depth of 1.5 inches (38 mm), shall not be less than 3 inches (76 mm) larger than the unit or the coil dimensions in width and length and shall be constructed of corrosion-resistant material. Metallic pans shall have a minimum thickness of not less than 0.0276-inch (0.7 mm) galvanized sheet metal. Nonmetallic pans shall have a minimum thickness of not less than 0.0625 inch (1.6 mm).

2. A separate overflow drain line shall be connected to the drain pan provided with the equipment. Such overflow drain shall discharge to a conspicuous point of disposal to alert occupants in the event of a stoppage of the primary drain. The overflow drain line shall connect to the drain pan at a higher level than the primary drain connection.

3. An auxiliary drain pan without a separate drain line shall be provided under the coils on which condensate will occur. Such pan shall be equipped with a water-level detection device that will shut off the equipment served prior to overflow of the pan. The auxiliary drain pan shall be constructed in accordance with Item 1 of this section.

4. A water level detection device shall be provided that will shut off the equipment served in the event that the primary drain is blocked. The device shall be installed in the primary drain line, the overflow drain line, or in the equipment-supplied drain pan, located at a point higher than the primary drain line connection and below the overflow rim of such pan.

   Exception: Fuel-fired appliances that automatically shut down operation in the event of a stoppage in the condensate drainage system.

CHAPTER 4
VENTILATION

Section 401.5.1 Change to read as shown: (M16-03/04 & M17-03/04)

401.5.1 Intake openings. Mechanical and gravity outdoor air intake openings shall be located a minimum of 10 feet (3048 mm) horizontally from any hazardous or noxious contaminant, such as vents, chimneys, plumbing vents, streets, alleys, parking lots and loading docks, except as otherwise specified in this code. Where a source of contaminant is located within 10 feet (3048 mm) horizontally of an intake opening, such opening shall be located a minimum of 2 feet (610 mm) below the contaminant source.

The exhaust from a bathroom or kitchen in a residential dwelling shall not be considered to be a hazardous or noxious contaminant.

Section 403.2 Change to read as shown: (M20-03/04)

403.2 Outdoor air required. The minimum ventilation rate of outdoor air shall be determined in accordance with Section 403.3.

Exception: Where the registered design professional demonstrates that an engineered ventilation system design will prevent the maximum concentration of contaminants from exceeding that obtainable by the rate of outdoor air ventilation determined in accordance with Section 403.3, the minimum required rate of outdoor air shall be reduced in accordance with such engineered system design.

Table 403.3 Add new entry and revise note d to read as shown: (M23-03/04 & M30-03/04)

TABLE 403.3
REQUIRED OUTDOOR VENTILATION AIR

OCCUPANCY CLASSIFICATION	ESTIMATED MAXIMUM OCCUPANT LOAD, PERSONS PER 1,000 SQUARE FEET [a]	OUTDOOR AIR (Cubic feet per Minute (cfm) Per person) UNLESS NOTED [e]
Public spaces Elevator car[g]	---	1.00 cfm/ft^2

(Portions of table not shown do not change)

(No change to notes a through c)
d. Ventilation systems in enclosed parking garages shall comply with Section 404.
(No change to notes e through g)

Section 404.1 Change to read as shown: (M35-03/04)

404.1 Enclosed parking garages. Mechanical ventilation systems for enclosed parking garages shall be permitted to operate intermittently where the system is arranged to operate automatically upon detection of vehicle operation or the presence of occupants by approved automatic detection devices.

CHAPTER 5
EXHAUST SYSTEMS

Section 501.3 Change to read as shown: (M36-03/04)

501.3 Pressure equalization. Mechanical exhaust systems shall be sized to remove the quantity of air required by this chapter to be exhausted. The system shall operate when air is required to be exhausted. Where mechanical exhaust is required in a room or space in other than occupancies in R-3, such space shall be maintained with a neutral or negative pressure. If a greater quantity of air is supplied by a mechanical ventilating supply system than is removed by a mechanical exhaust for a room, adequate means shall be provided for the natural or mechanical exhaust of the excess air supplied. If only a mechanical exhaust system is installed for a room or if a greater quantity of air is removed by a mechanical exhaust system than is supplied by a mechanical ventilating supply system for a room, adequate means shall be provided for the supply of the deficiency in the air supplied.

Sections 502.4, 502.4.1 and 502.4.2 Change to read as shown: (F39-03/04)

502.4 Stationary storage battery systems. Stationary storage battery systems, as regulated by Section 608 of the *International Fire Code*, shall be provided with ventilation in accordance with this chapter and Section 502.4.1 or 502.4.2.

502.4.1 Hydrogen limit in rooms. The ventilation system shall be designed to limit the maximum concentration of hydrogen to 1.0 percent of the total volume of the room.

502.4.2 Ventilation rate in rooms. Continuous ventilation shall be provided at a rate of not less than 1 cubic foot per minute per square foot (cfm/ft^2) [0.00508 m^3/(s • m^2)] of floor area of the room.

Section 502.5 Change to read as shown: (F39-03/04)

502.5 Valve-regulated lead-acid batteries in cabinets. Valve-regulated lead-acid (VRLA) batteries installed in cabinets, as regulated by Section 608 of the *International Fire Code*, shall be provided with ventilation in accordance with Section 502.5.1 or 502.5.2.

Section 502.5.1 Hydrogen limit in rooms and Section 502.5.2 Ventilation rate in rooms. Delete without substitution: (F39-03/04)

Sections 502.5.3 and 502.5.4 Renumber as Sections 501.5.1 and 501.5.2 and change to read as shown: (F39-03/04)

502.5.1 Hydrogen limit in cabinets. The cabinet ventilation system shall be designed to limit the maximum concentration of hydrogen to 1.0 percent of the total volume of the cabinet during the worst-case event of simultaneous boost charging of all batteries in the cabinet.

502.5.2 Ventilation rate in cabinets. Continuous cabinet ventilation shall be provided at a rate of not less than 1 cubic foot per minute per square foot (cfm/ft.^{2}) [0.00508 m^3/(s • m^2)] of the floor area covered by the cabinet. The room in which the cabinet is installed shall also be ventilated as required by Section 502.4.1 or 502.4.2.

Section 502.7.3.2 Change to read as shown: (F142-03/04)

502.7.3.2 Recirculation. Air exhausted from spraying operations shall not be recirculated.

Exceptions:

1. Air exhausted from spraying operations shall be permitted to be recirculated as makeup air for unmanned spray operations provided that:

    1.1. Solid particulate has been removed.

    1.2. The vapor concentration is less than 25 percent of the lower flammability limit (LFL).

    1.3. Approved equipment is used to monitor the vapor concentration.

    1.4. An alarm is sounded and spray operations are automatically shut down if the vapor concentration exceeds 25 percent of the LFL.

    1.5. The spray booths, spray spaces or spray rooms involved in any recirculation process shall be provided with mechanical ventilation that shall automatically exhaust 100 percent of the required air volume in the event of shutdown by approved equipment used to monitor vapor concentrations.

2. Air exhausted from spraying operations is allowed to be recirculated as makeup air to manned spraying operations where all of the conditions provided in Exception 1 are included in the installation and documents have been prepared to show that the installation does not pose a life safety hazard to personnel inside the spray booth, spray area or spray room.

2004 SUPPLEMENT TO THE IMC

Section 502.7.3.6 Change to read as shown: (F143-03/04)

502.7.3.6 Termination point. The termination point for exhaust ducts discharging to the atmosphere shall not be less than the following minimum distances:

1. Ducts conveying explosive or flammable vapors, fumes or dusts: 30 feet (9144 mm) from the property line; 10 feet (3048 mm) from openings into the building; 6 feet (1829 mm) from exterior walls and roofs; 30 feet (9144 mm) from combustible walls or openings into the building which are in the direction of the exhaust discharge; 10 feet (3048 mm) above adjoining grade.

2. Other product-conveying outlets: 10 feet (3048 mm) from the property line; 3 feet (914 mm) from exterior walls and roofs; 10 feet (3048 mm) from openings into the building; 10 feet (3048 mm) above adjoining grade.

Section 502.8.1.1 Change to read as shown: (F166-03/04)

502.8.1.1 System requirements. Exhaust ventilation systems shall comply with all of the following:

1. The installation shall be in accordance with this code.

2. Mechanical ventilation shall be provided at a rate of not less than 1 cfm/ft^2 [0.00508 m^3/(s · m^2)] of floor area over the storage area.

3. The systems shall operate continuously unless alternate designs are approved.

4. A manual shutoff control shall be provided outside of the room in a position adjacent to the access door to the room or in another approved location. The switch shall be a break-glass or other approved type and shall be labeled: VENTILATION SYSTEM EMERGENCY SHUTOFF.

5. The exhaust ventilation shall be designed to consider the density of the potential fumes or vapors released. For fumes or vapors that are heavier than air, exhaust shall be taken from a point within 12 inches (305 mm) of the floor. For fumes or vapors that are lighter than air, exhaust shall be taken from a point within 12 inches (305 mm) of the highest point of the room.

6. The location of both the exhaust and inlet air openings shall be designed to provide air movement across all portions of the floor or room to prevent the accumulation of vapors.

7. The exhaust air shall not be recirculated to occupied areas if the materials stored are capable of emitting hazardous vapors and contaminants have not been removed. Air contaminated with explosive or flammable vapors, fumes or dusts; flammable, highly toxic or toxic gases; or radioactive materials shall not be recirculated.

Section 502.8.4 Change to read as shown: (F144-03/04)

502.8.4 Indoor dispensing and use—point sources. Where gases, liquids or solids in amounts exceeding the maximum allowable quantity per control area and having a hazard ranking of 3 or 4 in accordance with NFPA 704 are dispensed or used, mechanical exhaust ventilation shall be provided to capture gases, fumes, mists or vapors at the point of generation.

Exception: Where it can be demonstrated that the gases, liquids or solids do not create harmful gases, fumes, mists or vapors.

Section 502.10.1 Change to read as shown: (F144-03/04)

502.10.1 Where required. Exhaust ventilation systems shall be provided in the following locations in accordance with the requirements of this section and the *International Building Code*.

1. Fabrication areas: Exhaust ventilation for fabrication areas shall comply with the *International Building Code*. Additional manual control switches shall be provided where required by the code official.

2. Workstations: A ventilation system shall be provided to capture and exhaust gases, fumes and vapors at workstations.

3. Liquid storage rooms: Exhaust ventilation for liquid storage rooms shall comply with Section 502.8.1.1 and the *International Building Code*.

4. HPM rooms: Exhaust ventilation for HPM rooms shall comply with Section 502.8.1.1 and the *International Building Code*.

5. Gas cabinets: Exhaust ventilation for gas cabinets shall comply with Section 502.8.2. The gas cabinet ventilation system is allowed to connect to a workstation ventilation system. Exhaust ventilation for gas cabinets containing highly toxic or toxic gases shall also comply with Sections 502.9.7 and 502.9.8.

6. Exhausted enclosures: Exhaust ventilation for exhausted enclosures shall comply with Section

502.7.2. Exhaust ventilation for exhausted enclosures containing highly toxic or toxic gases shall also comply with Sections 502.9.7 and 502.9.8.

7. Gas rooms: Exhaust ventilation for gas rooms shall comply with Section 502.8.2. Exhaust ventilation for gas cabinets containing highly toxic or toxic gases shall also comply with Sections 502.9.7 and 502.9.8.

Section 502.19 Add new section as shown: (M37-03/04)

502.19 Indoor firing ranges. Ventilation shall be provided in an approved manner in areas utilized as indoor firing ranges. Ventilation shall be designed to protect employees and the public in accordance with OSHA 29 CFR 1910.1025 where applicable.

Section 506.3.3.1 Add new section as shown: (M41-03/04)

506.3.3.1 Grease duct test. Prior to the use or concealment of any portion of a grease duct system, a leakage test shall be performed in the presence of the code official. Ducts shall be considered to be concealed where installed in shafts or covered by coatings or wraps that prevent the ductwork from being visually inspected on all sides. The permit holder shall be responsible to provide the necessary equipment and perform the grease duct leakage test. A light test or an approved equivalent test method shall be performed to determine that all welded and brazed joints are liquid tight. A light test shall be performed by passing a lamp having a power rating of not less than 100 watts through the entire section of duct work to be tested. The lamp shall be open so as to emit light equally in all directions perpendicular to the duct walls.

A test shall be performed for the entire duct system including the hood-to-duct connection. The ductwork shall be permitted to be tested in sections, provided that every joint is tested.

Section 506.3.4 Change to read as shown: (M45-03/04)

506.3.4 Air velocity. Grease duct systems serving a Type I hood shall be designed and installed to provide an air velocity within the duct system of not less than 500 feet per minute (2.5 m/s).

Exception: The velocity limitations shall not apply within duct transitions utilized to connect ducts to differently sized or shaped openings in hoods and fans, provided that such transitions do not exceed 3 feet (914 mm) in length and are designed to prevent the trapping of grease.

Section 506.3.12.3 Change to read as shown: (M44-03/04)

506.3.12.3 Termination location. Exhaust outlets shall be located not less than 10 feet (3048 mm) horizontally from parts of the same or contiguous buildings, adjacent buildings, adjacent property lines and air intake openings into any building and shall be located not less than 10 feet (3048 mm) above the adjoining grade level.

Exception: Exhaust outlets shall terminate not less than 5 feet (1524 mm) from parts of the same or contiguous building, an adjacent building, adjacent property line and air intake openings into a building where air from the exhaust outlet discharges away from such locations.

Section 506.5.5 Change to read as shown: (M47-03/04)

506.5.5 Termination Location. The outlet of exhaust equipment serving Type I hoods, shall be in accordance with Section 506.3.12.

Exception: The minimum horizontal distance between vertical discharge fans and parapet-type building structures shall be 2 feet (610 mm) provided that such structures are not higher than the top of the fan discharge opening.

Section 507.2.2 Change to read as shown: (M49-03/04)

507.2.2. Type II hoods. Type II hoods shall be installed where cooking or dishwashing appliances produce heat or steam and do not produce grease or smoke, such as steamers, kettles, pasta cookers and dishwashing machines.

Exceptions:

1. Under-counter-type commercial dishwashing machines.
2. A Type II hood is not required for dishwashers and potwashers that are provided with heat and water vapor exhaust systems that are supplied by the appliance manufacturer and are installed in accordance with the manufacturer's instructions.
3. A single light-duty electric convection, bread, retherm or microwave oven.

Section 507.2.4 Change to read as shown: (M51-03/04)

507.2.4 Extra-heavy-duty. Type I hoods for use over extra-heavy-duty cooking appliances shall not cover other appliances that require fire extinguishing equipment and such hoods shall discharge to an exhaust system that is independent of other exhaust systems.

2004 SUPPLEMENT TO THE IMC

Section 507.13 Change to read as shown: (M53-03/04)

507.13. Capacity of hoods. Commercial food service hoods shall exhaust a minimum net quantity of air determined in accordance with this section and Sections 507.13.1 through 507.13.4. The net quantity of exhaust air shall be calculated by subtracting any airflow supplied directly to a hood cavity from the total exhaust flow rate of a hood. Where any combination of heavy-duty, medium-duty and light-duty cooking appliances are utilized under a single hood, the exhaust rate required by this section for the heaviest duty appliance covered by the hood shall be used for the entire hood.

Section 507.13.5 Add new section as shown: (M52-03/04)

507.13.5. Dishwashing Appliances. The minimum net airflow for Type II hoods used for dishwashing appliances shall be 100 CFM per linear foot of hood length.

> **Exception:** Dishwashing appliances and equipment installed in accordance with Section 507.2.2, exception number 2.

Section 507.16.1 Change to read as shown: (M54-03/04)

507.16.1 Capture and containment test. The permit holder shall verify capture and containment performance of the exhaust of the exhaust system. This field test shall be conducted with all appliances under the hood at operating temperatures, with all sources of outdoor air providing makeup air for the hood operating and with all sources of recirculated air providing conditioning for the space in which the hood is located operating. Capture and containment shall be verified visually by observing smoke or steam produced by actual or simulated cooking, such as with smoke candles, smoke puffers, etc.

Section 510.1 Change to read as shown: (M55-03/04)

510.1 General. This section shall govern the design and construction of duct systems for hazardous exhaust and shall determine where such systems are required. Hazardous exhaust systems are systems designed to capture and control hazardous emissions generated from product handling or processes, and convey those emissions to the outdoors. Hazardous emissions include flammable vapors, gases, fumes, mists or dusts, and volatile or airborne materials posing a health hazard, such as toxic or corrosive materials For the purposes of this section, the health hazard rating of materials shall be as specified in NFPA 704.

For the purposes of the provisions of Section 510, a laboratory shall be defined as a facility where the use of chemicals is related to testing, analysis, teaching, research, or developmental activities. Chemicals are used or synthesized on a non-production basis, rather than in a manufacturing process.

Section 510.2 Change to read as shown: (M55-03/04)

510.2 Where required. A hazardous exhaust system shall be required wherever operations involving the handling or processing of hazardous materials, in the absence of such exhaust systems and under normal operating conditions, have the potential to create one of the following conditions:

1. A flammable vapor, gas, fume, mist or dust is present in concentrations exceeding 25 percent of the lower flammability limit of the substance for the expected room temperature.

2. A vapor, gas, fume, mist or dust with a health-hazard rating of 4 is present in any concentration.

3. A vapor, gas, fume, mist or dust with a health-hazard rating of 1,2, or 3 is present in concentrations exceeding 1 percent of the median lethal concentration of the substance for acute inhalation toxicity.

> **Exception:** Laboratories, as defined in section 510.1, except where the concentrations listed in Item 1 are exceeded or a vapor, gas, fume, mist or dust with a health-hazard rating of 1,2, 3, or 4 is present in concentrations exceeding 1 percent of the median lethal concentration of the substance for acute inhalation toxicity

Section 510.4 Change to read as shown: (M55-03/04; M56-03/04)

510.4 Independent system. Hazardous exhaust systems shall be independent of other types of exhaust systems. Incompatible materials, as defined in the *International Fire Code*, shall not be exhausted through the same hazardous exhaust system. Hazardous exhaust systems shall not share common shafts with other duct systems, except where such systems are hazardous exhaust systems originating in the same fire area.

> **Exception:** The provision of this section shall not apply to laboratory exhaust systems where all of the following conditions apply:
>
> 1. All of the hazardous exhaust ductwork and other laboratory exhaust within both the occupied space and the shafts is under negative pressure while in operation
>
> 2. The hazardous exhaust ductwork manifolded together within the occupied space must originate within the same fire area.

3. Each control branch has a flow regulating device

4. Perchloric acid hoods and connected exhaust shall be prohibited from manifolding

5. Radioisotope hoods are equipped with filtration and/or carbon beds where required by the registered design professional.

6. Biological safety cabinets are filtered.

7. Provision is made for continuous maintenance of negative static pressure in the ductwork

Contaminated air shall not be recirculated to occupied areas unless the contaminants have been removed. Air contaminated with explosive or flammable vapors, fumes or dusts; flammable, highly toxic or toxic gases; or radioactive material shall not be recirculated.

Section 510.7 Change to read as shown: (M55-03/04)

510.7 Suppression required. Ducts shall be protected with an approved automatic fire suppression system installed in accordance with the *International Building Code*.

Exceptions:

1. An approved automatic fire suppression system shall not be required in ducts conveying materials, fumes, mists, and vapors that are nonflammable and noncombustible under all conditions and at any concentrations

2. An approved automatic fire suppression system shall not be required in ducts where the largest cross-sectional diameter of the duct is less than 10 inches (254 mm).

3. For laboratories, as defined in Section 510.1, automatic fire protection systems shall not be required in laboratory hoods or exhaust systems.

Section 511.1.1 Change to read as shown: (M57-03/04)

511.1.1 Collectors and separators. Collectors and separators involving such systems as centrifugal separators, bag filter systems and similar devices, and associated supports shall be constructed of noncombustible materials and shall be located on the exterior of the building or structure. A collector or separator shall not be located nearer than 10 feet (3048 mm) to combustible construction or to an unprotected wall or floor opening, unless the collector is provided with a metal vent pipe that extends above the highest part of any roof with a distance of 30 feet (9144 mm).

2004 SUPPLEMENT TO THE IMC

Exception: Collectors such as "Point of Use" collectors, close extraction weld fume collectors, spray finishing booths, stationary grinding tables, sanding booths, and integrated or machine-mounted collectors shall be permitted to be installed indoors provided the installation is in accordance with the International Fire Code and the ICC Electrical Code.

Section 513.4.6 Change to read as shown: (FS117-03/04)

513.4.6 Duration of operation. All portions of active or passive smoke control systems shall be capable of continued operation after detection of the fire event for a period of not less than either 20 minutes or 1.5 times the calculated egress time, whichever is less.

Section 513.8.1 Change to read as shown: (FS118-03/04; FS119-03/04)

513.8.1 Exhaust rate. The height of the lowest horizontal surface of the accumulating smoke layer shall be maintained at least 6 feet (1829 mm) above any walking surface which forms a portion of a required egress system within the smoke zone. The required exhaust rate for the zone shall be the largest of the calculated plume mass flow rates for the possible plume configurations. Provisions shall be made for natural or mechanical supply of outside air from outside or adjacent smoke zones to make up for the air exhausted. Makeup airflow rates, when measured at the potential fire location, shall not increase the smoke production rate beyond the capabilities of the smoke control system. The temperature of the makeup air shall be such that it does not expose temperature-sensitive fire protection systems beyond their limits.

Section 513.8.3 Balcony spill plumes. Delete section without substitution: (FS120-03/04)

Section 513.8.4 Window plumes. Delete section without substitution: (FS121-03/04)

Section 513.9 Change to read as shown: (FS122-03/04)

513.9 Design fire. The design fire shall be based on a rational analysis performed by the registered design professional and approved by the code official. The design fire shall be based on the analysis in accordance with Section 513.4 and this section.

Section 513.9.2 Change to read as shown: (FS122-03/04)

513.9.2 Separation distance. Determination of the design fire shall include consideration of the type of fuel, fuel spacing and configuration.

IMC-7

CHAPTER 6
DUCT SYSTEMS

Section 601.2 Change to read as shown: (E3-03/04)

601.2 Air movement in egress elements. Corridors shall not serve as supply, return, exhaust, relief or ventilation air ducts.

Exceptions:

1. Use of a corridor as a source of makeup air for exhaust systems in rooms that open directly onto such corridors, including toilet rooms, bathrooms, dressing rooms, smoking lounges and janitor closets, shall be permitted provided that each such corridor is directly supplied with outdoor air at a rate greater than the rate of makeup air taken from the corridor.

2. Where located within a dwelling unit, the use of corridors for conveying return air shall not be prohibited.

3. Where located within tenant spaces of 1,000 square feet (93 m^2) or less in area, utilization of corridors for conveying return air is permitted.

Section 603.9 Change to read as shown: (M62-03/04)

603.9 Joints, seams and connections. All longitudal and transverse joints, seams and connections in metallic and nonmetallic ducts shall be constructed as specified in SMACNA HVAC *Duct Construction Standards- Metal and Flexible* and NAIMA *Fibrous Glass Duct Construction Standards.* All joints, longitudinal and transverse seams, and connections in ductwork shall be securely fastened and sealed with welds, gaskets, mastics (adhesives), mastic-plus-embedded-fabric systems or tapes. Tapes and mastics used to seal ductwork listed and labeled in accordance with UL 181A shall be marked "181A-P" for pressure-sensitive tape, "181 A-M" for mastic or "181 A-H" for heat-sensitive tape. Tapes and mastics used to seal flexible air ducts and flexible air connectors shall comply with UL 181B and shall be marked "181B-FX" for pressure-sensitive tape or "181B-M" for mastic. Duct connections to flanges of air distribution system equipment shall be sealed and mechanically fastened. Unlisted duct tape is not permitted as a sealant on any metal ducts.

Section 604.3 Change to read as shown: (M63-03/04)

604.3 Coverings and linings. Coverings and linings, including adhesives when used, shall have a flame spread index not more than 25 and a smoke-developed index not more than 50, when tested in accordance with ASTM E 84, using the specimen preparation and mounting procedures of ASTM E 2231. Duct coverings and linings shall not flame, glow, smolder or smoke when tested in accordance with ASTM C 411 at the temperature to which they are exposed in service. The test temperature shall not fall below 250°F (121°C).

Section 606.1 Change to read as shown: (M64-03/04)

606.1 Controls required. Air distribution systems shall be equipped with smoke detectors listed and labeled for installation in air distribution systems, as required by this section. Duct smoke detectors shall comply with UL 268A. Other smoke detectors shall comply with UL 268.

Section 607.2 Change to read as shown: (FS82-03/04)

607.2 Installation. Fire dampers, smoke dampers, combination fire/smoke dampers and ceiling radiation dampers located within air distribution and smoke control systems shall be installed in accordance with the requirements of this section, and the manufacturer's installation instructions and listing.

Section 607.3.2.1 Change to read as shown: (FS84-03/04)

607.3.2.1 Smoke damper actuation methods. The smoke damper shall close upon actuation of a listed smoke detector or detectors installed in accordance with Section 607 of this code and Sections 907.10 and 907.11 of the *International Building Code* and one of the following methods, as applicable:

1. Where a damper is installed within a duct, a smoke detector shall be installed in the duct within 5 feet (1524 mm) of the damper with no air outlets or inlets between the detector and the damper. The detector shall be listed for the air velocity, temperature and humidity anticipated at the point where it is installed. Other than in mechanical smoke control systems, dampers shall be closed upon fan shutdown where local smoke detectors require a minimum velocity to operate.

2. Where a damper is installed above smoke barrier doors in a smoke barrier, a spot-type detector listed for releasing service shall be installed on either side of the smoke barrier door opening.

3. Where a damper is installed within an unducted opening in a wall, a spot-type detector listed for releasing service shall be installed within 5 feet (1524mm) horizontally of the damper.

4. Where a damper is installed in a corridor wall or ceiling, the damper shall be permitted to be controlled by a smoke detection system installed in the corridor.

5. Where a total-coverage smoke detector system is provided within areas served by an HVAC system, dampers shall be permitted to be controlled by the smoke detection system.

Section 607.4 Change to read as shown: (FS85-03/04)

607.4 Access and identification. Fire and smoke dampers shall be provided with an approved means of access, large enough to permit inspection and maintenance of the damper and its operating parts. The access shall not affect the integrity of fire-resistance-rated assemblies. The access openings shall not reduce the fire-resistance rating of the assembly. Access points shall be permanently identified on the exterior by a label having letters not less than 0.5 inch (12.7 mm) in height reading: FIRE/SMOKE DAMPER, SMOKE DAMPER or FIRE DAMPER. Access doors in ducts shall be tight fitting and suitable for the required duct construction.

Section 607.5.2 Change to read as shown: (FS87-03/04)

607.5.2 Fire barriers. Duct penetrations and air transfer openings in fire barriers shall be protected with approved fire dampers installed in accordance with their listing.

> **Exceptions:** Fire dampers are not required at penetrations of fire barriers where any of the following apply:
>
> 1. Penetrations are tested in accordance with ASTM E119 as part of the fire-resistance-rated assembly.
>
> 2. Ducts are used as part of an approved smoke control system in accordance with Section 513 and where the use of a fire damper would interfere with the operation of the smoke control system.
>
> 3. Such walls are penetrated by ducted HVAC systems, have a required fire-resistance rating of 1 hour or less, are in areas of other than Group H and are in buildings equipped throughout with an automatic sprinkler system in accordance with Section 903.3.1.1 or 903.3.1.2 of the *International Building Code*. For the purposes of this exception, a ducted HVAC system shall be a duct system for the structure's HVAC system. Such a duct system shall be constructed of sheet metal not less than 26-gage (0.0217 inch) [0.55 mm] thickness and shall be continuous from the air-handling appliance or equipment to the air outlet and inlet terminals.

Section 607.5.3 Change to read as shown: (FS91-03/04)

607.5.3 Fire partitions. Duct penetrations in fire partitions shall be protected with approved fire dampers installed in accordance with their listing.

> **Exceptions:** In occupancies other than Group H, fire dampers are not required where any of the following apply:
>
> 1. The partitions are tenant separation or corridor walls in buildings equipped throughout with an automatic sprinkler system in accordance with Section 903.3.1.1 or 903.3.1.2 of the *International Building Code* and the duct is protected as a through penetration in accordance with Section 712 of the *International Building Code*.
>
> 2. The duct system is constructed of approved materials in accordance with this code and the duct penetrating the wall meets all of the following minimum requirements.
>
>   2.1. The duct shall not exceed 100 square inches (0.06 m^2).
>
>   2.2. The duct shall be constructed of steel a minimum of 0.0217-inch (0.55 mm) in thickness.
>
>   2.3. The duct shall not have openings that communicate the corridor with adjacent spaces or rooms.
>
>   2.4. The duct shall be installed above a ceiling.
>
>   2.5. The duct shall not terminate at a wall register in the fire-resistance-rated wall.
>
>   2.6. A minimum 12-inch-long (3048 mm) by 0.060-inch-thick (0.52 mm) steel sleeve shall be centered in each duct opening. The sleeve shall be secured to both sides of the wall and all four sides of the sleeve with minimum 1-1/2-inch by 1-1/2-inch by 0.060-inch (38 mm by 38 mm by 1.52 mm) steel retaining angles. The retaining angles shall be secured to the sleeve and the wall with No. 10 (M5) screws. The annular space between the steel sleeve and the wall opening shall be filled with rock (mineral) wool batting on all sides.

Section 607.5.5.1 Change to read as shown: (FS90-03/04)

607.5.5.1 Penetrations of shaft enclosures. Shaft enclosures that are permitted to be penetrated by ducts and air transfer openings shall be protected with approved fire and smoke dampers installed in accordance with their listing.

2004 SUPPLEMENT TO THE IMC

Exceptions:

1. Fire dampers are not required at penetrations of shafts where:

    1.1. Steel exhaust subducts extend at least 22 inches (559 mm) vertically in exhaust shafts provided there is a continuous airflow upward to the outside, or

    1.2. Penetrations are tested in accordance with ASTM E 119 as part of the fire-resistance-rated assembly, or

    1.3. Ducts are used as part of an approved smoke control system designed and installed in accordance with Section 909 of the *International Building Code,* and where the fire damper will interfere with the operation of the smoke control system, or

    1.4. The penetrations are in parking garage exhaust or supply shafts that are separated from other building shafts by not less than 2-hour fire-resistance-rated construction.

2. In Group B occupancies, equipped throughout with an automatic sprinkler system in accordance with Section 903.3.1.1 of the *International Building Code,* smoke dampers are not required at penetrations of shafts where bathroom and toilet room exhaust openings with steel exhaust subducts, having a wall thickness of at least 0.019 inch (0.48 mm) extend at least 22 inches (559 mm) vertically and the exhaust fan at the upper terminus is powered continuously in accordance with the provisions of Section 909.11 of the *International Building Code,* and maintains airflow upward to the outside.

3. Smoke dampers are not required at penetration of exhaust or supply shafts in parking garages that are separated from other building shafts by not less than 2-hour fire-resistance-rated construction.

4. Smoke dampers are not required at penetrations of shafts where ducts are used as part of an approved mechanical smoke control system designed in accordance with Section 909 and where the smoke damper will interfere with the operation of the smoke control system.

Section 607.6.2 Delete and substitute new Sections 607.6.2 and 607.6.2.1 as shown: (FS83-03/04)

607.6.2 Membrane penetrations. Duct systems constructed of approved materials, in accordance with this code, that penetrate the ceiling membrane of a fire-resistance-rated floor/ceiling or roof/ceiling assembly shall be protected with one of the following:

1. A fire resistance rated shaft enclosure in accordance with Sections 707 and 712.4 of the *International Building Code.*

2. An approved ceiling radiation damper installed at the ceiling line where the duct system penetrates the ceiling of a fire-resistance-rated floor/ceiling or roof/ceiling assembly.

3. An approved ceiling radiation damper installed at the ceiling line where a diffuser with no duct attached penetrates the ceiling of a fire-resistance-rated floor/ceiling or roof/ceiling assembly.

607.6.2.1 Ceiling radiation dampers. Ceiling radiation dampers shall be tested in accordance with UL 555C and installed in accordance with the manufacturer's installation instructions and listing. Ceiling radiation dampers are not required where either of the following apply:

1. ASTM E 119 fire tests have shown that ceiling radiation dampers are not necessary in order to maintain the fire-resistance rating of the assembly.

2. Where exhaust duct penetrations are protected in accordance with Section 712.4.2 of the *International Building Code* and the exhaust ducts are located within the cavity of a wall, and do not pass through another dwelling unit or tenant space.

Section 607.6.3 Change to read as shown: (FS92-03/04)

607.6.3 Nonfire-resistance-rated floor assemblies. Duct systems constructed of approved materials in accordance with this code that penetrate nonfire-resistance-rated floor assemblies shall be protected by any of the following methods:

1. A fire resistance rated shaft enclosure that meets the requirements of Section 712.4.3 of the *International Building Code.*

2. The duct connects not more than two stories, and the annular space around the penetrating duct is protected with an approved noncombustible material to resist the free passage of flame and the products of combustion

3. The duct connects not more than three stories, and the annular space around the penetrating duct is protected with an approved noncombustible

material to resist the free passage of flame and the products of combustion, and a fire damper is installed at each floor line.

Exception: Fire dampers are not required in ducts within individual residential dwelling units.

CHAPTER 8
CHIMNEYS AND VENTS

Section 804.3.5 Change to read as shown: (M67-03/04)

804.3.5 Vertical terminations. Vertical terminations shall comply with the following requirements:

1. Where located adjacent to walkways, the termination of mechanical draft systems shall be not less than 7 feet (2134 mm) above the level of the walkway.

2. Vents shall terminate at least 3 feet (914 mm) above any forced air inlet located within 10 feet (3048 mm) horizontally.

3. Where the vent termination is located below an adjacent roof structure, the termination point shall be located at least 3 feet (914 mm) from such structure.

4. The vent shall terminate at least 4 feet (1219 mm) below, 4 feet (1219 mm) horizontally from or 1 foot (305 mm) above any door, window or gravity air inlet for the building.

5. A vent cap shall be installed to prevent rain from entering the vent system.

6. The vent termination shall be located at least 3 feet (914 mm) horizontally from any portion of the roof structure.

CHAPTER 9
SPECIFIC APPLIANCES, FIREPLACES AND SOLID FUEL BURNING EQUIPMENT

Section 924 and 924.1 Change to read as shown: (M68-03/04)

SECTION 924
STATIONARY FUEL CELL POWER SYSTEMS

924.1 General. Stationary fuel cell power systems having a power output not exceeding 10 MW, shall be tested in accordance with ANSI CSA America FC 1 and shall be installed in accordance with the manufacturer's installation instructions and NFPA 853.

2004 SUPPLEMENT TO THE IMC

CHAPTER 10
BOILERS, WATER HEATERS AND PRESSURE VESSELS

Section 1002.2.2 Change to read as shown: (M69-03/04)

1002.2.2 Temperature Limitation. Where a combination potable water-heating and space-heating system requires water for space heating at temperatures higher than 140°F (60°C), a temperature actuated mixing valve that conforms to ASSE 1017 shall be provided to temper the water supplied to the potable hot water distribution system to a temperature of 140°F (60°C) or less.

CHAPTER 11
REFRIGERATION

Section 1105.9 Add new section as shown: (F36-03/04)

1105.9 Emergency pressure control system. Refrigeration systems containing more than 6.6 pounds (3 kg) of flammable, toxic or highly toxic refrigerant or ammonia shall be provided with an emergency pressure control system in accordance with the *International Fire Code*.

CHAPTER 12
HYDRONIC PIPING

Section 1203.3.7 Change to read as shown: (M73-03/04)

1203.3.7 Grooved and shouldered mechanical joints. Grooved and shouldered mechanical joints shall conform to the requirements of ASTM F 1476 and shall be installed in accordance with the manufacturer's installation instructions.

Section 1204.1 Change to read as shown: (M74-03/04)

1204.1 Insulation characteristics. Pipe insulation installed in buildings shall conform to the requirements of the *International Energy Conservation Code*; shall be tested in accordance with ASTM E 84, using the specimen preparation and mounting procedures of ASTM E 2231; and shall have a maximum flame spread index of 25 and a smoke-developed index not exceeding 450. Insulation installed in an air plenum shall comply with Section 602.2.1.

Exception: The maximum flame spread index and smoke-developed index shall not apply to one- and two-family dwellings.

2004 SUPPLEMENT TO THE IMC

Section 1206.9.1 Change to read as shown: (M75-03/04)

1206.9.1 Flood hazard. Piping located in a flood hazard area shall be capable of resisting hydrostatic and hydrodynamic loads and stresses, including the effects of buoyancy, during the occurrence of flooding to the design flood elevation.

CHAPTER 13
FUEL OIL PIPING AND STORAGE

Section 1301.5 Add new section as shown: (M76-03/04)

1301.5 Tanks abandoned or removed. All exterior above-grade fill piping shall be removed when tanks are abandoned or removed. Tank abandonment and removal shall be in accordance with the *International Fire Code*.

2004 SUPPLEMENT TO THE IMC

CHAPTER 15
REFERENCED STANDARDS

Change or add to read as shown: (M37-03/04; M63-03/04; M64-03/04; M68-03/04; M69-03/04; M73-03/04; M74-03/04; M77-03/04) (STANDARDS NOT SHOWN REMAIN UNCHANGED)

ANSI

American National Standards Institute
25 West 43rd Street - 4th Floor
New York, NY 10036

Standard reference number	Title	Referenced in code section number
Z21.8-1994 (R2002)	Installation of Domestic Gas Conversion Burners	919.1
Z21.83-1998	(DELETED)	
ANSI CSA America FC 1-2003	Standard Fuel Cell Power Systems	924.1

ARI

Air Conditioning and Refrigeration Institute
4100 North Fairfax Drive, Suite 200
Arlington, VA 22203

Standard reference number	Title	Referenced in code section number
700-99	Purity Specifications for Fluorocarbon Refrigerants	1102.2.2.3

ASME

American Society of Mechanical Engineers
Three Park Avenue
New York, NY 10016-5990

Standard reference number	Title	Referenced in code section number
B1.20.1-1983 (Reaffirmed 2001)	Pipe Threads, General Purpose (inch)	1203.3.5, 1303.3.5
B16.9-2001	Factory-Made Wrought Steel Buttwelding Fittings	Table 1202.5
B16.11-2001	Forged Fittings, Socket-Welding and Threaded	Table 1202.5
B16.18-2001	Cast Copper Alloy Solder Joint Pressure Fittings	513.13.1, Table 1202.5
B16.22-2001	Wrought Copper and Copper Alloy Solder Joint Pressure Fittings	513.13.1, Table 1202.5
B16.23-2002	Cast Copper Alloy Solder Joint Drainage Fittings: DWV	Table 1202.5
B16.24-2001	Cast Copper Alloy Pipe Flanges and Flanged Fittings: Class 150, 300, 400, 600, 900, 1500 and 2500	Table 1202.5
B16.29-2001	Wrought Copper and Wrought Copper Alloy Solder Joint Drainage Fittings - DWV	Table 1202.5
BPVC-2001	ASME Boiler & Pressure Vessel Code (2001 Edition) (Sections I, II, IV, V & VI)	1004.1, 1011.1
CSD-1-2002	Controls and Safety Devices for Automatically Fired Boilers	1004.1

2004 SUPPLEMENT TO THE IMC

ASSE

American Society of Sanitary Engineering
28901 Clemens Road, Suite A
Westlake, OH 44145

Standard reference number	Title	Referenced in code section number
1017-99	Performance Requirements for Temperature Actuated Mixing Valves for Hot Water Distribution Systems	1002.2.2

ASTM

ASTM International
100 Barr Harbor Drive
West Conshohocken, PA 19428-2859

Standard reference number	Title	Referenced in code section
A 53/A 53M-02	Specification for Pipe, Steel, Black and Hot Dipped, Zinc-Coated Welded and Seamless	Table 1202.4, Table 1302.3
A 106-02a	Specification for Seamless Carbon Steel Pipe for High-Temperature Service	Table 1202.4, Table 1302.3
A 254-97(2002)	Specification for Copper Brazed Steel Tubing	Table 1202.4, Table 1302.3
A 420/A 420M-02	Specification for Piping Fittings of Wrought Carbon Steel and Alloy Steel for Low-Temperature Service	Table 1202.5
B 32-00e01	Specification for Solder Metal	1203.3.3
B 42-02	Specification for Seamless Copper Pipe, Standard Sizes	513.3.1, 1107.4.2, Table 1202.4, Table 1302.3
B 68-02	Specification for Seamless Copper Tube, Bright Annealed	513.3.1
B 75-02	Specification for Seamless Copper Tube	Table 1202.4, Table 1302.3
B 88-02	Specification for Seamless Copper Water Tube	513.3.1, 1107.4.3, Table 1202.4, Table 1302.3
B 135-02	Specification for Seamless Brass Tube	Table 1202.4, Table 1302.3
B 251-02	Specification for General Requirements for Wrought Seamless Copper and Copper-Alloy Tube	513.3.1, Table 1202.4
B 280-02	Specification for Seamless Copper Tube for Air Conditioning and Refrigeration Field Service	513.3.1, 1107.4.3, Table 1302.3
B 302-02	Specification for Threadless Copper Pipe, Standard Sizes	Table 1202.4, Table 1302.3
C 315-02	Specification for Clay Flue Linings	801.16.1, Table 803.10.4
D 56-02a	Test Method for Flash Point by Tag Closed Tester	202
D 93-02	Test Method for Flash Point by Pensky-Martens Closed Cup Tester	202
D 2412-02	Test Method for Determination of External Loading Characteristics of Plastic Pipe by Parallel-Plate Loading	603.8.3
D 2466-02	Specification for Poly (Vinyl Chloride) (PVC) Plastic Pipe Fittings, Schedule 40	Table 1202.5
D 2467-02	Specification for Poly (Vinyl Chloride) (PVC) Plastic Pipe Fittings, Schedule 80	Table 1202.5
D 2513-03	Specification for Thermoplastic Gas Pressure Pipe, Tubing, and Fittings	Table 1202.4, 1203.15.3
D 2564-02	Specification for Solvent Cements for Poly (Vinyl Chloride) (PVC) Plastic Piping Systems	1203.3.4
D 2837-02	Test Method for Obtaining Hydrostatic Design Basis for Thermoplastic Pipe Materials	Table 1202.4
D 2996-01	Specification for Filament-Wound Fiberglass (Glass Fiber Reinforced Thermosetting Resin) Pipe	Table 1302.3
D 3350-02a	Specification for Polyethylene Plastics Pipe and Fittings Materials	Table 1202.4

2004 SUPPLEMENT TO THE IMC

ASTM (continued)

E 84-03	Test Method for Surface Burning Characteristics of Building Materials	202, 510.8, 602.2.1, 602.2.1.5
E 814-02	Test Method of Fire Tests of Through-Penetration Firestops	506.3.10
E 2231-02	Standard Practice for Specimen Preparation and Mounting of Pipe and Duct Insulation Materials to Assess Surface Burning Characteristics	604.3, 1204.1
F 438-02	Specification for Socket-Type Chlorinated Poly (Vinyl Chloride) (CPVC) Plastic Pipe Fittings, Schedule 40	Table 1202.5
F 439-02	Specification for Socket-Type Chlorinated Poly (Vinyl Chloride) (CPVC) Plastic Pipe Fittings, Schedule 80	Table 1202.5
F 441/F 441M-02	Specification for Chlorinated Poly (Vinyl Chloride) (CPVC) Plastic Pipe, Schedules 40 and 80	Table 1202.4
F 876-02e01	Specification for Crosslinked Polyethylene (PEX) Tubing	Table 1202.4
F 877-02e01	Specification for Crosslinked Polyethylene (PEX) Plastic Hot- and Cold-Water Distribution Systems	Table 1202.4, Table 1202.5
F 1281-02e02	Specification for Crosslinked Polyethylene/Aluminum/Crosslinked Polyethylene (PEX-AL-PEX) Pressure Pipe	Table 1202.4
F 1476-95a	Standard Specification for Performance of Gasketed Mechanical Couplings for Use in Piping Applications	1203.3.7
F 1974-02	Specification for Metal Insert Fittings for Polyethylene/Aluminum/Polyethylene and Crosslinked Polyethylene/Aluminum/Crosslinked Polyethylene Composite Pressure Pipe	Table 1202.5

CSA

Canadian Standards Association
5060 Spectrum Way, Suite 100
Mississauga, Ontario, Canada L4W 5N6

Standard reference number	Title	Referenced in code section
CAN/CSA B137.10-02	Crosslinked Polyethylene/Aluminum/Crosslinked Polyethylene Composite Pressure-Pipe Systems	Table 1202.4

DOL

Department of Labor
Occupational Safety and Health Administration
c/o Superintendent of Documents
US Government Printing Office
Washington, DC 20402-9325

Standard reference number	Title	Referenced in code section
29 CFR Part 1910.1025	Toxic and Hazardous Substances	502.19

MSS

Manufacturers Standardization Society of the
Valve & Fittings Industry, Inc.
127 Park Street, NE
Vienna, VA 22180-4602

Standard reference number	Title	Referenced in code section
SP-69-2002	Pipe Hangers and Supports - Selection and Application	305.4

NAIMA

North American Insulation Manufacturers
44 Canal Center Plaza, Suite 310
Alexandria, VA 22314

Standard reference number	Title	Referenced in code section
AH116-02	Fibrous Glass Duct Construction Standards, Fifth Edition	603.5, 603.9

NFPA

National Fire Protection Association
1 Batterymarch Park
Quincy, MA 02269-9101

Standard reference number	Title	Referenced in code section
37-02	Installation and Use of Stationary Combustion Engines and Gas Turbines	801.2.1, 801.18.1, 801.18.2, 920.2, 922.1, 1308.1
69-02	Explosion Prevention Systems	510.8.3
72-02	National Fire Alarm Code	606.3
262-02	Method of Test for Flame Travel and Smoke of Wires and Cables for Use in Air-Handling Spaces	602.2.1.1
704-01	System for the Identification of the Hazards of Materials for Emergency Response	502.8.4, Table 1103.1, 510.1

SMACNA

Sheet Metal & Air Conditioning Contractors
4021 Lafayette Center Road
Chantilly, VA 20151-1209

Standard reference number	Title	Referenced in code section
SMACNA-03	Fibrous Glass Duct Construction Standards (2003)	603.5, 603.9

UL

Underwriters Laboratories, Inc.
333 Pfingsten Road
Northbrook, IL 60062-2096

Standard reference number	Title	Referenced in code section
103-01	Factory-Built Chimneys, for Residential Type and Building Heating Appliances	805.2
181A.-95	Closure Systems for Use with Rigid Air Ducts and Air Connectors-with Revisions through December 1998	603.9

2004 SUPPLEMENT TO THE IMC

UL (continued)

181B-98	Closure Systems for Use with Flexible Air Ducts and Air Connectors-with Revisions through December 1998	603.9
197-03	Commercial Electric Cooking Appliances	507.1
207-01	Refrigerant-Containing Components and Accessories, Nonelectrical	1101.2
268-96	Smoke Detectors for Fire Protective Signaling Systems – with Revisions through January 1999	606.1
268A-98	Smoke Detectors for Duct Application – with Revisions through September 2001	606.1
343-97	Pumps for Oil-Burning Appliances - with Revisions through May 2002	1302.7
412-93	Refrigeration Unit Coolers - with Revisions through November 2001	1101.2
471-95	Commercial Refrigerators and Freezers - with Revisions through November 2001	1101.2
555-99	Fire Dampers - with Revisions through January 2002	607.3
555S-99	Smoke Dampers - with Revisions through January 2002	607.3, 607.3.1.1
726-98	Oil-Fired Boiler Assemblies - with Revisions through January 2001	916.1, 004.1
727-98	Oil-Fired Central Furnaces - with Revisions through January 1999	918.1
1240-94	Electric Commercial Clothes Drying Equipment - with Revisions through May 2000	913.1
1261-01	Electric Water Heaters for Pools and Tubs	916.1
2043-96	Fire Test for Heat and Visible Smoke Release for Discrete Products and their Accessories Installed in Air-Handling Spaces - with Revisions through June 2001	602.2.1.4
2162-01	Outline of Investigation for Commercial Wood-Fired Baking Ovens - Refractory Type	917.1

International Plumbing Code

2004 Supplement

INTERNATIONAL PLUMBING CODE 2004 SUPPLEMENT

SECTION 202
GENERAL DEFINITIONS

Change the definition of "Approved" to read as shown: (P1-03/04)

APPROVED. Acceptable to the code official or other authority having jurisdiction.

Replace the definition of "Branch Interval" to read as shown: (P2-03/04)

Branch Interval. A vertical measurement of distance, 8 feet (2438 mm) or more in developed length, between the connections of horizontal branches to a drainage stack. Measurements are taken down the stack from the highest horizontal branch connection.

Add new definition "Gridded Water Distribution System" to read as shown: (P51-03/04)

GRIDDED WATER DISTRIBUTION SYSTEM. A water distribution system where every water distribution pipe is interconnected so as to provide two or more paths to each fixture supply pipe.

CHAPTER 3
GENERAL REGULATIONS

Section 310.5 Add new section to read as shown: (P8-03/04)

310.5 Urinal privacy. Each urinal utilized by the public or employees shall occupy a separate area with walls or partitions to provide privacy. The construction of such walls or partitions shall incorporate waterproof, smooth, readily cleanable and nonabsorbent finish surfaces. The walls or partitions shall begin at a height not more than 12 inches (304.8 mm) from and extend not less than 60 inches (1524 mm) above the finished floor surface. The walls or partitions shall extend from the wall surface at each side of the urinal a minimum of 18 inches (457 mm) or to a point not less than 6 inches (152 mm) beyond the outermost front lip of the urinal measured from the finished back wall surface, whichever is greater.

> **Exception:** Urinal partitions shall not be required in a single occupant or unisex toilet room with a lockable door.

Section 312.5 Change to read as shown: (P9-03/04)

312.5 Water supply system test. Upon completion of a section of or the entire water supply system, the system, or portion completed shall be tested and proved tight under a water pressure not less than the working pressure of the system or, for piping systems other than plastic, by an air test of not less than 50 psi (344 kPa). This pressure shall be held for at least 15 minutes. The water utilized for tests shall be obtained from a potable source of supply. The required tests shall be performed in accordance with this section and Section 107.

Section 312.9.2 Change to read as shown: (P10-03/04)

312.9.2 Testing. Reduced pressure principle backflow preventer assemblies, double check-valve assemblies, pressure vacuum breaker assemblies, reduced pressure detector fire protection backflow prevention assemblies, double check detector fire protection backflow prevention assemblies, hose connection backflow preventers, and spill-proof vacuum breakers shall be tested at the time of installation, immediately after repairs or relocation and at least annually. The testing procedure shall be performed in accordance with one of the following standards: ASSE 5013, ASSE 5015, ASSE 5020, ASSE 5047, ASSE 5048, ASSE 5052, ASSE 5056, CSA B64.10, CSA B64.10.1.

Section 314.2.3 Change to read as shown: (M15-03/04)

307.2.3 Auxiliary and secondary drain systems. In addition to the requirements of Section 307.2.1, a secondary drain or auxiliary drain pan shall be required for each cooling or evaporator coil or fuel-fired appliance that produces condensate, where damage to any building components will occur as a result of overflow from the equipment drain pan or stoppage in the condensate drain piping. One of the following methods shall be used:

1. An auxiliary drain pan with a separate drain shall be provided under the coils on which condensation will occur. The auxiliary pan drain shall discharge to a conspicuous point of disposal to alert occupants in the event of a stoppage of the primary drain. The pan shall have a minimum depth of 1.5 inches (38 mm), shall not be less than 3 inches (76 mm) larger than the unit or the coil dimensions in width and length and shall be constructed of corrosion-resistant material. Metallic pans shall have a minimum thickness of not less than 0.0276-inch (0.7 mm) galvanized sheet metal. Nonmetallic pans shall have a minimum thickness of not less than 0.0625 inch (1.6 mm).

2. A separate overflow drain line shall be connected to the drain pan provided with the equipment. Such overflow drain shall discharge to a conspicuous point of disposal to alert occupants in the event of a stoppage of the primary drain. The overflow drain line shall connect to the drain pan at a higher level than the primary drain connection.

3. An auxiliary drain pan without a separate drain line shall be provided under the coils on which condensate will occur. Such pan shall be equipped with a water-level detection device that will shut off the equipment served prior to overflow of the pan. The auxiliary drain pan shall be constructed in accordance with Item 1 of this section.

4. A water level detection device shall be provided that will shut off the equipment served in the event that the primary drain is blocked. The device shall be installed in the primary drain line, the overflow drain line, or in the equipment-supplied drain pan, located at a point higher than the primary drain line connection and below the overflow rim of such pan.

   Exception: Fuel-fired appliances that automatically shut down operation in the event of a stoppage in the condensate drainage system.

2004 SUPPLEMENT TO THE IPC

CHAPTER 4
FIXTURES, FAUCETS AND FIXTURE FITTINGS

Table 403.1 Change table heading and entries to read as shown: (P11-03/04; P14-03/04; P15-03/04)

TABLE 403.1
MINIMUM NUMBER OF REQUIRED PLUMBING FIXTURES
(See Sections P2902.2 and P2902.3)

NO.	CLASSIFICATION	OCCUPANCY	DESCRIPTION	WATER CLOSETS (URINALS SEE SECTION 419.2) Male	WATER CLOSETS Female	LAVATORIES Male	LAVATORIES Female	BATHTUBS/ SHOWERS	DRINKING FOUNTAIN (SEE SECTION 410.1)	OTHER
1	Assembly (see Sections 403.2, 403.5 and 403.6)	A-1 [d]	Theaters usually with fixed seats and other buildings for the performing arts and motion pictures	1 per 125	1 per 65	1 per 200		—	1 per 500	1 service sink
		A-2 [d]	Nightclubs, bars, taverns, dance halls and buildings for similar purposes	1 per 40	1 per 40	1 per 75		—	1 per 500	1 service sink
			Restaurants, banquet halls and food courts	1 per 75	1 per 75	1 per 200		—	1 per 500	1 service sink
		A-3 [d]	Auditoriums without permanent seating, art galleries, exhibition halls, museums, lecture halls, libraries, arcades and gymnasiums	1 per 125	1 per 65	1 per 200		—	1 per 500	1 service sink
			Passenger terminals and transportation facilities	1 per 500	1 per 500	1 per 750		—	1 per 1000	1 service sink
			Places of worship and other religious services. Churches without assembly halls.	1 per 150	1 per 75	1 per 200		—	1 per 1000	1 service sink

IPC-3

2004 SUPPLEMENT TO THE IPC

Table 403.1 (continued)

NO.	CLASSIFICATION	OCCUPANCY	DESCRIPTION	WATER CLOSETS (URINALS SEE SECTION 419.2)		LAVATORIES		BATHTUBS/ SHOWERS	DRINKING FOUNTAIN (SEE SECTION 410.1)	OTHER
				Male	Female	Male	Female			
2	Business (see Sections 403.2, 403.4 and 403.6)	B	Buildings for the transaction of business, professional services, other services involving merchandise, office buildings, banks, light industrial and similar uses.	1 per 25 for the first 50 and 1 per 50 for the remainder exceeding 50		1 per 40 for the first 80 and 1 per 80 for the remainder exceeding 80		—	1 per 100	1 service sink
5	Institutional	I-4	Adult daycare and childcare	1 per 15		1 per 15		—	1 per 100	1 service sink

(Portions of table not shown do not change)

a. The fixtures shown are based on one fixture being the minimum required for the number of persons indicated or any fraction of the number of persons indicated. The number of occupants shall be determined by this code.
b. Toilet facilities for employees shall be separate from facilities for inmates or patients.
c. A single-occupant toilet room with one water closet and one lavatory serving not more than two adjacent patient rooms shall be permitted where such room is provided with direct access from each patient room and with provisions for privacy.
d. The occupant load for seasonal outdoor seating and entertainment areas shall be included when determining the minimum number of facilities required.
e. For attached one- and two-family dwellings, one automatic clothes washer connection shall be required per 20 dwelling units.

Section 403.2 Change exceptions to read as follows: (P16-03/04)

403.2 Separate facilities. Where plumbing fixtures are required, separate facilities shall be provided for each sex.

Exceptions:

1. Separate facilities shall not be required for dwelling units and sleeping units.

2. Separate employee facilities shall not be required in occupancies in which 15 or less people are employed.

3. Separate facilities shall not be required in structures or tenant spaces with a total occupant load, including both employees and customers, of 15 or less.

4. Separate facilities shall not be required in mercantile occupancies in which the maximum occupant load is 50 or less.

Section 403.6 Change to read as shown: (P19-03/04)

403.6 Public facilities. Customers, patrons and visitors shall be provided with public toilet facilities in structures and tenant spaces intended for public utilization. The accessible route to public facilities shall not pass through kitchens, storage rooms, closets or similar spaces. Public toilet facilities shall be located not more than one story above or below the space required to be provided with public toilet facilities and the path of travel to such facilities shall not exceed a distance of 500 feet (152 m).

Section 405.4 Change to read as shown: (P21-03/04)

405.4 Floor and wall drainage connections. Connections between the drain and floor outlet plumbing fixtures shall be made with a floor flange. The flange shall be attached to the drain and anchored to the structure. Connections between the drain and wall-hung water closets shall be made with an approved extension nipple or horn adaptor. The water closet shall be bolted to the hanger with corrosion-resistant bolts or screws. Joints shall be sealed with an approved elastomeric gasket, flange-to-fixture connection complying with ASME A112.4.3 or an approved setting compound.

Section 408.3 Add new section to read as shown: (P23-03/04)

408.3 Bidet water temperature. The discharge water temperature from a bidet fitting shall be limited to a maximum temperature of 110° F (43° C) by a water temperature limiting device conforming to ASSE 1070.

Section 412.2 Change to read as shown: (P25-03/04)

412.2 Floor drains. Floor drains shall have removeable strainers. The floor drain shall be constructed so that the drain is capable of being cleaned. Access shall be provided to the drain inlet.

Section 416.5 Add new section to read as shown: (P66-03/04)

416.5 Tempered water for public hand-washing facilities. Tempered water shall be delivered from public hand-washing facilities through an approved water temperature limiting device that conforms to ASSE 1070.

Section 419.1 Change to read as shown: (P27-03/04; P28-03/04)

419.1 Approval. Urinals shall conform to ANSI Z124.9, ASME A112.19.2M, CSA B45.1 or CSA B45.5. Urinals shall conform to the water consumption requirements of Section 604.4. Water supplied urinals shall conform to the hydraulic performance requirements of ASME A112.19.6, CSA B45.1 or CSA B45.5.

Section 421.2 Change to read as shown: (P32-03/04)

421.2 Installation. Whirlpool bathtubs shall be installed and tested in accordance with the manufacturer's installation instructions. The pump shall be located above the weir of the fixture trap.

Section 421.5 Add new section to read as shown: (P32-03/04)

421.5 Access to pump. Access shall be provided to circulation pumps in accordance with the fixture manufacturer's installation instructions. Where manufacturer's instructions do not specify the location and minimum size of field fabricated access openings, a 12"x 12" (304 mm x 304 mm) minimum size door or panel shall be installed to provide access to the circulation pump. Where pumps are located more than 2 feet (609 mm) from the access opening, a 18"x 18" (457 mm x 457 mm) minimum size door or panel shall be installed. In all cases access panel and door openings shall be unobstructed and large enough to permit the removal of the circulation pump.

Section 424.1.2 Change to read as shown: (P33-03/04)

424.1.2 Waste fittings. Waste fittings shall conform to ASME A112.18.2, ASTM F 409, CSA B125 or to one of the standards listed in Tables 702.1 and 702.4 for above-ground drainage and vent pipe and fittings.

Section 424.3 Change to read as shown: (P36-03/04; P37-03/04)

424.3 Individual shower valves. Individual shower and tub-shower combination valves shall be balanced-pressure, thermostatic or combination balanced-pressure/thermostatic valves that conform to the requirements of ASSE 1016 or CSA B125 and shall be installed at the point of use. Shower and tub-shower combination valves required by this section shall be equipped with a means to limit the maximum setting of the valve to 120°F (49°C), which shall be field adjusted in accordance with the manufacturer's instructions. In-line thermostatic valves shall not be utilized for compliance with this section.

Section 424.4 Add new section to read as shown and renumber subsequent sections: (P36-03/04)

424.4 Multiple (gang) showers supplied with a single tempered water supply pipe. Multiple (gang) showers supplied with a single tempered water supply pipe shall have the water supply for such showers controlled by an approved automatic temperature control mixing valve. Such valves shall be equipped with a means to limit the maximum setting of the valve to 120°F (49°C), which shall be field adjusted in accordance with the manufacturer's instructions.

Section 424.5 Change to read as shown: (P41-03/04)

424.5 Hose-connected outlets. Faucets and fixture fittings with hose-connected outlets shall conform to ASME A112.18.3M or CSA B125.

CHAPTER 5
WATER HEATERS

Sections 504.6, 504.6.1 and 504.6.2 Delete and substitute new Section 504.6 to read as shown: (P45/03/04)

504.6 Requirements for discharge piping. The discharge piping serving a pressure relief valve, temperature relief valve or combination thereof shall:

1. Not be directly connected to the drainage system.

2. Discharge through an air gap located in the same room as the water heater.

3. Not be smaller than the diameter of the outlet of the valve served and shall discharge full size to the air gap.

4. Serve a single relief device and shall not connect to piping serving any other relief device or equipment.

5. Discharge to the floor, to an indirect waste receptor, or to the outdoors. Where discharging to

2004 SUPPLEMENT TO THE IPC

the outdoors in areas subject to freezing, discharge piping shall be first piped to an indirect waste receptor through an air gap located in a conditioned area.

6. Discharge in a manner that does not cause personal injury or property damage.

7. Discharge to a termination point that is readily observable by the building occupants.

8. Not be trapped.

9. Be installed so as to flow by gravity.

10. Not terminate more than 6 inches (152 mm) above the floor or waste receptor.

11. Not have a threaded connection at the end of such piping.

12. Not have valves or tee fittings.

13. Be constructed of those materials listed in Section 605.4 or materials tested, rated and approved for such use in accordance with ASME A112.4.1.

Section 504.6.2 Change to read as shown: (P47-03/04)

504.6.2 Relief valve discharge. Piping shall be of those materials listed in Section 605.4 or shall be tested, rated and approved for such use in accordance with ASME A112.4.1.

Section 504.7.1 Change to read as shown: (P47-03/04)

504.7.1 Pan size and drain. The pan shall be not less than 1.5 inches (38 mm) deep and shall be of sufficient size and shape to receive all dripping or condensate from the tank or water heater. The pan shall be drained by an indirect waste pipe having a minimum diameter of 0.75 inch (19 mm). Piping for safety pan drains shall be of those materials listed in Table 605.4.

CHAPTER 6
WATER SUPPLY AND DISTRIBUTION

Section 604.5 Change to read as shown: (P51-03/04)

604.5 Size of fixture supply. The minimum size of a fixture supply pipe shall be as shown in Table 604.5. The fixture supply pipe shall not terminate more than 30 inches (762 mm) from the point of connection to the fixture. A reduced-size flexible water connector installed between the supply pipe and the fixture shall be of an approved type. The supply pipe shall extend to the floor or wall adjacent to the fixture. The minimum size of individual distribution lines utilized in gridded or parallel water distribution systems shall be as shown in Table 604.5.

Section 604.10 Change to read as shown: (P51-03/04)

604.10 Gridded and parallel water distribution system manifolds. Hot water and cold water manifolds installed with gridded or parallel connected individual distribution lines to each fixture or fixture fitting shall be designed in accordance with Sections 604.10.1 through 604.10.3.

Section 605.3 Change to read as shown: (P53-03/04)

605.3 Water service pipe. Water service pipe shall conform to NSF 61 and shall conform to one of the standards listed in Table 605.3. All water service pipe or tubing, installed underground and outside of the structure, shall have a minimum working pressure rating of 160 psi (1100 kPa) at 73.4°F (23°C). Where the water pressure exceeds 160 psi (1100 kPa), piping material shall have a minimum rated working pressure equal to the highest available pressure. Water service piping materials not third-party certified for water distribution shall terminate at or before the full open valve located at the entrance to the structure. All ductile iron water service piping shall be cement mortar lined in accordance with AWWA C104.

Table 605.3 Add entries to read as shown: (P56-03/04; P57-03/04)

TABLE 605.3
WATER SERVICE PIPE

MATERIAL	STANDARD
Cross-linked polyethylene/aluminum/high density polyethylene (PEX-AL-HDPE)	ASTM F1986
Polypropylene (PP-R) Plastic Pipe	CSA B137.11

(Portions of table not shown do not change)

Table 605.4 Add entries to read as shown: (P56-03/04; P57-03/04)

TABLE 605.4
WATER DISTRIBUTION PIPE

MATERIAL	STANDARD
Cross-linked polyethylene/aluminum/high density polyethylene (PEX-AL-HDPE)	ASTM F1986
Polypropylene (PP-R) Plastic Pipe	CSA B137.11

(Portions of table not shown do not change)

2004 SUPPLEMENT TO THE IPC

Table 605.5 Revise standards and add 3 entires (PEX-AL-HDPE, PB and PP-R) to read as shown: (P56-03/04; P57-03/04)

TABLE 605.5
PIPE FITTINGS

MATERIAL	STANDARD
Chlorinated polyvinyl chloride (CPVC) plastic	ASTM F437, ASTM F438, ASTM F439, CSA B137.6
Cross-linked polyethylene/aluminum/high density polyethylene (PEX-AL-HDPE)	ASTM F1986
Fittings for cross-linked polyethylene (PEX) plastic tubing	ASTM F1807, ASTM F1960, ASTM F2080, CSA B137.5
Polybutylene (PB) plastic	CSA B137.8
Polyethylene (PE) plastic	ASTM D2609, CSA B137.1
Polypropylene (PP-R) plastic	CSA B137.11

(Portions of table not shown do not change)

Section 607.1 Change to read as shown: (P66-03/04)

607.1 Where required. In residential occupancies, hot water shall be supplied to all plumbing fixtures and equipment utilized for bathing, washing, culinary purposes, cleansing, laundry or building maintenance. In nonresidential occupancies, hot water shall be supplied for culinary purposes, cleansing, laundry or building maintenance purposes. In nonresidential occupancies, hot water or tempered water shall be supplied for bathing and washing purposes. Tempered water shall be supplied through a water temperature limiting device that conforms to ASSE 1070 and shall limit the tempered water to a maximum of 110° F (43°C). This provision shall not supersede the requirement for protective shower valves in accordance with Section 424.3.

Section 607.4 Change to read as shown: (P69-03/04)

607.4 Flow of hot water to fixtures. Fixture fittings, faucets and diverters shall be installed and adjusted so that the flow of hot water from the fittings corresponds to the left-hand side of the fixture fitting.

Exception: Shower and tub/shower mixing valves conforming to ASSE 1016 or CSA B125, where the flow of hot water corresponds to the markings on the device.

Table 608.1 Revise entries to read as shown: (P71-03/04)

TABLE 608.1
APPLICATION FOR BACKFLOW PREVENTERS

DEVICE	DEGREE OF HAZARD	APPLICATION [b]	APPLICABLE STANDARDS
Reduced pressure principle backflow preventer and reduced pressure principle fire protection backflow preventer	High or low hazard	Backpressure or backsiphonage Sizes 3/8"-16"	CSA B64.4, CSA B64.4.1 ASSE 1013 AWWA C511
Double check backflow prevention assembly and double check fire protection backflow prevention assembly	Low hazard	Backpressure or backsiphonage Sizes 3/8"-16"	CSA B64.5, CSA B64.5.1 ASSE 1015 AWWA C510
Backflow preventer with intermediate atmospheric vents	Low hazard	Backpressure or backsiphonage Sizes 1/4"-3/4"	CSA B64.3 ASSE 1012
Backflow preventer for carbonated beverage machines	Low hazard	Backpressure or backsiphonage Sizes 1/4"-3/8"	CSA B64.3.1 ASSE 1022
Pipe-applied atmospheric-type vacuum breaker	High or low hazard	Backsiphonage only Sizes 1/4"-4"	CSA B64.1.1 ASSE 1001
Pressure vacuum breaker assembly	High or low hazard	Backsiphonage only Sizes 1/2"-2"	CSA B64.1.2 ASSE 1020
Hose-connection vacuum breaker	High or low hazard	Low head backpressure or backsiphonage Sizes ½", 3/4", 1"	CSA B64.2, CSA B64.2.1 ASSE 1011
Vacuum breaker wall hydrants, frost-resistant, automatic draining type	High or low hazard	Low head backpressure or backsiphonage Sizes 3/4", 1"	CSA B64.2.2 ASSE 1019
Hose connection backflow preventer	High or low hazard	Low head backpressure, rated working pressure backpressure or backsiphonage Sizes 1/2"-1"	CSA B64.2.1.1 ASSE 1052

(Portions of table not shown do not change)

Section 608.2 Change to read as shown: (P72-03/04)

608.2 Plumbing fixtures. The supply lines and fittings for every plumbing fixture shall be installed as to prevent backflow. Plumbing fixture fittings shall provide backflow protection in accordance with ASME A112.18.1.

Section 608.13.2 Change to read as shown: (P73-03/04)

IPC-7

2004 SUPPLEMENT TO THE IPC

608.13.2 Reduced pressure principle backflow preventers. Reduced pressure principle backflow preventers shall conform to ASSE 1013, AWWA C511, CSA B64.4 or CSA B64.4.1. Reduced pressure detector assembly backflow preventers shall conform to ASSE 1047. These devices shall be permitted to be installed where subject to continuous pressure conditions. The relief opening shall discharge by air gap and shall be prevented from being submerged.

Section 608.13.5 Change to read as shown: (P74-03/04)

608.13.5 Pressure-type vacuum breakers. Pressure-type vacuum breakers shall conform to ASSE 1020 or CSA B64.1.2 and spillproof vacuum breakers shall comply with ASSE 1056. These devices are designed for installation under continuous pressure conditions when the critical level is installed at the required height. Pressure-type vacuum breakers shall not be installed in locations where spillage could cause damage to the structure.

Section 608.13.6 Change to read as shown: (P75-03/04)

608.13.6 Atmospheric-type vacuum breakers. Pipe-applied atmospheric-type vacuum breakers shall conform to ASSE 1001 or CSA B64.1.1. Hose-connection vacuum breakers shall conform to ASSE 1011, ASSE 1019, ASSE 1035, ASSE 1052, CSA B64.2, CSA B64.2.1, CSA B64.2.1.1, CSA B64.2.2 or CSA B64.7. These devices shall operate under normal atmospheric pressure when the critical level is installed at the required height.

Section 608.13.7 Change to read as shown: (P76-03/04)

608.13.7 Double check-valve assemblies. Double check valve assemblies shall conform to ASSE 1015, CSA B64.5, CSA B64.5.1 or AWWA C510. Double-detector check-valve assemblies shall conform to ASSE 1048. These devices shall be capable of operating under continuous pressure conditions.

Section 608.16.1 Change to read as shown: (P78-03/04)

608.16.1 Beverage dispensers. The water supply connection to beverage dispensers shall be protected against backflow by a backflow preventer conforming to ASSE 1022 or CSA B64.3.1 or by air gap. The backflow preventer device and the piping downstream therefrom shall not be affected by carbon dioxide gas.

CHAPTER 7
SANITARY DRAINAGE

Table 702.2 Revise entries to read as shown: (P83-03/04)

TABLE 702.2
UNDERGROUND BUILDING DRAINAGE AND VENT PIPE

MATERIAL	STANDARD
Polyolefin pipe	ASTM F1412, CSA B181.3

(Portions of table not shown do not change)

Table 702.4 Add entry to read as shown: (P83-03/04)

TABLE 702.4
PIPE FITTINGS

MATERIAL	STANDARD
Polyolefin	ASTM F1412, CSA B181.3

(Portions of table not shown do not change)

Section 705.5.3 Change to read as shown: (P85-03/04)

705.5.3 Mechanical joint coupling. Mechanical joint couplings for hubless pipe and fittings shall comply with CISPI 310, ASTM C1277 or ASTM C1540. The elastomeric sealing sleeve shall conform to ASTM C564 or CSA B602 and shall be provided with a center stop. Mechanical joint couplings shall be installed in accordance with the manufacturer's installation instructions.

Sections 705.16, 705.16.1 and 705.16.2 Add new sections to read as shown and renumber subsequent sections: (P86-03/04)

705.16 Polyethylene plastic pipe. Joints between polyethylene plastic pipe and fittings shall comply with Sections 705.16.1 and 705.16.2.

705.16.1 Heat-fusion joints. Joint surfaces shall be clean and free from moisture. All joint surfaces shall be heated to melting temperature and joined. The joint shall be undisturbed until cool. Joints shall be made in accordance with ASTM D 2657.

705.16.2 Mechanical joints. Mechanical joints shall be installed in accordance with the manufacturer's instructions.

Sections 705.17, 705.17.1 and 705.17.2 Add new sections to read as shown and renumber subsequent sections: (P87-03/04)

705.17 Polyolefin plastic. Joints between polyolefin plastic pipe and fittings shall comply with Sections 705.17.1 and 705.17.2.

705.17.1 Heat-fusion joints. Heat fusion joints for polyolefin pipe and tubing joints shall be installed with socket-type heat-fused polyolefin fittings or electrofusion polyolefin fittings. Joint surfaces shall be clean and free from moisture. The joint shall be undisturbed until cool. Joints shall be made in accordance with ASTM F 1412 or CSA B181.3.

705.17.2 Mechanical and compression sleeve joints. Mechanical and compression sleeve joints shall be installed in accordance with the manufacturer's instructions.

Section 706.4 Add new section to read as shown: (P88-03/04)

706.4 Heel-or-side-inlet quarter bends. Heel-inlet quarter bends shall be an acceptable means of connection, except where the quarter bend serves a water closet. A low-heel inlet shall not be used as a wet-vented connection. Side-inlet quarter bends shall be an acceptable means of connection for drainage, wet venting and stack venting arrangements.

Table 709.1 Change table to read as shown: (P89-03/04; P90-03/04)

TABLE 709.1
DRAINAGE FIXTURE UNITS FOR FIXTURES AND GROUPS

FIXTURE TYPE	DRAINAGE FIXTURE UNIT VALUE AS LOAD FACTORS	MINIMUM SIZE OF TRAP (inches)
Service Sink	2	1½
Urinal, non-water supplied	0.5	Footnote d

(Portions of table not shown do not change)

Section 712.2 Delete exception: (P91-03/04)

712.2 Valves required. A check valve and full open valve, located on the discharge side of the check valve, shall be installed in the pump or ejector discharge piping between the pump or ejector and the gravity drainage system. Access shall be provided to such valves. Such valves shall be located above the sump cover required by Section 712.1 or, where the discharge pipe from the ejector is below grade, the valves shall be accessibly located outside the sump below grade in an access pit with a removable access cover.

CHAPTER 9
VENTS

Section 903 Change title to read as shown: (P93-03/04)

SECTION 903
OUTDOOR VENT EXTENSION

Sections 903.1 and 903.1.1 Delete and substitute new sections 903.1, 903.1.1 and 903.1.2 to read as shown: (P93-03/04)

903.1 Required vent extension. The vent system serving each building drain shall have at least one vent pipe that extends to the outdoors.

903.1.1 Installation. The required vent shall be a dry vent that connects to the building drain or an extension of a drain that connects to the building drain. Such vent shall not be an island fixture vent as allowed by Section 913.

903.1.2 Size. The required vent shall be sized in accordance with Section 916.2 based on the required size of the building drain.

Section 903.3 Change to read as shown: (P95-03/04)

903.3 Vent termination. Every vent stack or stack vent shall terminate outdoors to the open air or to a stack-type air admittance valve in accordance with Section 917.

Section 906.1 Change to read as shown: (P97-03/04)

906.1 Distance of trap from vent. Each fixture trap shall have a protecting vent located so that the slope and the developed length in the fixture drain from the trap weir to the vent fitting are within the requirements set forth in Table 906.1.

Exception: The developed length of the fixture drain from the trap weir to the vent fitting for self-siphoning fixtures such as water closets, shall not be limited.

Table 906.1 Delete table in its entirety and substitute as follows: (P98-03/04)

TABLE 906.1
MAXIMUM DISTANCE OF
FIXTURE TRAP FROM VENT

SIZE OF TRAP (inches)	SLOPE (inch per foot)	DISTANCE FROM TRAP (feet)
1¼	¼	5
1½	¼	6
2	¼	8
3	⅛	12
4	⅛	16

Section 906.2 Change to read as shown: (P99-03/04)

906.2 Venting of fixture drains. The total fall in a fixture drain due to pipe slope shall not exceed the diameter of the fixture drain, nor shall the vent connection to a fixture drain, except for water closets, be below the weir of the trap.

2004 SUPPLEMENT TO THE IPC

Section 917.1 Change to read as shown: (P95-03/04)

917.1 General. Vent systems utilizing air admittance valves shall comply with this section. Stack type air admittance valves shall conform to ASSE 1050. Individual and branch type air admittance valves shall conform to ASSE 1051.

Section 917.3 Change to read as shown: (P95-03/04)

917.3 Where permitted. Individual, branch and circuit vents shall be permitted to terminate with a connection to an individual or branch-type air admittance valve. Stack vents and vent stacks shall be permitted to terminate to stack-type air admittance valves. Individual and branch-type air admittance valves shall vent only fixtures that are on the same floor level and connect to a horizontal branch drain. The horizontal branch drain having individual and branch-type air admittance valves shall conform to Section 917.3.1 or 917.3.2. Stack type air admittance valves shall conform to Section 917.3.3.

Section 917.3.2 Change to read as shown: (P95-03/04)

917.3.2 Relief vent. Where the horizontal branch is located more than four branch intervals from the top of the stack, the horizontal branch shall be provided with a relief vent that shall connect to a vent stack, or stack vent or extend outdoors to the open air. The relief vent shall connect to the horizontal branch drain between the stack and the most downstream fixture drain connected to the horizontal branch drain. The relief vent shall be sized in accordance with Section 916.2 and installed in accordance with Section 905. The relief vent shall be permitted to serve as the vent for other fixtures

Section 917.3.3 Add new section to read as shown: (P95-03/04)

917.3.3 Stack. Stack-type air admittance valves shall not serve as the vent terminal for vent stacks or stack vents that serve drainage stacks exceeding six branch intervals.

Section 917.4 Change to read as shown: (P95-03/04)

917.4 Location. Individual and branch-type air admittance valves shall be located a minimum of 4 inches (102 mm) above the horizontal branch drain or fixture drain being vented. Stack-type air admittance valves shall be located not less than 6 inches (152 mm) above the flood level rim of the highest fixture being vented. The air admittance valve shall be located within the maximum developed length permitted for the vent. The air admittance valve shall be installed a minimum of 6 inches (152 mm) above insulation materials.

CHAPTER 10
TRAPS, INTERCEPTORS AND SEPARATORS

Section 1003.4 Change to read as shown: (P104-03/04)

1003.4 Oil separators required. At repair garages, carwashing facilities and factories where oily and flammable liquid wastes are produced, separators shall be installed into which all oil-bearing, grease-bearing or flammable wastes shall be discharged before emptying into the building drainage system or other point of disposal.

CHAPTER 11
STORM DRAINAGE

Table 1102.4 Change entries to read as shown: (P105-03/04)

TABLE 1102.4
BUILDING STORM SEWER PIPE

MATERIAL	STANDARD
Acrylonitrile butadiene styrene (ABS) plastic pipe	ASTM D2661; ASTM D2751; ASTM F628; CSA B181.1; CSA B182.1
Polyvinyl chloride (PVC) plastic pipe (Type DWV, SDR26, SDR35, SDR41, PS50, PS100)	ASTM D2665; ASTM D3034; ASTM F891; CSA B182.4; CSA B181.2; CSA B182.2,

(Portions of table not shown do not change)

Table 1102.5 Change entries to read as shown: (P106-03/04)

TABLE 1102.5
SUBSOIL DRAIN PIPE

MATERIAL	STANDARD
Polyethylene (PE) plastic pipe	ASTM F405; CSA B182.1; CSA B182.6; CSA B182.8
Polyvinyl chloride (PVC) Plastic pipe (type sewer pipe, PS25, PS50 or PS100)	ASTM D2729; ASTM F891; CSA B182.2; CSA B182.4

(Portions of table not shown do not change)

Table 1102.7 Change entire table to read as shown: (P33-03/04; P107-03/04)

TABLE 1102.7
PIPE FITTINGS

MATERIAL	STANDARD
Acrylonitrile butadiene styrene (ABS) plastic	ASTM D2661; ASTM D3311; CSA B181.1
Cast-iron	ASME B16.4; ASME B16.12; ASTM A888; CISPI 301; ASTM A74
Coextruded composite ABS DWV Schedule 40 IPS pipe (solid or cellular core)	ASTM D2661; ASTM D3311; ASTM F628
Coextruded composite PVC DWV Schedule 40 IPS-DR, PS140, PS200 (solid or cellular core)	ASTM D2665; ASTM D3311; ASTM F891
Coextruded composite ABS sewer and drain DR-PS in PS35, PS50, PS100, PS140, PS200	ASTM D2751
Coextruded composite PVC sewer and drain DR-PS in PS35, PS50, PS100, PS140, PS200	ASTM D3034
Copper or copper alloy	ASME B16.15; ASME B16.18; ASME B16.22; ASME B16.23; ASME B16.26; ASME B16.29
Gray iron and ductile iron	AWWA C110
Malleable iron	ASME B16.3
Plastic, general	ASTM F409
Polyvinyl chloride (PVC) plastic	ASTM D2665; ASTM D 3311; ASTM F1866
Steel	ASME B16.9; ASME B16.11; ASME B16.28
Stainless steel drainage Systems, Type 316L	ASME A112.3.2

2004 SUPPLEMENT TO THE IPC

CHAPTER 13
REFERENCED STANDARDS

Change, delete or add the following referenced standards to read as shown: (P10-03/04; P23; P27; P33; P41; P56; P57; P59; P66; P69; P71; P72; P73; P74; P75; P76; P78; P83; P85; P86; P87; P95; P105; P106; P107; P114) (STANDARDS NOT SHOWN REMAIN UNCHANGED)

ANSI
American National Standards Institute
25 West 43rd Street, Fourth Floor
New York, NY 10036

Standard reference number	Title	Referenced In code section number
Z21.22—99	Relief Valves for Hot Water Supply Systems — with Addenda Z21.22a-2000 and Z21.22b-2001	504.2, 504.5
Z124.9—94	Plastic Urinal Fixtures	419.1

ARI
Air Conditioning and Refrigeration Institute
4100 North Fairfax Drive, Suite 200
Arlington, VA 22203

Standard reference number	Title	Referenced In code section number
1010—02	Self-Contained, Mechanically-Refrigerated Drinking-Water Coolers	410.1

ASME
American Society of Mechanical Engineers
Three Park Avenue
New York, NY 10016-5990

Standard reference number	Title	Referenced In code section number
A112.1.2—1991 (R 2002)	Air Gaps in Plumbing Systems	Table 608.1
A112.4.1—1993 (R 2002)	Water Heater Relief Valve Drain Tubes	504.6.2
A112.6.1M—1997 (R 2002)	Floor-Affixed Supports for Off-the-Floor Plumbing Fixtures for Public Use	405.4.3
A112.18.1—2000	Plumbing Fixture Fittings	424.1, 608.2
A112.18.2—2002	Plumbing Fixtures Waste Fittings	424.1.2
A112.19.1M—1994 (R 1999)	Enameled Cast Iron Plumbing Fixtures — with 1998 and 2000 Supplements	407.1, 410.1, 415.1, 416.1, 418.1
A112.19.2M—1998 (R 2002)	Vitreous China Plumbing Fixtures — with 2000 Supplement	401.2, 405.9, 408.1, 410.1, 416.1, 418.1, 419.1, 420.1
A112.19.3M—2000	Stainless Steel Plumbing Fixtures (Designed for Residential Use) - with 2002 Supplement	405.9, 415.1, 416.1, 418.1
A112.19.9M—1991 (R1998)	Non-Vitreous Ceramic Plumbing Fixtures — with 2002 Supplement	407.1, 408.1, 410.1, 415.1, 416.1, 417.1, 418.1, 420.1
A112.36.2M—1991 (R 2002)	Cleanouts	708.2

ASME (continued)

B1.20.1—1983 (R 2001)	Pipe Threads, General Purpose (inch)	605.10.3, 605.12.3, 605.14.4, 605.16.3, 605.18.1, 705, 705.2.3, 705.4.3
B16.3—1998	Malleable Iron Threaded Fittings Classes 150 and 300	Table 605.5, Table 702.4, Table 1102.7
B16.9—2001	Factory-Made Wrought Steel Buttwelding Fittings	Table 605.5, Table 702.4, Table 1102.7
B16.11—2001	Forged Fittings, Socket-Welding and Threaded	Table 605.5, Table 702.4, Table 1102.7
B16.18—2001	Cast Copper Alloy Solder Joint Pressure Fittings	Table 605.5, Table 702.4, Table 1102.7
B16.22—2001	Wrought Copper and Copper Alloy Solder Joint Pressure Fittings	Table 605.5, Table 702.4, Table 1102.7
B16.23—2002	Cast Copper Alloy Solder Joint Drainage Fittings: DWV	Table 605.5, Table 702.4, Table 1102.7
B16.29—2001	Wrought Copper and Wrought Copper Alloy Solder Joint Drainage Fittings - DWV	Table 605.5, Table 702.4, Table 1102.7

ASSE

American Society of Sanitary Engineering
901 Canterbury Road, Suite A
Westlake, OH 44145

Standard reference number	Title	Referenced In code section number
1001—02	Performance Requirements for Atmospheric Type Vacuum Breakers	425.2, Table 608.1, 608.13.6
1003—01	Performance Requirements for Water Pressure Reducing Valves	604.8
1010—96	Performance Requirements for Water Hammer Arresters	604.9
1011—93	Performance Requirements for Hose Connection Vacuum Breakers	Table 608.1, 608.13.6
1012—02	Performance Requirements for Backflow Preventers with Intermediate Atmospheric Vent	Table 608.1, 608.13.3, 608.16.2
1018—01	Performance Requirements for Trap Seal Primer Valves - Potable Water Supplied	1002.4
1022—03	Performance Requirements for Backflow Preventer for Carbonated Beverage Machines	Table 608.1, 608.16.1
1024—03	Performance Requirements for Dual Check Valve Backflow Preventers	605.3.1, Table 608.1
1035—02	Performance Requirements for Laboratory Faucet Blackflow Preventers	Table 608.1, 608.13.6
1044—01	Performance Requirements for Trap Seal Primer Devices - Drainage Types and Electronic Design Types	1002.4
1050—02	Performance Requirements for Stack Air Admittance Valves for Sanitary Drainage Systems	917.1
1051—02	Performance Requirements for Individual and Branch Type Air Admittance Valves for Sanitary Drainage Systems	917.1
1052—93	Performance Requirements for Hose Connection Backflow Preventers	Table 608.1, 608.13.6
1056—01	Performance Requirements for Spill Resistant Vacuum Breaker	Table 608.1, 608.13.5, 608.13.8
1070-04	Performance Requirements for Water Temperature Limiting Devices	408.3, 416.5, 607.1

ASTM

ASTM International
100 Barr Harbor Drive
West Conshohocken, PA 19428-2959

Standard reference number	Title	Referenced In code section number
A 53/A 53M—02	Specification for Pipe, Steel, Black and Hot Dipped, Zinc-Coated Welded and Seamless	Table 605.3, Table 605.4, Table 702.1
A 74—03	Specification for Cast Iron Soil Pipe and Fittings	Table 702.1, Table 702.2, Table 702.3, Table 702.4, 708.2, Table 1102.4, Table 1102.5, Table 1102.7
A 312/A 312M—02	Specification for Seamless and Welded Austenitic Stainless Steel Pipes	Table 605.4, Table 605.5, Table 605.6, 605.23.2
B 32—00e01	Specification for Solder Metal	605.14.3, 605.15.4, 705.9.3, 705.10.3
B 42—02	Specification for Seamless Copper Pipe, Standard Sizes	Table 605.3, Table 605.4, Table 702.1
B 75—02	Specification for Seamless Copper Tube	Table 605.3, Table 605.4, Table 702.1, Table 702.2, Table 702.3, Table 1102.4
B 88—02	Specification for Seamless Copper Water Tube	Table 605.3, Table 605.4, Table 702.1, Table 702.2, Table 702.3, Table 1102.4
B 251—02	Specification for General Requirements for Wrought Seamless Copper and Copper-Alloy Tube	Table 605.3, Table 605.4, Table 702.1, Table 702.2, Table 702.3, Table 1102.4
B 302—02	Specification for Threadless Copper Pipe, Standard Sizes	Table 605.3, Table 605.4, Table 702.1
B 306—02	Specification for Copper Drainage Tube (DWV)	Table 702.1, Table 702.2, Table 1102.4
B 447—02	Specification for Welded Copper Tube	Table 605.3, Table 605.4
B 828—02	Practice for Making Capillary Joints by Soldering of Copper and Copper Alloy Tube and Fittings	605.14.3, 605.15.4, 705.9.3, 705.10.3
C 4—03	Specification for Clay Drain Tile and Perforated Clay Drain Tile	Table 702.3, Table 1102.4, Table 1102.5
C 76—02	Specification for Reinforced Concrete Culvert, Storm Drain, and Sewer Pipe	Table 702.3, Table 1102.4
C 425—02	Specification for Compression Joints for Vitrified Clay Pipe and Fittings	705.15, 705.16
C 428—97 (2002)	Specification for Asbestos-Cement Nonpressure Sewer Pipe	Table 702.2, Table 702.3, Table 1102.4
C 443—02a	Specification for Joints for Concrete Pipe and Manholes, Using Rubber Gaskets	705.6, 705.16
C 564—03	Specification for Rubber Gaskets for Cast Iron Soil Pipe and Fittings	705.5.2, 705.5.3, 705.16
C 700—02	Specification for Vitrified Clay Pipe, Extra Strength, Standard Strength, and Perforated	Table 702.3, Table 1102.4, Table 1102.5
C 1173—02	Specification for Flexible Transition Couplings for Underground Piping System	705.7.1, 705.14.1, 705.16
C 1277—03	Specification for Shielded Coupling Joining Hubless Cast Iron Soil Pipe and Fittings	705.5.3
C 1440—99e01	Specification for Thermoplastic Elastomeric (TPE) Gasket Materials for Drain, Waste, and Vent (DWV), Sewer, Sanitary and Storm Plumbing Systems	705.16

ASTM (continued)

C 1461—02	Specification for Mechanical Couplings Using Thermoplastic Elastomeric (TPE) Gaskets for Joining Drain, Waste, and Vent (DWV) Sewer, Sanitary and Storm Plumbing Systems for Above and Below Ground Use	705.16
C 1540—02	Standard Specification for Heavy Duty Shielded Couplings Joining Hubless Cast Iron Soil Pipe and Fittings	705.5.3
D 2466—02	Specification for Poly (Vinyl Chloride) (PVC) Plastic Pipe Fittings, Schedule 40	Table 605.5, Table 1102.7
D 2467—02	Specification for Poly (Vinyl Chloride) (PVC) Plastic Pipe Fittings, Schedule 80	Table 605.5, Table 1102.7
D 2468—96a	Specification for Acrylonitrile-Butadiene-Styrene (ABS) Plastic Pipe Fittings, Schedule 40	Table 605.5
D 2564—02	Specification for Solvent Cements for Poly (Vinyl Chloride) (PVC) Plastic Piping Systems	605.21.2, 705.8.2, 705.14.2
D 2657—97	Standard Practice for Heat Fusion-Joining of Polyolefin Pipe and Fitting	605.19.2, 605.20.2, 705.16.1
D 2661—02	Specification for Acrylonitrile-Butadiene-Styrene (ABS) Schedule 40 Plastic Drain, Waste, and Vent Pipe and Fittings	Table 702.1, Table 702.2, Table 702.3, Table 702.4, 705.2.2, 705.7.2, Table 1102.4, Table 1102.7
D 2665—02a	Specification for Poly (Vinyl Chloride) (PVC) Plastic Drain, Waste, and Vent Pipe and Fittings	Table 702.1, Table 702.2, Table 702.3, Table 702.4, Table 1102.4, Table 1102.7
D 2751—96a	Specification for Acrylonitrile-Butadiene-Styrene (ABS) Sewer Pipe and Fittings	Table 702.3, Table 1102.4, Table 1102.7
D 2855—96 (2002)	Standard Practice for Making Solvent-Cemented Joints with Poly (Vinyl Chloride) (PVC) Pipe and Fittings	605.21.2, 705.8.2, 705.14.2
D 2949—01a	Specification for 3.25-In Outside Diameter Poly (Vinyl Chloride) (PVC) Plastic Drain, Waste, and Vent Pipe and Fittings	Table 702.1, Table 702.2, Table 702.3
D 3034—00	Specification for Type PSM Poly (Vinyl Chloride) (PVC) Sewer Pipe and Fittings	Table 702.3, Table 702.4, Table 1102.4, Table 1102.7
D 3311—02	Specification for Drain, Waste and Vent (DWV) Plastic Fitting Patterns	Table 702.4, Table 1102.7
F 409—02	Specification for Thermoplastic Accessible and Replaceable Plastic Tube and Tubular Fittings	424.1.2, Table 1102.7.
F 438—02	Specification for Socket-Type Chlorinated Poly (Vinyl Chloride) (CPVC) Plastic Pipe Fittings, Schedule 40	Table 605.5, Table 1102.7
F 439—02	Specification for Socket-Type Chlorinated Poly (Vinyl Chloride) (CPVC) Plastic Pipe Fittings, Schedule 80	Table 605.5, Table 1102.7
F 441/F 441M—02	Specification for Chlorinated Poly (Vinyl Chloride) (CPVC) Plastic Pipe, Schedules 40 and 80	Table 605.3, Table 605.4, Table 605.5
F 477—02e01	Specification for Elastomeric Seals (Gaskets) for Joining Plastic Pipe	605.22, 705.16
F 628—01	Specification for Acrylonitrile-Butadiene-Styrene (ABS) Schedule 40 Plastic Drain, Waste, and Vent Pipe With a Cellular Core	Table 702.1, Table 702.2, Table 702.3, Table 702.4, 705.2.2, 705.7.2, Table 1102.4, Table 1102.7
F 656—02	Specification for Primers for Use in Solvent Cement Joints of Poly (Vinyl Chloride) (PVC) Plastic Pipe and Fittings	605.21.2, 705.8.2, 705.14.2
F 714—01	Specification for Polyethylene (PE) Plastic Pipe (SDR-PR) Based on Outside Diameter	Table 702.3
F 876—02e01	Specification for Cross-linked Polyethylene (PEX) Tubing	Table 605.3

2004 SUPPLEMENT TO THE IPC

ASTM (continued)

F 877—02e01	Specification for Cross-linked Polyethylene (PEX) Plastic Hot and Cold Water Distribution Systems	Table 605.3, Table 605.4
F 891—00e01	Specification for Coextruded Poly (Vinyl Chloride) (PVC) Plastic Pipe with a Cellular Core	Table 702.1, Table 702.2, Table 702.3, Table 702.4, Table 1102.4, Table 1102.5, Table 1102.7
F 1281—02e02	Specification for Cross-Linked Polyethylene/Aluminum/Cross-Linked Polyethylene (PEX-AL-PEX) Pressure Pipe	Table 605.3, Table 605.4
F 1282—02e02	Specification for Polyethylene/Aluminum/Polyethylene (PE-AL-PE) Composite Pressure Pipe	Table 605.3, Table 605.4
F 1412—01	Specification for Polyolefin Pipe and Fittings for Corrosive Waste Drainage	Table 702.2, Table 702.4, 705.17.1
F 1488—00e01	Specification for Coextruded Composite Pipe	Table 702.1, Table 702.2, Table 702.3
F 1807—02a	Specification for Metal Insert Fittings Utilizing a Copper Crimp Ring for SDR9 Cross-linked Polyethylene (PEX) Tubing	Table 605.5, 605.17.2
F 1960—03	Specification for Cold Expansion Fittings with PEX Reinforcing Rings for use with Cross-linked Polyethylene (PEX) Tubing	Table 605.5
F 1974—02	Specification for Metal Insert Fittings for Polyethylene/Aluminum/ Polyethylene and Cross-linked Polyethylene/Aluminum/Cross-linked Polyethylene Composite Pressure Pipe	Table 605.5
F 1986—00a	Specification for Multilayer Pipe, Type 2, Compression Fittings, and Compression Joints for Hot and Cold Drinking Water Systems	Table 605.3, Table 605.4, Table 605.5
F 2080—02	Specifications for Cold-Expansion Fittings with Metal Compression-Sleeves for Cross-linked Polyethylene (PEX) Pipe	Table 605.5

AWWA

American Water Works Association
6666 West Quincy Avenue
Denver, CO 80235

Standard reference number	Title	Referenced In code section number
C151/A21.51—02	Standard for Ductile-Iron Pipe, Centrifugally Cast for Water	Table 605.3
C652—02	Disinfection of Water-Storage Facilities	610.1

CSA

Canadian Standards Association
5060 Spectrum Way, Suite 100
Mississauga, Ontario, Canada L4W 5N6

Standard reference number	Title	Referenced In code section number
B45.1—02	Ceramic Plumbing Fixtures	408.1, 416.1, 418.1, 419.1, 420.1
B45.2—02	Enameled Cast-Iron Plumbing Fixtures	407.1, 415.1, 416.1, 418.1
B45.3—02	Porcelain Enameled Steel Plumbing Fixtures	407.1, 416.1, 418.1
B45.4—02	Stainless-Steel Plumbing Fixtures	415.1, 416.1, 418.1, 420.1
B45.5—02	Plastic Plumbing Fixtures	407.1, 416.2, 417.1, 419.1, 420.1, 421.1
B45.9—02	Macerating Systems and Related Components	712.4.1
B64.1.1—01	Vacuum Breakers, Atmospheric Type (AVB)	Table 608.1

2004 SUPPLEMENT TO THE IPC

CSA (continued)

B64.1.2—01	Vacuum Breakers, Pressure Type (PVB)	Table 608.1, 608.13.5
B64.2—01	Vacuum Breakers, Hose Connection Type (HCVB)	Table 608.1
B64.2.1—01	Vacuum Breakers, Hose Connection Type (HCVB) with Manual Draining Feature	Table 608.1, 608.13.6
B64.2.2—01	Vacuum Breakers, Hose Connection Type (HCVB) with Automatic Draining Feature	Table 608.1
B64.2.1.1—01	Vacuum Breakers, Hose Connection Dual Check Type (HCDVB)	Table 608.1, 608.13.6
B64.3—01	Backflow Preventers, Dual Check Valve Type with Atmospheric Port (PCAP)	Table 608.1
B64.3.1—01	Backflow Preventers, Dual Check Valve Type with Atmospheric Port for Carbonators (DCAPC)	Table 608.1, 608.16.1
B64.4—01	Backflow Preventers, Reduced Pressure Principle Type (RP)	Table 608.1
B64.4.1—01	Backflow Preventers, Reduced Pressure Principle Type for Fire Systems (RPF)	Table 608.1, 608.13.2
B64.5—01	Backflow Preventers, Double Check Valve Type (DCVA)	Table 608.1, 608.13.7
B64.5.1—01	Backflow Preventers, Double Check Valve Type for Fire Systems (DCVAF)	Table 608.1, 608.13.7
B64.7—01	Vacuum Breakers, Laboratory Faucet Type (LFVB)	Table 608.1, 608.13.6
B64.10/B64.10—01	Manual for the Selection and Installation of Backflow Prevention Devices/ Manual for the Maintenance and Field Testing of Backflow Prevention Devices	.312.9.2
B125—01	Plumbing Fittings	424.1, 424.3, 425.3.1 424.4, 424.5, 607.4, Table 608.1
B137.1—02	Polyethylene Pipe, Tubing and Fittings for Cold Water Pressure Services	Table 605.3
B137.2—02	PVC Injection-Moulded Gasketed Fittings for Pressure Applications	Table 605.5, Table 1102.7
B137.3—02	Rigid Poly (Vinyl Chloride) (PVC) Pipe for Pressure Applications	Table 605.3, 605.21.2, 705.8.2, 705.14.2
B137.5—02	Cross-Linked Polyethylene (PEX) Tubing Systems for Pressure Applications	Table 605.3, Table 605.4
B137.6—02	CPVC Pipe, Tubing and Fittings for Hot and Cold Water Distribution Systems	Table 605.3, Table 605.4
B137.8—02	Polybutylene (PB) Piping for Pressure Applications	Table 605.3, Table 605.4, 605.19.2, 605.19.3
B137.9M—02	Polyethylene/Aluminum/Polyethylene Composite Pressure Pipe Systems	Table 605.3
B137.10M—02	Cross-linked Polyethylene/Aluminum/Cross-linked Polyethylene Composite Pressure Pipe Systems	Table 605.3, Table 605.4
B137.11—99	Polypropylene (PP-R) Pipe and Fittings for Pressure Applications	Table 605.3, Table 605.4, Table 605.5
B181.1—02	ABS Drain, Waste, and Vent Pipe and Pipe Fittings	Table 702.1, Table 702.2, Table 702.4, 705.2.2, 705.7.2, 715.2, Table 1102.4, Table 1102.7
B181.2—02	PVC Drain, Waste, and Vent Pipe and Pipe Fittings	Table 702.1, Table 702.2, 705.8.2, 705.14.2, 715.2, Table 1102.4
B181.3—02	Polyolefin Laboratory Drainage Systems	Table 702.1, Table 702.2, Table 702.4, 705.17.1
B182.1—02	Plastic Drain and Sewer Pipe and Pipe Fittings	705.8.2, 705.14.2, Table 1102.4

2004 SUPPLEMENT TO THE IPC

CSA (continued)

B182.2—02	PVC Sewer Pipe and Fittings (PSM Type)	Table 702.3, Table 1102.4, Table 1102.5
B182.4—02	Profile PVC Sewer Pipe and Fittings	Table 702.3, Table 1102.4, Table 1102.5
B182.6—02	Profile Polyethylene Sewer Pipe and Fittings for Leak-Proof Sewer Applications	Table 1102.5
B182.8—02	Profile Polyethylene Storm Sewer and Drainage Pipe and Fittings	Table 1102.5
B602—02	Mechanical Couplings for Drain, Waste, and Vent Pipe and Sewer Pipe	705.2.1, 705.5.3, 705.6, 705.7.1, 705.14.1, 705.15, 705.16

FS

Federal Specifications
1941 Jefferson Davis Highway, Suite 104
Arlington, VA 22202

Standard reference number	Title	Referenced In code section number

(DELETE ENTIRE FS SPECIFICATIONS)

TT P1536A (1975)	Federal Specification for Plumbing Fixture Setting Compound	

ISEA

International Safety Equipment Association
1901 North Moore Street, Suite 808
Arlington, VA 22209

Standard reference number	Title	Referenced In code section number
Z358.1—03	Emergency eyewash and shower equipment	411.1

NFPA

National Fire Protection Association
1 Batterymarch Park
Quincy, MA 02269-9101

Standard reference number	Title	Referenced In code section number
51—02	Design and Installation of Oxygen-Fuel Gas Systems for Welding, Cutting, and Allied Processes	1203.1

NSF

NSF International
789 North Dixboro Road
Ann Arbor, MI 48105

Standard reference number	Title	Referenced In code section number
3—2001	Commercial Warewashing Equipment	409.1
14—2003	Plastic Piping System Components and Related Materials	303.3, 611.3
18—1996	Manual Food and Beverage Dispensing Equipment	426.1

2004 SUPPLEMENT TO THE IPC

NSF (continued)

42—2002	Drinking Water Treatment Units — Aesthetic Effects	611.1, 611.3
44—2002	Residential Cation Exchange Water Softeners	611.1, 611.3
53—2002	Drinking Water Treatment Units—Health Effects	611.1, 611.3
58—2002	Reverse Osmosis Drinking Water Treatment Systems	611.2
61—2002	Drinking Water System Components — Health Effects	424.1, 605.3, 605.4, 605.5, 611.3

PDI

Plumbing and Drainage Institute
45 Bristol Drive, Suite 101
South Easton, MA 02375

Standard reference number	Title	Referenced In code section number
G101(2003)	Testing and Rating Procedure for Grease Interceptors with Appendix of Sizing and Installation Data	1003.3.4

2004 SUPPLEMENT TO THE IPC

International Private Sewage Disposal Code

2004 Supplement

INTERNATIONAL PRIVATE SEWAGE DISPOSAL CODE 2004 SUPPLEMENT

CHAPTER 1
ADMINISTRATION

Section 107.4 Coordination of inspection. Delete section without substitution: (PSD1-03/04)

SECTION 202
GENERAL DEFINITIONS

Add new definition to read as shown: (PSD2-03/04)

AIR BREAK (Drainage System). A piping arrangement in which a drain from a fixture, appliance or device discharges indirectly into another fixture, receptacle or interceptor at a point below the flood level rim and above the trap seal.

CHAPTER 3
GENERAL REGULATIONS

Section 302.6 Change to read as shown: (PSD3-03/04)

302.6 Water softener and iron filter backwash. Water softener or iron filter discharge shall be indirectly connected by means of an air gap to the private sewage disposal system or discharge onto the ground surface, provided a nuisance is not created.

CHAPTER 5
MATERIALS

SECTION 505
PIPE, JOINTS AND CONNECTIONS

Sections 505.3.1 and 505.3.2 Change to read as shown: (PSD4-03/04)

505.3.1 Mechanical joints. Mechanical joints on drainage pipes shall be made with an elastomeric seal conforming to ASTM C 1173, ASTM D 3212 or CSA B 602. Mechanical joints shall be installed only in underground systems, unless otherwise approved. Joints shall be installed in accordance with the manufacturer's instructions.

505.3.2 Solvent cementing. Joint surfaces shall be clean and free from moisture. Solvent cement conforming to ASTM D 2235 or CSA 181.1 shall be applied to all joint surfaces. The joint shall be made while the cement is wet. Joints shall be made in accordance with ASTM D 2235, ASTM D 2661, ASTM F 628 or CSA 181.1. Solvent cement joints shall be permitted above or below ground.

Section 505.4 Add new section to read as shown: (PSD4-03/04)

505.4 Coextruded composite ABS pipe and joints. Joints between coextruded composite pipe with an ABS outer layer or ABS fittings shall comply with Sections 505.4.1 and 505.4.2.

505.4.1 Mechanical joints. Mechanical joints on drainage pipe shall be made with an elastomeric seal conforming to ASTM C 1173, ASTM D 3212, or CSA B 602. Mechanical joints shall not be installed in above-ground systems, unless otherwise approved. Joints shall be installed in accordance with the manufacturer's instructions.

505.4.2 Solvent cementing. Joint surfaces shall be clean and free from moisture. Solvent cement conforming to ASTM D 2235 or CSA B 181.1 shall be applied to all joint surfaces. The joint shall be made while the cement is wet. Joints shall be made in accordance with ASTM D 2235, ASTM D 2661, ASTM F 628 or CSA B181.1. Solvent cement joints shall be permitted above or below ground.

(Renumber subsequent sections)

Section 505.5 Bituminized fiber pipe. Delete section without substitution: (PSD4-03/04)

Sections 505.6.1, 505.6.2 and 505.6.3 Change to read as shown: (PSD 4-03/04)

505.6.1 Caulked joints. Joints for hub and spigot pipe shall be firmly packed with oakum or hemp. Molten lead shall be poured in one operation to a depth of not less than 1 inch (25 mm). The lead shall not recede more than 0.125 inch (3.2 mm) below the rim of the hub, and shall be caulked tight. Paint, varnish or other coatings shall not be applied to the joining material until after the joint has been tested and approved. Lead shall be run in one pouring and shall be caulked tight. Acid-resistant rope and acidproof cement shall be permitted.

505.6.2 Mechanical compression joints. Compression gaskets for hub and spigot pipe and fittings shall conform to ASTM C 564. Gaskets shall be compressed when the pipe is fully inserted.

505.6.3 Mechanical joint coupling. Mechanical joint couplings for hubless pipe and fittings shall comply with CISPI 310 or ASTM C 1277. The elastomeric sealing sleeve shall conform to ASTM C 564 or CSA B 602 and shall be provided with a center stop. Mechanical joint couplings shall be installed in accordance with the manufacturer's installation instructions.

2004 SUPPLEMENT TO THE IPSDC

Section 505.7 Change to read as shown: (PSD4-03/04)

505.7 Concrete pipe. Joints between concrete pipe or fittings shall be made by the use of an elastomeric seal conforming to ASTM C 443, ASTM C 1173, CSA A 257.3M or CSA B 602.

Section 505.8 Change to read as shown: (PSD4-03/04)

505.8 Copper or copper-alloy tubing or pipe. Joints between copper or copper-alloy tubing, pipe or fittings shall be in accordance with Sections 505.8.1 and 505.8.2.

505.8.1 Mechanical joints. Mechanical joints shall be installed in accordance with the manufacturer's instructions.

505.8.2 Soldered joints. Solder joints shall be made in accordance with the methods of ASTM B 828. All cut ends shall be reamed to the full inside diameter of the tube end. All joint surfaces shall be cleaned. A flux conforming to ASTM B 813 shall be applied. The joint shall be soldered with a solder conforming to ASTM B 32.

Section 505.10 Change to read as shown: (PSD4-03/04)

505.10 PVC plastic pipe. Joints between polyvinyl chloride (PVC) plastic pipe and fittings shall be in accordance with Sections 505.10.1 and 505.10.2.

505.10.1 Mechanical joints. Mechanical joints shall be made with an elastomeric seal conforming to ASTM C1173, ASTM D3212 or CSA B602. Mechanical joints shall not be installed in above-ground systems, unless otherwise approved. Joints shall be installed in accordance with the manufacturer's instructions.

505.10.2 Solvent cementing. Joint surfaces shall be clean and free from moisture. A purple primer that conforms to ASTM F 656 shall be applied. Solvent cement not purple in color and conforming to ASTM D2564, CSA B137.3, CSA B181.2 or CSA B182.1 shall be applied to all joint surfaces. The joint shall be made while the cement is wet, and shall be in accordance with ASTM D2855. Solvent cement joints shall be permitted above or below ground.

Section 505.11 Add new section to read as shown: (PSD4-03/04)

505.11 Coextruded composite PVC pipe. Joints between coextruded composite pipe with a PVC outerlayer or PVC fittings shall comply with Section 505.11.1 and 505.11.2.

505.11.1 Mechanical joints. Mechanical joints on drainage pipe shall be made with an elastomeric seal conforming to ASTM D3212. Mechanical joints shall not be installed in above-ground systems, unless otherwise approved. Joints shall be installed in accordance with the manufacturer's instructions.

505.11.2 Solvent cementing. Joint surfaces shall be clean and free from moisture. A purple primer that conforms to ASTM F656 shall be applied. Solvent cement not purple in color and conforming to ASTM D2564, CSA B137.3, CSA B181.2 or CSA B 182.1 shall be applied to all joint surfaces. The joint shall be made while the cement is wet, and shall be in accordance with ASTM D2855. Solvent cement joints shall be permitted above or below ground.

(Renumber subsequent sections)

Section 505.12 Change to read as shown: (PSD4-03/04)

505.12 Different piping materials. Joints between different piping materials shall be made with a mechanical joint of the compression or mechanical-sealing type conforming to ASTM C1173, ASTM C1460 or ASTM C1461. Connectors or adapters shall be approved for the application and such joints shall have an elastomeric seal conforming to ASTM C425, ASTM C443, ASTM C564, ASTM C1440, ASTM D1869, ASTM F477, CSA A257.3M or CSA B602, or as required in Sections 505.12.1 and 505.12.2. Joints shall be installed in accordance with the manufacturer's instructions.

2004 SUPPLEMENT TO THE IPSDC

CHAPTER 14
REFERENCED STANDARDS

Change, delete or add the following referenced standards to read as shown: (PSD4-03/04, PSD5-03/04) (STANDARDS NOT SHOWN REMAIN UNCHANGED)

ASTM

ASTM International
100 Barr Harbor Drive
West Conshohocken, PA 19428-2959

Standard reference number	Title	Referenced In code section number
A74—03	Specification for Cast Iron Soil Pipe and Fittings	Table 505.1
B32—00e01	Specification for Solder Metal	505.8.2
B75—02	Specification for Seamless Copper Tube	Table 505.1
B88—02	Specification for Seamless Copper Water Tube	Table 505.1
B251—02	Specification for General Requirements for Wrought Seamless Copper and Copper-Alloy Tube	Table 505.1
B828-02	Practice for Making Capillary Joints by Soldering of Copper and Copper Alloy Tube and Fittings	505.8.2
C4—03	Specification for Clay Drain Tile and Perforated Clay Drain Tile	Table 505.1
C76—02	Specification for Reinforced Concrete Culvert, Storm Drain, and Sewer Pipe	Table 505.1
C425—02	Specification for Compression Joints for Vitrified Clay Pipe and Fittings	505.11, 505.12
C428—97 (2002)	Specification for Asbestos–Cement Nonpressure Sewer Pipe	Table 505.1
C443—02a	Specification for Joints for Concrete Pipe and Manholes, Using Rubber Gaskets	505.7, 505.12
C564—03	Specification for Rubber Gaskets for Cast Iron Soil Pipe and Fittings	505.6.2, 505.6.3, 505.12
C700—02	Specification for Vitrified Clay Pipe, Extra Strength, Standard Strength, and Perforated	Table 505.1
C913—02	Specification for Precast Concrete Water and Waste water Structures	504.2
C1173—02	Specification for Flexible Transition Couplings for Underground Piping Systems	505.3.1, 505.7, 505.10.1, 505.11, 505.13
C1277-03	Specification for Shielding Coupling Joining Hubless Cast Iron Soil Pipe and Fittings	505..6.3
C1440-99e01	Specification for Thermoplastic Elastomeric (TPE) Gasket Materials for Drain, Waste, and Vent (DWV), Sewer, Sanitary and Storm Plumbing Systems	505.13
C1460-00	Specification for Shielded Transition Couplings for Use with Dissimilar DWV Pipe and Fittings Above Ground	505.13
C1461-02	Specification for Mechanical Couplings Using Thermoplastic Elastomeric (TPE) Gaskets for Joining Drain, Waste, and Vent (DWV) Sewer, Sanitary and Storm Plumbing Systems for Above and Below Ground Use	505.13
D2564—02	Specification for Solvent Cements for Poly (Vinyl Chloride) (PVC) Plastic Piping Systems	505.10.2, 505.11.2
D2661—02	Specification for Acrylonitrile–Butadiene–Styrene (ABS) Schedule 40 Plastic Drain, Waste, and Vent Pipe and Fittings	Table 505.1, 505.3.2
D2665—02a	Specification for Poly (Vinyl Chloride) (PVC) Plastic Drain, Waste, and Vent Pipe and Fittings	Table 505.1
D2855—96 (2002)	Standard Practice for Making Solvent–Cemented Joints with Poly (Vinyl Chloride) (PVC) Pipe and Fittings	505.10.2, 505.11.2
D2949—01a	Specification for 3.25–In. Outside Diameter Poly (Vinyl Chloride) (PVC) Plastic Drain, Waste, and Vent Pipe and Fittings	Table 505.1

IPSDC-3

2004 SUPPLEMENT TO THE IPSDC

ASTM (continued)

D3212-96a	Specification for Joints for Drain and Sewer Plastics Using Flexible Elastomeric Seals	505.3.1, 505.10.1, 505.11.1
D4021—92	DELETED	
F477—02e01	Specification for Elastomeric Seals (Gaskets) for Joining Plastic Pipe	505.12
F656—02	Specification for Primers for Use in Solvent Cement Joints of Poly (Vinyl Chloride) (PVC) Plastic Pipe and Fittings	505.10.2
F891—00e01	Specification for Coextruded Poly (Vinyl Chloride) (PVC) Plastic Pipe with a Cellular Core	Table 505.1
F1488—00e01	Specification for Coextruded Composite Pipe	Table 505.1, Table 505.1.1

CISPI

Cast Iron Soil Pipe Institute
5959 Shallowford Road, Suite 419
Chattanooga, TN 37421

Standard reference number	Title	Referenced In code section number
310-97	Specification for Coupling for Use in Connection with Hubless Cast Iron Soil Pipe and Fittings for Sanitary and Storm Drain, Waste and Vent Piping Applications	505.6.3

CSA

Canadian Standards Association
5060 Spectrum Way, Suite 100
Mississauga, Ontario, Canada L4W 5N6

Standard reference number	Title	Referenced In code section number
B137.3—02	Rigid Poly Vinyl Chloride (PVC) Pipe for Pressure Applications	505.10.2, 505.11.2
B181.1—02	ABS Drain, Waste, and Vent Pipe and Pipe Fittings	505.3.2
B181.2—02	PVC Drain, Waste, and Vent Pipe and Pipe Fittings	505.10.2, 505.11.2
B182.1—02	Plastic Drain and Sewer Pipe and Pipe Fittings	505.10.2, 505.11.2
B182.2—02	PVC Sewer Pipe and Fittings (PSM Type)	Table 505.1
B182.4—02	Profile PVC Sewer Pipe and Fittings	Table 505.1
B602—02	Mechanical Couplings for Drain, Waste, and Vent Pipe and Sewer Pipe	505.3.1, 505.6.3, 505.7, 505.10.1, 505.11, 505.12

UL

Underwriters Laboratories, Inc.
333 Pfingsten Road
Northbrook, IL 60062-2096

Standard reference number	Title	Referenced In code section number
70—01	Septic Tanks, Bituminous Coated Material	504.3

International Property Maintenance Code

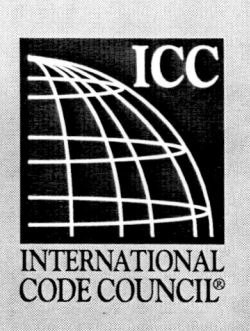

2004 Supplement

INTERNATIONAL PROPERTY MAINTENANCE CODE 2004 SUPPLEMENT

Chapter 1
ADMINISTRATION

Section 104.8 Coordination of inspections Delete without substitution: (PM1-03/04)

Section 106.3 Change to read as shown: (PM2-03/04)

106.3 Prosecution of violation. Any person failing to comply with a notice of violation or order served in accordance with Section 107 shall be deemed guilty of a misdemeanor or civil infraction as determined by the local municipality, and the violation shall be deemed a strict liability offense. If the notice of violation is not complied with, the code official shall institute the appropriate proceeding at law or in equity to restrain, correct or abate such violation, or to require the removal or termination of the unlawful occupancy of the structure in violation of the provisions of this code or of the order or direction made pursuant thereto. Any action taken by the authority having jurisdiction on such premises shall be charged against the real estate upon which the structure is located and shall be a lien upon such real estate.

CHAPTER 3
GENERAL REQUIREMENTS

Section 304.14 Change to read as shown: (PM3-03/04)

304.14 Insect screens. During the period from [DATE] to [DATE], every door, window and other outside opening required for ventilation of habitable rooms, food preparation areas, food service areas, or any areas where products to be included or utilized in food for human consumption are processed, manufactured, packaged or stored, shall be supplied with approved tightly fitting screens of not less than 16 mesh per inch (16 mesh per 25 mm) and every screen door used for insect control shall have a self-closing device in good working condition.

> **Exception:** Screens shall not be required where other approved means, such as air curtains or insect repellent fans, are employed.

Section 403.3 Change to read as shown: (PM5-03/04)

403.3 Cooking facilities. Unless approved through the certificate of occupancy, cooking shall not be permitted in any rooming unit or dormitory unit, and a cooking facility or appliance shall not be permitted to be present in the rooming unit or dormitory unit.

> **Exceptions:**
> 1. Where specifically approved in writing by the code official.
> 2. Devices such as coffee pots and microwave ovens shall not be considered cooking appliances.

Sections 404.4 and 404.4.1 Delete and substitute as shown: (PM6-03/04)

404.4 Bedroom and living room requirements. Every bedroom and living room shall comply with the requirements of Sections 404.4.1 through 404.4.5.

404.4.1 Room area. Every living room shall contain at least 120 square feet (11.2 m^2) and every bedroom shall contain at least 70 square feet (6.5 m^2).

Section 404.5 Delete and substitute as shown: (PM6-03/04)

404.5 Overcrowding. The number of persons occupying a dwelling unit shall not create conditions that, in the opinion of the building official, endanger the life, health, safety, or welfare of the occupants.

Table 405 Minimum Area Requirements Delete table without substitution: (PM6-03/04)

CHAPTER 6
MECHANICAL AND ELECTRICAL REQUIREMENTS

Section 605.3 Change to read as shown: (EL3-03/04)

605.3 Luminaires. Every public hall, interior stairway, toilet room, kitchen, bathroom, laundry room, boiler room and furnace room shall contain at least one electric luminaire.

International Residential Code

For One- and Two-Family Dwellings

2004 Supplement

INTERNATIONAL RESIDENTIAL CODE 2004 SUPPLEMENT

PART I - Administrative

CHAPTER 1
ADMINISTRATION

Section R101.2 Change to read as shown: (RB1-03/04)

R101.2 Scope. The provisions of the *International Residential Code for One- and Two-family Dwellings* shall apply to the construction, alteration, movement, enlargement, replacement, repair, equipment, use and occupancy, location, removal and demolition of detached one- and two-family dwellings and multiple single-family dwellings (townhouses) not more than three stories above-grade in height with a separate means of egress and their accessory structures.

Exception: Existing buildings undergoing repair, alteration or additions, and change of occupancy shall be permitted to comply with the *International Existing Building Code.*

Section R105.2 Change to read as shown: (RB5-03/04, RB6-03/04 & RB8-03/04)

R105.2 Work exempt from permit. Permits shall not be required for the following. Exemption from permit requirements of this code shall not be deemed to grant authorization for any work to be done in any manner in violation of the provisions of this code or any other laws or ordinances of this jurisdiction.

Building:

1. One-story detached accessory structures used as tool and storage sheds, playhouses and similar uses, provided the floor area does not exceed 120 square feet (11.15 m^2).

2. Fences not over 6 feet (1829 mm) high.

3. Retaining walls that are not over 4 feet (1219 mm) in height measured from the bottom of the footing to the top of the wall, unless supporting a surcharge.

4. Water tanks supported directly upon grade if the capacity does not exceed 5,000 gallons (18 927 L) and the ratio of height to diameter or width does not exceed 2 to 1.

5. Sidewalks and driveways.

6. Painting, papering, tiling, carpeting, cabinets, counter tops and similar finish work.

7. Prefabricated swimming pools that are less than 24 inches (610 mm) deep.

8. Swings and other playground equipment.

9. Window awnings supported by an exterior wall which do not project more than 54 inches (1372 mm) from the exterior wall and do not require additional support.

Electrical:

Repairs and maintenance: A permit shall not be required for minor repair work, including the replacement of lamps or the connection of approved portable electrical equipment to approved permanently installed receptacles.

Gas:

1. Portable heating, cooking or clothes drying appliances.

2. Replacement of any minor part that does not alter approval of equipment or make such equipment unsafe.

3. Portable-fuel-cell appliances that are not connected to a fixed piping system and are not interconnected to a power grid.

Mechanical:

1. Portable heating appliance.

2. Portable ventilation appliances.

3. Portable cooling units.

4. Steam, hot or chilled water piping within any heating or cooling equipment regulated by this code.

5. Replacement of any minor part that does not alter approval of equipment or make such equipment unsafe.

6. Portable evaporative coolers.

7. Self-contained refrigeration systems containing 10 pounds (4.54 kg) or less of refrigerant or that are actuated by motors of 1 horsepower (746 W) or less.

8. Portable-fuel-cell appliances that are not connected to a fixed piping system and are not interconnected to a power grid.

The stopping of leaks in drains, water, soil, waste or vent pipe; provided, however, that if any concealed trap, drainpipe, water, soil, waste or vent pipe becomes defective and it becomes necessary to remove and replace the same with new material, such work shall be considered as new work and a permit shall be obtained and inspection made as provided in this code.

The clearing of stoppages or the repairing of leaks in pipes, valves or fixtures, and the removal and reinstallation of water closets, provided such repairs do not involve or require the replacement or rearrangement of valves, pipes or fixtures.

Section R105.3.1.1 Change heading to read as shown: (CCC)

R105.3.1.1 Determination of substantially improved or substantially damaged existing buildings in flood hazard areas. For applications for reconstruction, rehabilitation, addition or other improvement of existing buildings or structures located in an area prone to flooding as established by Table R301.2(1), the building official shall examine or cause to be examined the construction documents and shall prepare a finding with regard to the value of the proposed work. For buildings that have sustained damage of any origin, the value of the proposed work shall include the cost to repair the building or structure to its predamage condition. If the building official finds that the value of proposed work equals or exceeds 50 percent of the market value of the building or structure before the damage has occurred or the improvement is started, the finding shall be provided to the board of appeals for a determination of substantial improvement or substantial damage. Applications determined by the board of appeals to constitute substantial improvement or substantial damage shall meet the requirements of Section R323.

Section R106.1.3 Change to read as shown: (RB10-03/04)

R106.1.3 Information for construction in flood hazard areas. For buildings and structures located in whole or in part in flood hazard areas as established by Table R301.2(1), construction documents shall include:

1. Delineation of flood hazard areas, floodway boundaries and flood zones and the design flood elevation, as appropriate;

2. The elevation of the proposed lowest floor, including basement; in areas of shallow flooding (AO zones), the height of the proposed lowest floor, including basement, above the highest adjacent grade; and

3. The elevation of the bottom of the lowest horizontal structural member in coastal high hazard areas (V Zone); and

4. If design flood elevations are not included on the community's Flood Insurance Rate Map (FIRM), the building official and the applicant shall obtain and reasonably utilize any design flood elevation and floodway data available from other sources.

Section 106.3.1 Change to read as shown: (RB11-03/04)

R106.3.1 Approval of construction documents. When the building official issues a permit, the construction documents shall be approved, in writing or by a stamp which states "APPROVED PLANS PER IRC R106.3.1." One set of construction documents so reviewed shall be retained by the building official. The other set shall be returned to the applicant, shall be kept at the site of work and shall be open to inspection by the building official or his or her authorized representative.

Section R108.2 Change to read as shown: (G11-03/04)

108.2 Schedule of permit fees. On buildings, structures, electrical, gas, mechanical and plumbing systems or alterations requiring a permit, a fee for each permit shall be paid as required, in accordance with the schedule as established by the applicable governing authority. See Appendix L.

Section R109.1.1 Change to read as shown: (RB14-03/04)

R109.1.1 Foundation inspection. Inspection of the foundation shall be made after poles or piers are set or trenches or basement areas are excavated and any required forms erected and any required reinforcing steel is in place and supported prior to the placing of concrete. The foundation inspection shall include excavations for thickened slabs intended for the support of bearing walls, partitions, structural supports, or equipment and special requirements for wood foundation.

Section R110.1 Change to read as shown: (RB15-03/04)

R110.1 Use and occupancy. No building or structure shall be used or occupied, and no change in the existing occupancy classification of a building or structure or portion thereof shall be made until the building official has issued a certificate of occupancy therefor as provided herein. Issuance of a certificate of occupancy shall not be construed as an approval of a violation of the provisions of this code or of other ordinances of the jurisdiction. Certificates presuming to give authority to violate or cancel the provisions of this code or other ordinances of the jurisdiction shall not be valid.

Exceptions:

1. Certificates of occupancy are not required for work exempt from permits under Section R105.2.

2. Accessory buildings or structures.

Section R110.3 Change to read as shown: (RB16-03/04)

R110.3 Certificate issued. After the building official inspects the building or structure and finds no violations of the provisions of this code or other laws that are enforced by the department of building safety, the building official shall issue a certificate of occupancy which shall contain the following:

1. The building permit number.
2. The address of the structure.
3. The name and address of the owner.
4. A description of that portion of the structure for which the certificate is issued.
5. A statement that the described portion of the structure has been inspected for compliance with the requirements of this code.
6. The name of the building official.
7. The edition of the code under which the permit was issued.
8. If an automatic sprinkler system is provided and whether the sprinkler system is required.
9. Any special stipulations and conditions of the building permit.

Part II—Definitions

CHAPTER 2
DEFINITIONS

Section R202 Change the definition of "Approved" to read as shown: (RB19-03/04)

APPROVED. Approved refers to approval by the building official as the result of investigation and tests conducted by him or her, or by reason of accepted principles or tests by an approved agency.

Section R202 Change the definition of "Accessory Structure" to read as shown: (RB17-03/04)

ACCESSORY STRUCTURE. A structure not greater than 3,000 square feet (279 m^2) in floor area, and not over two stories in height, the use of which is customarily accessory to and incidental to that of the dwelling(s) and which is located on the same lot.

Section R202 Change the definition of "Branch Interval" to read as shown: (P2-03/04

BRANCH INTERVAL. A vertical measurement of distance, 8 feet (2438 mm) or more in developed length, between the connections of horizontal branches to a drainage stack. Measurements are taken down the stack from the highest horizontal branch connection.

Section R202 Change definition of "Exterior Wall" to read as shown: (RB21-03/04)

EXTERIOR WALL. An above-grade wall that defines the exterior boundaries of a building. Includes between-floor spandrels, peripheral edges of floors, roof and basement knee walls, dormer walls, gable end walls, walls enclosing a mansard roof and basement walls with an average below-grade wall area that is less than 50 percent of the total opaque and nonopaque area of that enclosing side.

Section R202 Add new definition to read as shown: (P51-03/04)

GRIDDED WATER DISTRIBUTION SYSTEM. A water distribution system where every water distribution pipe is interconnected so as to provide two or more paths to each fixture supply pipe.

Section R202 Delete definition of "RESIDENTIAL BUILDING TYPE" without substitution. (EC48-03/04)

Section R202 Change the definition of "Sunroom" to read as shown: (EC48-03/04)

SUNROOM. A one-story structure attached to a dwelling with a glazing area in excess of 40 percent of the gross area of the structure's exterior walls and roof.

Section R202 Change the definition of "Thermal Isolation" to read as shown: (EC48-03/04)

THERMAL ISOLATION. Physical and space conditioning separation from conditioned space(s). The conditioned space(s) shall be controlled as separate zones for heating and cooling or conditioned by separate equipment.

Section R202 Add new definition to read as shown: (RB25-03/04)

WALL, RETAINING. A wall not laterally supported at the top, that resists lateral soil load and other imposed loads.

Part III—Building Planning and Construction

CHAPTER 3
BUILDING PLANNING

Section R301.2.1 Change to read as follows: (RB28-03/04 & RB30-03/04)

R301.2.1. Wind limitations. Buildings and portions thereof shall be limited by wind speed, as defined in Table R301.2(1) and construction methods in accordance with this code. Basic wind speeds shall be determined from Figure R301.2(4). Where different construction methods and structural materials are used for various portions of a building, the applicable requirements of this section for each portion shall apply. Where loads for wall coverings, curtain walls, roof coverings, exterior windows, skylights, garage doors and exterior doors are not otherwise specified, the loads listed in Table R301.2(2) adjusted for height and exposure using Table R301.2(3) shall be used to determine design load performance requirements for wall coverings, curtain walls, roof coverings, exterior windows, skylights, garage doors and exterior doors. Asphalt shingles shall be designed for wind speeds in accordance with Section R905.2.6.

Section R301.2.1.2 Change to read as shown: (RB34-03/04, RB35-03/04, RB36-03/04 & S17-03/04)

R301.2.1.2 Protection of Openings. Windows in buildings located in wind-borne debris regions shall have glazed openings protected from wind-borne debris. Glazed opening protection for wind-borne debris shall meet the requirements of the Large Missile Test of an approved impact resisting standard or ASTM E 1996 and ASTM E 1886 referenced therein.

> **Exception:** Wood structural panels with a minimum of 7/16 inch (11.1 mm) and a maximum span of 8 feet (2438 mm) shall be permitted for opening protection in one- and-two-story buildings. Panels shall be precut so that they shall be attached to the framing surrounding the opening containing the product with the glazed opening. Panels shall be secured with the attachment hardware provided. Attachments shall be designed to resist the component and cladding loads determined in accordance with either Table R301.2(2) or Section 1609.6.5 of the *International Building Code*. Attachment in accordance with Table R301.2.1.2 is permitted for buildings with a mean roof height of 33 feet (10 058 mm) or less where wind speeds do not exceed 130 miles per hour (58 m/s).

2004 SUPPLEMENT TO THE IRC

Figure R301.2(2) Change figure to read as shown: (RB31-03/04) (CCC)

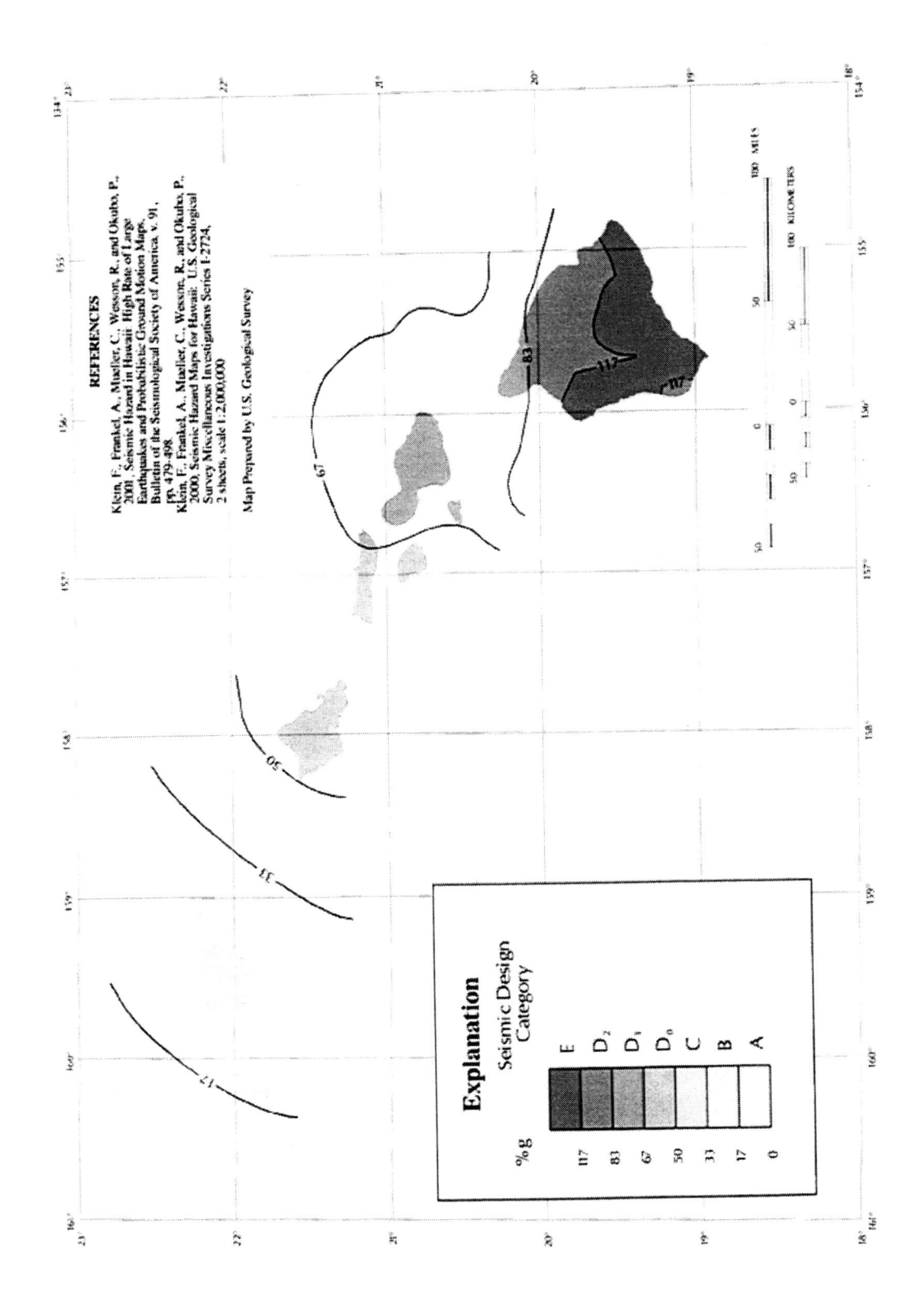

FIGURE R301.2(2)
SEISMIC DESIGN CATEGORIES—SITE CLASS D

IRC-5

2004 SUPPLEMENT TO THE IRC

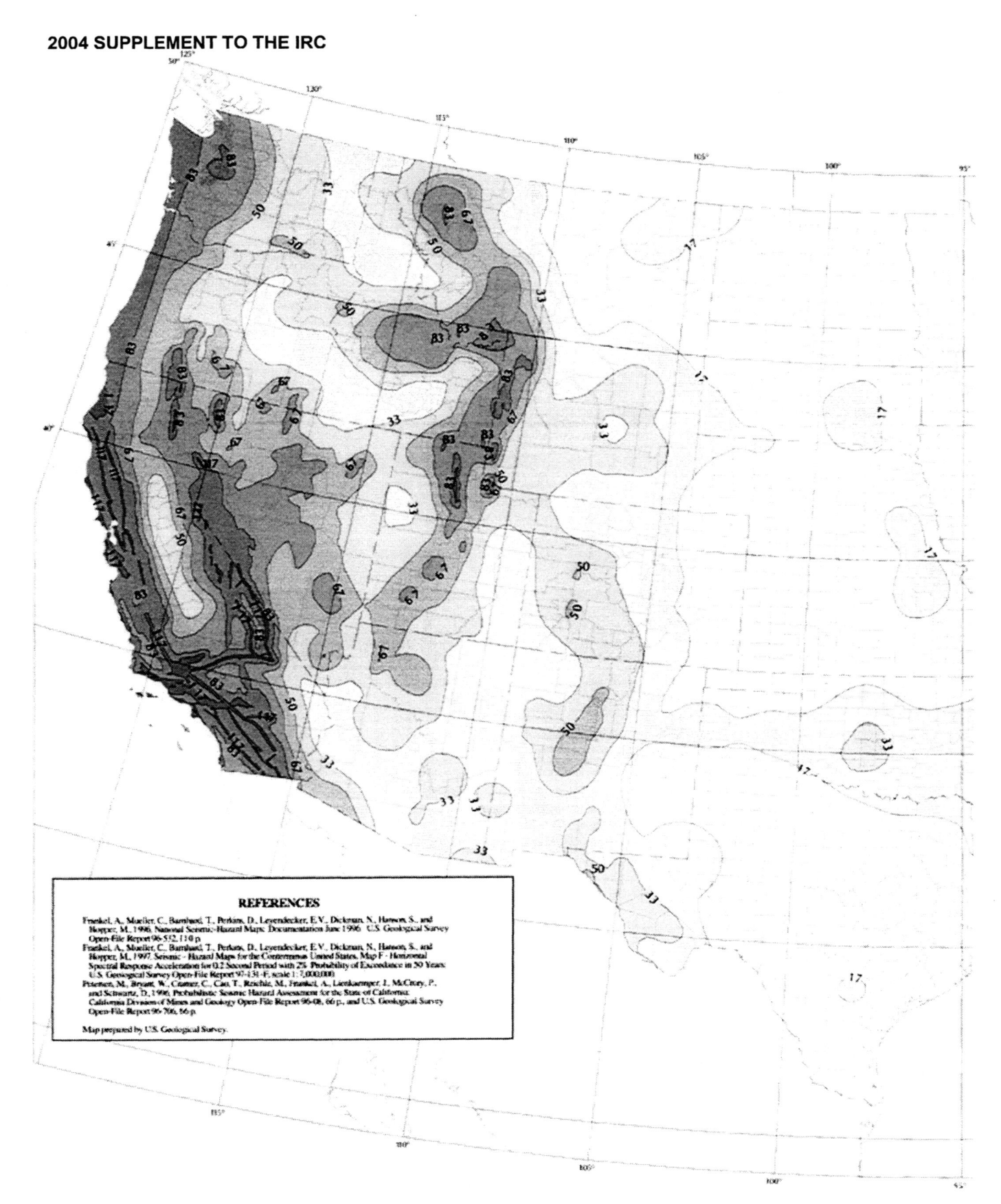

FIGURE R301.2(2)—continued
SEISMIC DESIGN CATEGORIES—SITE CLASS D

2004 SUPPLEMENT TO THE IRC

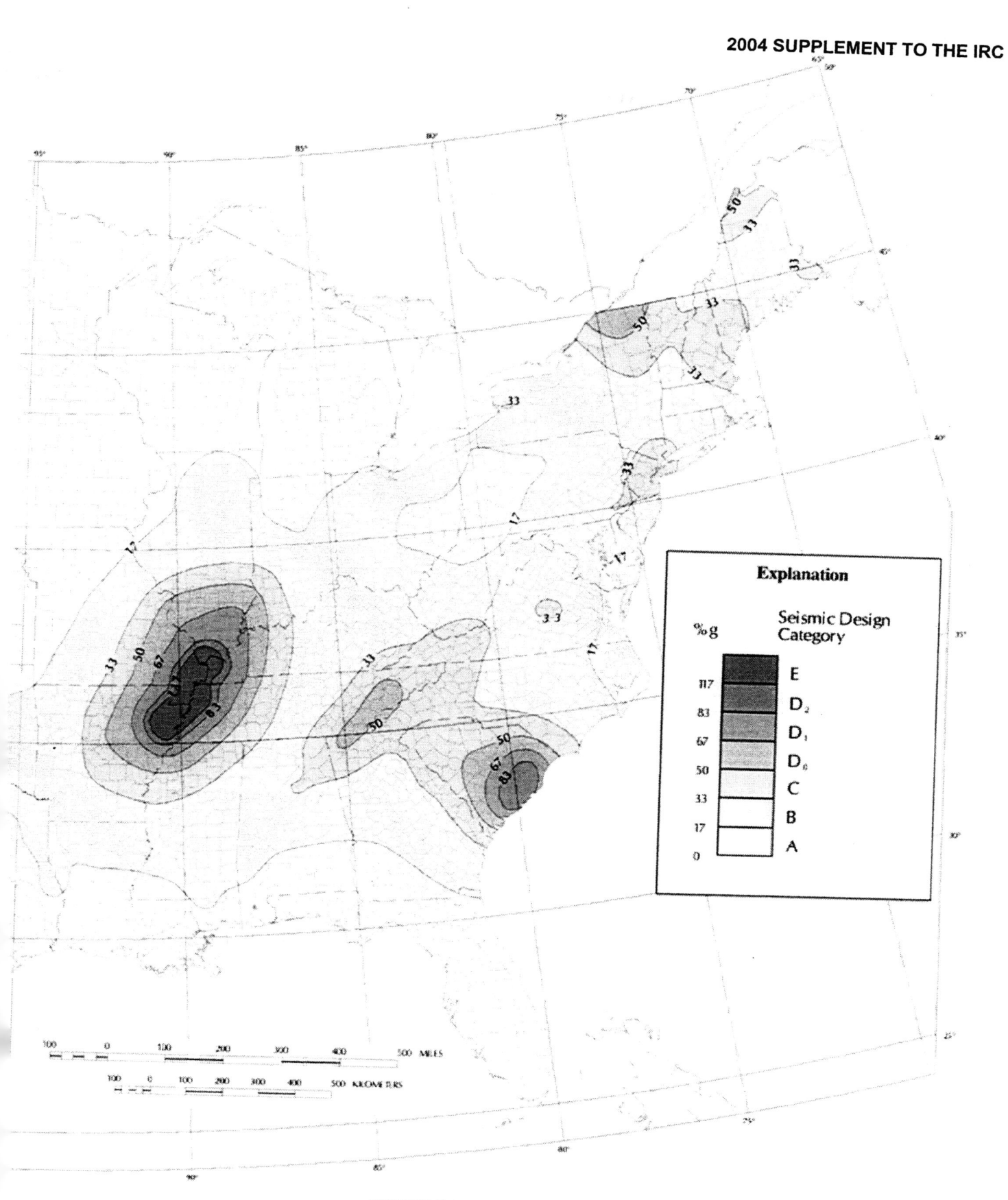

FIGURE R301.2(2)—continued
SEISMIC DESIGN CATEGORIES—SITE CLASS D

IRC-7

2004 SUPPLEMENT TO THE IRC

FIGURE R301.2(2)
SEISMIC DESIGN CATEGORIES—SITE CLASS D

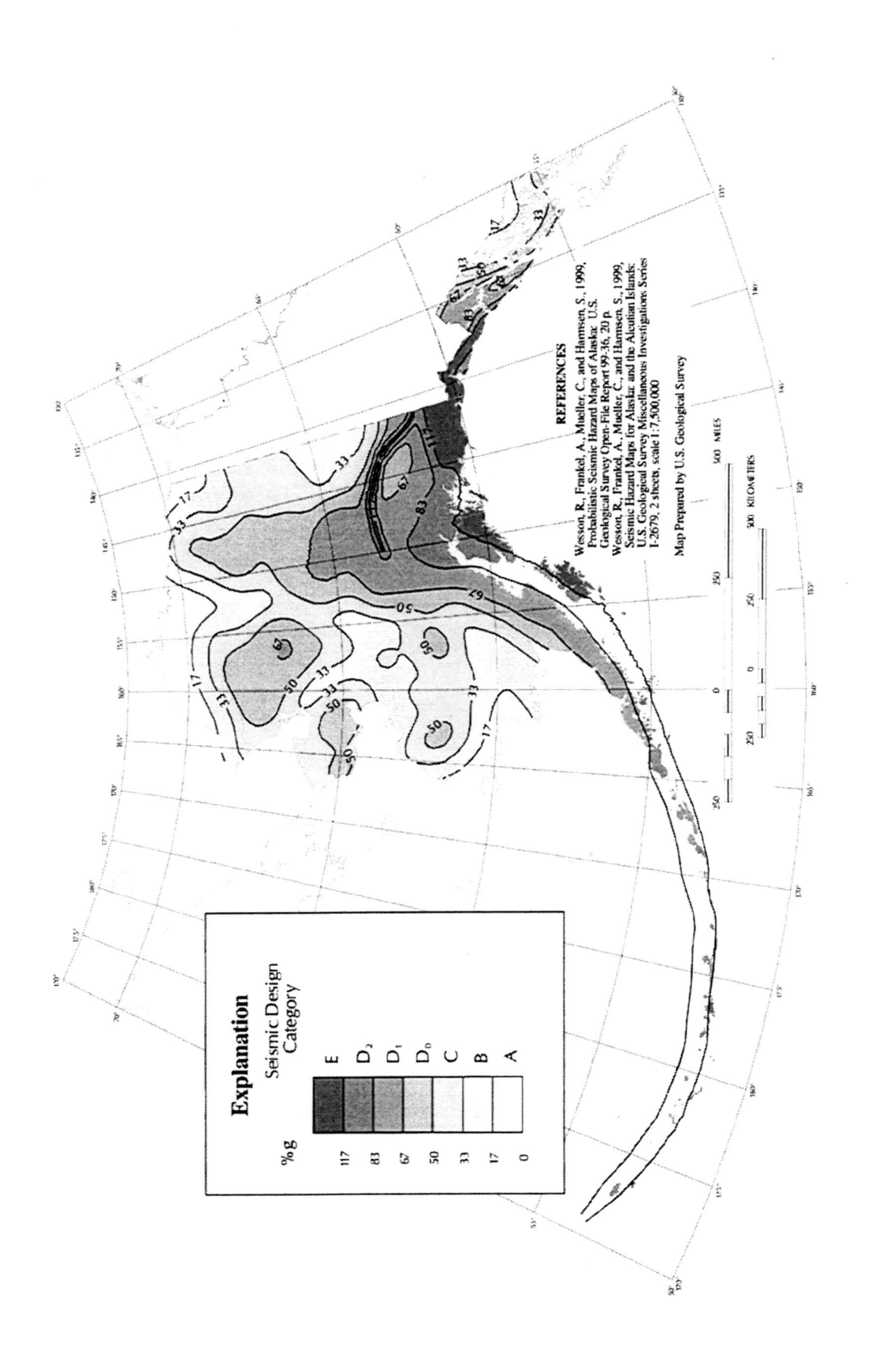

301.2.1.2 Change table text and table notes to read as shown: (S17-03/04)

TABLE R301.2.1.2
WINDBORNE DEBRIS PROTECTION FASTENING
SCHEDULE FOR WOOD STRUCTURAL PANELS [a,b,c,d]

FASTENER TYPE	FASTENER SPACING (inches)		
	Panel Span ≤ 4 feet	4 feet < Panel Span ≤ 6 feet	6 feet < Panel Span ≤ 8 feet
No. 6 Screws	16	12	9
No. 8 Screws	16	16	12

a. This table is based on 130 mph and a 33-foot mean roof height.
b. Fasteners shall be installed at opposing ends of the wood structural panel. Fasteners shall be located a minimum of 1 inch from the edge of the panel.
c. Fasteners shall be long enough to penetrate through the exterior wall covering and a minimum of 1 1/4 inch into wood wall framing and a minimum of 1 1/4 inch into concrete block or concrete, and into steel framing a minimum of 3 exposed threads. Fasteners shall be located a minimum of 2 ½ inch from the edge of concrete block or concrete.
d. Where screws are attached to masonry or masonry/stucco, they shall be attached utilizing vibration-resistant anchors having a minimum ultimate withdrawal capacity of 490 pounds.

Section R301.2.2 Change to read as shown: (RB31-03/04)

R301.2.2 Seismic provisions. The seismic provisions of this code shall apply to buildings constructed in Seismic Design Categories C, D_0, D_1 and D_2, as determined in accordance with this section. Buildings in Seismic Design Category E shall be designed in accordance with the *International Building Code*, except when the Seismic Design Category is reclassified to a lower Seismic Design Category in accordance with Section R301.2.2.1.

Exception: Detached one- and two-family dwellings located in Seismic Design Category C are exempt from the seismic requirements of this code.

The weight and irregularity limitations of Section R301.2.2.2 shall apply to buildings in all Seismic Design Categories regulated by the seismic provisions of this code. Buildings in Seismic Design Category C shall be constructed in accordance with the additional requirements of Sections R301.2.2.3. Buildings in Seismic Design Categories D_0, D_1, and D_2 shall be constructed in accordance with the additional requirements of Section R301.2.2.4

Table R301.2.2.1.1 Change to read as shown: (RB31-03/04)

TABLE R301.2.2.1.1
SEISMIC DESIGN CATEGORY DETERMINATION

CALCULATED S_{DS}	SEISMIC DESIGN CATEGORY
S_{DS} ≤ 0.17 g	A
0.17g < S_{DS} ≤ 0.33 g	B
0.33g < S_{DS} ≤ 0.50 g	C
0.50g < S_{DS} ≤ 0.67 g	D_0
0.67g < S_{DS} ≤ 0.83 g	D_1
0.83g < S_{DS} ≤ 1.17 g	D_2
1.17g < S_{DS}	E

2004 SUPPLEMENT TO THE IRC

Section R301.2.2.2.2 Change to read as shown: (RB31-03/04)

R301.2.2.2.2 Irregular buildings. Concrete construction complying with Section R611 or R612 and conventional light-frame construction shall not be used in irregular portions of structures in Seismic Design Categories C, D_0, D_1 and D_2. Only such irregular portions of structures shall be designed in accordance with accepted engineering practice to the extent such irregular features affect the performance of the conventional framing system. A portion of a building shall be considered to be irregular when one or more of the following conditions occur:

1. When exterior shear wall lines or braced wall panels are not in one plane vertically from the foundation to the uppermost story in which they are required.

    Exception: For wood light-frame construction, floors with cantilevers or setbacks not exceeding four times the nominal depth of the wood floor joists are permitted to support braced wall panels that are out of plane with braced wall panels below provided that:

    1. Floor joists are nominal 2 inches by 10 inches (51 mm by 254 mm) or larger and spaced not more than 16 inches (406 mm) on center.
    2. The ratio of the back span to the cantilever is at least 2 to 1.
    3. Floor joists at ends of braced wall panels are doubled.
    4. For wood-frame construction, a continuous rim joist is connected to ends of all cantilever joists. When spliced, the rim joists shall be spliced using a galvanized metal tie not less than 0.058 inch (1.47 mm) (16 gage) and 1/2 inches (38 mm) wide fastened with six 16d nails on each side of the splice or a block of the same size as the rim joist of sufficient length to fit securely between the joist space at which the splice occurs fastened with eight 16d nails on each side of the splice; and
    5. Gravity loads carried at the end of cantilevered joists are limited to uniform wall and roof load and the reactions from headers having span of 8 feet (2438 mm) or less.

2. When a section of floor or roof is not laterally supported by shear walls or braced wall lines on all edges.

    Exception: Portions of floors that do not support shear walls or braced wall panels above, or roofs, shall be permitted to extend no more than 6 feet (1829 mm) beyond a shear wall or braced wall line.

3. When the end of a braced wall panel occurs over an opening in the wall below and ends at a horizontal distance greater than 1 foot (305 mm) from the edge of the opening. This provision is applicable to shear walls and braced wall panels offset in plane and to braced wall panels offset out of plane as permitted by the exception to Item 1 above.

    Exception: For wood light-frame wall construction, one end of a braced wall panel shall be permitted to extend more than 1 foot (305 mm) over an opening of not more than 8 feet (2438 mm) in width in the wall below provided that the opening includes a header in accordance with the following:

    1. The building width, loading condition and member species limitations of Table R502.5(1) shall apply and
    2. Not less than 1-2x12 or 2-2x10 for an opening not more than 4 feet in width or
    3. Not less than 2-2x12 or 3-2x10 for an opening not more than 6 feet (1829 mm) in width or
    4. Not less than 3-2x12 or 4-2x10 for an opening not more than 8 feet (2438 mm) in width and
    5. The entire length of the braced wall panel shall not occur over an opening in the wall below.

4. When an opening in a floor or roof exceeds the lesser of 12 feet (3657 mm) or 50 percent of the least floor or roof dimension.

5. When portions of a floor level are vertically offset.

    Exceptions:

    1. Framing supported directly by continuous foundations at the perimeter of the building.
    2. For wood light-frame construction, floors shall be permitted to be vertically offset when the floor framing is lapped or tied together as required by Section R502.6.1.

6. When shear walls and braced wall lines do not occur in two perpendicular directions.

7. When stories above-grade partially or completely braced by wood wall framing in accordance with Section R602 or steel wall framing in accordance with Section R603 include masonry or concrete construction.

   Exception: Fireplaces, chimneys and masonry veneer as permitted by this code.

   When this irregularity applies, the entire story shall be designed in accordance with accepted engineering practice.

Sections R301.2.2.4 Change to read as shown: (RB31-03/04)

R301.2.2.4 Seismic Design Categories D_0, D_1 and D_2. Structures assigned to Seismic Design Categories D_0, D_1 and D_2 shall conform to the requirements for Seismic Design Category C and the additional requirements of this section.

Section R301.2.2.4.2 and R301.2.2.4.3 Change to read as shown: (RB31-03/04)

R301.2.2.4.2 Anchored stone and masonry veneer. Buildings with anchored stone and masonry veneer shall be designed in accordance with accepted engineering practice.

Exceptions:

1. In Seismic Design Categories D_0 and D_1, exterior masonry veneer with a maximum nominal thickness of 4 inches (102 mm) is permitted in accordance with Section R703.7, Exception 3.

2. In Seismic Design Category D_2, exterior masonry veneer with a maximum actual thickness of 3 inches (76 mm) is permitted in accordance with Section R703.7, Exception 4.

R301.2.2.4.3 Masonry construction. Masonry construction in Seismic Design Categories D_0 and D_1 shall comply with the requirements of Section R606.11.3. Masonry construction in Seismic Design Category D_2 shall comply with the requirements of Section R606.11.4.

Section R301.2.2.4.5 Change to read as shown: (RB31-03/04)

R301.2.2.4.5 Cold-formed steel framing in Seismic Design Categories D_0, D_1 and D_2. In Seismic Design Categories D_0, D_1 and D_2 in addition to the requirements of this code, cold-formed steel framing shall comply with the requirements of COFS/PM.

Section R301.2.4 Change to read as shown: (RB10-03/04)

R301.2.4 Floodplain construction. Buildings and structures constructed in whole or in part in flood hazard areas (including A or V Zones) as established in Table R301.2(1) shall be designed and constructed in accordance with Section R323.

Exception: All buildings and structures located in whole or in part in identified floodways as established in Table R301.2(1) shall be designed and constructed as stipulated in the *International Building Code*.

Table R301.5 Change Footnote b and add new table footnotes as shown: (RB42-03/04, S13-03/04 and S14-03/04)

TABLE R301.5
MINIMUM UNIFORMLY DISTRIBUTED
LIVE LOADS
(in pounds per square foot)

USE	LIVE LOAD
Attics with limited storage [b, g, h]	20
Attics without storage [b]	10
Guardrails and handrails [d]	200[i]
Guardrails in-fill components [i]	50[i]

(Portions of table not shown do not change)

a. (No change to current text)
b. Attics without storage are those where the maximum clear height between joist and rafter is less than 42 inches, or where there are not two or more adjacent trusses with the same web configuration capable of containing a rectangle 42 inches high by 2 feet wide, or greater, located within the plane of the truss. For attics without storage, this live load need not be assumed to act concurrently with any other live load requirements.
c through f (No change to current text)
g. For attics with limited storage and constructed with trusses, this live load need be applied only to those portions of the bottom chord of not less than two adjacent trusses with the same web configuration containing a rectangle 42 inches high or greater by 2 feet wide or greater, located within the plane of the truss. The rectangle shall fit between the top of the bottom chord and the bottom of any other truss member, provided that each of the following criteria is met:
   1. The attic area is accessible by a pull-down stairway or framed opening in accordance with Section R807.1; and
   2. The truss shall have a bottom chord pitch less than 2:12.
h. Attic spaces served by a fixed stair shall be designed to support the minimum live load specified for sleeping rooms.
i. Glazing used in handrail assemblies and guards shall be designed with a safety factor of 4. The safety factor shall be applied to each the concentrated load-applied to the top of the rail, and to the load on the in-fill components. These loads shall be determined independent of one another, and loads are assumed not to occur with any other live load.

2004 SUPPLEMENT TO THE IRC

Sections R302.1, R302.2 and R302.3 Change to read as shown: (G128-03/04)

R302.1 Exterior walls. Exterior walls with a fire separation distance less than 5 feet (1524 mm) shall have not less than a 1-hour fire-resistance rating with exposure from both sides. Projections shall not extend to a point closer than 4 feet (1219 mm) from the line used to determine the fire separation distance.

> **Exception:** Detached garages accessory to a dwelling located within 2 feet (610 mm) of a lot line may have roof eave projections not exceeding 4 inches (102 mm).

Projections extending into the fire separation distance shall have not less than 1-hour fire-resistant construction on the underside. The above provisions shall not apply to walls that are perpendicular to the line used to determine the fire separation distance.

> **Exception:** Detached tool and storage sheds, playhouses and similar structures exempted from permits are not required to provide wall protection based on location on the lot. Projections beyond the exterior wall shall not extend over the lot line.

R302.2 Openings. Openings shall not be permitted in the exterior wall of a dwelling or accessory building with a fire separation distance less than 3 feet (914 mm). Openings in excess of 25 percent of the area of the wall shall not be permitted in the exterior wall of a dwelling or accessory building with a fire separation distance between 3 and 5 feet (914 and 1524 mm). This distance shall be measured perpendicular to the line used to determine the fire separation distance.

> **Exceptions:**
> 1. Openings shall be permitted in walls that are perpendicular to the line used to determine the fire separation distance.
> 2. Foundation vents installed in compliance with this code are permitted.

R302.3 Penetrations. Penetrations located in the exterior wall of a dwelling with a fire separation distance less than 5 feet (1524 mm) shall be protected in accordance with Section R317.3.

> **Exception:** Penetrations shall be permitted in walls that are perpendicular to the line used to determine the fire separation distance.

Section R303.1 Change to read as shown (RB47-03/04):

R303.1 Habitable Rooms. All habitable rooms shall be provided with aggregate glazing area of not less than 8 percent of the floor area of such rooms. Natural ventilation shall be through windows, doors, louvers or other approved openings to the outdoor air. Such openings shall be provided with ready access or shall otherwise be readily controllable by the building occupants. The minimum openable area to the outdoors shall be 4 percent of the floor area being ventilated.

> **Exceptions:**
> 1. The glazed areas need not be openable where the opening is not required by Section R310 and an approved mechanical ventilation system is provided capable of producing 0.35 air change per hour in the room or a whole-house mechanical ventilation system is installed capable of supplying outdoor ventilation air of 15 cubic feet per minute (cfm) (7.08 L/s) per occupant computed on the basis of two occupants for the first bedroom and one occupant for each additional bedroom.
> 2. The glazed areas need not be provided in rooms where Exception 1 above is satisfied and artificial light is provided capable of producing an average illumination of 6 footcandles (64.6 lux) over the area of the room at a height of 30 inches (762 mm) above the floor level.
> 3. Sunroom additions and patio covers, as defined in Section R202, shall be permitted to be used for natural ventilation if in excess of 40 percent of the exterior sunroom walls are open, or are enclosed only by insect screening.

Section R303.5.2 Add new section as shown: (RB47-03/04)

R303.5.2 Sunroom additions. Required glazed openings shall be permitted to open into sunroom additions or patio covers, that abut a street, yard or court if in excess of 40 percent of the exterior sunroom walls are open, or are enclosed only by insect screening, and the ceiling height of the sunroom is not less than 7 feet (2134 mm).

Section R307.1 Change to read as shown: (CCC)

R307.1 Space required. Fixtures shall be spaced as per Figure R307.1.

Figure R307.2 Change title as shown: (CCC)

FIGURE R307.1
MINIMUM FIXTURE CLEARANCES

Section R308.4 Change to read as shown: (S101-03/04)

R308.4 Hazardous locations. The following shall be considered specific hazardous locations for the purposes of glazing:

1. Glazing in swinging doors except jalousies.

2. Glazing in fixed and sliding panels of sliding door assemblies and panels in sliding and bifold closet door assemblies.

3. Glazing in storm doors.

4. Glazing in all unframed swinging doors.

5. Glazing in doors and enclosures for hot tubs, whirlpools, saunas, steam rooms, bathtubs and showers. Glazing in any part of a building wall enclosing these compartments where the bottom exposed edge of the glazing is less than 60 inches (1524 mm) measured vertically above any standing or walking surface.

6. Glazing, in an individual fixed or operable panel adjacent to a door where the nearest vertical edge is within a 24-inch (610 mm) arc of the door in a closed position and whose bottom edge is less than 60 inches (1524 mm) above the floor or walking surface.

7. Glazing in an individual fixed or operable panel, other than those locations described in Items 5 and 6 above, that meets all of the following conditions:

    7.1. Exposed area of an individual pane greater than 9 square feet (0.836 m^2).

    7.2. Bottom edge less than 18 inches (457 mm) above the floor.

    7.3. Top edge greater than 36 inches (914 mm) above the floor.

    7.4. One or more walking surfaces within 36 inches (914 mm) horizontally of the glazing.

8. All glazing in railings regardless of an area or height above a walking surface. Included are structural baluster panels and nonstructural infill panels.

9. Glazing in walls and fences enclosing indoor and outdoor swimming pools, hot tubs and spas where the bottom edge of the glazing is less than 60 inches (1524 mm) above a walking surface and within 60 inches (1524 mm) horizontally of the water's edge. This shall apply to single glazing and all panes in multiple glazing.

10. Glazing adjacent to stairways, landings and ramps within 36 inches (914 mm) horizontally of a walking surface when the exposed surface of the glass is less than 60 inches (1524 mm) above the plane of the adjacent walking surface.

11. Glazing adjacent to stairways within 60 inches (1524 mm) horizontally of the bottom tread of a stairway in any direction when the exposed surface of the glass is less than 60 inches (1524 mm) above the nose of the tread.

Exception: The following products, materials and uses are exempt from the above hazardous locations:

1. Openings in doors through which a 3-inch (76 mm) sphere is unable to pass.

2. Decorative glass in Items 1, 6 or 7.

3. Glazing in Section R308.4, Item 6, when there is an intervening wall or other permanent barrier between the door and the glazing.

4. Glazing in Section R308.4, Item 6, in walls perpendicular to the plane of the door in a closed position or where access through the door is to a closet or storage area 3 feet (914 mm) or less in depth. Glazing in these applications shall comply with Section R308.4, Item 7.

5. Glazing in Section R308.4, Items 7 and 10, when a protective bar is installed on the accessible side(s) of the glazing 36 inches ± 2 inches (914 mm ± 51 mm) above the floor. The bar shall be capable of withstanding a horizontal load of 50 pounds per linear foot (730 N/m) without contacting the glass and be a minimum of 1 1/2 inches (38 mm) in height.

6. Outboard panes in insulating glass units and other multiple glazed panels in Section R308.4, Item 7, when the bottom edge of the glass is 25 feet (7620 mm) or more above-grade, a roof, walking surfaces, or other horizontal [within 45 degrees (0.79 rad) of horizontal] surface adjacent to the glass exterior.

7. Louvered windows and jalousies complying with the requirements of Section R308.2.

8. Mirrors and other glass panels mounted or hung on a surface that provides a continuous backing support.

9. Safety glazing in Section R308.4, items 10 and 11, is not required where:

    9.1 The side of a stairway, landing or ramp has a guardrail or handrail, including balusters or in-fill panels, complying with the provisions of Sections 1012 and 1607.7 of the *International Building Code;* and

    9.2 The plane of the glass is greater than 18

2004 SUPPLEMENT TO THE IRC

inches (457 mm) from the railing or,

9.3 When a solid wall or panel extends from the plane of the adjacent walking surface to 34 inches (863 mm) to 36 inches (914 mm) above the floor and the construction at the top of that wall or panel is capable of withstanding the same horizontal load as the protective bar.

Section R308.6.1 Change definition of "Skylights and Sloped Glazing" to read as shown: (EC5-03/04)

SKYLIGHTS AND SLOPED GLAZING. Glass or other transparent or translucent glazing material installed at a slope of 15 degrees (0.26 rad) or more from vertical. Glazing materials in skylights, including unit skylights, solariums, sunrooms, roofs and sloped walls are included in this definition.

Section R309.2 Change to read as shown: (RB56-03/04)

R309.2 Separation required. The garage shall be separated from the residence and its attic area by not less than 1/2-inch (12.7 mm) gypsum board applied to the garage side. Garages beneath habitable rooms shall be separated from all habitable rooms above by not less than 5/8-inch (15.9 mm) Type X gypsum board or equivalent. Where the separation is a floor-ceiling assembly, the structure supporting the separation shall also be protected by not less than 1/2-inch (12.7 mm) gypsum board or equivalent. Garages located less than 3 feet (914 mm) from a dwelling unit on the same lot shall be protected with not less than 1/2-inch (12.7 mm) gypsum board applied to the interior side of exterior walls that are within this area. Openings in these walls shall be regulated by Section R309.1. This provision does not apply to garage walls that are perpendicular to the adjacent dwelling unit wall.

Section R310.1 Change to read as shown: (RB67-03/04)

R310.1 Emergency escape and rescue required. Basements and every sleeping room shall have at least one operable emergency and rescue opening. Such opening shall open directly into a public street, public alley, yard or court. Where basements contain one or more sleeping rooms, emergency egress and rescue openings shall be required in each sleeping room, but shall not be required in adjoining areas of the basement. Where emergency escape and rescue openings are provided they shall have a sill height of not more than 44 inches (1118 mm) above the floor. Where a door opening having a threshold below the adjacent ground elevation serves as an emergency escape and rescue opening and is provided with a bulkhead enclosure, the bulkhead enclosure shall comply with Section R310.3. The net clear opening dimensions required by this section shall be obtained by the normal operation of the emergency escape and rescue opening from the inside. Emergency escape and rescue openings with a finished sill height below the adjacent ground elevation shall be provided with a window well in accordance with Section R310.2.

Exception: Basements used only to house mechanical equipment and not exceeding total floor area of 200 square feet (18.58 m^2).

Section R310.1.4 Change to read as shown: (RB75-03/04)

R310.1.4 Operational Constraints. Emergency escape and rescue openings shall be operational from the inside of the room without the use of keys, tools or special knowledge

Section R310.4 Change to read as shown: (RB78-03/04)

R310.4 Bars, grills, covers and screens. Bars, grills, covers, screens or similar devices are permitted to be placed over emergency escape and rescue openings, bulkhead enclosures, or window wells that serve such openings, provided the minimum net clear opening size complies with Sections R310.1.1 to R310.1.3, and such devices shall be releasable or removable from the inside without the use of a key, tool, special knowledge or force greater than that which is required for normal operation of the escape and rescue opening.

Section R310.5 Add new section to read as shown: (RB79-03/04)

R310.5 Emergency escape windows under decks and porches. Emergency escape windows are allowed to be installed under decks and porches provided the location of the deck allows the emergency escape window to be fully opened and provides a path not less than 36 inches (914 mm) in height to a yard or court.

Section R311.4.3 Change to read as shown: (RB80-03/04)

R311.4.3 Landings at doors. There shall be a floor or landing on each side of each exterior door. The landing shall be permitted to have a slope not to exceed 0.25 unit vertical in 12 units horizontal (2-percent).

Exception: Where a stairway of two or fewer risers is located on the exterior side of a door, other than the required exit door, a landing is not required for the exterior side of the door.

The floor or landing at the exit door required by Section R311.4.1 shall not be more than 1.5 inches (38 mm) lower than the top of the threshold. The floor or landing at exterior doors other than the exit door required by Section R311.4.1 shall not be required to comply with this requirement but shall have a rise no greater than that permitted in Section R311.5.3.

> **Exception:** The landing at an exterior doorway shall not be more than 7 3/4 inches (196 mm) below the top of the threshold, provided the door, other than an exterior storm or screen door does not swing over the landing.

The width of each landing shall not be less than the door served. Every landing shall have a minimum dimension of 36 inches (914mm) measured in the direction of travel.

Section R311.5.4 Change to read as shown: (RB87-03/04)

R311.5.4 Landings for stairways. There shall be a floor or landing at the top and bottom of each stairway.

> **Exception:** A floor or landing is not required at the top of an interior flight of stairs, including stairs in an enclosed garage, provided a door does not swing over the stairs.

A flight of stairs shall not have a vertical rise greater than 12 feet (3658 mm) between floor levels or landings.

The width of each landing shall not be less than the stairway served. Every landing shall have a minimum dimension of 36 inches (914 mm) measured in the direction of travel.

Section R312.1 Change to read as shown: (RB90-03/04)

R312.1 Guards. Porches, balconies, ramps or raised floor surfaces located more than 30 inches (762 mm) above the floor or grade below shall have guards not less than 36 inches (914 mm) in height. Open sides of stairs with a total rise of more than 30 inches (762 mm) above the floor or grade below shall have guards not less than 34 inches (864 mm) in height measured vertically from the nosing of the treads.

Porches and decks which are enclosed with insect screening shall be provided with guards where the walking surface is located more than 30 inches (762 mm) above the floor or grade below.

Section R313.1 Change to read as shown: (F116-03/04)

R313.1 Smoke alarms. Smoke alarms shall be installed in the following locations:

1. In each sleeping room.
2. Outside each separate sleeping area in the immediate vicinity of the bedrooms.
3. On each additional story of the dwelling, including basements but not including crawl spaces and uninhabitable attics. In dwellings or dwelling units with split levels and without an intervening door between the adjacent levels, a smoke alarm installed on the upper level shall suffice for the adjacent lower level provided that the lower level is less than one full story below the upper level.

When more than one smoke alarm is required to be installed within an individual dwelling unit the alarm devices shall be interconnected in such a manner that the actuation of one alarm will activate all of the alarms in the individual unit. The alarm shall be clearly audible in all bedrooms over background noise levels with all intervening doors closed.

All smoke alarms shall be listed in accordance with UL 217 and installed in accordance with the provisions of this code and the household fire warning equipment provisions of NFPA 72.

Section R314.3 Change to read as shown: (RB102-03/04)

R314.3 Specific approval. Plastic foam not meeting the requirements of Sections R314.1 and R314.2 may be specifically approved on the basis of one of the following approved tests: ASTM E 84, FM 4880, UL 1040, NFPA 286 or UL 1715 or fire tests related to actual end-use configurations. The specific approval may be based on the end use, quantity, location and similar considerations where such tests would not be applicable or practical.

Section R316.2 Change to read as shown: (FS96-03/04)

R316.2 Loose-fill insulation. Loose-fill insulation materials that cannot be mounted in the ASTM E 84 apparatus without a screen or artificial supports shall comply with the flame spread and smoke-developed limits of Sections R316.1 and R316.4 when tested in accordance with CAN/ULC S102.2.

> **Exception:** Cellulose loose-fill insulation shall not be required to comply with the flame spread index requirement of CAN/ULC S102.2, provided such insulation complies with the requirements of Section R316.3.

Section R317.1 Change to read as shown: (RB104-03/04)

R317.1 Two-Family dwellings. Dwelling units in two-family dwellings shall be separated from each other by wall and/or floor assemblies having not less than a 1-hour fire-resistance rating when tested in accordance with ASTM E 119. Fire-resistance-rated floor-ceiling and wall assemblies shall extend to and be tight against the exterior wall, and wall assemblies shall extend to the underside of the roof sheathing.

2004 SUPPLEMENT TO THE IRC

Exceptions:

1. A fire-resistance rating of ½ hour shall be permitted in buildings equipped throughout with an automatic sprinkler system installed in accordance with NFPA 13.

2. Wall assemblies need not extend through attic spaces when the ceiling is protected by not less than 5/8-inch (15.9 mm) Type X gypsum board and an attic draft stop constructed as specified in Section R502.12.1 is provided above and along the wall assembly separating the dwellings. The structural framing supporting the ceiling shall also be protected by not less than ½-inch (12.7 mm) gypsum board or equivalent.

Section R317.2.2 Change to read as shown: (RB105-03/04)

R317.2.2 Parapets. Parapets constructed in accordance with Section R317.2.3 shall be provided for townhouses as an extension of exterior walls or common walls in accordance with the following:

1. Where roof surfaces adjacent to the wall or walls are at the same elevation, the parapet shall extend not less than 30 inches (762 mm) above the roof surfaces.

2. Where roof surfaces adjacent to the wall or walls are at different elevations and the higher roof is not more than 30 inches (762 mm) above the lower roof, the parapet shall extend not less than 30 inches (762 mm) above the lower roof surface.

   Exception: A parapet is not required in the two cases above when the roof is covered with a minimum class C roof covering, and the roof decking or sheathing is of noncombustible materials or approved fire retardant-treated wood for a distance of 4 feet (1219 mm) on each side of the wall or walls, or one layer of 5/8-inch (15.9 mm) Type X gypsum board is installed directly beneath the roof decking or sheathing, supported by a minimum of nominal 2-inch (51 mm) ledgers attached to the sides of the roof framing members, for a minimum distance of 4 feet (1220 mm) on each side of the wall or walls.

3. A parapet is not required where roof surfaces adjacent to the wall or walls are at different elevations and the higher roof is more than 30 inches (762 mm) above the lower roof. The common wall construction from the lower roof to the underside of the higher roof deck shall not have less than a 1-hour fire-resistance rating. The wall shall be rated for exposure from both sides.

Section R318.1 Change to read as shown: (EC48-03/04)

R318.1 Moisture control. In all framed walls, floors and roof/ceilings comprising elements of the building thermal envelope, a vapor retarder shall be installed on the warm-in-winter side of the insulation.

Exceptions:

1. In construction where moisture or freezing will not damage the materials.

2. Where the framed cavity or space is ventilated to allow moisture to escape.

3. In counties identified as in climate zones 1 through 4 in Table N1101.2

Sections R319.1 Change to read as shown: (S51-03/04)

R319.1 Location required. In areas subject to decay damage as established by Table R301.2(1), the following locations shall require the use of an approved species and grade of lumber, pressure treated in accordance with AWPA U1 for the species, product, preservative and end use or of the decay-resistant heartwood of redwood, black locust or cedars. Preservatives shall conform to AWPA P1/13, P2, P3 or P5.

Section R319.1.5 Add new section to read as shown: (RB107-03/04)

R319.1.5 Exposed glued-laminated timbers. The portions of glued-laminated timbers that form the structural supports of a building or other structure and are exposed to weather and not properly protected by a roof, eave or similar covering shall be pressure treated with preservative, or be manufactured from naturally durable or preservative-treated wood.

Section R323.1 Change to read as shown: (RB10-03/04)

R323.1 General. Buildings and structures constructed in whole or in part in flood hazard areas (including A or V Zones) as established in Table R301.2(1) shall be designed and constructed in accordance with the provisions contained in this section.

Exception: All buildings and structures located in whole or in part in identified floodways as established in Table R301.2(1) shall be designed and constructed as stipulated in the *International Building Code*.

Section R323.1.3.1 Add new section to read as shown: (RB111-03/04)

R323.1.3.1 Determination of design flood elevations. If design flood elevations are not specified, the building official is authorized to require the applicant to:

1. Obtain and reasonably utilize data available from a federal, state or other source, or

2. Determine the design flood elevation in accordance with accepted hydrologic and hydraulic engineering practices used to define special flood hazard areas. Determinations shall be undertaken by a registered design professional who shall document that the technical methods used reflect currently accepted engineering practice. Studies, analyses and computations shall be submitted in sufficient detail to allow thorough review and approval.

Section R323.1.3.2 Add new section to read as shown: (RB112-03/04)

R323.1.3.2 Determination of impacts. In riverine flood hazard areas where design flood elevations are specified but floodways have not been designated, the applicant shall demonstrate that the effect of the proposed buildings and structures on design flood elevations, including fill, when combined with all other existing and anticipated flood hazard area encroachments, will not increase the design flood elevation more than 1 foot (305 mm) at any point within the jurisdiction.

Section R323.1.7 Change to read as shown: (RB113-03/04)

R323.1.7 Flood Resistant Materials. Building materials used below the design flood elevation shall comply with the following:

1. All wood, including floor sheathing, shall be pressure-preservative treated in accordance with AWPA U1 for the species, product, preservative and end use or be the decay-resistant heartwood of redwood, black locust or cedars. Preservatives shall conform to AWPA P1/13, P2, P3 or P5.

2. Materials and installation methods used for flooring and interior and exterior walls and wall coverings shall conform to the provisions of FEMA/FIA-TB-2.

Section R323.2 Change to read as shown: (RB10-03/04)

R323.2 Flood hazard areas (including A Zones). All areas that have been determined to be prone to flooding but not subject to high velocity wave action shall be designated as flood hazard areas. All buildings and structures constructed in whole or in part in flood hazard areas shall be designed and constructed in accordance with Sections R323.2.1 and R323.2.3.

Section R323.2.2 Change to read as shown: (RB114-03/04)

R323.2.2 Enclosed area below design flood elevation. Enclosed areas, including crawl spaces, that are below the design flood elevation shall:

1. Be used solely for parking of vehicles, building access or storage.

2. Be provided with flood openings which shall meet the following criteria:

    2.1. There shall be a minimum of two openings on different sides of each enclosed area; if a building has more than one enclosed area below the design flood elevation, each area shall have openings on exterior walls.

    2.2. The total net area of all openings shall be at least 1 square inch (645 mm²) for each square foot (0.093 m²) of enclosed area, or the openings shall be designed and the construction documents shall include a statement that the design and installation will provide for equalization of hydrostatic flood forces on exterior walls by allowing for the automatic entry and exit of floodwaters.

    2.3. The bottom of each opening shall be 1 foot (305 mm) or less above the adjacent ground level.

    2.4. Openings shall be at least 3 inches (76 mm) in diameter.

    2.5. Any louvers, screens or other opening covers shall allow the automatic flow of floodwaters into and out of the enclosed area.

    2.6. Openings installed in doors and windows, that meet requirements 2.1 through 2.5, are acceptable; however, doors and windows without installed openings do not meet the requirements of this section.

Section R323.3 Change to read as shown: (RB10-03/04)

R323.3 Coastal high hazard areas (including V Zones). Areas that have been determined to be subject to wave heights in excess of 3 feet (914 mm) or subject to high velocity wave action or wave-induced erosion shall be designated as coastal high hazard areas. All buildings

2004 SUPPLEMENT TO THE IRC

and structures constructed in whole or in part in coastal high hazard areas shall be designated and constructed in accordance with Sections R323.3.1 through R323.3.6.

Section R323.3.3 Change to read as shown: (RB117-03/04)

R323.3.3 Foundations. All buildings and structures erected in coastal high hazard areas shall be supported on pilings or columns and shall be adequately anchored to such pilings or columns. Pilings shall have adequate soil penetrations to resist the combined wave and wind loads (lateral and uplift). Water loading values used shall be those associated with the design flood. Wind loading values shall be those required by this code. Pile embedment shall include consideration of decreased resistance capacity caused by scour of soil strata surrounding the piling. Pile systems design and installation shall be certified in accordance with Section R323.3.6. Mat, raft or other foundations that support columns shall not be permitted where soil investigations that are required in accordance with Section R401.4 indicate that soil material under the mat, raft or other foundation is subject to scour or erosion from wave-velocity flow conditions. Slabs, pools, pool decks and walkways shall be located and constructed to be structurally independent of buildings and structures and their foundations to prevent transfer of flood loads to the buildings and structures during conditions of flooding, scour or erosion from wave-velocity flow conditions, unless the buildings and structures and their foundation are designed to resist the additional flood load.

CHAPTER 4
FOUNDATIONS

Section R401.1 Change to read as shown: (RB31-03/04 and RB122-03/04)

R401.1 Application. The provisions of this chapter shall control the design and construction of the foundation and foundation spaces for all buildings. In addition to the provisions of this chapter, the design and construction of foundations in areas prone to flooding as established by Table R301.2(1) shall meet the provisions of Section R323. Wood foundations shall be designed and installed in accordance with AF&PA Report No. 7.

> **Exception**: The provisions of this chapter shall be permitted to be used for wood foundations only in the following situations:
>
> 1. In buildings that have no more than two floors and a roof.
>
> 2. When interior basement and foundation walls are provided at intervals not exceeding 50 feet (15 240 mm).

Wood foundations in Seismic Design Categories D_0, D_1 or D_2 shall be designed in accordance with accepted engineering practice.

Section R401.3 Change to read as shown: (S44-03/04)

R401.3 Drainage. Surface drainage shall be diverted to a storm sewer conveyance or other approved point of collection so as to not create a hazard. Lots shall be graded so as to drain surface water away from foundation walls. The grade shall fall a minimum of 6 inches (152 mm) within the first 10 feet (3048 mm).

> **Exception:** Where lot lines, walls, slopes or other physical barriers prohibit 6 inches (152 mm) of fall within 10 feet (3048 mm), the final grade shall slope away from the foundation at a minimum slope of 5 percent and the water shall be directed to drains or swales to ensure drainage away from the structure. Swales shall be sloped a minimum of 2 percent when located within 10 feet (3048 mm) of the building foundation. Impervious surfaces within 10 feet (3048 mm) of the building foundation shall be sloped a minimum of 2 percent away from the building.

Section R401.4.2 Renumber 401.5 to R401.4.2 and change to read as shown: (RB123-03/04)

R401.4.2 Compressible or shifting soil. In lieu of a complete geotechnical evaluation, when top or subsoils are compressible or shifting, such soils shall be removed to a depth and width sufficient to assure stable moisture content in each active zone and shall not be used as fill or stabilized within each active zone by chemical, dewatering, or presaturation.

Section R401.5 relocated and revised to Section R401.4.2

Section R402.1.2 Change to read as shown: (RB124-03/04)

R402.1.2 Wood Treatment. All lumber and plywood shall be pressure-preservative treated and dried after treatment in accordance with AWPA U1 (Commodity Specification A, Use Category 4B and section 5.2), and shall bear the label of an accredited agency. Where lumber and/or plywood is cut or drilled after treatment, the treated surface shall be field treated with copper naphthenate, the concentration of which shall contain a minimum of 2 percent copper metal, by repeated brushing, dipping or soaking until the wood absorbs no more preservative.

Section R402.2 Change to read as shown: (CCC)

R402.2 Concrete. Concrete shall have a minimum specified compressive strength of f'_c, as shown in Table R402.2. Concrete subject to weathering as indicated in

Table R301.2(1) shall be air entrained as specified in Table R402.2. The maximum weight of fly ash, other pozzolans, silica fume, or slag that is included in concrete mixtures for garage floor slabs and for exterior porches, carport slabs, and steps that will be exposed to deicing chemicals shall not exceed the percentages of the total weight of cementitious materials specified in ACI 318. Materials used to produce concrete and testing thereof shall comply with the applicable standards listed in Chapter 3 of ACI 318.

Sections R403.1.2 and R403.1.3 Change to read as shown: (RB31-03/04)

R403.1.2 Continuous footing in Seismic Design Categories D_0, D_1 and D_2. The braced wall panels at exterior walls of buildings located in Seismic Design Categories D_0, D_1 and D_2 shall be supported by continuous footings. All required interior braced wall panels in buildings with plan dimensions greater than 50 feet (15 240 mm) shall also be supported by continuous footings.

R403.1.3 Seismic reinforcing. Concrete footings located in Seismic Design Categories D_0, D_1 and D_2, as established in Table R301.2(1), shall have minimum reinforcement. Bottom reinforcement shall be located a minimum of 3 inches (76 mm) clear from the bottom of the footing.

In Seismic Design Categories D_0, D_1 and D_2 where a construction joint is created between a concrete footing and stem wall, a minimum of one No. 4 bar shall be provided at not more than 4 feet (1219 mm) on center. The vertical bar shall extend to 3 inches (76 mm) clear of the bottom of the footing, have a standard hook and extend a minimum of 14 inches (357 mm) into the stem wall.

In Seismic Design Categories D_0, D_1 and D_2 where a grouted masonry stem wall is supported on a concrete footing and stem wall, a minimum of one No. 4 bar shall be provided at not more than 4 feet on center. The vertical bar shall extend to 3 inches (76 mm) clear of the bottom of the footing and have a standard hook.

In Seismic Design Categories D_0, D_1 and D_2 masonry stem walls without solid grout and vertical reinforcing shall not be permitted.

> **Exception:** In detached one- and two-family dwellings which are three stories or less in height and constructed with stud bearing walls, plain concrete footings without longitudinal reinforcement supporting walls and isolated plain concrete footings supporting columns or pedestals are permitted.

Section R403.1.4.1 Change to read as shown: (S49-03/04)

R403.1.4.1 Frost Protection. Except where otherwise protected from frost, foundation walls, piers and other permanent supports of buildings and structures shall be protected from frost by one or more of the following methods:

1. Extended below the frost line specified in Table R301.2.(1);
2. Constructing in accordance with Section R403.3;
3. Constructing in accordance with ASCE 32-01; or
4. Erected on solid rock.

> **Exceptions:**
>
> 1. Freestanding accessory structures with an area of 600 square feet (56 m^2) or less, of light-framed construction, with an eave height of 10 feet (3080 mm) or less shall not be required to be protected.
>
> 2. Freestanding accessory structures with an area of 400 square feet (37 m^2) or less, of other than light-framed construction, with an eave height of 10 feet (3080 mm) or less shall not be required to be protected.
>
> 3. Decks not supported by a dwelling need not be provided with footings that extend below the frost line.

Footings shall not bear on frozen soil unless such frozen condition is of a permanent character.

Section R403.1.4.2 Change to read as shown: (RB31-03/04 and RB132-03/04)

R403.1.4.2 Seismic conditions. In Seismic Design Categories D_0, D_1 and D_2, interior footings supporting bearing or bracing walls and cast monolithically with a slab on grade shall extend to a depth of not less than 12 inches (305 mm) below the top of the slab.

Section R403.1.6 Change to read as shown: (RB31-03/04 and RB133-03/04)

R403.1.6 Foundation Anchorage. When braced wall panels are supported directly on continuous foundations, the wall wood sill plate or cold-formed steel bottom track shall be anchored to the foundation in accordance with this section.

The wood sole plate at exterior walls on monolithic slabs and wood sill plate shall be anchored to the foundation with anchor bolts spaced a maximum of 6 feet (1829 mm) on center. There shall be a minimum of two bolts per plate section with one bolt located not more than

12 inches (305 mm) or less than seven bolt diameters from each end of the plate section. In Seismic Design Categories D_0, D_1 and D_2, anchor bolts shall be spaced at 6 feet (1829 mm) on center and located within 12 inches (305 mm) from the ends of each plate section at interior braced wall lines when required by Section R602.10.9 to be supported on a continuous foundation. Bolts shall be at least ½ inch (12.7 mm) in diameter and shall extend a minimum of 7 inches (178 mm) into masonry or concrete. Interior bearing wall sole plates on monolithic slab foundation shall be positively anchored with approved fasteners. A nut and washer shall be tightened on each bolt of the plate. Sills and sole plates shall be protected against decay and termites where required by Sections R319 and R320. Cold-formed steel framing systems shall be fastened to the wood sill plates or anchored directly to the foundation as required in Section R505.3.1 or R603.1.1.

Exceptions:

1. Foundation anchor straps, spaced as required to provide equivalent anchorage to ½-inch-diameter (12.7 mm) anchor bolts.

2. Walls 24 inches (610 mm) total length or shorter connecting offset braced wall panels shall be anchored to the foundation with a minimum of one anchor bolt located in the center third of the plate section and shall be attached to adjacent braced wall panels per Figure R602.10.5 at corners.

3. Walls 12 inches (305 mm) total length or shorter connecting offset braced wall panels shall be permitted to be connected to the foundation without anchor bolts. The wall shall be attached to adjacent braced wall panels per Figure R602.10.5 at corners.

Section R403.1.6.1 Change to read as shown: (RB31-03/04 & S83-03/04)

R403.1.6.1 Foundation anchorage in Seismic Design Categories C, D_0, D_1 and D_2. In addition to the requirements of Section R403.1.6, the following requirements shall apply to wood light-frame structures in Seismic Design Categories D_0, D_1 and D_2 and wood light-frame townhouses in Seismic Design Category C.

1. Plate washers conforming to Section R602.11.1 shall be provided for all anchor bolts over the full length of required braced wall lines in a braced wall line. Properly sized cut washers shall be permitted for anchor bolts in wall lines not containing braced wall panels.

2. Interior braced wall plates shall have anchor bolts spaced at not more than 6 feet (1829 mm) on center and located within 12 inches (305 mm) from the ends of each plate section when supported on a continuous foundation.

3. Interior bearing wall sole plates shall have anchor bolts spaced at not more than 6 feet (1829 mm) on center and located within 12 inches (305 mm) from the ends of each plate section when supported on a continuous foundation.

4. The maximum anchor bolt spacing shall be 4 feet (1219 mm) for buildings over two stories in height.

5. Stepped cripple walls shall conform to Section R602.11.3.

6. Where continuous wood foundations in accordance with Section R404.2 are used, the force transfer shall have a capacity equal to or greater than the connections required by Section R602.11.1 or the braced wall panel shall be connected to the wood foundations in accordance with the braced wall panel-to-floor fastening requirements of Table R602.3(1).

Section R404 Change title to read as shown: (RB138-03/04)

SECTION R404
FOUNDATION AND RETAINING WALLS

Sections R404.1.1 and R404.1.2 Change to read as shown: (RB31-03/04)

R404.1.1 Masonry foundation walls. Concrete masonry and clay masonry foundation walls shall be constructed as set forth in Tables R404.1.1(1), R404.1.1(2), R404.1.1(3) and R404.1.1(4) and shall also comply with the provisions of this section and the applicable provisions of Sections R606, R607 and R608. In Seismic Design Categories D_0, D_1 and D_2, concrete masonry and clay masonry foundation walls shall comply with Section R404.1.4. Rubble stone masonry foundation walls shall be constructed in accordance with Sections R404.1.8 and R607.2.2. Rubble stone masonry walls shall not be used in Seismic Design Categories D_0, D_1 and D_2.

R404.1.2 Concrete foundation walls. Concrete foundation walls shall be constructed as set forth in Tables R404.1.1(1), R404.1.1(2), R404.1.1(3) and R404.1.1(4), and shall also comply with the provisions of this section and the applicable provisions of Section R402.2. In Seismic Design Categories D_0, D_1 and D_2, concrete foundation walls shall comply with Section R404.1.4.

Sections R404.1.4 Change to read as shown (RB31-03/04)

R404.1.4 Seismic Design Categories D_0, D_1 and D_2. In addition to the requirements of Table R404.1.1(1), plain concrete and plain masonry foundation walls located in

Seismic Design Categories D_0, D_1 and D_2, as established in Table R301.2(1), shall comply with the following.

1. Minimum reinforcement shall consist of one No. 4 (No. 13) horizontal bar located in the upper 12 inches (305 mm) of the wall,

2. Wall height shall not exceed 8 feet (2438 mm),

3. Height of unbalanced backfill shall not exceed 4 feet (1219 mm), and

4. A minimum thickness of 7.5 inches (191 mm) is required for plain concrete foundation walls except that a minimum thickness of 6 inches (152 mm) shall be permitted for plain concrete foundation walls with a maximum height of 4 feet, 6 inches (1372 mm).

5. Plain masonry foundation walls shall be a minimum of 8 inches (203 mm) thick.

Vertical reinforcement for masonry stem walls shall be tied to the horizontal reinforcement in the footings. Masonry stem walls located in Seismic Design Categories D_0, D_1 and D_2 shall have a minimum vertical reinforcement of one No. 3 bar located a maximum of 4 feet (1220 mm) on center in grouted cells.

Foundations walls located in Seismic Design Categories D_0, D_1 and D_2, as established in Table R301.2(1), supporting more than 4 feet (1219 mm) of unbalanced backfill or exceeding 8 feet (2438 mm) in height shall be constructed in accordance with Table R404.1.1(2), R404.1.1(3) or R404.1.1(4) and shall have two No. 4 (No. 13) horizontal bars located in the upper 12 inches (305 mm) of the wall.

R404.1.5.1 Change to read as shown (RB31-03/04)

R404.1.5.1 Pier and curtain wall foundations. Pier and curtain wall foundations shall be permitted to be used to support light-frame construction not more than two stories in height, provided the following requirements are met:

1. All load-bearing walls shall be placed on continuous concrete footings placed integrally with the exterior wall footings.

2. The minimum actual thickness of a load-bearing masonry wall shall be not less than 4 inches (102 mm) nominal or 3 3/8 inches (92 mm) actual thickness, and shall be bonded integrally with piers spaced in accordance with Section R606.8.

3. Piers shall be constructed in accordance with Section R606.5 and Section R606.5.1, and shall be bonded into the load-bearing masonry wall in accordance with Section R608.1.1 or Section R608.1.1.2.

4. The maximum height of a 4-inch (102 mm) load-bearing masonry foundation wall supporting wood framed walls and floors shall not be more than 4 feet (1219 mm).

5. Anchorage shall be in accordance with Section R403.1.6, Figure R404.1.5(1), or as specified by engineered design accepted by the building official.

6. The unbalanced fill for 4-inch (102 mm) foundation walls shall not exceed 24 inches (610 mm) for solid masonry or 12 inches (305 mm) for hollow masonry.

7. In Seismic Design Categories D_0, D_1 and D_2, prescriptive reinforcement shall be provided in the horizontal and vertical direction. Provide minimum horizontal joint reinforcement of two No.9 gage wires spaced not less than 6 inches (152 mm) or one 1/4 inch (6.4 mm) diameter wire at 10 inches (254 mm) on center vertically. Provide minimum vertical reinforcement of one No. 4 bar at 48 inches (1220 mm) on center horizontally grouted in place.

Section R404.1.8 Change to read as shown: (RB31-03/04 and RB140-03/04)

R404.1.8 Rubble stone masonry. Rubble stone masonry foundation walls shall have a minimum thickness of 16 inches (406 mm), shall not support an unbalanced backfill exceeding 8 feet (2438 mm) in height, shall not support a soil pressure greater than 30 pounds per square foot per foot (4.71 kPa/m), and shall not be constructed in Seismic Design Categories D_0, D_1, D_2 or townhouses in Seismic Design Category C, as established in Figure R301.2(2).

Section R404.5 Add new section to read as shown: (RB138-03/04)

R404.5 Retaining walls. Retaining walls that are not laterally supported at the top and that retain in excess of 24 inches (610 mm) of unbalanced fill shall be designed to ensure stability against overturning, sliding, excessive foundation pressure and water uplift. Retaining walls shall be designed for a safety factor of 1.5 against lateral sliding and overturning.

Section R406.1 Change to read as shown: (RB142-03/04 and RB143-03/04)

R406.1 Concrete and masonry foundation dampproofing. Except where required by Section R406.2

to be waterproofed, foundation walls that retain earth and enclose interior spaces and floors below grade shall be dampproofed from the top of the footing to the finished grade. Masonry walls shall have not less than 3/8 inch (9.5 mm) portland cement parging applied to the exterior of the wall. The parging shall be dampproofed in accordance with one of the following:

1. Bituminous coating.
2. 3 pounds per square yard (1.63 kg/m^2) of acrylic modified cement.
3. 1/8-inch (3.2 mm) coat of surface-bonding cement complying with ASTM C 887.
4. Any material permitted for waterproofing in Section R406.2.
5. Other approved methods or materials.

   Exception: Parging of unit masonry walls is not required where a material is approved for direct application to the masonry.

Concrete walls shall be dampproofed by applying any one of the above listed dampproofing materials or any one of the waterproofing materials listed in Section R406.2 to the exterior of the wall.

Section R406.2 Change to read as shown: (RB145-03/04 and RB146-03/04)

R406.2 Concrete and masonry foundation waterproofing. In areas where a high water table or other severe soil-water conditions are known to exist, exterior foundation walls that retain earth and enclose interior spaces and floors below grade shall be waterproofed from the top of the footing to the finished grade. Walls shall be waterproofed in accordance with one of the following:

1. 2-ply hot-mopped felts.
2. 55 pound (25 kg) roll roofing.
3. 6-mil (0.15 mm) polyvinyl chloride.
4. 6-mil (0.15 mm) polyethylene.
5. 40-mil (1 mm) polymer-modified asphalt.
6. 60-mil (1.5 mm) flexible polymer cement.
7. 1/8 inch cement-based, fiber-reinforced, waterproof coating.
8. 8.60-mil (0.22 mm) solvent-free liquid-applied synthetic rubber.

   Exception: Organic-solvent-based products such as hydrocarbons, chlorinated hydrocarbons, ketones and esters shall not be used for ICF walls with expanded polystyrene form material. Plastic roofing cements, acrylic coatings, latex coatings, mortars and pargings are permitted to be used to seal ICF walls. Cold-setting asphalt or hot asphalt shall conform to type C of ASTM D 449. Hot asphalt shall be applied at a temperature of less than 200° F (93°C).

All joints in membrane waterproofing shall be lapped and sealed with an adhesive compatible with the membrane.

Section R408.2 Delete the "exceptions" without substitution. (EC48-03/04)

R408.2 Openings for under-floor ventilation. The minimum net area of ventilation openings shall not be less than 1 square foot (0.0929 m^2) for each 150 square feet (14 m^2) of under-floor area. One such ventilating opening shall be within 3 feet (914 mm) of each corner of the building. Ventilation openings shall be covered for their height and width with any of the following materials provided that the least dimension of the covering shall not exceed 1/4 inch (6.4 mm):

1. Perforated sheet metal plates not less than 0.070 inch (1.8 mm) thick.
2. Expanded sheet metal plates not less than 0.047 inch (1.2 mm) thick.
3. Cast iron grill or grating.
4. Extruded load-bearing brick vents.
5. Hardware cloth of 0.035 inch (0.89 mm) wire or heavier.
6. Corrosion-resistant wire mesh, with the least dimension being 1/8 inch (3.2 mm).

Section R408.3 Add new section as shown and renumber existing text: (EC48-03/04)

R408.3 Unvented crawl space. Ventilation openings in under-floor spaces specified in Sections R408.1 and R408.2 shall not be required where:

1. Exposed earth is covered with a continuous vapor retarder. All joints of the vapor retarder shall

overlap by 6 inches (153 mm) and shall be sealed or taped. The edges of the vapor retarder shall extend at least 6 inches (153 mm) up the stem wall and shall be attached and sealed to the stem wall,

2. And one of the following is provided for the under-floor space:

   a. Continuously operated mechanical exhaust ventilation at a rate equal to 1 cfm (0.47 L/s) for each 50 ft^2 (4.7 m^2) of crawlspace floor area, including an air pathway to the common area (such as a duct or transfer grille), and perimeter walls insulated in accordance with Section N1102.2.8, or

   b. Conditioned air supply sized to deliver at a rate equal to 1 cfm (0.47 L/s) for each 50 ft^2 (4.7 m^2) of under-floor area, including a return air pathway to the common area (such as a duct or transfer grille), and perimeter walls insulated in accordance with Section N1102.2.8, or

   c. Plenum complying with Section M1601.4, if under-floor space is used as a plenum.

CHAPTER 5
FLOORS

Table R502.3.3(1) Change table note "f" as shown (no change to table): RB31-03/04

(No change to table)

a. through e. (No change to current text)
f. See Section R301.2.2.2.2, Item 1, for additional limitations on cantilevered floor joists for detached one- and two-family dwellings in Seismic Design Categories D_0, D_1 or D_2 and townhouses in Seismic Design Categories C, D_0, D_1 or D_2.
g. and h. (No change to current text)

2004 SUPPLEMENT TO THE IRC

Table R502.5(1) Change table to read as shown: (S81-03/04)

TABLE R502.5(1)
GIRDER SPANS[a] AND HEADER SPANS[a] FOR EXTERIOR BEARING WALLS
(Maximum header spans for douglas fir-larch, hem fir, southern pine and spruce pine fir[b] and required number of jack studs)

GIRDERS AND HEADERS SUPPORTING	SIZE	\multicolumn{6}{c	}{GROUND SNOW LOAD (psf)[e]}											
		\multicolumn{6}{c	}{30}	\multicolumn{6}{c	}{50}	\multicolumn{6}{c	}{70}
		\multicolumn{18}{c	}{Building Width[c] (feet)}
		\multicolumn{2}{c|}{20}	\multicolumn{2}{c|}{28}	\multicolumn{2}{c|}{36}	\multicolumn{2}{c|}{20}	\multicolumn{2}{c|}{28}	\multicolumn{2}{c|}{36}	\multicolumn{2}{c|}{20}	\multicolumn{2}{c|}{28}	\multicolumn{2}{c|}{36}
		Span	NJ[d]	Span	NJ[d]	Span	NJ[d]	Span	NJ[d]	Span	NJ[d]	Span	NJ[d]	Span	NJ[d]	Span	NJ[d]	Span	NJ[d]
Roof and ceiling	2-2x4	3-6	1	3-2	1	2-10	1	3-2	1	2-9	1	2-6	1	2-10	1	2-6	1	2-3	1
	2-2x6	5-5	1	4-8	1	4-2	1	4-8	1	4-1	1	3-8	2	4-2	1	3-8	2	3-3	2
	2-2x8	6-10	1	5-11	2	5-4	2	5-11	2	5-2	2	4-7	2	5-4	2	4-7	2	4-1	2
	2-2x10	8-5	2	7-3	2	6-6	2	7-3	2	6-3	2	5-7	2	6-6	2	5-7	2	5-0	2
	2-2x12	9-9	2	8-5	2	7-6	2	8-5	2	7-3	2	6-6	2	7-6	2	6-6	2	5-10	3
	3-2x8	8-4	1	7-5	1	6-8	1	7-5	1	6-5	2	5-9	2	6-8	1	5-9	2	5-2	2
	3-2x10	10-6	1	9-1	2	8-2	2	9-1	2	7-10	2	7-0	2	8-2	2	7-0	2	6-4	2
	3-2x12	12-2	2	10-7	2	9-5	2	10-7	2	9-2	2	8-2	2	9-5	2	8-2	2	7-4	2
	4-2x8	9-2	1	8-4	1	7-8	1	8-4	1	7-5	1	6-8	1	7-8	1	6-8	1	5-11	2
	4-2x10	11-8	1	10-6	1	9-5	2	10-6	1	9-1	2	8-2	2	9-5	2	8-2	2	7-3	2
	4-2x12	14-1	1	12-2	2	10-11	2	12-2	2	10-7	2	9-5	2	10-11	2	9-5	2	8-5	2
Roof, ceiling, and one center-bearing floor	2-2x4	3-1	1	2-9	1	2-5	1	2-9	1	2-5	1	2-2	1	2-7	1	2-3	1	2-0	1
	2-2x6	4-6	1	4-0	1	3-7	2	4-1	1	3-7	2	3-3	2	3-9	2	3-3	2	2-11	2
	2-2x8	5-9	2	5-0	2	4-6	2	5-2	2	4-6	2	4-1	2	4-9	2	4-2	2	3-9	2
	2-2x10	7-0	2	6-2	2	5-6	2	6-4	2	5-6	2	5-0	2	5-9	2	5-1	2	4-7	3
	2-2x12	8-1	2	7-1	2	6-5	2	7-4	2	6-5	2	5-9	3	6-8	2	5-10	3	5-3	3
	3-2x8	7-2	1	6-3	2	5-8	2	6-5	2	5-8	2	5-1	2	5-11	2	5-2	2	4-8	2
	3-2x10	8-9	2	7-8	2	6-11	2	7-11	2	6-11	2	6-3	2	7-3	2	6-4	2	5-8	2
	3-2x12	10-2	2	8-11	2	8-0	2	9-2	2	8-0	2	7-3	2	8-5	2	7-4	2	6-7	2
	4-2x8	8-1	1	7-3	1	6-7	1	7-5	1	6-6	1	5-11	2	6-10	1	6-0	2	5-5	2
	4-2x10	10-1	1	8-10	2	8-0	2	9-1	2	8-0	2	7-2	2	8-4	2	7-4	2	6-7	2
	4-2x12	11-9	2	10-3	2	9-3	2	10-7	2	9-3	2	8-4	2	9-8	2	8-6	2	7-7	2
Roof, ceiling, and one clear span floor	2-2x4	2-8	1	2-4	1	2-1	1	2-7	1	2-3	1	2-0	1	2-5	1	2-1	1	1-10	1
	2-2x6	3-11	1	3-5	2	3-0	2	3-10	2	3-4	2	3-0	2	3-6	2	3-1	2	2-9	2
	2-2x8	5-0	2	4-4	2	3-10	2	4-10	2	4-2	2	3-9	2	4-6	2	3-11	2	3-6	2
	2-2x10	6-1	2	5-3	2	4-8	2	5-11	2	5-1	2	4-7	3	5-6	2	4-9	2	4-3	3
	2-2x12	7-1	2	6-1	3	5-5	3	6-10	2	5-11	3	5-4	3	6-4	2	5-6	3	5-0	3
	3-2x8	6-3	2	5-5	2	4-10	2	6-1	2	5-3	2	4-8	2	5-7	2	4-11	2	4-5	2
	3-2x10	7-7	2	6-7	2	5-11	2	7-5	2	6-5	2	5-9	2	6-10	2	6-0	2	5-4	2
	3-2x12	8-10	2	7-8	2	6-10	2	8-7	2	7-5	2	6-8	2	7-11	2	6-11	2	6-3	3
	4-2x8	7-2	1	6-3	2	5-7	2	7-0	1	6-1	2	5-5	2	6-6	1	5-8	2	5-1	2
	4-2x10	8-9	2	7-7	2	6-10	2	8-7	2	7-5	2	6-7	2	7-11	2	6-11	2	6-2	2
	4-2x12	10-2	2	8-10	2	7-11	2	9-11	2	8-7	2	7-8	2	9-2	2	8-0	2	7-2	2
Roof, ceiling, and two center-bearing floors	2-2x4	2-7	1	2-3	1	2-0	1	2-6	1	2-2	1	1-11	1	2-4	1	2-0	1	1-9	1
	2-2x6	3-9	2	3-3	2	2-11	2	3-8	2	3-2	2	2-10	2	3-5	2	3-0	2	2-8	2
	2-2x8	4-9	2	4-2	2	3-9	2	4-7	2	4-0	2	3-8	2	4-4	2	3-9	2	3-5	2
	2-2x10	5-9	2	5-1	2	4-7	3	5-8	2	4-11	2	4-5	3	5-3	2	4-7	3	4-2	3
	2-2x12	6-8	2	5-10	3	5-3	3	6-6	2	5-9	3	5-2	3	6-1	3	5-4	3	4-10	3
	3-2x8	5-11	2	5-2	2	4-8	2	5-9	2	5-1	2	4-7	2	5-5	2	4-9	2	4-3	2
	3-2x10	7-3	2	6-4	2	5-8	2	7-1	2	6-2	2	5-7	2	6-7	2	5-9	2	5-3	2
	3-2x12	8-5	2	7-4	2	6-7	2	8-2	2	7-2	2	6-5	3	7-8	2	6-9	2	6-1	3
	4-2x8	6-10	1	6-0	2	5-5	2	6-8	1	5-10	2	5-3	2	6-3	2	5-6	2	4-11	2
	4-2x10	8-4	2	7-4	2	6-7	2	8-2	2	7-2	2	6-5	2	7-7	2	6-8	2	6-0	2
	4-2x12	9-8	2	8-6	2	7-8	2	9-5	2	8-3	2	7-5	2	8-10	2	7-9	2	7-0	2
Roof, ceiling, and two clear span floors	2-2x4	2-1	1	1-8	1	1-6	2	2-0	1	1-8	1	1-5	2	2-0	1	1-8	1	1-5	2
	2-2x6	3-1	2	2-8	2	2-4	2	3-0	2	2-7	2	2-3	2	2-11	2	2-7	2	2-3	2
	2-2x8	3-10	2	3-4	2	3-0	3	3-10	2	3-4	2	2-11	3	3-9	2	3-3	2	2-11	3
	2-2x10	4-9	2	4-1	3	3-8	3	4-8	2	4-0	3	3-7	3	4-7	3	4-0	3	3-6	3
	2-2x12	5-6	3	4-9	3	4-3	3	5-5	3	4-8	3	4-2	3	5-4	3	4-7	3	4-1	4
	3-2x8	4-10	2	4-2	2	3-9	2	4-9	2	4-1	2	3-8	2	4-8	2	4-1	2	3-8	2
	3-2x10	5-11	2	5-1	2	4-7	3	5-10	2	5-0	2	4-6	3	5-9	2	4-11	2	4-5	3
	3-2x12	6-10	2	5-11	3	5-4	3	6-9	2	5-10	3	5-3	3	6-8	2	5-9	3	5-2	3
	4-2x8	5-7	2	4-10	2	4-4	2	5-6	2	4-9	2	4-3	2	5-5	2	4-8	2	4-2	2
	4-2x10	6-10	2	5-11	2	5-3	2	6-9	2	5-10	2	5-2	2	6-7	2	5-9	2	5-1	2
	4-2x12	7-11	2	6-10	2	6-2	3	7-9	2	6-9	2	6-0	3	7-8	2	6-8	2	5-11	3

No change to table notes.

Section R502.7 Change 'exception' to read as shown: (RB31-03/04)

R502.7 Lateral restraint at supports. Joists shall be supported laterally at the ends by full-depth solid blocking not less than 2 inches (51 mm) nominal in thickness; or by attachment to a header, band or rim joist, or to an adjoining stud or shall be otherwise provided with lateral support to prevent rotation.

Exception: In Seismic Design Categories D_0, D_1 and D_2, lateral restraint shall also be provided at each intermediate support.

Section R502.8.2 Change to read as shown: (S82-03/04)

R502.8.2 Engineered wood products. Cuts, notches and holes bored in trusses, structural composite lumber, structural glue-laminated members or I-joists are not permitted unless allowed by the manufacturer's recommendations or unless the effects of such penetrations are specifically considered in the design of the member by a registered design professional.

Section R502.13 Change to read as shown: (RB153-03/04)

R502.13 Fireblocking required. Fireblocking shall be provided in accordance with Section R602.8.

Table R503.2.1.1(1) Change table heading to reference Footnote d as shown: (CCC)

TABLE R503.2.1.1(1)
ALLOWABLE SPANS AND LOADS FOR WOOD STRUCTURAL PANELS FOR ROOF AND SUBFLOOR SHEATHING AND COMBINATION SUBFLOOR UNDERLAYMENT [a, b, c]

SPAN RATING	MINIMUM NOMINAL PANEL THICKNESS (inch)	MAXIMUM SPAN (Inches) With edge support[d]	MAXIMUM SPAN (Inches) Without edge support	LOAD (pounds per square foot, at maximum span) Total load	LOAD (pounds per square foot, at maximum span) Live load	MAXIMUM SPAN (inches)

(Portions of table not shown do not change)

Section R504.3 Change to read as shown: (RB154-03/04)

R504.3 Materials. All framing materials, including sleepers, joists, blocking and plywood subflooring, shall be pressure- preservative treated and dried after treatment in accordance with AWPA U1 (Commodity Specification A, Use Category 4B and section 5.2), and shall bear the label of an accredited agency.

Section R505.1.1 Change to read as shown: (RB155-03/04)

R505.1.1 Applicability limits. The provisions of this section shall control the construction of steel floor framing for buildings not greater than 60 feet (18 288 mm) in length perpendicular to the joist span, not greater than 36 feet (10 973 mm) in width parallel to the joist span, and not greater than two stories in height. Steel floor framing constructed in accordance with the provisions of this section shall be limited to sites subjected to a maximum design wind speed of 110 miles per hour (49 m/s) Exposure A, B or C and a maximum ground snow load of 70 psf (3.35 kN/m²).

2004 SUPPLEMENT TO THE IRC

Section R505.2.1 Change to read as shown: (RB157-03/04)

R505.2.1 Material. Load-bearing members utilized in steel floor construction shall be cold-formed to shape from structural quality sheet steel complying with the requirements of one of the following:

1. ASTM A 653: Grades 33, 37, 40 and 50 (Class 1 and 3).

2. ASTM A 792: Grades 33, 37, 40 and 50A.

3. ASTM A 875: Grades 33, 37, 40 and 50 (Class 1 and 3).

4. ASTM A 1003: Grades 33, 37, 40 and 50

Section R506.2.4 Add new section to read as shown: (RB159-03/04)

R506.2.4 Reinforcement Support. Where provided in slabs on ground, welded wire fabric shall be supported to remain in place from the center to upper one third of the slab, for the duration of the concrete placement.

2004 SUPPLEMENT TO THE IRC

Table R505.3.1(2) Change third column to read as shown: (RB158-03/04)

TABLE R505.3.1(2)
FLOOR FASTENING SCHEDULE [a]

DESCRIPTION OF BUILDING ELEMENTS	NUMBER AND SIZE OF FASTENERS	SPACING OF FASTENERS
Floor joist to track of an interior load-bearing wall per Figures R505.3.1(7) and R505.3.1(8)	2 No. 8 screws	Each joist
Floor joist to track and end of joist	2 No. 8 screws	One per flange or two per bearing stiffener
Subfloor to floor joists	No. 8 screws	6" o.c. on edges and 12" o.c. at intermediate supports

For SI: 1 inch = 25.4 mm
a. All screw sizes shown are minimum.

CHAPTER 6
WALL CONSTRUCTION

Table R602.3(1) Change entire table to read as shown: (RB163-03/04, RB168-03/04, S62-03/04)

TABLE R602.3(1)
FASTENER SCHEDULE FOR STRUCTURAL MEMBERS

DESCRIPTION OF BUILDING ELEMENTS	NUMBER AND TYPE OF FASTENER [a,b,c,d]	SPACING OF FASTENERS
Joist to sill or girder, toe nail	3-8d (2-1/2" x 0.113")	-
1" x 6" subfloor or less to each joist, face nail	2-8d (2-1/2" x 0.113")	-
	2 staples, 1 3/4"	-
2" subfloor to joist or girder, blind and face nail	2-16d (3-1/2" x 0.135")	-
Sole plate to joist or blocking, face nail	16d (3-1/2" x 0.135")	16" o.c.
Top or sole plate to stud, end nail	2-16d (3-1/2" x 0.135")	-
Stud to sole plate, toe nail	3-8d (2-1/2" x 0.113") or 2-16d (3-1/2" x 0.135")	-
Double studs, face nail	10d (3" x 0.128")	24" o.c.
Double top plates, face nail	10d (3" x 0.128")	24" o.c.
Sole plate to joist or blocking at braced wall panels	3-16d (3-1/2" x 0.135")	16" o.c.
Double top plates, minimum 24-inch offset of end joints, face nail in lapped area	8-16d (3-1/2" x 0.135")	-
Blocking between joists or rafters to top plate, toe nail	3-8d (2-1/2" x 0.113")	-
Rim joist to top plate, toe nail	8d (2-1/2" x 0.113")	6" o.c.
Top plates, laps at corners and intersections, face nail	2-10d (3" x 0.128")	-
Built-up header, two pieces with 1/2" spacer	16d (3-1/2" x 0.135")	16" o.c. along each edge
Continued header, two pieces	16d (3-1/2" x 0.135")	16" o.c. along each edge
Ceiling joists to plate, toe nail	3-8d (2-1/2" x 0.113")	-
Continuous header to stud, toe nail	4-8d (2-1/2" x 0.113")	-
Ceiling joist, laps over partitions, face nail	3-10d (3" x 0.128")	-
Ceiling joist to parallel rafters, face nail	3-10d (3" x 0.128")	-
Rafter to plate, toe nail	2-16d (3-1/2" x 0.135")	-
1" brace to each stud and plate, face nail	2-8d (2-1/2" x 0.113")	-
	2 staples, 1 3/4"	-
1" x 6" sheathing to each bearing, face nail	2-8d (2-1/2" x 0.113")	-
	2 staples, 1 3/4"	-
1" x 8" sheathing to each bearing, face nail	2-8d (2-1/2" x 0.113")	-
	3 staples, 1 3/4"	-
Wider than 1" x 8" sheathing to each bearing, face nail	3-8d (2-1/2" x 0.113")	-
	4 staples, 1 3/4"	-
Built-up corner studs	10d (3" x 0.128")	24" o.c.
Built-up girders and beams, 2-inch lumber layers	10d (3" x 0.128")	Nail each layer as follows: 32" o.c. at top and bottom and staggered. Two nails at ends and at each splice.
2" planks	2-16d (3-1/2" x 0.135")	At each bearing
Roof rafters to ridge, valley or hip rafters:		
toe nail	4-16d (3-1/2" x 0.135")	-
face nail	3-16d (3-1/2" x 0.135")	-
Rafter ties to rafters, face	3-8d (2-1/2" x 0.113")	-
Collar tie to rafter, face nail, or 1-1/4" x 20 gage ridge strap	3-10d (3" x 0.128")	-

IRC-27

2004 SUPPLEMENT TO THE IRC

TABLE 602.2(1)-continued
FASTENER SCHEDULE FOR STRUCTURAL MEMBERS

DESCRIPTION OF BUILDING MATERIALS	DESCRIPTION OF FASTENER[a,c,d,e]	SPACING OF FASTENERS	
		Edges (inches)[f]	Intermediate supports[c,e] inches
Wood structural panels, subfloor, roof and wall sheathing to framing, and particleboard wall sheathing to framing			
5/16"-½"	6d common (2" x 0.113") nail (subfloor, wall) 8d common (2-1/2" x 0.131") nail (roof)[f]	6	12[g]
19/32"-1"	8d common nail (2-1/2" x 0.131")	6	12[g]
1 1/8"-1 1/4"	10d common (3" x 0.148") nail or 8d (2-1/2" x 0.131") deformed nail	6	12
Other wall sheathing[h]			
½" regular cellulosic fiberboard sheathing	1 1/2" galvanized roofing nail 6d common (2" x 0.113") nail staple 16 ga., 1 1/2 long	3	6
½" structural cellulosic fiberboard sheathing	1 1/2" galvanized roofing nail 8d common (2-1/2" x 0.131") nail staple 16 ga., 1 1/2 long	3	6
25/32" structural cellulosic fiberboard sheathing	1 3/4" galvanized roofing nail 8d common (2-1/2" x 0.131") nail staple 16 ga., 1 3/4 long	3	6
½" gypsum sheathing	1 1/2" galvanized roofing nail; 6d common (2" x 0.113") nail; staple galvanized, 1 1/2" long; 1 1/4" screws, Type W or S	4	8
5/8" gypsum sheathing	1 3/4" galvanized roofing nail; 8d common (2-1/2" x 0.131") nail; staple galvanized, 1 5/8" long; 1 5/8" screws, Type W or S	4	8
Wood structural panels, combination subfloor underlayment to framing			
3/4" and less	6d deformed (2" x 0.120") nail or 8d common (2-1/2" x 0.131") nail	6	12
7/8"-1"	8d common (2-1/2" x 0.131") nail or 8d deformed (2-1/2" x 0.120") nail	6	12
1 1/8"- 1 1/4"	10d common (3" x 0.148") nail or 8d deformed (2-1/2" x 0.120") nail	6	12

For SI: 1 inch = 25.4 mm, 1 foot = 304.8 mm, 1 mile per hour = 0.447 m/s.
a. through d. (No change to current text)
e. Spacing of fasteners not included in this table shall be based on Table R602.3(2) ALTERNATE ATTACHMENTS.
f. For regions having basic wind speed of 110 mph or greater, 8d deformed (2-1/2" x 0.120") nails shall be used for attaching plywood and wood structural panel roof sheathing to framing within minimum 48-inch distance from gable end walls, if mean roof height is more than 25 feet, up to 35 feet maximum.
g. through l. (No change to current text)

Table R602.3.1 Change title to read as shown: (RB31-03/04)

TABLE R602.3.1
MAXIMUM ALLOWABLE LENGTH OF WOOD WALL STUDS EXPOSED TO WIND SPEEDS OF 100 MPH OR LESS IN SEISMIC DESIGN CATEGORIES A, B, C, D_0, D_1, and D_2[b,c]

(Portions of table not shown do not change)

Section R602.6 Change to read as shown: (RB169-03/04)

R602.6 Drilling and notching – studs. Drilling and notching of studs shall be in accordance with the following:

1. Notching. Any stud in an exterior wall or bearing partition may be cut or notched to a depth not exceeding 25 percent of its width. Studs in nonbearing partitions may be notched to a depth not to exceed 40 percent of a single stud width.

2. Drilling. Any stud may be bored or drilled, provided that the diameter of the resulting hole is no greater than 60 percent of the stud width, the edge of the hole is no greater than 5/8 inch (16 mm) to the edge of the stud, and the hole is not located in the same section as a cut or notch. Studs located in exterior walls or bearing partitions drilled over 40 percent and up to 60 percent shall also be doubled with no more than two successive doubled studs bored. See Figures R602.6(1) and R602.6(2).

Exception: Use of approved stud shoes is permitted when they are installed in accordance with the manufacturer's recommendations.

Section R602.10 Change to read as shown: (RB31-03/04)

R602.10 Wall bracing. All exterior walls shall be braced in accordance with this section. In addition, interior braced wall lines shall be provided in accordance with Section R602.10.1.1. For buildings

IRC-28

in Seismic Design Categories D_0, D_1 and D_2, walls shall be constructed in accordance with the additional requirements of Sections R602.10.9, R602.10.11, and R602.11.

Section R602.10.1 Change to read as shown: (RB171-03/04)

R602.10.1 Braced wall lines. Braced wall lines shall consist of braced wall panel construction in accordance with Section R602.10.3. The amount and location of bracing shall be in accordance with Table R602.10.1 and the amount of bracing shall be the greater of that required by the Seismic Design Category or the design wind speed. Braced wall panels shall begin no more than 12.5 feet (3810 mm) from each end of a braced wall line. Braced wall panels that are counted as part of a braced wall line shall be in line, except that offsets out-of-plane of up to 4 feet (1219 mm) shall be permitted provided that the total out-to-out offset dimension in any braced wall line is not more than 8 feet (2438 mm).

A designed collector shall be provided if the bracing begins more than 12.5 feet (3810 mm) from each end of a braced wall line.

Table R602.10.1 Change first column and Footnote 'c' to read as shown: (RB31-03/04 and S80-03/04)

(Portions of table not shown do not change)

SEISMIC DESIGN CATEGORY OR WIND SPEED
(unchanged)
(unchanged)
Categories D_0 and D_1 ($S_s \leq 1.25g$ and $S_{ds} \leq 0.83g$) or less than 110 mph
(unchanged)

a. and b. (No change to current text)
c. Methods of bracing shall be as described in Section R602.10.3. The alternate braced wall panels described in Sections R602.10.6.1 or R602.10.6.2 shall also be permitted.
d. and e. (No change to current text)

Section R602.10.3 Change to read as shown: (S80-03/04)

R602.10.3 Braced wall panel construction methods. The construction of braced wall panels shall be in accordance with one of the following methods:

1. Nominal 1-inch-by 4-inch (25.4 mm by 102 mm) continuous diagonal braces let in to the top and bottom plates and the intervening studs or approved metal strap devices installed in accordance with the manufacturer's specifications. The let-in bracing shall be placed at an angel not more than 60 degrees (1.06 rad) or less than 45 degrees (0.79 rad) from the horizontal.

2. Wood boards of 5/8 inch (15.9 mm) net minimum thickness applied diagonally on studs spaced a maximum of 24 inches (610 mm). Diagonal boards shall be attached to studs in accordance with Table R602.3(1).

3. Wood structural panel sheathing with a thickness not less than 5/16 inch (7.9 mm) for 16-inch (406 mm) stud spacing and less than 3/8 inch (9.5 mm) for 24-inch (610 mm) stud spacing. Wood structural panels shall be installed in accordance with Table R602.3(3).

4. One-half-inch (12.7 mm) or 25/32 -inch (19.8mm) thick structural fiberboard sheating applied vertically or horizontally on studs spaced a maximum of 16 inches (406 mm) on center. Structural fiberboard sheathing shall be installed in accordance with Table R602.3(1).

5. Gypsum board with minimum ½-inch (12.7 mm) thickness placed on studs spaced a maximum of 24 inches (610 mm) on center and fastened at 7 inches (178 mm) on center with the size nails specified in Table R602.3(1) for sheating and Table R702.3.5 for interior gypsum board.

6. Particleboard wall sheating panels installed in accordance with Table R602.3(4).

7. Portland cement plaster on studs spaced a maximum of 16 inches (406 mm) on center and installed in accordance with Section R703.6.

8. Hardboard panel siding when installed in accordance with Table R703.4.

Exception: Alternate braced wall panels constructed in accordance with Sections R602.10.6.1 or R602.10.6.2 shall be permitted to replace any of the above methods of braced wall panels.

Section R602.10.4 Change to read as shown: (S80-03/04)

R602.10.4 Length of braced panels. For Methods 2, 3, 4, 6, 7 and 8 above, each braced wall panel shall be at least 48 inches (1219 mm) in length, covering a minimum of three stud spaces where studs are spaced 16 inches (406 mm) on center and covering a minimum of two stud spaces where studs are spaced 24 inches (610 mm) on center. For Method 5 above, each braced wall panel shall be at least 96 inches (2438 mm) in length where applied to one face of a braced wall panel and at least 48 inches (1219 mm) where applied to both faces.

Exceptions:

1. Lengths of braced wall panels for continuous wood structural panel sheathing shall be in accordance with Section R602.10.5.

2. Lengths of alternate braced wall panels shall be in accordance with Section R602.10.6.1 or Section R602.10.6.2.

Table R602.10.5 Add new footnote 'c' to read as shown: (RB178-03/04)

(No change to table)

(No change to footnotes a and b)

c. Walls on either or both sides of openings in garages attached to fully sheathed dwellings shall be permitted to be built in accordance with Section R602.10.6.2 and Figure R602.10.6.2 except that a single bottom plate shall be permitted and two anchor bolts shall be placed at 1/3 points. In addition, tie-down devices shall not be required and the vertical wall segment shall have a maximum 6:1 height-to-width ratio (with height being measured from top of header to the bottom of the sill plate). This option shall be permitted for the first story of two-story applications in Seismic Design Categories A through C.

(No change to footnotes d and e)

Section R602.10.6.1 Change to read as shown: (S80-03/04)

R602.10.6.1 Alternate braced wall panels. Alternate braced wall lines constructed in accordance with one of the following provisions shall be permitted to replace each 4 feet (1219 mm) of braced wall panel as required by Section R602.10.4:

1. In one-story buildings, each panel shall have a length of not less than 2 feet, 8 inches (813 mm) and a height of not more than 10 feet (3048 mm). Each panel shall be sheathed on one face with 3/8-inch-minimum-thick-ness (9.5 mm) wood structural panel sheathing nailed with 8d common or galvanized box nails in accordance with Table R602.3(1) and blocked at all wood structural panel sheathing edges. Two anchor bolts installed in accordance with Figure R403.1(1) shall be provided in each panel. Anchor bolts shall be placed at panel quarter points. Each panel end stud shall have a tie-down device fastened to the foundation, capable of providing an uplift capacity of at least 1,800 pounds (816.5 kg). The tie-down device shall be installed in accordance with the manufacturer's recommendations. The panels shall be supported directly on a foundation or on floor framing supported directly on a foundation which is continuous across the entire length of the braced wall line. This foundation shall be reinforced with not less than one No. 4 bar top and bottom. When the continuous foundation is required to have a depth greater than 12 inches (305 mm), a minimum 12-inch-by-12-inch (305 mm by 305 mm) continuous footing or turned down slab edge is permitted at door openings in the braced wall line. This continuous footing or turned down slab edge shall be reinforced with not less than one No. 4 bar top and bottom. This reinforcement shall be lapped 15 inches (381 mm) with the reinforcement required in the continuous foundation located directly under the braced wall line.

2. In the first story of two-story buildings, each braced wall panel shall be in accordance with Item 1 above, except that the wood structural panel sheathing shall be provided on both faces, sheathing edge nailing spacing shall not exceed four inches on center, at least three anchor bolts shall be placed at one-fifth points, and tie-down device uplift capacity shall not be less than 3,000 pounds (1360.8 kg).

Section R602.10.6.2 Add new section to read as shown: (S80-03/04)

R602.10.6.2 Alternate bracing wall panel adjacent to a door or window opening. Alternate braced wall panels constructed in accordance with one of the following provisions are also permitted to replace each 4 feet (1219 mm) of braced wall panel as required by Section R602.10.4 for use adjacent to a window or door opening with a full-length header:

1. In one-story buildings, each panel shall have a length of not less than 16 inches (406 mm) and

a height of not more than 10 feet (3048 mm). Each panel shall be sheathed on one face with a single layer of 3/8-inch-minimum-thickness (9.5 mm) wood structural panel sheathing nailed with 8d common or galvanized box nails in accordance with Figure R602.10.6.2. The wood structural panel sheathing shall extend up over the solid sawn or glued-laminated header and shall be nailed in accordance with Figure R602.10.6.2. A built-up header consisting of at least two 2 x 12s and fastened in accordance with Table R602.3(1) shall be permitted to be used. A spacer, if used, shall be placed on the side of the built-up beam opposite the wood structural panel sheathing. The header shall extend between the inside faces of the first full-length outer studs of each panel. The clear span of the header between the inner studs of each panel shall be not less than 6 feet (1829 mm) and not more than 18 feet (5486 mm) in length. A strap with an uplift capacity of not less than 1000 pounds (4448 N) shall fasten the header to the side of the inner studs opposite the sheathing. One anchor bolt not less than 5/8-inch -diameter (16 mm) and installed in accordance with Section R403.1.6 shall be provided in the center of each sill plate. The studs at each end of the panel shall have a tie-down device fastened to the foundation with an uplift capacity of not less than 4,200 pounds (18683 N).

Where a panel is located on one side of the opening, the header shall extend between the inside face of the first full-length stud of the panel and the bearing studs at the other end of the opening. A strap with an uplift capacity of not less than 1000 pounds (4448 N) shall fasten the header to the bearing studs. The bearing studs shall also have a tie-down device fastened to the foundation with an uplift capacity of not less than 1000 pounds (4448 N).

The tie-down devices shall be an embedded-strap type, installed in accordance with the manufacturer's recommendations. The panels shall be supported directly on a foundation which is continuous across the entire length of the braced wall line. The foundation shall be reinforced with not less than one No. 4 bar top and bottom.

Where the continuous foundation is required to have a depth greater than 12 inches (305 mm), a minimum 12-inch-by-12-inch (305 mm by 305 mm) continuous footing or turned down slab edge is permitted at door openings in the braced wall line. This continuous footing or turned down slab edge shall be reinforced with not less than one No. 4 bar top and bottom. This reinforcement shall be lapped not less than 15 inches (381 mm) with the reinforcement required in the continuous foundation located directly under the braced wall line.

2. In the first story of two-story buildings, each wall panel shall be braced in accordance with Item 1 above, except that each panel shall have a length of not less than 24 inches (610 mm).

2004 SUPPLEMENT TO THE IRC

Figure R602.10.6.2 Add new figure as shown: (S80-03/04)

EXTENT OF HEADER
DOUBLE PORTAL FRAME (TWO BRACED WALL PANELS)

EXTENT OF HEADER
SINGLE PORTAL FRAME (ONE BRACED WALL PANEL)

MIN. 3" X 11.25" NET HEADER

6' TO 18'

FASTEN TOP PLATE TO HEADER WITH TWO ROWS OF 16D SINKER NAILS AT 3" O.C. TYP.

1000 LB STRAP OPPOSITE SHEATHING

FASTEN SHEATHING TO HEADER WITH 8D COMMON OR GALVANIZED BOX NAILS IN 3" GRID PATTERN AS SHOWN AND 3" O.C. IN ALL FRAMING (STUDS, BLOCKING, AND SILLS) TYP.

MIN. WIDTH = 16" FOR ONE STORY STRUCTURES
MIN. WIDTH = 24" FOR USE IN THE FIRST OF TWO STORY STRUCTURES

MIN. 2x4 FRAMING

3/8" MIN. THICKNESS WOOD STRUCTURAL PANEL SHEATHING

MIN. 4200 LB TIE-DOWN DEVICE (EMBEDDED INTO CONCRETE AND NAILED INTO FRAMING)

SEE SECTION R602.10.6.2

MAX. HEIGHT 10'

1000 LB STRAP

MIN. DOUBLE 2x4 POST

MIN. 1000 LB TIE DOWN DEVICE

TYPICAL PORTAL FRAME CONSTRUCTION

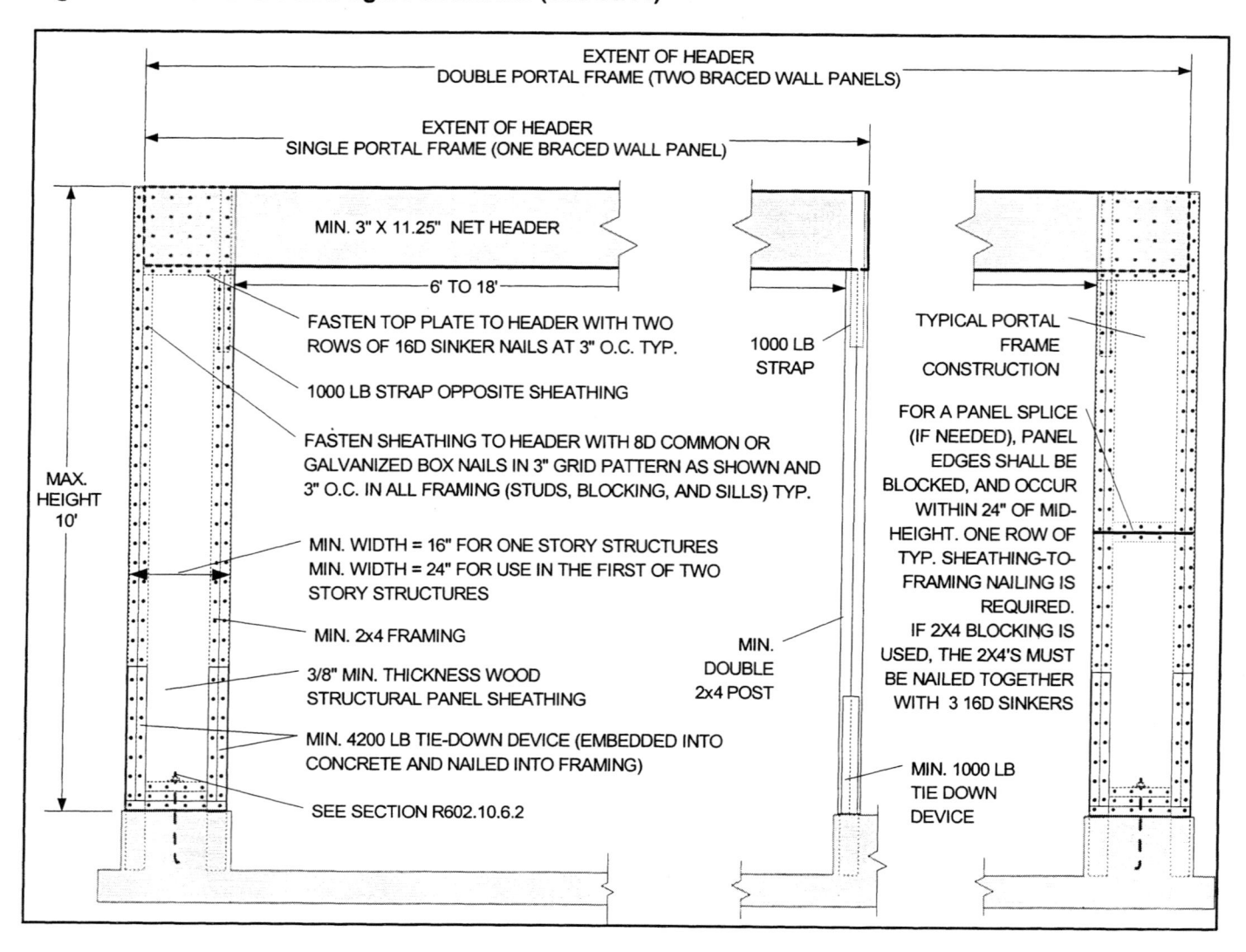

FOR A PANEL SPLICE (IF NEEDED), PANEL EDGES SHALL BE BLOCKED, AND OCCUR WITHIN 24" OF MID-HEIGHT. ONE ROW OF TYP. SHEATHING-TO-FRAMING NAILING IS REQUIRED. IF 2X4 BLOCKING IS USED, THE 2X4'S MUST BE NAILED TOGETHER WITH 3 16D SINKERS

**FIGURE R602.10.6.2
ALTERNATE BRACED WALL PANEL ADJACENT TO A DOOR OR WINDOW OPENING**

Section R602.10.7 Change to read as shown: (RB181-03/04)

R602.10.7 Panel joints. All vertical joints of panel sheathing shall occur over, and be fastened to, common studs. Horizontal joints in braced wall panels shall occur over, and be fastened to common blocking of a minimum 1-1/2 inch (38 mm) thickness.

> **Exception:** Blocking is not required behind horizontal joints in Seismic Design Categories A and B and detached dwellings in Seismic Design Category C when constructed in accordance with Section R602.10.3, braced-wall-panel construction method 3 and Table R602.10.1, method 3, or where permitted by the manufacturer's installation requirements for the specific sheathing material.

Section R602.10.11 Change to read as shown: (RB31-03/04 and RB171-03/04)

R602.10.11 Bracing in Seismic Design Categories D_0, D_1 and D_2. Structures located in Seismic Design Categories D_0, D_1 and D_2 shall be provided with exterior and interior braced wall lines. Spacing between braced wall lines in each story shall not exceed 25 feet (7620 mm) on center in both the longitudinal and transverse directions.

> **Exception:** In one- and two-story buildings, spacing between braced wall lines shall not exceed 35 feet (10 363 mm) on center in order to accommodate one single room not exceeding 900 square feet (84 m²) in each dwelling unit. The length of wall bracing in braced wall lines spaced greater or less than 25 feet (7620 mm) apart shall be the length required by Table R602.10.1 multiplied by the appropriate adjustment factor from Table R602.10.11.

Exterior braced wall lines shall have a braced wall panel located at each end of the braced wall line.

> **Exception:** For braced wall panel construction Method 3 of Section R602.10.3, the braced wall panel shall be permitted to begin no more than 8 feet (2438 mm) from each end of the braced wall line provided one of the following is satisfied:
>
> 1. A minimum 24-inch-wide (610 mm) panel is applied to each side of the building corner and the two 24-inch-wide (610 mm) panels at the corner shall be attached to framing in accordance with Figure R602.10.5 or,
>
> 2. The end of each braced wall panel closest to the corner shall have a tie-down device fastened to the stud at the edge of the braced wall panel closest to the corner and to the foundation or framing below. The tie-down device shall be capable of providing an uplift allowable design value of at least 1,800 pounds (8007 N). The tie-down device shall be installed in accordance with the manufacturer's recommendations.

A designed collector shall be provided if the bracing is not located at each end of a braced wall line as indicated above or more than 8 feet (2438 mm) from each end of a braced wall line as indicated in the exception.

Section R602.10.11.2 Change to read as shown: (RB31-03/04)

R602.10.11.2 Sheathing attachment. Adhesive attachment of wall sheathing shall not be permitted in Seismic Design Categories C, D_0, D_1 and D_2.

Section R602.11 Change to read as shown: (RB31-03/04 and RB 171-03/04)

R602.11 Framing and connections for Seismic Design Categories D_0, D_1 and D_2. The framing and connections details of buildings located in Seismic Design Categories D_0, D_1 and D_2 shall be in accordance with Sections R602.11.1 through R602.11.3.

Section R602.11.1 Change to read as shown: (S83-03/04 & RB31-03/04)

R602.11.1 Wall anchorage. Braced wall line sills shall be anchored to concrete or masonry foundations in accordance with Sections R403.1.6 and R602.11. For all buildings in Seismic Design Categories D_0, D_1 and D_2 and townhouses in Seismic Design Category C, plate washers, a minimum of 1/4 inch by 3 inches by 3 inches (6.4 mm by 76 mm by 76 mm) in size, shall be provided between the foundation sill plate and the nut. The hole in the plate washer is permitted to be diagonally slotted with a width of up to 3/16 inch (5 mm) larger than the bolt diameter and a slot length not to exceed 1 3/4 inches (44 mm), provided a standard cut washer is placed between the plate washer and the nut.

Section R603.1.1 Change to read as shown: (RB155-03/04)

R603.1.1 Applicability limits. The provisions of this section shall control the construction of exterior steel wall framing and interior load-bearing steel wall framing for buildings not greater than 60 feet (18 288 mm) in length perpendicular to the joist or truss span,

2004 SUPPLEMENT TO THE IRC

not greater than 36 feet (10 973 mm) in width parallel to the joist or truss span, and not greater than two stories in height. All exterior walls installed in accordance with the provisions of this section shall be considered as load-bearing walls. Steel walls constructed in accordance with the provisions of this section shall be limited to sites subjected to a maximum design wind speed of 110 miles per hour (49 m/s) Exposure A, B or C and a maximum ground snow load of 70 psf (3.35 kN/m^2).

Section R603.2.1 Change to read as shown: (RB157-03/04)

R603.2.1 Material. Load-bearing steel framing members shall be cold-formed to shape from structural quality sheet steel complying with the requirements of one of the following:

1. ASTM A 653: Grades 33, 37, 40 and 50 (Class 1 and 3).
2. ASTM A 792: Grades 33, 37, 40 and 50A.
3. ASTM A 875: Grades 33, 37, 40 and 50 (Class 1 and 3).
4. ASTM A 1003: Grades 33, 37, 40 and 50

Section R603.3.2 Change to read as shown: (RB194-03/04)

R603.3.2 Load-bearing walls. Steel studs shall comply with Tables R603.3.2(2) through R603.3.2(7) for steels with minimum yield strength of 33 ksi (227.7 MPa) and Tables R603.3.2(8) through R603.3.2(13) for steels with minimum yield strength of 50 ksi (345 MPa). Where the second floor has rooms other than sleeping rooms (See Table R301.5), live loads of 40 psf (1.92 kN/m^2) shall be permitted, provided that the next higher snow load column is used to select the stud size from Tables R603.3.2(2) through R603.3.2(13). Fastening requirements shall be in accordance with Section R603.2.4 and Table R603.3.2(1). Tracks shall have the same minimum thickness as the wall studs. Exterior walls with a minimum of ½-inch (12.7 mm) gypsum board installed in accordance with Section R702 on the interior surface and wood structural panels of minimum 7/16-inch thick (11.1 mm) oriented-strand board or 15/32-inch thick (11.9 mm) plywood installed in accordance with Table R603.3.2(1) on the outside surface shall be permitted to use the next thinner stud from Tables R603.3.2(2) through R603.3.2(13) but not less than 33 mils (0.84 mm). Interior load bearing walls with a minimum ½-inch (12.7 mm) gypsum board installed in accordance with Section R702 on both sides of the wall shall be permitted to use the next thinner stud from Tables R603.3.2(2) through R603.3.2(13) but not less than 33 mils (0.84 mm).

Table R603.6(5) Change second column to read as shown: (RB31-03/04)

HEADER SPAN (feet)	BASIC WIND SPEED (mph), EXPOSURE		
	85 A/B or Seismic Design Categories A,B,C and D$_0$, D$_1$, and D$_2$	85 C or less than 110 A/B	Less than 110 C

(No change to portions of table not shown)

Section R603.7 Change to read as shown: (RB195-03/04)

R603.7 Structural sheathing. In areas where the basic wind speed is less than 110 miles per hour (49 m/s), wood structural panel sheathing shall be installed on all exterior walls of buildings in accordance with this section. Wood structural panel sheathing shall consist of minimum 7/16-inch thick (11.1 mm) oriented-strand board or 15/32-inch thick (11.9 mm) plywood and shall be installed on all exterior wall surfaces in accordance with Section R603.7.1 and Figure R603.3. The minimum length of full height sheathing on exterior walls shall be determined in accordance with Table R603.7, but shall not be less than 20 percent of the braced wall length in any case. The minimum percentage of full height sheathing in Table R603.7 shall include only those sheathed wall sections, uninterrupted by openings, which are a minimum of 48 inches (1120 mm) wide. The minimum percentage of full-height structural sheathing shall be multiplied by a 1.10 for 9-foot-high (2743 mm) walls and multiplied by 1.20 for 10-foot-high (3048 mm) walls. In addition, structural sheathing shall:

1. Be installed with the long dimension parallel to the stud framing and shall cover the full vertical height of studs, from the bottom of the bottom track to the top of the top track of each story.

2. Be applied to each end (corners) of each of the exterior walls with a minimum 48-inch-wide (1220 mm) panel.

Section R603.7.1 Change to read as shown: (RB196-03/04)

R603.7.1 Structural sheathing fastening. All edges and interior areas of wood structural panel sheathing shall be fastened to a framing member and tracks in accordance with Table R603.3.2(1).

2004 SUPPLEMENT TO THE IRC

Section R606.2.4 Change to read as shown: (RB31-03/04)

R606.2.4 Parapet walls. Unreinforced solid masonry parapet walls shall not be less than 8 inches (203 mm) in thickness and their height shall not exceed four times their thickness. Unreinforced hollow unit masonry parapet walls shall be not less than 8 inches (203 mm) in thickness, and their height shall not exceed three times their thickness. Masonry parapet walls in areas subject to wind loads of 30 pounds per square foot (1.44 kN/m^2) located in Seismic Design Category D_0, D_1 or D_2, or on townhouses in Seismic Design Category C shall be reinforced in accordance with Section R606.11

Section R606.3 Change to read as shown: (RB197-03/04)

R606.3 Corbeled masonry. Solid masonry units shall be used for corbelling. The maximum corbelled projection beyond the face of the wall shall not be more than one-half of the wall thickness or one-half the wythe thickness for hollow walls; the maximum projection of one unit shall not exceed one-half the height of the unit or one-third the thickness at right angles to the wall. When corbelled masonry is used to support floor or roof-framing members, the top course of the corbel shall be a header course or the top course bed joint shall have ties to the vertical wall. The hollow space behind the corbelled masonry shall be filled with mortar or grout.

Section R606.3.1 Support conditions. Delete (relocated to new Section R606.4) (RB197-03/04)

Section R606.4 Add new section to read as shown: (RB197-03/04)

R606.4 Support conditions.

R606.4.1 Bearing on support. Each masonry wythe shall be supported by at least two-thirds of the wythe thickness.

R606.4.2 Support at foundation. Cavity wall or masonry veneer construction may be supported on an 8-inch (203 mm) foundation wall, provided the 8-inch (203 mm) wall is corbelled with solid masonry to the width of the wall system above. The total horizontal projection of the corbel shall not exceed 2 inches (51 mm) with individual corbels projecting not more than one-third the thickness of the unit or one-half the height of the unit.

(Renumber subsequent sections)

Section R606.11 Change to read as shown: (RB31-03/04)

R606.11 Seismic requirements. The seismic requirements of this section shall apply to the design of masonry and the construction of masonry building elements located in Seismic Design Category D_0, D_1 or D_2. Townhouses in Seismic Design Category C shall comply with the requirements of Section R606.11.2. These requirements shall not apply to glass unit masonry conforming to Section R610 or masonry veneer conforming to Section R703.7.

Section R606.11.1.1 Change to read as shown: (RB31-03/04 and RB198-03/04)

R606.11.1.1 Floor and roof diaphragm construction. Floor and roof diaphragms shall be constructed of wood structural panels attached to wood framing in accordance with Table R602.3(1) or to cold-formed steel floor framing in accordance with Table R505.3.1(2) or to cold-formed steel roof framing in accordance with Table R804.3. Additionally, sheathing panel edges perpendicular to framing members shall be backed by blocking, and sheathing shall be connected to the blocking with fasteners at the edge spacing. For Seismic Design Categories C, D_0, D_1 and D_2, where the width-to-thickness dimension of the diaphragm exceeds 2-to-1, edge spacing of fasteners shall be 4 inches (102 mm) on center.

Figure R606.10(3) Change figure title to read as shown: (RB31-03/04)

**FIGURE R606.10(3)
REQUIREMENTS FOR REINFORCED
MASONRY CONSTRUCTION IN SEISMIC
DESIGN CATEGORY D_0, D_1 or D_2**

(No change to figure or note)

Section R606.11.3 Change to read as shown: (RB31-03/04)

R606.11.3 Seismic Design Category D_0 and D_1. Structures in Seismic Design Categories D_0 or D_1 shall comply with the requirements of Seismic Design Category C and the additional requirements of this section.

Table R606.11.3.2 Change title to read as shown: (RB31-03/04)

**TABLE R606.11.3.2
MINIMUM DISTRIBUTED WALL
REINFORCEMENT FOR BUILDINGS ASSIGNED
TO SEISMIC DESIGN CATEGORY D_0 OR D_1**

(No change to table or notes)

2004 SUPPLEMENT TO THE IRC

Table R606.14.1 Change to read as shown: (RB199-03/04)

TABLE R606.14.1
MINIMUM CORROSION PROTECTION

MASONRY METAL ACCESSORY	STANDARD
Sheet metal ties or anchors completely embedded in mortar or grout	ASTM A 653, Coating Designation G60

(Portions of table not shown do not change)

Section R607.1.3 Change to read as shown: (RB31-03/04)

R607.1.3 Masonry in Seismic Design Categories D_0, D_1 and D_2. Mortar for masonry serving as the lateral-force-resisting system in Seismic Design Categories D_0, D_1 and D_2 shall be type M or S portland cement-lime or mortar cement mortar.

Section R611.2 Change to read as shown: (RB31-03/04)

R611.2 Applicability limits. The provisions of this section shall apply to the construction of insulating concrete form walls for buildings greater than 60 feet (18 288 mm) in plan dimensions, and floors not greater than 32 feet (9754 mm) or roofs not greater than 40 feet (12 192 mm) in clear span. Buildings shall not exceed two stories in height above-grade. Insulating concrete form (ICF) walls shall comply with the requirements in Table R611.2. Walls constructed in accordance with the provisions of this section shall be limited to buildings subjected to a maximum design wind speed of 150 miles per hour (67 m/s), and Seismic Design Categories A, B, C, D_0, D_1 and D_2. The provisions of this section shall not apply to the construction of insulating concrete form walls for buildings or portions of buildings considered irregular as defined in Section R301.2.2.2.2.

2004 SUPPLEMENT TO THE IRC

Table R611.2 Change table headings to read as shown: (RB201-03/04)

TABLE R611.2
REQUIREMENTS FOR ICF WALLS [b]

WALL TYPE AND SIZE	MAXIMUM WALL WEIGHT (PSF) [c]	MINIMUM WIDTH OF VERTICAL CORE (INCHES) [a]	MINIMUM THICKNESS OF VERTICAL CORE (INCHES) [a]	MAXIMUM SPACING OF VERTICAL CORES (INCHES)	MAXIMUM SPACING OF HORIZONTAL CORES (INCHES)	MINIMUM WEB THICKNESS (INCHES)

(Portions of table not shown do not change)

No change to tabular values or footnotes.

For townhouses in Seismic Design Category C and all buildings in Seismic Design Category D_0, D_1 or D_2, the provisions of this section shall apply only to buildings meeting the following requirements.

Sections R611.4 and R611.5 Change to read as shown: (CCC)

Table R611.3(2) Change table notes to read as shown: (RB31-03/04)

TABLE R611.3(2)
MINIMUM VERTICAL WALL REINFORCEMENT FOR FLAT ICF ABOVE-GRADE WALLS[a, b, c, d]

(No change to table)
a. This table is based on reinforcing bars with a minimum yield strength of 40,000 psi (276 MPa) and concrete with a minimum specified compressive strength of 2,500 psi (17.2MPa). For Seismic Design Categories D_0, D_1, and D_2, reinforcing bars shall have a minimum yield strength of 60,000 psi (414 MPa). See Section R611.6.2.
b. through e. (No change to current text)
f. See Section R611.7.1.2 for limitations on maximum spacing of vertical reinforcement in Seismic Design Categories C, D_0, D_1, and D_2.
g. (No change to current text)

R611.4 Waffle-grid insulating concrete form wall systems. Waffle-grid wall systems shall comply with Figure R611.4 and shall have reinforcement in accordance with Tables R611.3(1) and R611.4(1) and Section R611.7. The minimum core dimensions shall comply with Table R611.2.

Table R611.4(1) Change table notes to read as shown: (RB31-03/04)

TABLE R611.4(1)
MINIMUM VERTICAL WALL REINFORCEMENT FOR WAFFLE-GRID ICF ABOVE-GRADE WALLS[a, b, c]

(No change to table)

a. This table is based on reinforcing bars with a minimum yield strength of 40,000 psi (276 MPa) and concrete with a minimum specified compressive strength of 2,500 psi (17.2 MPa). For Seismic Design Categories D_0, D_1 and D_2, reinforcing bars shall have a minimum yield strength of 60,000 psi (414 MPa). See Section R611.6.2.
b. through d. (No change to current text)
e. See Section R611.7.1.2 for limitations on maximum spacing of vertical reinforcement in Seismic Design Categories C, D_0, D_1 and D_2.

Table R611.4(2) Delete Table R611.4(2) without substitution. (CCC)

R611.5 Screen-grid insulating concrete form wall systems. Screen-grid ICF wall systems shall comply with Figure R611.5 and shall have reinforcement in accordance with Tables R611.3(1) and R611.5 and Section R611.7. The minimum core dimensions shall comply with Table R611.2.

Section R611.6.1 Change to read as shown: (RB31-3/04)

R611.6.1 Concrete material. Ready-mixed concrete for insulating concrete form walls be in accordance with Section R402.2. Maximum slump shall not be greater than 6 inches (152 mm) as determined in accordance with ASTM C 143. Maximum aggregate size shall not be larger than 3/4 inch (19.1 mm).

Exception: Concrete mixes conforming to the ICF manufacturer's recommendations.

1. Rectangular buildings with a maximum building aspect ratio of 2:1. The building aspect ratio shall be determined by dividing the longest dimension of the building by the shortest dimension of the building.

2. Walls are aligned vertically with the walls below.

3. Cantilever and setback construction shall not be permitted.

2004 SUPPLEMENT TO THE IRC

4. The weight of interior and exterior finishes applied to ICF walls shall not exceed 8 psf (0.38 kN/m^2).

5. The gable portion of ICF walls shall be constructed of light-frame construction.

In Seismic Design Categories D_0, D_1 and D_2, the minimum concrete compressive strength shall be 3,000 psi (20.5 MPa).

Section R611.6.2 Change to read as shown: (RB31-03/04)

R611.6.2 Reinforcing steel. Reinforcing steel shall meet the requirements of ASTM A 615, A 706, or A 996. Except in Seismic Design Categories D_0, D_1 and D_2, the minimum yield strength of reinforcing steel shall be 40,000 psi (Grade 40) (276 MPa). In Seismic Design Categories D_0, D_1 and D_2, reinforcing steel shall meet the requirements of ASTM A 706 for low-alloy steel with a minimum yield strength of 60,000 psi (Grade 60) (414 Mpa).

Section R611.7.1.2 Change to read as shown: (RB31-03/04)

R611.7.1.2 Vertical steel. Above-grade concrete walls shall have reinforcement in accordance with Sections R611.3, R611.4, or R611.5 and R611.7.2. All vertical reinforcement in the top-most ICF story shall terminate with a bend or standard hook and be provided with a minimum lap splice of 24 inches (610 mm) with the top horizontal reinforcement.

For townhouses in Seismic Design Category C, The minimum vertical reinforcement shall be one No. 5 bar at 24 inches (610 mm) on center or one No. 4 at 16 inches (407 mm) on center. For all buildings in Seismic Design Categories D_0, D_1 and D_2, the minimum vertical reinforcement shall be one No. 5 bar at 18 inches (457 mm) on center or one No. 4 at 12 inches (305 mm) on center.

Above-grade ICF walls shall be supported on concrete foundations reinforced as required for the above-grade wall immediately above, or in accordance with Tables R404.4(1) through R404.4(5), whichever requires the greater amount of reinforcement.

Vertical reinforcement shall be continuous from the bottom of the foundation wall to the roof. Lap splices, if required, shall comply with Section R611.7.1.5. Where vertical reinforcement in the above-grade wall is not continuous with the foundation wall reinforcement, dowel bars with a size and spacing to match the vertical ICF wall reinforcement shall be embedded $40d_b$ into the foundation wall and shall be lap spliced with the above-grade wall reinforcement. Alternatively, for No. 6 and larger bars, the portion of the bar embedded in the foundation wall shall be embedded 24 inches in the foundation wall and shall have a standard hook.

Section R611.7.1.3 Change to read as shown: (RB31-03/04)

R611.7.1.3 Horizontal reinforcement. Concrete walls with a minimum thickness of 4 inches (102 mm) shall have a minimum of one continuous No. 4 horizontal reinforcing bar placed at 32 inches (812 mm) on center with one bar within 12 inches (305 mm) of the top of the wall story. Concrete walls 5.5 inches (140 mm) thick or greater shall have a minimum of one continuous No. 4 horizontal reinforcing bar placed at 48 inches (1219 mm) on center with one bar located within 12 inches (305 mm) of the top of the wall story.

For townhouses in Seismic Design Category C, the minimum horizontal reinforcement shall be one No. 5 bar at 24 inches (610 mm) on center or one No. 4 at 16 inches (407 mm) on center. For all buildings in Seismic Design Categories D_0, D_1 and D_2, the minimum horizontal reinforcement shall be one No. 5 bar at 18 inches (457 mm) on center or one No. 4 at 12 inches (305 mm) on center.

Horizontal reinforcement shall be continuous around building corners using corner bars or by bending the bars. In either case, the minimum lap splice shall be 24 inches (610 mm). For townhouses in Seismic Design Category C and for all buildings in Seismic Design Categories D_0, D_1 and D_2, each end of all horizontal reinforcement shall terminate with a standard hook or lap splice.

Table R611.5 Change table notes to read as shown: (RB31-03/04)

TABLE R611.5
MINIMUM VERTICAL WALL REINFORCEMENT FOR SCREEN-GRID ICF ABOVE-GRADE WALLS[a, b, c]

(No change to table)

a. This table is based on reinforcing bars with a minimum yield strength of 40,000 psi (276 MPa) and concrete with a minimum specified compressive strength of 2,500 psi (17.2 MPa). For Seismic Design Categories D_0, D_1 and D_2, reinforcing bars shall have a minimum yield strength of 60,000 psi (414 MPa). See Section R611.6.2.
b. through d. (No change to current text)
e. See Section R611.7.1.2 for limitations on maximum spacing of vertical reinforcement in Seismic Design Categories C, D_0, D_1 and D_2.

Table R611.7(1) Change third column to read as shown: (RB31-03/04)

**TABLE R611.7(1)
MINIMUM WALL OPENING REINFORCEMENT REQUIREMENTS IN ICF WALLS**

MINIMUM VERTICAL OPENING REINFORCEMENT
(no change)
In locations with wind speeds less than or equal to 110 mph (177km/hr) or in Seismic Design Categories A and B, provide one No.4 bar for the full height of the wall story within 12 inches (305mm) of each side of the opening
In locations with wind speeds greater than 110 mph (49 m/s), townhouses in Seismic Design Category C, or all buildings in Seismic Design Categories D_0, D_1 and D_2, provide two No. 4 bars or one No. 5 bar for the full height of the wall story within 12 inches (305 mm) of each side of the opening.

(Portions of table not shown do not change)
(No change to footnote)

Tables R611.7(4) and R611.7(5) Change Footnote g to read as shown: (CCC)

a through f (no change)
g. For actual wall lintel width, refer to Table R611.2.
h. (no change)

Tables R611.7(6) and R611.7(7) Change Footnote h to read as shown: (CCC)

a through g (no change)
h. For actual wall lintel width, refer to Table R611.2.
I. (no change)

Table R611.7(9) Change Footnote g to read as shown: (CCC)

a through f (no change)
g. Actual thickness is shown for flat lintels while nominal thickness is given for waffle-grid and screen-grid lintels. Lintel thickness corresponds to the nominal waffle-grid and screen-grid ICF wall thickness. Refer to Table R611.2 for actual wall thickness.
h. (no change)

Table R611.7(11) Change title to read as shown: (RB31-03/04)

2004 SUPPLEMENT TO THE IRC

**TABLE R611.7(11)
MINIMUM PERCENTAGE OF SOLID WALL LENGTH ALONG EXTERIOR WALL LINES FOR TOWNHOUSES IN SEISMIC DESIGN CATEGORY C AND ALL BUILDINGS IN SEISMIC DESIGN CATEGORIES D_0, D_1 and D_2[a,b]**

(No change to table or notes)

Section R611.7.4 Change to read as shown: (RB31-03/04)

R611.7.4 Minimum length of wall without openings. The wind velocity pressures of Table R611.7.4 shall be used to determine the minimum amount of solid wall length in accordance with Tables R611.7(9A) through R611.7(10B) and Figure R611.7.4. Table R611.7(11) shall be used to determine the minimum amount of solid wall length for townhouses in Seismic Design Category C, and all buildings in Seismic Design Categories D_0, D_1 and D_2 for all types of ICF walls. The greater amount of solid wall length required by wind loading or seismic loading shall apply. The minimum percentage of solid wall length shall include only those solid wall segments that are a minimum of 24 inches (610 mm) in length. The maximum distance between wall segments included in determining solid wall length shall not exceed 18 feet (5486 mm). A minimum length of 24 inches (610 mm) of solid wall segment, extending the full height of each wall story, shall occur at all interior and exterior corners of exterior walls.

Section R611.8.1.1 Change to read as shown: (RB31-03/04)

R611.8.1.1 Top bearing requirements for Seismic Design Categories C, D_0, D_1 and D_2. For townhouses in Seismic Design Category C, wood sill plates attached to ICF walls shall be anchored with Grade A 307, 3/8-inch-diameter (9.5 mm) headed anchor bolts embedded a minimum of 7 inches (178 mm) and placed at a maximum spacing of 36 inches (914 mm) on center. For all buildings in Seismic Design Category D_0 or D_1, wood sill plates attached to ICF walls shall be anchored with ASTM A 307, Grade A, 3/8-inch-diameter (9.5 mm) headed anchor bolts embedded a minimum of 7 inches (178 mm) and placed at a maximum spacing of 24 inches (610 mm) on center. For all buildings in Seismic Design Category D_2, wood sill plates attached to ICF walls shall be anchored with ASTM A 307, Grade A, 3/8-inch-diameter (9.5 mm) headed anchor bolts embedded a minimum of 7 inches (178 mm) and placed at a maximum spacing of 16 inches (406 mm) on center. Larger diameter bolts than specified herein shall not be used.

For townhouses in Seismic Design Category C, each floor joist perpendicular to an ICF wall shall be attached to the sill plate with an 18-gage [(0.0478 in.) (1.2 mm)] angle bracket using 3 - 8d common nails per leg in accordance with Figure R611.8(1). For all buildings in Seismic Design Category D_0 or D_1, each floor joist perpendicular to an ICF wall shall be attached to the sill plate with an 18-gage [(0.0478 in..) (1.2 mm)] angle bracket using 4 - 8d common nails per leg in accordance with Figure R611.8(1). For all buildings in Seismic Design Category D_2, each floor joist perpendicular to an ICF wall shall be attached to the sill plate with an 18-gage [(0.0478 in.) (1.2 mm)] angle bracket using 6 - 8d common nails per leg in accordance with Figure R611.8(1).

For ICF walls parallel to floor framing in townhouses in Seismic Design Category C, full depth blocking shall be placed at 24 inches (610 mm) on center and shall be attached to the sill plate with an 18-gage [(0.0478 in.) (1.2 mm)] angle bracket using 5 - 8d common nails per leg in accordance with Figure R611.8(6). For ICF walls parallel to floor framing for all buildings in Seismic Design Category D_0 or D_1, full depth blocking shall be placed at 24 inches (610 mm) on center and shall be attached to the sill plate with an 18-gage [(0.0478 in.) (1.2 mm)] angle bracket using 6 - 8d common nails per leg in accordance with Figure R611.8(6). For ICF walls parallel to floor framing for all buildings in Seismic Design Category D_2, full depth blocking shall be placed at 24 inches (610 mm) on center and shall be attached to the sill plate with an 18-gage [(0.0478 in.) (1.2 mm)] angle bracket using 9 - 8d common nails per leg in accordance with Figure R611.8(6).

Section R611.8.2.1 Change to read as shown: (RB31-03/04)

R611.8.2.1 Ledger bearing requirements for Seismic Design Categories C, D_0, D_1 and D_2. Additional anchorage mechanisms connecting the wall to the floor system shall be installed at a maximum spacing of 6 feet (1829 mm) on center for townhouses in Seismic Design Category C and 4 feet (1220 mm) on center for all buildings in Seismic Design Categories D_0, D_1 and D_2. The additional anchorage mechanisms shall be attached to the ICF wall reinforcement and joist rafters or blocking in accordance with Figures R611.8(1) through R611.8(7). The additional anchorage shall be installed through an oversized hole in the ledger board that is ½ inch (12.7 mm) larger than the anchorage mechanism diameter to prevent combined tension and shear in the mechanism. The blocking shall be attached to floor or roof sheathing in accordance with edge fastener spacing. Such additional anchorage shall not be accomplished by the use of toe nails or nails subject to withdrawal nor shall such anchorage mechanisms induce tension stresses perpendicular to grain in ledgers or nailers. The capacity of such anchors shall result in connections capable of resisting the design values listed in Table R611.8(2). The diaphragm sheathing fasteners applied directly to a ledger shall not be considered effective in providing the additional anchorage required by this section.

Where the additional anchorage mechanisms consist of threaded rods with hex nuts or headed bolts complying with ASTM A 307, Grade A or ASTM F 1554, Grade 36, the design tensile strengths shown in Table R611.9 shall be equal to or greater than the product of the design values listed in Table R611.8(2) and the spacing of the bolts in feet (mm). Anchor bolts shall be embedded as indicated in Table R611.9. Bolts with hooks shall not be used.

Section R611.8.3 Change to read as shown: (RB202-03/04)

R611.8.3 Floor and roof diaphragm construction. Floor and roof diaphragms shall be constructed of wood structural panel sheathing attached to wood framing in accordance with Table R602.3(1) or Table R602.3(2) or to cold-formed steel floor framing in accordance with Table R505.3.1(2) or to cold-formed steel roof framing in accordance with Table R804.3.

Section R611.8.3.1 Change to read as shown: (RB31-03/04)

R611.8.3.1 Floor and roof diaphragm construction requirements in Seismic Design Categories D_0, D_1 and D_2. The requirements of this section shall apply in addition to those required by Section R611.8.3. Edge spacing of fasteners in floor and roof sheathing shall be 4 inches (102 mm) on center for Seismic Design Category D_0 or D_1 and 3 inches (76 mm) on center for Seismic Design Category D_2. In Seismic Design Categories D_0, D_1 and D_2, all sheathing edges shall be attached to framing or blocking. Minimum sheathing fastener size shall be 0.113 inch (2.8 mm) diameter with a minimum penetration of 13/8-inches (35 mm) into framing members supporting the sheathing. Minimum wood structural panel thickness shall be 7/16 inch (11 mm) for roof sheathing and 23/32 inch (18 mm) for floor sheathing. Vertical offsets in floor framing shall not be permitted.

Section R611.9.1 Change to read as shown: (RB31-03/04)

R611.9.1 ICF wall to top sill plate (roof) connections for Seismic Design Categories C, D_0, D_1 and D_2. The requirements of this section shall apply in addition to those required by Section R611.9. The top of an ICF wall at a gable shall be attached to an attic floor in accordance with Section R611.8.1.1. For townhouses in Seismic Design Category C, attic floor diaphragms shall be constructed of structural wood sheathing panels attached to wood framing in accordance with Table R602.3(1) or Table R602.3(2). Edge spacing of fasteners in attic floor sheathing shall be 4 inches (102 mm) on center for Seismic Design Category D_0 or D_1 and 3 inches (76 mm) on center for Seismic Design Category D_2. In Seismic Design Categories D_0, D_1 and D_2, all sheathing edges shall be attached to framing or blocking. Minimum sheathing fastener size shall be 0.113 inch (2.8 mm) diameter with a minimum penetration of 1 3/8 inches (35 mm) into framing members supporting the sheathing. Minimum wood structural panel thickness shall be 7/16 inch (11 mm) for the attic floor sheathing. Where hipped roof construction is used, the use of a structural attic floor is not required.

For townhouses in Seismic Design Category C, wood sill plates attached to ICF walls shall be anchored with ASTM A 307, Grade A, 3/8-inch (9.5 mm) diameter anchor bolts embedded a minimum of 7 inches (178 mm) and placed at a maximum spacing of 36 inches (914 mm) on center. For all buildings in Seismic Design Category D_0 or D_1, wood sill plates attached to ICF walls shall be anchored with ASTM A 307, Grade A, 3/8-inch (9.5 mm) diameter anchor bolts embedded a minimum of 7 inches (178 mm) and placed at a maximum spacing of 16 inches (406 mm) on center. For all buildings in Seismic Design Category D_2, wood sill plates attached to ICF walls shall be anchored with ASTM A 307, Grade A, 3/8-inch (9.5 mm) diameter anchor bolts embedded a minimum of 7 inches (178 mm) and placed at a maximum spacing of 16 inches (406 mm) on center.

For townhouses in Seismic Design Category C, each floor joist shall be attached to the sill plate with an 18-gage [(0.0478 in.) (1.2 mm)] angle bracket using 3 - 8d common nails per leg in accordance with Figure R611.8(1). For all buildings in Seismic Design Category D_0 or D_1, each floor joist shall be attached to the sill plate with an 18-gage [(0.0478 in.) (1.2 mm)] angle bracket using 4 - 8d common nails per leg in accordance with Figure R611.8(1). For all buildings in Seismic Design Category D_2, each floor joist shall be attached to the sill plate with an 18-gage [(0.0478 in.) (1.2 mm)] angle bracket using 6-8d common nails per leg in accordance with Figure R611.8(1).

Where hipped roof construction is used without an attic floor, the following shall apply. For townhouses in Seismic Design Category C, each rafter shall be attached to the sill plate with an 18-gage [(0.0478 in.) (1.2 mm)] angle bracket using 3 - 8d common nails per leg in accordance with Figure R611.9. For all buildings in Seismic Design Category D_0 or D_1, each rafter shall be attached to the sill plate with an 18-gage [(0.0478 in.) (1.2 mm)] angle bracket using 4 - 8d common nails per leg in accordance with Figure R611.9. For all buildings in Seismic Design Category D_2, each rafter shall be attached to the sill plate with an 18-gage [(0.0478 in.) (1.2 mm)] angle bracket using 6-8d common nails per leg in accordance with Figure R611.9.

Table R611.8(2) Change title to read as shown: (RB31-03/04)

TABLE R611.8(2)
DESIGN VALUES (plf) FOR FLOOR JOIST-TO-WALL ANCHORS REQUIRED FOR TOWNHOUSES IN SEISMIC DESIGN CATEGORY C AND ALL BUILDINGS IN SEISMIC DESIGN CATEGORIES D_0, D_1 and D_2[a,b]

(No change to table or notes)

Figures R611.8(6) and R611.8(7) Change title to read as shown: (RB31-03/04)

FIGURE R611.8(6)
ANCHORAGE REQUIREMENTS FOR TOP BEARING WALLS FOR TOWNHOUSES IN SEISMIC DESIGN CATEGORY C AND ALL BUILDINGS IN SEISMIC DESIGN CATEGORIES D_0, D_1 and D_2 FOR FLOOR FRAMING PARALLEL TO WALL

(No change to figure)

FIGURE R611.8(7)
ANCHORAGE REQUIREMENTS FOR LEDGER BEARING WALLS FOR TOWNHOUSES IN SEISMIC DESIGN CATEGORY C AND ALL BUILDINGS IN SEISMIC DESIGN CATEGORIES D_0, D_1 and D_2 FOR FLOOR FRAMING PARALLEL TO WALL

(No change to figure)

Figure R611.9 Change text box to read as shown: (RB31-03/04)

CLIP ANGLE AT EACH ROOF FRAMING MEMBER IN SEISMIC DESIGN CATEGORIES C, D_0, D_1 and D_2 PER SECTION R611.9.1

Section R613.1 Change to read as shown: (RB203-03/04)

R613.1 General. This section prescribes performance and construction requirements for exterior window systems installed in wall systems. Windows shall be installed and flashed in accordance with the manufacturer's written installation instructions. Each window shall be provided with written installation instructions provided by the manufacturer of the product.

2004 SUPPLEMENT TO THE IRC

Section R613.2 Add new section to read as shown: (RB205-03/04)

R613.2 Window sills. In dwelling units, where the rough opening for the sill portion of an operable window is located more than 72 inches above the ground or other surface below, the rough opening for the sill portion of the window shall be a minimum of 24 inches above the finished floor of the room in which the window is located.

> **Exception.** Windows whose openings will not allow a 4 inch diameter sphere to pass through the opening when the opening is in its largest opened position.

Section R613.4 Change to read as shown: (CCC)

R613.4 Windborne debris protection. Protection of exterior windows and glass doors in buildings located in windborne debris regions shall be in accordance with Section R301.2.1.2.

Section R613.4.1 Add new text to read as shown: (RB207-03/04)

R613.4.1 Fenestration testing and labeling. Fenestration shall be tested by an approved independent laboratory, listed by an approved entity, and bear a label identifying manufacturer, performance characteristics, and approved inspection agency to indicate compliance with the requirements of the following specifications:

ASTM E 1886 and ASTM E 1996 or   AAMA 506

Section R613.6 and R613.6.1 Change to read as shown: (RB208-03/04 and RB209-03/04)

R613.6 Mullions occurring between individual window and glass door assemblies.

R613.6.1 Mullions. Mullions shall be tested by an approved testing laboratory in accordance with AAMA 450, or be engineered in accordance with accepted engineering practice. Mullions tested as stand alone units or qualified by engineering shall use performance criteria cited in Sections R613.6.2, R613.6.3 and R613.6.4. Mullions qualified by an actual test of an entire assembly shall comply with Sections R613.6.2 and R613.6.4.

CHAPTER 7
WALL COVERING

Section R702.3.7 Add new text to read as shown: (RB210-03/04)

R702.3.7 Horizontal Gypsum Board Diaphragm Ceilings. Use of gypsum board shall be permitted on wood joists to create a horizontal diaphragm in accordance with Table R702.3.7. Gypsum board shall be installed perpendicular to ceiling framing members. End joints of adjacent courses of board shall not occur on the same joist. The maximum allowable diaphragm proportions shall be 1 1/2:1 between shear resisting elements. Rotation or cantilever conditions shall not be permitted. Gypsum board shall not be used in diaphragm ceilings to resist lateral forces imposed by masonry or concrete construction. All perimeter edges shall be blocked using wood members not less than 2-inch (51 mm) by 6-inch (159 mm) nominal dimension. Blocking material shall be installed flat over the top plate of the wall to provide a nailing surface not less than 2 inches (51 mm) in width for the attachment of the gypsum board.

Section R702.3.8 and R702.3.8.1 Add new text to read as shown: (RB211-03/04)

R702.3.8 Water-resistant gypsum backing board. Gypsum board used as the base or backer for adhesive application of ceramic tile or other required nonabsorbent finish material shall conform to ASTM C 630 or C 1178. Use of water-resistant gypsum backing board shall be permitted on ceilings where framing spacing does not exceed 12 inches (305 mm) on center for 1/2-inch-thick (12.7 mm) or 16 inches (406 mm) for 5/8-inch-thick (15.9 mm) gypsum board. Water-resistant gypsum board shall not be installed over a vapor retarder in a shower or tub compartment. All cut or exposed edges, including those at wall intersections, shall be sealed as recommended by the manufacturer.

R702.3.8.1 Limitations. Water resistant gypsum backing board shall not be used where there will be direct exposure to water, or in areas subject to continuous high humidity.

Table R702.3.7 Add new table to read as shown: (RB210-03/04)

TABLE R702.3.7
SHEAR CAPACITY FOR HORIZONTAL WOOD-FRAMED
GYPSUM BOARD DIAPHRAGM CEILING ASSEMBLIES

MATERIAL	THICKNESS OF MATERIAL (min.) (in.)	SPACING OF FRAMING MEMBERS (max.) (in.)	SHEAR VALUE[a,b] (plf OF CEILING)	MINIMUM FASTENER SIZE[c,d]
Gypsum Board	1/2	16 o.c.	90	5d cooler or wallboard nail; 1 5/8-inch long; 0.086 inch shank; 15/64 inch head
Gypsum Board	1/2	24 o.c.	70	5d cooler or wallboard nail; 1 5/8-inch long; 0.086 inch shank; 15/64 inch head

For SI: 1 inch = 25.4 mm, 1 pound per linear foot = 1.488 kg/m.

[a] Values are not cumulative with other horizontal diaphragm values and are for short-term loading due to wind or seismic loading. Values shall be reduced 25 percent for normal loading.

[b] Values shall be reduced 50 percent in Seismic Design Categories D_0, D_1, D_2, and E.

[c] 1-1/4"; #6 Type S or W screws may be substituted for the listed nails.

[d] Fasteners shall be spaced not more than 7 inches on center at all supports, including perimeter blocking, and not less than 3/8 inch form the edges and ends of the gypsum board.

Sections R702.4.2 and R702.4.3 are deleted without substitution. (RB211-03/04)

Section R702.4.2 Add new section to read as shown: (RB212-03/04)

R702.4.2. Cement, fiber-cement, and glass mat gypsum backers. Cement, fiber-cement or glass mat gypsum backers in compliance with ASTM C 1288, C 1325 or C 1178 and installed in accordance with manufacturers' recommendations shall be used as backers for wall tile in tub and shower areas and wall panels in shower areas.

Section R703.1 Change to read as shown: (RB213-03/04)

R703.1 General. Exterior walls shall provide the building with a weather-resistant exterior wall envelope. The exterior wall envelope shall include flashing as described in Section R703.8. The exterior wall envelope shall be designed and constructed in such a manner as to prevent the accumulation of water within the wall assembly by providing a water-resistant barrier behind the exterior veneer as required by Section R703.2. and a means of draining water that enters the assembly to the exterior. Protection against condensation in the exterior wall assembly shall be provided in accordance with Chapter 11 of this code.

Exceptions:

1. A weather-resistant exterior wall envelope shall not be required over concrete or masonry walls designed in accordance with Chapter 6 and flashed according to Section R703.7 or R703.8.

2. Compliance with the requirements for a means of drainage, and the requirements of Section 703.2 and Section 703.8, shall not be required for an exterior wall envelope that has been demonstrated to resist wind-driven rain through testing of the exterior wall envelope, including joints, penetrations and intersections with dissimilar materials, in accordance with ASTM E 331 under the following conditions:

   2.1 Exterior wall envelope test assemblies shall include at least one opening, one control joint, one wall/eave interface and one wall sill. All tested openings and penetrations shall be representative of the intended end-use configuration.

   2.2 Exterior wall envelope test assemblies shall be at least 4 feet (1219 mm) by 8 feet (2438 mm) in size.

- 2.3 Exterior wall assemblies shall be tested at a minimum differential pressure of 6.24 pounds per square foot (299 Pa).

- 2.4 Exterior wall envelope assemblies shall be subjected to a minimum test exposure duration of 2 hours.

The exterior wall envelope design shall be considered to resist wind-driven rain where the results of testing indicate that water did not penetrate control joints in the exterior wall envelope, joints at the perimeter of openings penetration or intersections of terminations with dissimilar materials.

Section R703.2 Change to read as shown: (RB214-03/04)

R703.2 Weather-resistant sheathing paper. One layer of No. 15 asphalt felt, free from holes and breaks, complying with ASTM D 226 for Type 1 felt or other approved weather-resistive materials shall be applied over studs or sheathing of all exterior walls. Such felt or material shall be applied horizontally, with the upper layer lapped over the lower layer not less than 2 inches (51 mm). Where joints occur, felt shall be lapped not less than 6 inches (152 mm). Such felt or other approved material shall be continuous to the top of walls and terminated at penetrations and building appendages in a manner to meet the requirements of the exterior wall envelope as described in Section R703.1.

Exceptions: Omission of such felt or material is permitted in the following situations:

1. In detached accessory buildings.

2. Under exterior wall finish materials as permitted in Table R703.4.

3. Under paperbacked stucco lath when the paper backing is an approved weather-resistive sheathing paper.

Section R703.4 Change to read as shown: (RB216-03/04)

R703.4 Attachments. Unless specified otherwise, all wall coverings shall be securely fastened in accordance with Table R703.4 or with other approved aluminum, stainless steel, zinc-coated or other approved corrosion-resistive fasteners. Where the basic wind speed per Figure R301.2(4) is 110 miles per hour (49 m/s) or greater, the attachment of wall coverings shall be designed to resist the component and cladding loads specified in Table R301.2(2), adjusted for height and exposure in accordance with Table R301.2(3).

2004 SUPPLEMENT TO THE IRC

Table R703.4 Change entire table to read as shown: (RB218-03/04, RB219-03/04, RB220-03/04 and S62-03/04)

TABLE R703.4
WEATHER-RESISTANT SIDING ATTACHMENT AND MINIMUM THICKNESS

Siding Material		Nominal Thickness[a] (inches)	Joint Treatment	Weather Resistant Barrier Required	Wood or wood structural panel sheathing	Fiberboard sheathing into stud	Gypsum sheathing into stud	Foam plastic sheathing into stud	Direct to Studs	Number or spacing of fasteners
Horizontal aluminum[e]	Without insulation	0.019[f]	Lap	Yes	0.120 nail 1½" long	0.120 nail 2" long	0.120 nail 2" long	0.120 nail[y]	Not allowed	Same as stud spacing
		0.024	Lap	yes	0.120 nail 1½" long	0.120 nail 2" long	0.120 nail 2" long	0.120 nail[y]	Not allowed	
	With insulation	0.019	Lap	Yes	0.120 nail 1½" long	0.120 nail 2½" long	0.120 nail 2½" long	0.120 nail[z]	0.120 nail 1½" long	
Brick veneer[z] Concrete masonry veneer[z]		2 2	Section R703	Yes (note l)	See Section R703 and Figure R703.7[g]					
Hardboard[k] Panel siding-vertical		7/16		Yes	Note n	Note n	Note n	Note n	Note n	6" panel edges 12" inter.sup.[o]
Hardboard[k] Lap-siding-horizontal		7/16	Note q	Yes	Note p	Note p	Note p	Note p	Note p	Same as stud spacing 2 per bearing
Steel[h]		29 ga.	Lap	Yes	0.113 nail 1¾" Staple-1¾"	0.113 nail 2¾" Staple-2½"	0.013 nail 2½" Staple- 2¼"	0.113 nail[y] Staple[y]	Not allowed	Same as stud spacing
Stone veneer		2	Section R703	Yes (Note l)	See Section R703 and Figure R703.7[g]					
Particleboard panels		3/8 - ½		Yes	6d box (2" x 0.099") nail	6d box (2" x 0.099") nail	6d box (2" x 0.099") nail	box nail[y]	6d box (2" x 0.099") nail, 3/8 not allowed	6" panel edge 12" inter. sup.
		5/8		yes	6d box (2" x 0.099") nail	8d box (2½" x 0.113") nail	8d box (2½" x 0.113") nail	box nail[y]	6d box (2" x 0.099") nail	
Plywood panel[i] (exterior grade)		3/8		Yes	0.099 nail-2"	0.113 nail-2½"	0.099 nail-2"	0.113 nail[y]	0.099 nail-2"	6" panel edges, 12" inter. sup.
Vinyl siding[m]		0.035	Lap	Yes	0.120 nail 1½" Staple-1¾	0.120 nail 2" Staple-2½	0.120 nail 2" Staple-2½	0.120 nail[y] staple[y]	Not allowed	Same as stud spacing
Wood[j] Rustic, drop		3/8 Min	Lap	Yes	Fastener penetration into stud-1"				0.113 nail-2½" Staple-2"	Face nailing up to 6" widths, 1 nail per bearing; 8" widths and over 2 nail per bearing
Shiplap		no change	Lap	Yes						
Bevel		no change								
Butt tip		no change	Lap	Yes						

IRC-45

2004 SUPPLEMENT TO THE IRC

TABLE R703.4 (continued)

Fiber cement panel siding[r]	5/16	Note s	Yes Note x	6d corrosion resistant nail[t]	6d corrosion resistant nail[t]	6d corrosion resistant nail[t]	—	4d corrosion resistant nail[u]	6" oc on edges, 12" oc on intermed. studs
Fiber cement lap siding[r]	no change	Note v	Yes Note x	6d corrosion resistant nail[t]	6d corrosion resistant nail[t]	6d corrosion resistant nail[t]	—	6d corrosion resistant nail[w]	Note w

For SI: 1 inch = 25.4 mm.

a. Based on stud spacing of 16 inches on center where studs are spaced 24 inches, siding shall be applied to sheathing approved for that spacing.
b. Nail is a general description and shall be T-head, modified round head, or round head with smooth or deformed shanks.
c. Staples shall have a minimum crown width of 7/16-inch outside diameter and be manufactured of minimum 16 gage wire.
d. Nails or staples shall be aluminum, galvanized, or rust-preventative coated and shall be driven into the studs for fiberboard or gypsum backing.
e. Aluminum nails shall be used to attach aluminum siding.
f. Aluminum (0.019 inch) shall be unbacked only when the maximum panel width is 10 inches and the maximum flat area is 8 inches. The tolerance for aluminum siding shall be +0.002 inch of the nominal dimension.
g. All attachments shall be coated with a corrosion-resistive coating.
h. Shall be of approved type.
i. Three-eights-inch plywood shall not be applied directly to studs spaced greater than 16 inches on center when long dimension is parallel to studs. Plywood 1/2-inch or thinner shall not be applied directly to studs spaced greater than 24 inches on center. The stud spacing shall not exceed the panel span rating provided by the manufacturer unless the panels are installed with the face grain perpendicular to the studs or over sheathing approved for that stud spacing.
j. Wood board sidings applied vertically shall be nailed to horizontal nailing strips or blocking set 24 inches on center. Nails shall penetrate 1 1/2 inches into studs, studs and wood sheathing combined, or blocking. A weather-resistant membrane shall be installed weatherboard fashion under the vertical siding unless the siding boards are lapped or battens are used.
k. Hardboard siding shall comply with AHA A135.6.
l. For Masonry veneer, a weather-resistant sheathing paper is not required over a sheathing that performs as a weather-resistive barrier when a 1-inch air space is provided between the veneer and the sheathing. When the 1-inch space is filled with mortar, a weather-resistant sheathing paper is required over studs or sheathing.
m. Vinyl siding shall comply with ASTM D 3679.
n. Minimum shank diameter of 0.092 inch, minimum head diameter of 0.025 inch, and nail length must accommodate sheathing and penetrate framing 1 1/2 inches.
o. When used to resist shear forces, the spacing must be 4 inches at panel edges and 8 inches on interior supports.
p. Minimum shank diameter of 0.099 inch, minimum head diameter of 0.240 inch, and nail length must accommodate sheathing and penetrate framing 1 1/2 inches.
q. Vertical end joints shall occur at studs and shall be covered with a joint cover or shall be caulked.
r. Fiber cement siding shall comply with the requirements of ASTM C 1186.
s. See R703.10.1.
t. Minimum 0.102" smooth shank, 0.255" round head.
u. Minimum 0.099" smooth shank, 0.250" round head.
v. See R703.10.2.
w. Face nailing: 2 nails at each stud. Concealed nailing: one 11 gage 1-1/2 galv. roofing nail (0.371" head diameter, 0.120" shank) or 6d galv. box nail at each stud.
x. See R703.2 Exceptions.
y. Minimum nail length must accommodate sheathing and penetrate framing 1 1/2 inches.
z. Adhered masonry veneer shall comply with the requirements in Sections 6.1 and 6.3 of ACI 530/ASCE 5/TMS-402.

Section R703.6 Change to read as shown: (RB222-03/04)

R703.6 Exterior plaster. Installation of these materials shall be in compliance with ASTM C 926 and ASTM C 1063 and the provisions of this code.

Section R703.6.2 Change to read as shown: (RB223-03/04)

R703.6.2 Plaster. Plastering with portland cement plaster shall be not less than three coats when applied over metal lath or wire lath and shall be not less than two coats when applied over masonry, concrete, pressure preservatively treated wood or decay resistive wood as specified in Section R323.1 or gypsum backing. If the plaster surface is completely covered by veneer or other facing material or is completely concealed, plaster application need be only two coats, provided the total thickness is as set forth in Table 702.1(1).

On wood-frame construction with an on-grade floor slab system, exterior plaster shall be applied in such a manner as to cover, but not extend below, lath, paper and screed.

The proportion of aggregate to cementitious materials shall be as set forth in Table R702.1(3).

Section R 703.6.3 A dd n ew t ext t o r ead a s s hown: (RB224-03/04)

R703.6.3 Weather-resistant barriers. Weather-resistant barriers shall be installed as required in Section R703.2 and, where applied over wood-based sheathing, shall include a weather-resistive vapor-permeable barrier with a performance at least equivalent to two layers of Grade D paper.

> **Exception:** Where the weather-resistant barrier that is applied over wood-based sheathing has a water resistance equal to or greater than that of 60 minute Grade D paper and is separated from the stucco by an intervening, substantially nonwater-absorbing layer or designed drainage space.

Sections R703.7 Change to read as shown: (RB31-03/04)

R703.7 Stone and masonry veneer, general. All stone and masonry veneer shall be installed in accordance with this chapter, Table R703.4 and Figure R703.7. Such veneers installed over a backing of wood or cold-formed steel shall be limited to the first story above-grade and shall not exceed 5 inches (127 mm) in thickness.

Exceptions:

1. In Seismic Design Categories A and B, exterior masonry veneer with a backing of wood or cold-formed steel framing shall not exceed 30 feet (9144 mm) in height above the noncombustible foundation, with an additional 8 feet (2348 mm) permitted for gabled ends.

2. In Seismic Design Category C, exterior masonry veneer with a backing of wood or cold-formed steel framing shall not exceed 30 feet (9144 mm) in height above the noncombustible foundation, with an additional 8 feet (2348 mm) permitted for gabled ends. In other than the topmost story, the length of bracing shall be 1.5 times the length otherwise required in Chapter 6.

3. For detached one- or two-family dwellings with a maximum nominal thickness of 4 inches (102 mm) of exterior masonry veneer with a backing of wood frame located in Seismic Design Categories D_0 or D_1, the masonry veneer shall not exceed 20 feet (6096 mm) in height above a noncombustible foundation, with an additional 8 feet (2438 mm) permitted for gabled ends, or 30 feet (9144 mm) in height with an additional 8 feet (2438 mm) permitted for gabled ends where the lower 10 feet (3048 mm) has a backing of concrete or masonry wall, provided the following criteria are met:

   3.1. Braced wall panels shall be constructed with a minimum of 7/16-inch thick (11.1 mm) sheathing fastened with 8d common nails at 4 inches (102 mm) on center on panel edges and at 1 2 i nches (305 m m) o n c enter o n intermediate supports.

   3.2. The bracing of the top story shall be located at each end and at least every 25 feet (7620 mm) on center but not less than 45 percent of the braced wall line. The bracing of the first story shall be as provided in Table R602.10.1.

   3.3. Hold down connectors shall be provided at the ends of braced walls for the second floor to first floor wall assembly with an allowable design load of 2100 lb (9.34 kN). Hold down connectors shall be provided at the ends of each wall segment of the braced walls for the first floor to foundation assembly with an allowable design of 3700 lb (16.46 kN). In all cases, the hold down connector force shall be transferred to the foundation.

   3.4. Cripple walls shall not be permitted.

4. For detached one- and two-family dwellings with a maximum actual thickness of 3 inches (76 mm) of exterior masonry veneer with a backing of wood frame located in Seismic Design Category D_2, the masonry veneer shall not exceed 20 feet (6096 mm) in height above a noncombustible foundation, with an additional 8 feet (2438 mm) permitted for gabled ends, or 30 feet (9144 mm) in height with

2004 SUPPLEMENT TO THE IRC

an additional 8 feet (2438 mm) permitted for gabled ends where the lower 10 feet (3048 mm) has a backing of concrete or masonry wall, provided the following criteria are met:

4.1. Braced wall panels shall be constructed with a minimum of 7/16 inch (11.1 mm) thick sheathing fastened with 8d common nails at 4 inches (102 mm) on center on panel edges and at 12 inches (305 mm) on center on intermediate supports.

4.2. The bracing of the top story shall be located at each end and at least every 25 feet (7620 mm) on center but not less than 55% of the braced wall line. The bracing of the first story shall be as provided in Table R602.10.1.

4.3. Hold down connectors shall be provided at the ends of braced walls for the second floor to first floor wall assembly with an allowable design of 2300 lbs. (1043 kg). Hold down connectors shall be provided at the ends of each wall segment of the braced walls for the first floor to foundation assembly with an allowable design of 3900 lbs. (1769 kg). In all cases, the hold down connector force shall be transferred to the foundation.

4.4. Cripple walls shall not be permitted.

Section R703.7.2 Change to read as shown: (RB31-03/04)

R703.7.2 Exterior veneer support. Except in Seismic Design Categories D_0, D_1 and D_2, exterior masonry veneers having an installed weight of 40 pounds per square foot (195 kg/m^2) or less shall be permitted to be supported on wood or cold-formed steel construction. When masonry veneer supported by wood or cold-formed steel construction adjoins masonry veneer supported by the foundation, there shall be a movement joint between the veneer supported by the wood or cold-formed steel construction and the veneer supported by the foundation. The wood or cold-formed steel construction supporting the masonry veneer shall be designed to limit the deflection to 1/600 of the span for the supporting members. The design of the wood or cold-formed steel construction shall consider the weight of the veneer and any other loads.

Section R703.7.4.1 Change to read as shown: (RB31-03/04)

R703.7.4.1 Size and spacing. Veneer ties, if strand wire, shall not be less in thickness than No. 9 U.S. gage [(0.148 in.)(3.8 mm)] wire and shall have a hook embedded in the mortar joint, or if sheet metal, shall be not less than No. 22 U.S. gage by [(0.0299 in.)(0.76 mm)] 7/8 inch (22.3 mm) corrugated. Each tie shall be spaced not more than 24 inches (610 mm) on center horizontally and vertically and shall support not more than 2.67 square feet (0.248 m^2) of wall area.

Exception: In Seismic Design Categories D_0, D_1 or D_2 and townhouses in Seismic Design Category C and in wind areas of more than 30 pounds per square foot pressure (1.44 kN/m^2), each tie shall support not more than 2 square feet (0.186 m^2) of wall area.

CHAPTER 8
ROOF-CEILING CONSTRUCTION

Section R802.3.1 Change to read as shown: (RB168-03/04)

R802.3.1 Ceiling joist and rafter connections. Ceiling joists and rafters shall be nailed to each other in accordance with Table R802.5.1(9), and the rafter shall be nailed to the top wall plate in accordance with Table R602.3(1). Ceiling joists shall be continuous or securely joined in accordance with Table 802.5.1(9) where they meet over interior partitions and nailed to adjacent rafters to provide a continuous tie across the building when such joists are parallel to the rafters.

Where ceiling joists are not connected to the rafters at the top wall plate, joists connected higher in the attic shall be installed as rafter ties, or rafter ties shall be installed to provide a continuous tie. Where ceiling joists are not parallel to rafters, rafter ties shall be installed. Rafter ties shall be a minimum of 2-inch by 4-inch (51 mm by 102 mm) (nominal), installed in accordance with the connection requirements in Table R802.5.1(9), or connections of equivalent capacities shall be provided. Where ceiling joists or rafter ties are not provided, the ridge formed by these rafters shall be supported by a wall or girder designed in accordance with accepted engineering practice.

Collar ties or ridge straps to resist wind uplift shall be connected in the upper third of the attic space in accordance with Table R602.3(1).

Collar ties shall be a minimum of 1-inch by 4-inch (25.4 mm by 102 mm) (nominal), spaced not more than 4 feet (1219 mm) on center.

Tables R802.5.1(1) through R802.5.1(8) Change footnote a to read as shown: (RB229-03/04)

a. The tabulated rafter spans assume that ceiling joists are located at the bottom of the attic space or that some other method of resisting the outward push of the rafters on the bearing walls, such as rafter ties, is provided at that location. When ceiling joists or rafter ties are located higher in the attic space, the rafter spans shall be multiplied by the factors given below:

H_c/H_R	Rafter Span Adjustment Factor
1/3	0.67
1/4	0.76
1/5	0.83
1/6	0.90
1/7.5 or less	1.00

where:
H_c = Height of ceiling joists or rafter ties measured vertically above the top of the rafter support walls.
H_R = Height of roof ridge measured vertically above the top of the rafter support walls.

Table 802.5.1(9) Add footnote 'g' to title and notes as shown: (RB168-03/04)

TABLE R802.5.1(9)
RAFTER/CEILING JOIST HEEL CONNECTIONS[a, b, c, d, e, f, g]

(Portions of table and notes not shown do not change)

g. Tabulated heel joint connection requirements assume that ceiling joists or rafter ties are located at the bottom of the attic space. When ceiling joists or rafter ties are located higher in the attic, heel joint connection requirements shall be increased by the following factors:

H_c/H_R	Heel Joint Connection Adjustment Factor
1/3	1.5
1/4	1.33
1/5	1.25
1/6	1.2
1/10 or less	1.11

where:
H_c = Height of ceiling joists or rafter ties measured vertically above the top of the rafter support walls.
H_R = Height of roof ridge measured vertically above the top of the rafter support walls.

Section R802.7.2 Change to read as shown: (S82-03/04)

R802.7.2 Engineered wood products. Cuts, notches and holes bored in trusses, structural composite lumber, structural glue-laminated members or I-joists are not permitted unless allowed by the manufacturer's recommendations or unless the effects of such penetrations are specifically considered in the design of the member by a registered design professional.

Section R804.1.1 Change to read as shown: (RB155-03/04)

R804.1.1 Applicability limits. The provisions of this section shall control the construction of steel roof framing for buildings not greater than 60 feet (18 288 mm) in length perpendicular to the joist, rafter or truss span, not greater than 36 feet (10 973 mm) in width parallel to the joist span or truss, not greater than two stories in height and roof slopes not smaller than 3:12 (25 percent slope) or greater than 12:12 (100 percent slope). Steel roof framing constructed in accordance with the provisions of this section shall be limited to sites subjected to a maximum design wind speed of 110 miles per hour (49 m/s) Exposure A, B, or C and a maximum ground snow load of 70 psf (3.35 kN/m^2).

Section R804.2.1 Change to read as shown: (RB157-03/04)

R804.2.1 Material. Load-bearing steel framing members shall be cold-formed to shape from structural quality sheet steel complying with the requirements of one of the following:

1. ASTM A 653: Grades 33, 37, 40 and 50 (Class 1 and 3).
2. ASTM A 792: Grades 33, 37, 40 and 50A.
3. ASTM A 875: Grades 33, 37, 40 and 50 (Class 1 and 3).
4. ASTM A 1003: Grades 33, 37, 40 and 50

Section R806.2 Change to read as shown: (RB231-03/04)

R806.2 Minimum area. The total net free ventilating area shall not be less than 1/150 of the area of the space ventilated except that the total area is permitted to be reduced to 1/300, provided that at least 50 percent and not more than 80 percent of the required ventilating area is provided by ventilators located in the upper portion of the space to be ventilated at least 3 feet (914 mm) above the eave or cornice vents with the balance of the required ventilation provided by eave or cornice vents. As an alternative, the net free cross-ventilation area may be reduced to 1/300 when a vapor barrier having a transmission rate not exceeding 1 perm (57.4 mg/s.m^2.Pa) is installed on the warm-in-winter side of the ceiling.

Section R806.4 Add new section as shown: (EC48-03/04)

R806.4 Conditioned attic assemblies: Unvented conditioned attic assemblies (spaces between the ceiling joists of the top story and the roof rafters) are permitted under the following conditions:

1. No interior vapor retarders are installed on the ceiling side (attic floor) of the unvented attic assembly.

2. An air-impermeable insulation is applied in direct contact to the underside/interior of the structural roof deck. "Air-impermeable" shall be defined by ASTM E 283.

    Exception: In zones 2B and 3B, insulation is not required to be air impermeable.

3. In the warm humid locations as defined in Section N1101.2.1:

    a. For asphalt roofing shingles: A 1-perm (57.4 mg/s · m^2 · Pa) or less vapor retarder (determined using Procedure B of ASTM E 96) is placed to the exterior of the structural roof deck; i.e., just above the roof structural sheathing.

    b. For wood shingles and shakes: a minimum continuous 1/4-inch (6 mm) vented air space separates the shingles/shakes and the roofing felt placed over the structural sheathing.

4. In zones 3 through 8 as defined in Section N1101.2, sufficient insulation is installed to maintain the monthly average temperature of the condensing surface above 45°F (7°C). The condensing surface is defined as either the structural roof deck or the interior surface of an air-impermeable insulation applied in direct contact to the underside/interior of the structural roof deck. "Air-impermeable" is quantitatively defined by ASTM E 283. For calculation purposes, an interior temperature of 68°F (20° C) is assumed. The exterior temperature is assumed to be the monthly average outside temperature.

Section R808.1 Change to read as shown: (EC48-03/04; EL3-03/04)

R808.1 Combustible insulation. Combustible insulation shall be separated a minimum of 3 inches (76 mm) from recessed luminaires, fan motors and other heat-producing devices.

   Exception: Where heat-producing devices are listed for lesser clearances, combustible insulation complying with the listing requirements shall be separated in accordance with the conditions stipulated in the listing.

   Recessed luminaires installed in the building thermal envelope shall meet the requirements of Section N1102.4.3.

CHAPTER 9
ROOF ASSEMBLIES

Section R905.1 Change to read as shown: (RB233-03/04)

R905.1 Roof covering application. Roof coverings shall be applied in accordance with the applicable provisions of this section and the manufacturer's installation instructions. Unless otherwise specified in this section, roof coverings shall be installed to resist the component and cladding loads specified in Table R301.2(2), adjusted for height and exposure in accordance with Table R301.2(3).

Section R905.2.3 Change to read as shown: (RB234-03/04)

R905.2.3 Underlayment. Unless otherwise noted, required underlayment shall conform to ASTM D 226 Type I, ASTM D 4869 Type I, or ASTM D 6757.

Section R905.2.6 Change to read as shown: (RB236-03/04)

R905.2.6 Attachment. Asphalt shingles shall have the minimum number of fasteners required by the manufacturer. For normal application, asphalt shingles shall be secured to the roof with not less than four fasteners per strip shingle or two fasteners per individual shingle. Where the roof slope exceeds 20 units vertical in 12 units horizontal (167 percent slope), special methods of fastening are required. For roofs located where the basic wind speed per Figure 301.2(4) is 110 mph (49 m/s) or greater, special methods of fastening are required. Special fastening methods shall be tested in accordance with ASTM D 3161 Class F. Asphalt shingle wrappers shall bear a label indicating compliance with ASTM D 3161 Class F.

Shingles classified using ASTM D 3161 are acceptable for use in wind zones less than 110 mph (49 m/s). Shingles classified using ASTM D 3161 Class F are acceptable for use in all cases where special fastening is required.

Section R905.2.7 Change to read as shown: (RB237-03/04)

R905.2.7 Underlayment application. For roof slopes from two units vertical in 12 units horizontal (17-percent slope), up to four units vertical in 12 units horizontal (33-percent slope), underlayment shall be two layers applied in the following manner. Apply a 19-inch (483 mm) strip of underlayment felt parallel to and starting at the eaves, fastened sufficiently to hold in place. Starting at the eave, apply 36-inch-wide (914 mm) sheets of underlayment, overlapping successive sheets 19 inches (483 mm), and fastened sufficiently to hold in place. Distortions in the underlayment shall not interfere with the ability of the shingles to seal. For roof slopes of four units vertical in 12 units horizontal (33-percent slope) or greater, underlayment shall be one layer applied in the following manner. Underlayment shall be applied shingle fashion, parallel to and starting from the eave and lapped 2 inches (51 mm), fastened sufficiently to hold in place. Distortions in the underlayment shall not interfere with the ability of the shingles to seal. End laps shall be offset by 6 feet (1829 mm).

Section R905.2.8.2 Change to read as shown: (RB239-03/04 and RB240-03/04)

R905.2.8.2 Valleys. Valley linings shall be installed in accordance with the manufacturer's installation instructions before applying shingles. Valley linings of the following types shall be permitted:

1. For open valley (valley lining exposed) lined with metal, the valley lining shall be at least 24 inches (610 mm) wide and of any of the corrosion-resistant metals in Table R905.2.8.2.

2. For open valleys, valley lining of two plies of mineral surfaced roll roofing, complying with ASTM D 3909 or ASTM D 6380 Class M, shall be permitted. The bottom layer shall be 18 inches (457mm) and the top layer a minimum of 36 inches (914 mm) wide.

3. For closed valleys (valley covered with shingles), valley lining of one ply of smooth roll roofing complying with ASTM D 6380 Class S Type III, Class M Type II, or ASTM D 3909 and at least 36 inches wide (914mm) or valley lining as described in Items 1 and 2 above shall be permitted. Specialty underlayment complying with ASTM D 1970 may be used in lieu of the lining material.

Section R905.3.3 Change to read as shown: (RB241-03/04)

R905.3.3 Underlayment. Unless otherwise noted, required underlayment shall conform to ASTM D 226 Type II; ASTM D 2626 Type I; or ASTM D 6380 Class M mineral surfaced roll roofing.

Section R905.4.3 Change to read as shown: (RB242-03/04)

R905.4.3 Underlayment. In areas where the average daily temperature in January is 25° F (-4°C) or less, or when Table R301.2 criteria so designate, an ice barrier that consists of at least two layers of underlayment cemented together or of a self-adhering polymer-modified bitumen sheet shall be used in lieu of normal underlayment and extend from the eave's edge to a point at least 24 inches (610 mm) inside the exterior wall line of the building. Underlayment shall comply with ASTM D 226, Type I, or ASTM D 4869, Type I or II.

Exception: Detached accessory structures that contain no conditioned floor area.

Section R905.5.3 Change to read as shown: (RB243-03/04)

R905.5.3 Underlayment. In areas where the average daily temperature in January is 25° F (-4°C) or less, or when Table R301.2 criteria so designate, an ice barrier that consists of at least two layers of underlayment cemented together or of a self-adhering polymer-modified bitumen sheet shall be used in lieu of normal underlayment and extend from the eave's edge to a point at least 24 inches (610 mm) inside the exterior wall line of the building. Underlayment shall comply with ASTM D 226, Type I, or ASTM D 4869, Type I or II.

Exception: Detached accessory structures that contain no conditioned floor area.

Section R905.5.4 Change to read as shown: (RB244-03/04)

R905.5.4 Material standards. Mineral-surfaced roll roofing shall conform to ASTM D 3909 or ASTM D 6380, Class M.

2004 SUPPLEMENT TO THE IRC

Section R905.6.3 Change to read as shown: (RB245-03/04)

R905.6.3 Underlayment. In areas where the average daily temperature in January is 25° F (-4°C) or less, or when Table R301.2 criteria so designate, an ice barrier that consists of at least two layers of underlayment cemented together or of a self-adhering polymer-modified bitumen sheet shall be used in lieu of normal underlayment and extend from the eave's edge to a point at least 24 inches (610 mm) inside the exterior wall line of the building. Underlayment shall comply with ASTM D 226, Type I, or ASTM D 4869, Type I or II.

Exception: Detached accessory structures that contain no conditioned floor area.

Section R905.7.3 Change to read as shown: (RB246-03/04)

R905.7.3 Underlayment. In areas where the average daily temperature in January is 25°F (-4°C) or less, or when Table R301.2 criteria so designate, an ice barrier that consists of at least two layers of underlayment cemented together or of a self-adhering polymer-modified bitumen sheet shall be used in lieu of normal underlayment and extend from the eave's edge to a point at least 24 inches (610 mm) inside the exterior wall line of the building. Underlayment shall comply with ASTM D 226, Type I, or ASTM D 4869, Type I or II.

Exception: Detached accessory structures that contain no conditioned floor area.

Section R905.8.3 Change to read as shown: (RB247-03/04)

R905.8.3 Underlayment. In areas where the average daily temperature in January is 25°F (-4°C) or less, or when Table R301.2 criteria so designate, an ice barrier that consists of at least two layers of underlayment cemented together or of a self-adhering polymer-modified bitumen sheet shall be used in lieu of normal underlayment and extend from the eave's edge to a point at least 24 inches (610 mm) inside the exterior wall line of the building. Underlayment shall comply with D226, Type I, or ASTM D 4869, Type I or II.

Exception: Detached accessory structures that contain no conditioned floor area.

Table R905.8.5 Change to read as shown: (FS169-03/04)

TABLE R905.8.5
WOOD SHAKE MATERIAL REQUIREMENTS

Material	Minimum Grades	Applicable Grading Rules
Preservative-treated taper sawn shakes of Southern pine treated in accordance with AWPA Standard U1 (Commodity Specification A, Use Category 3B and Section 5.6)	1 or 2	Forest Products Laboratory of the Texas Forest Services

(Portions of table not shown do not change)

Section R905.10.2 Change to read as shown: (RB248-03/04)

R905.10.2 Slope. Minimum slopes for metal roof panels shall comply with the following:

1. The minimum slope for lapped, nonsoldered seam metal roofs without applied lap sealant shall be three units vertical in 12 units horizontal (25-percent slope).

2. The minimum slope for lapped, nonsoldered seam metal roofs with applied lap sealant shall be one-half vertical unit in 12 units horizontal (4-percent slope). Lap sealants shall be applied in accordance with the approved manufacturer's installation instructions.

3. The minimum slope for standing seam roof systems shall be one-quarter unit vertical in 12 units horizontal (2-percent slope).

Table R905.10.3 Add two new entries for Aluminum and change existing entry for Copper as shown: (RB249-03/04, RB250-03/04 and RB251-03/04)

**TABLE R905.10.3
METAL ROOF COVERINGS STANDARDS**

ROOF COVERINGS TYPE	STANDARD APPLICATION RATE/THICKNESS
Aluminum-coated steel	ASTM A 463 T2 65
Aluminum Alloy-coated steel	ASTM A 875 GF60
Copper	ASTM B 370, 16 oz. per sq. ft. for metal sheet roof covering systems; 12 oz. per sq. ft. for preformed metal shingle systems.

(Portions of table not shown do not change)

Section R905.10.4 Change to read as shown: (RB252-03/04)

R905.10.4 Attachment. Metal roof panels shall be secured to the supports in accordance with this chapter and the manufacturer's installation instructions. In the absence of manufacturer's installation instructions, the following fasteners shall be used:

1. Galvanized fasteners shall be used for steel roofs.

2. Three hundred series stainless steel fasteners shall be used for copper roofs.

3. Stainless steel fasteners are acceptable for metal roofs.

Section R905.12.2 Change to read as shown: (RB253-03/04)

R905.12.2 Material Standards. Thermoset single-ply roof coverings shall comply with ASTM D 4637, ASTM D 5019 or CGSB 37-GP-52M.

Section R905.13.2 Change to read as shown: (RB254-03/04 & RB256-03/04)

R905.13.2 Material Standards. Thermoplastic single-ply roof coverings shall comply with ASTM D 4434, ASTM D 6754, ASTM D 6878, or CSGB 37-GP-54M.

Section R905.15.2 Change to read as shown: (RB255-03/04)

Section R905.15.2 Material standards. Liquid-applied roof coatings shall comply with ASTM C 836, C 957, D 1227, D 3468, D 6083 or D 6694.

2004 SUPPLEMENT TO THE IRC

Section R907.3 Add new exception No. 3 as shown: (RB258-03/04)

R907.3 Recovering versus replacement. New roof coverings shall not be installed without first removing existing roof coverings where any of the following conditions occur:

1. Where the existing roof or roof covering is water-soaked or has deteriorated to the point that the existing roof or roof covering is not adequate as a base for additional roofing.

2. Where the existing roof covering is wood shake, slate, clay, cement or asbestos-cement tile.

3. Where the existing roof has two or more applications of any type of roof covering.

Exceptions:

1. Complete and separate roofing systems, such as standing-seam metal roof systems, that are designed to transmit the roof loads directly to the building's structural system and that do not rely on existing roofs and roof coverings for support, shall not require the removal of existing roof coverings.

2. Metal panel, metal shingle, and concrete and clay tile roof coverings shall be permitted to be installed over existing wood shake roofs when applied in accordance with Section R907.4.

3. The application of new protective coating over existing spray polyurethane foam roofing systems shall be permitted without tear-off of existing roof coverings.

**CHAPTER 10
CHIMNEYS AND FIREPLACES**

Section R1001.1 Change to read as shown: (RB31-03/04)

R1001.1 General. Masonry chimneys shall be constructed, anchored, supported and reinforced as required in this chapter and the applicable provisions of Chapters 3, 4 and 6. In Seismic Design Category D_0, D_1 or D_2 masonry and concrete chimneys shall be reinforced and anchored as detailed in Section R1003 for chimneys serving fireplaces. In Seismic Design Category A, B or C, reinforcement and seismic anchorage is not required. Chimneys shall be structurally sound, durable, smoke-tight and capable of conveying flue gases to the exterior safely.

2004 SUPPLEMENT TO THE IRC

Section R1001.9 Change to read as shown: (RB259-03/04)

R1001.9 Clay flue lining (installation). Clay flue liners shall be installed in accordance with ASTM C 1283 and extend from a point not less than 8 inches (203 mm) below the lowest inlet or, in the case of fireplaces, from the top of the smoke chamber to a point above the enclosing walls. The lining shall be carried up vertically, with a maximum slope no greater than 30 degrees from the vertical.

Clay flue liners shall be laid in medium-duty refractory mortar conforming to ASTM C 199 with tight mortar joints left smooth on the inside and installed to maintain an air space or insulation not to exceed the thickness of the flue liner separating the flue liners from the interior face of the chimney masonry walls. Flue liners shall be supported on all sides. Only enough mortar shall be placed to make the joint and hold the liners in position.

Section R1003.3 Change to read as shown: (RB31-03/04)

R1003.3 Seismic reinforcing. Masonry or concrete chimneys in Seismic Design Category D_0, D_1 or D_2 shall be reinforced. Reinforcing shall conform to the requirements set forth in Table R1003.1 and Section R609, Grouted Masonry.

Section R1003.4 Change to read as shown: (RB31-03/04)

R1003.4 Seismic anchorage. Masonry or concrete chimneys in Seismic Design Categories D_0, D_1 or D_2 shall be anchored at each floor, ceiling or roof line more than 6 feet (1829 mm) above grade, except where constructed completely within the exterior walls. Anchorage shall conform to the requirements of Section R1003.4.1.

Section R1003.12 Mantel and Trim. Delete section without substitution. (RB261-03/04)

Section R1006.2 Change to read as shown: (RB31-03/04)

R1006.2 Seismic Reinforcing. Masonry heaters shall be anchored and reinforced as required in this chapter. All masonry heaters shall maintain a minimum clearance of 4 inches (102 mm) to adjacent framing from the body of the masonry heater. In Seismic Design Categories A, B and C, reinforcement and seismic anchorage shall not be required. In Seismic Design Categories D_0, D_1 and D_2, masonry heaters shall be anchored to the foundation. Where the masonry chimney shares a common wall with the facing of the masonry heater, the chimney portion of the structure shall be reinforced in accordance with Section R1003.3.

Part IV—Energy Conservation

CHAPTER 11
ENERGY EFFICIENCY

Chapter 11 Delete Chapter 11 and replace with the following: (RB264-03/04, EC27-03/04, EC48-03/04 & EL3-03/04)

CHAPTER 11
ENERGY EFFICIENCY

SECTION N1101
GENERAL

N1101.1 Scope. This chapter regulates the energy efficiency for the design and construction of buildings regulated by this code.

Exception: Portions of the building envelope that do not enclose conditioned space.

N1101.2 Compliance. Compliance shall be demonstrated by either meeting the requirements of the *International Energy Conservation Code* or meeting the requirements of this chapter. Climate zones from Figure N1101.2 or Table N1101.2 shall be used in determining the applicable requirements from this chapter.

N1101.2.1 Warm humid counties. Warm humid counties are listed in Table N1101.2.1.

N1101.3 Identification. Materials, systems and equipment shall be identified in a manner that will allow a determination of compliance with the applicable provisions of this chapter.

N1101.4 Building thermal envelope insulation. An *R*-value identification mark shall be applied by the manufacturer to each piece of building thermal envelope insulation 12 inches (305 mm) or greater in width. Alternately, the insulation installers shall provide a certification listing the type, manufacturer and *R*-value of insulation installed in each element of the building thermal envelope. For blown or sprayed insulation, the initial installed thickness, settled thickness, settled *R*-value, installed density, coverage area and number of bags installed shall be listed on the certification. The insulation installer shall sign, date and post the certification in a conspicuous location on the job site.

N1101.4.1 Blown or sprayed roof/ceiling insulation. The thickness of blown in or sprayed roof/ceiling insulation shall be written in inches (mm) on markers that are installed at least one for every 300 square foot (28 m^2) throughout the attic space. The markers shall be affixed to the trusses or joists and marked with the minimum initial installed thickness with numbers a minimum of 1 inch (25 mm) in height. Each marker shall face the attic access opening.

N1101.4.2 Insulation mark installation. Insulating materials shall be installed such that the manufacturer's R-value mark is readily observable upon inspection.

N1101.5 Fenestration product rating. U-factors of fenestration products (windows, doors and skylights) shall be determined in accordance with NFRC 100 by an accredited, independent laboratory, and labeled and certified by the manufacturer. Products lacking such a labeled U-factor shall be assigned a default U-factor from Tables N1101.5(1) and N1101.5(2). The solar heat gain coefficient (SHGC) of glazed fenestration products (windows, glazed doors and skylights) shall be determined in accordance with NFRC 200 by an accredited, independent laboratory, and labeled and certified by the manufacturer. Products lacking such a labeled SHGC shall be assigned a default SHGC from Table 1101.5(3).

TABLE N1101.5(1)
DEFAULT GLAZED FENESTRATION U-FACTORS

FRAME TYPE	SINGLE PANE	DOUBLE PANE	SKYLIGHT SINGLE	SKYLIGHT DOUBLE
Metal	1.20	0.80	2.00	1.30
Metal with Thermal Break	1.10	0.65	1.90	1.10
Nonmetal or Metal Clad	0.95	0.55	1.75	1.05
Glazed Block	0.60			

TABLE N1101.5(2)
DEFAULT DOOR U-FACTORS

DOOR TYPE	U-FACTOR
Uninsulated Metal	1.20
Insulated Metal	0.60
Wood	0.50
Insulated, nonmetal edge, max 45% glazing, any glazing double pane	0.35

TABLE N1101.5(3)
DEFAULT GLAZED FENESTRATION SHGC

| SINGLE GLAZED || DOUBLE GLAZED || GLAZED BLOCK |
Clear	Tinted	Clear	Tinted	
0.7	0.6	0.6	0.5	0.6

N1101.6 Installation. All materials, systems and equipment shall be installed in accordance with the manufacturer's installation instructions and the provisions of this code.

2004 SUPPLEMENT TO THE IRC

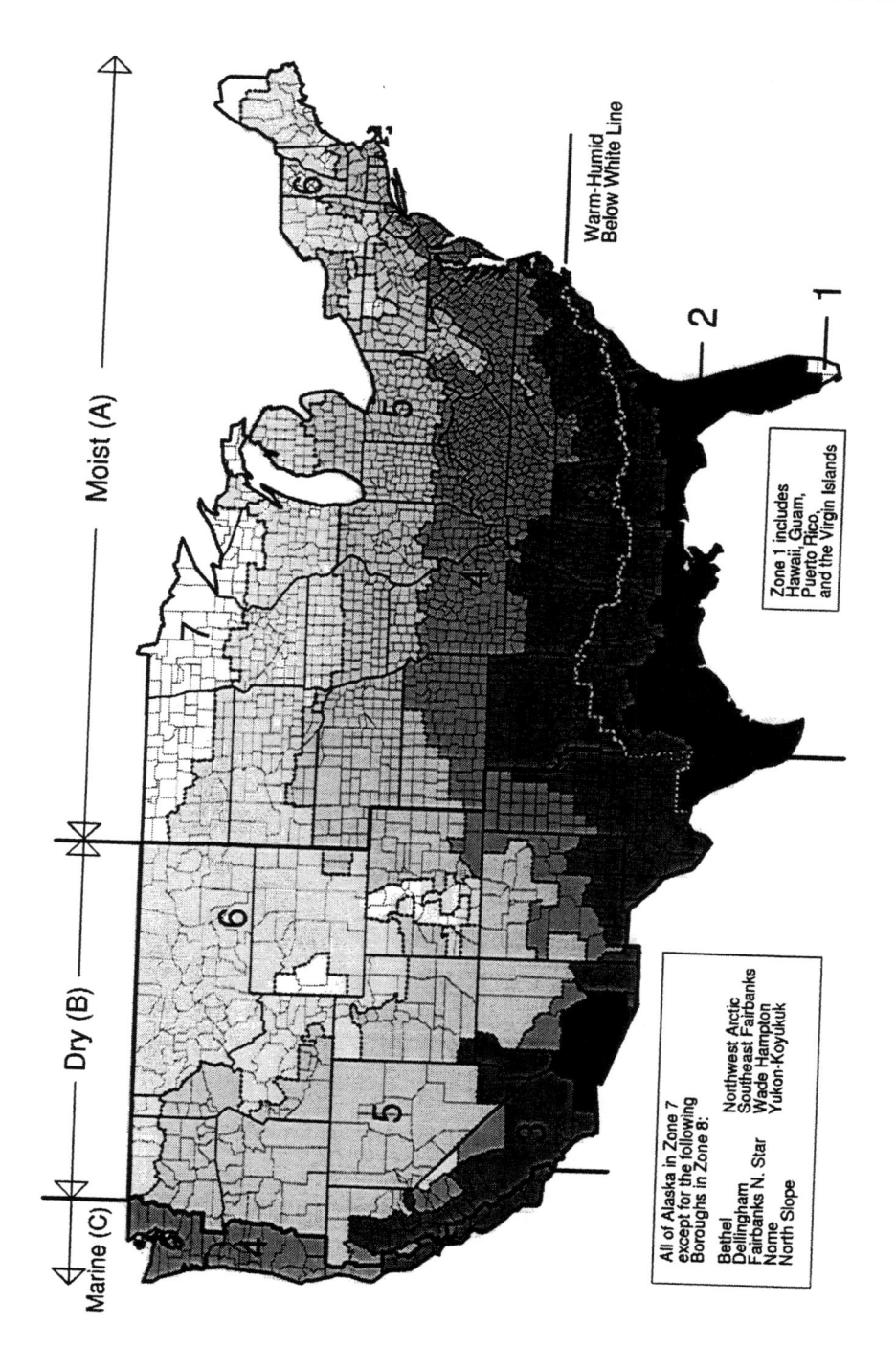

Figure N1101.2. Climate Zones

IRC-57

TABLE N1101.2
CLIMATE ZONES BY STATES AND COUNTIES

Alabama
Zone 3 except
Zone 2
Baldwin
Mobile

Alaska
Zone 7 except
Zone 8
Bethel
Dellingham
Fairbanks North Star
Nome
North Slope
Northwest Arctic
Southeast Fairbanks
Wade Hampton
Yukon-Koyukuk

Arizona
Zone 3 except
Zone 2
La Paz
Maricopa
Pima
Pinal
Yuma
Zone 4
Gila
Yavapai
Zone 5
Apache
Coconino
Navajo

Arkansas
Zone 3 except
Zone 4
Baxter
Benton
Boone
Carroll
Fulton
Izard
Madison
Marion
Newton
Searcy
Stone
Washington

California
Zone 3 Dry except
Zone 2
Imperial
Zone 3 Marine
Alameda
Marin
Mendocino
Monterey
Napa
San Benito
San Francisco
San Luis Obispo
San Mateo
Santa Barbara
Santa Clara
Santa Cruz
Sonoma
Ventura
Zone 4 Dry
Amador
Calaveras
El Dorado
Inyo
Lake
Mariposa
Trinity
Tuolumne
Zone 4 Marine
Del Norte
Humboldt
Zone 5
Lassen
Modoc
Nevada
Plumas
Sierra
Siskiyou
Zone 6
Alpine
Mono

Colorado
Zone 5 except
Zone 4
Baca
Las Animas
Otero
Zone 6
Alamosa
Archuleta
Chaffee
Conejos
Costilla
Custer
Dolores
Eagle
Moffat
Ouray
Rio Blanco
Saguache
San Miguel
Zone 7
Clear Creek
Grand
Gunnison
Hinsdale
Jackson
Lake
Mineral
Park
Pitkin
Rio Grande
Routt
San Juan
Summit

Connecticut
Zone 5

Delaware
Zone 4

Dist Of Columbia
Zone 4

Florida
Zone 2 except
Zone 1
Broward
Dade
Monroe

Georgia
Zone 3 except
Zone 2
Appling
Atkinson
Bacon
Baker
Berrien
Brantley
Brooks
Bryan
Camden
Charlton
Chatham
Clinch
Colquitt
Cook
Decatur
Echols
Effingham
Evans
Glynn
Grady
Jeff Davis
Lanier
Liberty
Long
Lowndes
McIntosh
Miller
Mitchell
Pierce
Seminole
Tattnall
Thomas
Toombs
Ware
Wayne
Zone 4
Banks
Catoosa
Chattooga
Dade
Dawson
Fannin
Floyd
Franklin
Gilmer
Gordon
Habersham
Hall
Lumpkin
Murray
Pickens
Rabun
Stephens
Towns
Union
Walker
White
Whitfield

Hawaii
Zone 1

Idaho
Zone 6 except
Zone 5
Ada
Benewah
Canyon
Cassia
Clearwater
Elmore
Gem
Gooding
Idaho
Jerome
Kootenai
Latah
Lewis
Lincoln
Minidoka
Nez Perce
Owyhee
Payette
Power
Shoshone
Twin Falls
Washington

Illinois
Zone 5 except
Zone 4
Alexander
Bond
Christian
Clay
Clinton
Crawford
Edwards
Effingham
Fayette
Franklin
Gallatin
Hamilton
Hardin
Jackson
Jasper
Jefferson
Johnson
Lawrence
Macoupin
Madison
Marion
Massac
Monroe
Montgomery
Perry
Pope
Pulaski
Randolph
Richland
Saline
Shelby
St Clair
Union
Wabash
Washington
Wayne
White
Williamson

Indiana
Zone 5 except
Zone 4
Brown
Clark
Crawford
Daviess
Dearborn
Dubois
Floyd
Gibson
Greene
Harrison
Jackson
Jefferson
Jennings
Knox
Lawrence
Martin
Monroe
Ohio
Orange
Perry
Pike
Posey
Ripley
Scott
Spencer
Sullivan
Switzerland
Vanderburgh
Warrick
Washington

Iowa
Zone 5 except
Zone 6
Allamakee
Black Hawk
Bremer
Buchanan
Buena Vista
Butler
Calhoun
Cerro Gordo
Cherokee
Chickasaw
Clay
Clayton
Delaware
Dickinson
Emmet
Fayette
Floyd
Franklin
Grundy
Hamilton
Hancock
Hardin
Howard
Humboldt
Ida

2004 SUPPLEMENT TO THE IRC

Kossuth
Lyon
Mitchell
O'Brien
Osceola
Palo Alto
Plymouth
Pocahontas
Sac
Sioux
Webster
Winnebago
Winneshiek
Worth
Wright

Kansas
Zone 4 except
Zone 5
Cheyenne
Cloud
Decatur
Ellis
Gove
Graham
Greeley
Hamilton
Jewell
Lane
Logan
Mitchell
Ness
Norton
Osborne
Phillips
Rawlins
Republic
Rooks
Scott
Sheridan
Sherman
Smith
Thomas
Trego
Wallace
Wichita

Kentucky
Zone 4

Louisiana
Zone 2 except
Zone 3
Bienville
Bossier
Caddo
Caldwell
Catahoula
Claiborne
Concordia
De Soto
East Carroll
Franklin
Grant
Jackson
La Salle
Lincoln
Madison
Morehouse
Natchitoches
Ouachita

Red River
Richland
Sabine
Tensas
Union
Vernon
Webster
West Carroll
Winn

Maine
Zone 6 except
Zone 7
Aroostook

Maryland
Zone 4 except
Zone 5 Garrett

Massachusetts
Zone 5

Michigan
Zone 5 except
Zone 6
Alcona
Alger
Alpena
Antrim
Arenac
Benzie
Charlevoix
Cheboygan
Clare
Crawford
Delta
Dickinson
Emmet
Gladwin
Grand Traverse
Huron
Iosco
Isabella
Kalkaska
Lake
Leelanau
Manistee
Marquette
Mason
Mecosta
Menominee
Missaukee
Montmorency
Newaygo
Oceana
Ogemaw
Oscoda
Otsego
Presque Isle
Roscommon
Sanilac
Wexford
Zone 7
Baraga
Chippewa
Gogebic
Houghton
Iron
Keweenaw
Luce
Mackinac
Ontonagon

Schoolcraft

Minnesota
Zone 6 except
Zone 7
Aitkin
Becker
Beltrami
Carlton
Cass
Clay
Clearwater
Cook
Crow Wing
Grant
Hubbard
Itasca
Kanabec
Kittson
Koochiching
Lake Of The Woods
Mahnomen
Marshall
Mille Lacs
Norman
Otter Tail
Pennington
Pine
Polk
Red Lake
Roseau
St Louis
Wadena
Wilkin

Mississippi
Zone 3 except
Zone 2
Hancock
Harrison
Jackson
Pearl River
Stone

Missouri
Zone 4 except
Zone 5
Adair
Andrew
Atchison
Buchanan
Caldwell
Chariton
Clark
Clinton
Daviess
De Kalb
Gentry
Grundy
Harrison
Holt
Knox
Lewis
Linn
Livingston
Macon
Marion
Mercer
Nodaway
Pike
Putnam
Ralls

Schuyler
Scotland
Shelby
Sullivan
Worth

Montana
Zone 6

Nebraska
Zone 5

Nevada
Zone 5 except
Zone 3
Clark

New Hampshire
Zone 6 except
Zone 5
Cheshire
Hillsborough
Rockingham
Strafford

New Jersey
Zone 4 except
Zone 5
Bergen
Hunterdon
Mercer
Morris
Passaic
Somerset
Sussex
Warren

New Mexico
Zone 4 except
Zone 3
Chaves
Dona Ana
Eddy
Hidalgo
Lea
Luna
Otero
Zone 5
Catron
Cibola
Colfax
Harding
Los Alamos
McKinley
Mora
Rio Arriba
San Juan
San Miguel
Sandoval
Santa Fe
Taos
Torrance

New York
Zone 5 except
Zone 4
Bronx
Kings
Nassau
New York
Queens
Richmond

Suffolk
Westchester
Zone 6
Allegany
Broome
Cattaraugus
Chenango
Clinton
Delaware
Essex
Franklin
Fulton
Hamilton
Herkimer
Jefferson
Lewis
Madison
Montgomery
Oneida
Otsego
Schoharie
Schuyler
St Lawrence
Steuben
Sullivan
Tompkins
Ulster
Warren
Wyoming

North Carolina
Zone 3 except
Zone 4
Alamance
Alexander
Bertie
Buncombe
Burke
Caldwell
Caswell
Catawba
Chatham
Cherokee
Clay
Cleveland
Davie
Durham
Forsyth
Franklin
Gates
Graham
Granville
Guilford
Halifax
Harnett
Haywood
Henderson
Hertford
Iredell
Jackson
Lee
Lincoln
Macon
Madison
McDowell
Nash
Northampton
Orange
Person
Polk
Rockingham

IRC-59

2004 SUPPLEMENT TO THE IRC

Rutherford
Stokes
Surry
Swain
Transylvania
Vance
Wake
Warren
Wilkes
Yadkin
Zone 5
Alleghany
Ashe
Avery
Mitchell
Watauga
Yancey

North Dakota
Zone 7 except
Zone 6
Adams
Billings
Bowman
Burleigh
Dickey
Dunn
Emmons
Golden Valley
Grant
Hettinger
La Moure
Logan
McIntosh
McKenzie
Mercer
Morton
Oliver
Ransom
Richland
Sargent
Sioux
Slope
Stark

Ohio
Zone 5 except
Zone 4
Adams
Brown
Clermont
Gallia
Hamilton
Lawrence
Pike
Scioto
Washington

Oklahoma
Zone 3 Moist except
Zone 4 Dry
Beaver
Cimarron
Texas

Oregon
Zone 4 Marine except
Zone 5 Dry
Baker
Crook
Deschutes
Gilliam

Grant
Harney
Hood River
Jefferson
Klamath
Lake
Malheur
Morrow
Sherman
Umatilla
Union
Wallowa
Wasco
Wheeler

Pennsylvania
Zone 5 except
Zone 4
Bucks
Chester
Delaware
Montgomery
Philadelphia
York
Zone 6
Cameron
Clearfield
Elk
McKean
Potter
Susquehanna
Tioga
Wayne

Rhode Island
Zone 5

South Carolina
Zone 3

South Dakota
Zone 6 except
Zone 5
Bennett
Bon Homme
Charles Mix
Clay
Douglas
Gregory
Hutchinson
Jackson
Mellette
Todd
Tripp
Union
Yankton

Tennessee
Zone 4 except
Zone 3
Chester
Crockett
Dyer
Fayette
Hardeman
Hardin
Haywood
Henderson
Lake
Lauderdale
Madison
McNairy

Shelby
Tipton

Texas
Zone 2 Moist except
Zone 2 Dry
Bandera
Dimmit
Edwards
Kinney
La Salle
Maverick
Medina
Real
Uvalde
Val Verde
Webb
Zapata
Zavala
Zone 3 Dry
Andrews
Baylor
Borden
Brewster
Callahan
Childress
Coke
Coleman
Collingsworth
Concho
Cottle
Crane
Crockett
Crosby
Culberson
Dawson
Dickens
Ector
El Paso
Fisher
Foard
Gaines
Garza
Glasscock
Hall
Hardeman
Haskell
Hemphill
Howard
Hudspeth
Irion
Jeff Davis
Jones
Kent
Kerr
Kimble
King
Knox
Loving
Lubbock
Lynn
Martin
Mason
Mcculloch
Menard
Midland
Mitchell
Motley
Nolan
Pecos
Presidio

Reagan
Reeves
Runnels
Schleicher
Scurry
Shackelford
Sterling
Stonewall
Sutton
Taylor
Terrell
Terry
Throckmorton
Tom Green
Ward
Wheeler
Wilbarger
Winkler
Zone 3 Moist
Archer
Blanco
Bowie
Brown
Burnet
Camp
Cass
Clay
Collin
Comanche
Cooke
Dallas
Delta
Denton
Eastland
Ellis
Erath
Fannin
Franklin
Gillespie
Grayson
Gregg
Hamilton
Harrison
Henderson
Hood
Hopkins
Hunt
Jack
Johnson
Kaufman
Kendall
Lamar
Lampasas
Llano
Montague
Stephens
Wichita
Wise
Young
Marion
Mills
Morris
Nacogdoches
Navarro
Palo Pinto
Panola
Parker
Rains
Red River

Rockwall
Rusk
Sabine
San Augustine
San Saba
Shelby
Smith
Somervell
Tarrant
Titus
Upshur
Van Zandt
Wood
Zone 4
Armstrong
Bailey
Briscoe
Carson
Castro
Cochran
Dallam
Deaf Smith
Donley
Floyd
Gray
Hale
Hansford
Hartley
Hockley
Hutchinson
Lamb
Lipscomb
Moore
Ochiltree
Oldham
Parmer
Potter
Randall
Roberts
Sherman
Swisher
Yoakum

Utah
Zone 5 except
Zone 3
Washington
Zone 6
Box Elder
Cache
Carbon
Daggett
Duchesne
Morgan
Rich
Summit
Uintah
Wasatch

Vermont
Zone 6

Virginia
Zone 4

Washington
Zone 4 Marine except
Zone 5 Dry
Adams
Asotin
Benton

Chelan
Columbia
Douglas
Franklin
Garfield
Grant
Kittitas
Klickitat
Lincoln
San Juan
Skamania
Spokane
Walla Walla
Whitman
Yakima
Zone 6 Dry
Ferry
Okanogan
Pend
Oreille
Stevens

West Virginia
Zone 5 except
Zone 4
Berkeley
Boone
Braxton
Cabell
Calhoun
Clay
Gilmer
Jackson
Jefferson
Kanawha
Lincoln
Logan
Mason
McDowell
Mercer
Mingo
Monroe
Morgan
Pleasants
Putnam
Ritchie
Roane
Tyler
Wayne
Wirt
Wood
Wyoming

Wisconsin
Zone 6 except
Zone 7
Ashland
Bayfield
Burnett
Douglas
Florence
Forest
Iron
Langlade
Lincoln
Oneida
Price
Sawyer
Taylor
Vilas
Washburn

Wyoming
Zone 6 except
Zone 5
Goshen
Platte
Zone 7
Lincoln
Sublette
Teto

TABLE N1101.2.1
WARM HUMID COUNTIES

Alabama
Autauga
Baldwin
Barbour
Bullock
Butler
Choctaw
Clarke
Coffee
Conecuh
Covington
Crenshaw
Dale
Dallas
Elmore
Escambia
Geneva
Henry
Houston
Lowndes
Macon
Marengo
Mobile
Monroe
Montgomery
Perry
Pike
Russell
Washington
Wilcox

Arkansas
Columbia
Hempstead
Lafayette
Little River
Miller
Sevier
Union

Florida
All

Georgia
All in Zone 2
Plus
Ben Hill
Bleckley
Bulloch
Calhoun
Candler
Chattahoochee
Clay
Coffee
Crisp
Dodge
Dooly
Dougherty
Early
Emanuel
Houston
Irwin
Jenkins
Johnson
Laurens
Lee
Macon

Marion
Montgomery
Peach
Pulaski
Quitman
Randolph
Schley
Screven
Stewart
Sumter
Taylor
Telfair
Terrell
Tift
Treutlen
Turner
Twiggs
Webster
Wheeler
Wilcox
Worth

Louisiana
All in Zone 2
Plus
Bienville
Bossier
Caddo
Caldwell
Catahoula
Claiborne
Concordia
De Soto
Franklin
Grant
Jackson
La Salle
Lincoln
Madison
Natchitoches
Ouachita
Red River
Richland
Sabine
Tensas
Union
Vernon
Webster
Winn

Mississippi
All in Zone 2
Plus
Adams
Amite
Claiborne
Copiah
Covington
Forrest
Franklin
George
Greene
Hinds
Jefferson
Jefferson Davis
Jones
Lamar

Lawrence
Lincoln
Marion
Perry
Pike
Rankin
Simpson
Smith
Walthall
Warren
Wayne
Wilkinson

North Carolina
Brunswick
Carteret
Columbus
New Hanover
Onslow
Pender

South Carolina
Allendale
Bamberg
Barnwell
Beaufort
Berkeley
Charleston
Colleton
Dorchester
Georgetown
Hampton
Horry
Jasper

Texas
All in Zone 2
Plus
Blanco
Bowie
Brown
Burnet
Camp
Cass
Collin
Comanche
Dallas
Delta
Denton
Ellis
Erath
Franklin
Gillespie
Gregg
Hamilton
Harrison
Henderson
Hood
Hopkins
Hunt
Johnson
Kaufman
Kendall
Lamar
Lampasas
Llano
Marion
Mills

Morris
Nacogdoches
Navarro
Palo
Pinto
Panola
Parker
Rains
Red
River
Rockwall
Rusk
Sabine
San Augustine
San Saba
Shelby
Smith
Somervell
Tarrant
Titus
Upshur
Van Zandt
Wood

N1101.6.1 Protection of exposed foundation insulation. Insulation applied to the exterior of basement walls, crawl space walls, and the perimeter of slab-on-grade floors shall have a rigid, opaque and weather-resistant protective covering to prevent the degradation of the insulation's thermal performance. The protective covering shall cover the exposed exterior insulation and extend a minimum of 6 inches (153 mm) below grade.

N1101.7 Above code programs. The building official or other authority having jurisdiction shall be permitted to deem a national, state or local energy efficiency program to exceed the energy efficiency required by this chapter. Buildings approved in writing by such an energy efficiency program shall be considered in compliance with this chapter.

N1101.8 Certificate. A permanent certificate shall be posted on or in the electrical distribution panel. The certificate shall be completed by the builder or registered design professional. The certificate shall list the predominant R-values of insulation installed in or on ceiling/roof, walls, foundation (slab, basement wall, crawlspace wall and/or floor) and ducts outside conditioned spaces; U-factors for fenestration; and, where requirements apply, the solar heat gain coefficient (SHGC) of fenestration. Where there is more than one value for each component, the certificate shall list the value covering the largest area. The certificate shall list the type and efficiency of heating, cooling and service water heating equipment.

SECTION N1102
BUILDING THERMAL ENVELOPE

N1102.1 Insulation and fenestration criteria. The building thermal envelope shall meet the requirements of Table N1102.1 based on the climate zone specified in Table N1101.2.

(See page 66 for Table N1102.1)

N1102.1.1 R-value computation. Insulation material used in layers, such as framing cavity insulation and insulating sheathing, shall be summed to compute the component R-value. The manufacturer's settled R-value shall be used for blown insulation. Computed R-values shall not include an R-value for other building materials or air films.

N1102.1.2 U-factor alternative. An assembly with a U-factor equal to or less than that specified in Table N1102.1.2 shall be permitted as an alternative to the R-value in Table N1102.1.

Exception: For mass walls not meeting the criterion for insulation location in Section N1102.2.3, the U-factor shall be permitted to be:

a) U-factor of 0.17 in Climate Zone 1

b) U-factor of 0.14 in Climate Zone 2

c) U-factor of 0.12 in Climate Zone 3

(See page 66 for Table N1102.1.2)

N1102.1.3 Total UA alternative. If the total building thermal envelope UA (sum of U-factor times assembly area) is less than or equal to the total UA resulting from using the U-factors in Table N1102.1.2, the building shall be considered in compliance with Table N1102.1. The UA calculation shall be done using a method consistent with the ASHRAE *Handbook of Fundamentals* and shall include the thermal bridging effects of framing materials. The SHGC requirements shall be met in addition to UA compliance.

N1102.2 Specific insulation requirements.

N1102.2.1 Ceilings with attic spaces. When Section N1102.1 would require R-38 in the ceiling, R-30 shall be deemed to satisfy the requirement for R-38 wherever the full height of uncompressed R-30 insulation extends over the wall top plate at the eaves. Similarly R-38 shall be deemed to satisfy the requirement for R-49 wherever the full height of uncompressed R-38 insulation extends over the wall top plate at the eaves.

N1102.2.2 Ceilings without attic spaces. Where Section N1102.1 would require insulation levels above R-30 and the design of the roof/ceiling assembly does not allow sufficient space for the required insulation, the minimum required insulation for such roof/ceiling assemblies shall be R-30. This reduction of insulation from the requirements of Section N1102.1 shall be limited to 500 ft^2 (46 m^2) of ceiling area.

N1102.2.3 Mass walls. Mass walls, for the purposes of this chapter, shall be considered walls of concrete block, concrete, insulated concrete form (ICF), masonry cavity, brick (other than brick veneer), earth (adobe, compressed earth block, rammed earth) and solid timber/logs. The provisions of Section N1102.1 for mass walls shall be applicable when at least 50 percent of the required insulation R-value is on the exterior of, or integral to, the wall. Walls that do not meet this criterion for insulation placement shall meet the wood frame wall insulation requirements of Section N1102.1.

Exception: For walls that do not meet this criterion for insulation placement the minimum added insulation R-value shall be permitted to be:

a) R-value of 4 in Climate Zone 1

b) R-value of 6 in Climate Zone 2

c) R-value of 8 in Climate Zone 3

N1102.2.4 Steel-frame ceilings, walls and floors. Steel-frame ceilings, walls and floors shall meet the insulation requirements of Table N1102.2.4 or shall meet the *U*-factor requirements in Table N1102.1.2. The calculation of the *U*-factor for a steel-frame envelope assembly shall use a series-parallel path calculation method.

(See page 67 for Table N1102.2.4)

N1102.2.5 Floors. Floor insulation shall be installed to maintain permanent contact with the underside of the subfloor decking.

N1102.2.6 Basement walls. Exterior walls associated with conditioned basements shall be insulated from the top of the basement wall down to 10 feet (3048 mm) below grade or to the basement floor, whichever is less. Walls associated with unconditioned basements shall meet this requirement unless the floor overhead is insulated in accordance with Sections N1102.1 and N1102.2.5.

N1102.2.7 Slab-on-grade floors. Slab-on-grade floors with a floor surface less than 12 inches below grade shall be insulated in accordance with Table N1102.1. The insulation shall extend downward from the top of the slab on the outside or inside of the foundation wall. Insulation located below grade shall be extended the distance provided in Table N1102.1 by any combination of vertical insulation, insulation extending under the slab or insulation extending out from the building. Insulation extending away from the building shall be protected by pavement or by a minimum of 10 inches (254 mm) of soil. The top edge of the insulation installed between the exterior wall and the edge of the interior slab shall be permitted to be cut at a 45-degree (0.79 rad) angle away from the exterior wall. Slab-edge insulation is not required in jurisdictions designated by the code official as having a very heavy termite infestation.

N1102.2.8 Crawl space walls. As an alternative to insulating floors over crawl spaces, crawl space walls shall be permitted to be insulated when the crawl space is not vented to the outside. Crawl space wall insulation shall be permanently fastened to the wall and extend downward from the floor to the finished grade level and then vertically and/or horizontally for at least an additional 24 inches (610 mm). Exposed earth in unvented crawl space foundations shall be covered with a continuous vapor retarder. All joints of the vapor retarder shall overlap by 6 inches (153 mm) and be sealed or taped. The edges of the vapor retarder shall extend at least 6 inches (153 mm) up the stem wall and shall be attached to the stem wall.

N1102.2.9 Masonry veneer. Insulation shall not be required on the horizontal portion of the foundation that supports a masonry veneer.

N1102.2.10 Thermally isolated sunroom insulation. The minimum ceiling insulation *R*-values shall be R-19 in zones 1 through 4 and R-24 in zones 5 though 8. The minimum wall *R*-value shall be R-13 in all zones. New wall(s) separating the sunroom from conditioned space shall meet the building thermal envelope requirements.

N1102.3 Fenestration.

N1102.3.1 *U*-factor. An area-weighted average of fenestration products shall be permitted to satisfy the *U*-factor requirements.

N1102.3.2 Glazed fenestration SHGC. An area-weighted average of fenestration products more than 50 percent glazed shall be permitted to satisfy the solar heat gain coefficient (SHGC) requirements.

N1102.3.3 Glazed fenestration exemption. Up to 15 ft^2 (1.4 m^2) of glazed fenestration per dwelling unit shall be permitted to be exempt from *U*-factor and solar heat gain coefficient (SHGC) requirements in Section N1102.1.

N1102.3.4 Opaque door exemption. One opaque door assembly is exempted from the *U*-factor requirement in Section N1102.1.

N1102.3.5 Thermally isolated sunroom *U*-factor. For zones 4 through 8 the maximum fenestration *U*-factor shall be 0.50 and the maximum skylight *U*-factor shall be 0.75. New windows and doors separating the sunroom from conditioned space shall meet the building thermal envelope requirements.

N1102.3.6 Replacement fenestration. Where some or all of an existing fenestration unit is replaced with a new fenestration product, including frame, sash and glazing, the replacement fenestration unit shall meet the applicable requirements for *U*-factor and solar heat gain coefficient (SHGC) in Table N1102.1.

N1102.4 Air leakage.

N1102.4.1 Building thermal envelope. The building thermal envelope shall be durably sealed to limit infiltration. The sealing methods between dissimilar materials shall allow for differential expansion and contraction. The following shall be caulked, gasketed, weatherstripped or otherwise sealed with an air barrier material, suitable film or solid material.

1. All joints, seams and penetrations.

2. Site-built windows, doors and skylights.

3. Openings between window and door assemblies and their respective jambs and framing.

4. Utility penetrations.

5. Dropped ceilings or chases adjacent to the thermal envelope.

6. Knee walls.

7. Walls and ceilings separating the garage from conditioned spaces.

8. Behind tubs and showers on exterior walls.

9. Common walls between dwelling units.

10. Other sources of infiltration.

N1102.4.2 Fenestration air leakage. Windows, skylights and sliding-glass doors shall have an air infiltration rate of no more than 0.3 cfm/ft^2 [1.5(L/s)/m^2], and swinging doors no more than 0.5 cfm/ft^2 [2.5(L/s)/m^2], when tested according to NFRC 400, 101/I.S.2, or 101/I.S.2/NAFS by an accredited, independent laboratory, and listed and labeled by the manufacturer.

Exception: Site-built windows, skylights and doors.

N1102.4.3 Recessed lighting. Recessed luminaires installed in the building thermal envelope shall be sealed to limit air leakage between conditioned and unconditioned spaces by being:

1. IC-rated and labeled with enclosures that are sealed or gasketed to prevent air leakage to the ceiling cavity or unconditioned space; or

2. IC-rated and labeled as meeting ASTM E 283 when tested at 1.57 psi (75 Pa) pressure differential with no more than 2.0 cfm (0.944 L/s) of air movement from the conditioned space to the ceiling cavity; or

3. Located inside an airtight sealed box with clearances of at least 0.5 inch (12.7 mm) from combustible material and 3 inches (76 mm) from insulation.

N1102.5 Moisture control. The building design shall not create conditions of accelerated deterioration from moisture condensation. Above-grade frame walls, floors and ceilings not ventilated to allow moisture to escape shall be provided with an approved vapor retarder. The vapor retarder shall be installed on the warm-in-winter side of the thermal insulation.

Exceptions:

1. In construction where moisture or its freezing will not damage the materials.

2. Frame walls, floors and ceilings in jurisdictions in Zones 1 through 4. (Crawl space floor vapor retarders are not exempted.)

3. Where other approved means to avoid condensation are provided.

N1102.5.1 Maximum fenestration *U*-factor. The area weighted average maximum fenestration *U*-factor permitted using trade offs from Section N1102.1.3 in Zones 6 through 8 shall be 0.55.

SECTION N1103
SYSTEMS

N1103.1 Controls. At least one thermostat shall be provided for each separate heating and cooling system.

N1103.2 Ducts.

N1103.2.1 Insulation. Supply and return ducts shall be insulated to a minimum of R-8. Ducts in floor trusses shall be insulated to a minimum of R-6.

Exception: Ducts or portions thereof located completely inside the building thermal envelope.

N1103.2.2 Sealing. All ducts, air handlers, filter boxes and building cavities used as ducts shall be sealed. Joints and seams shall comply with M1601.3.1.

N1103.2.3 Building cavities. Building framing cavities shall not be used as supply ducts.

N1103.3 Mechanical system piping insulation. Mechanical system piping capable of carrying fluids above 105°F (40°C) or below 55°F (13°C) shall be insulated to a minimum of R-2.

N1103.4 Circulating hot water systems. All circulating service hot water piping shall be insulated to at least R-2. Circulating hot water systems shall include an automatic or readily accessible manual switch that can turn off the hot water circulating pump when the system is not in use.

N1103.5 Mechanical ventilation. Outdoor air intakes and exhausts shall have automatic or gravity dampers that close when the ventilation system is not operating.

N1103.6 Equipment sizing. Heating and cooling equipment shall be sized as specified in Section M1401.3.

TABLE N1102.1
INSULATION AND FENESTRATION REQUIREMENTS BY COMPONENT [a]

CLIMATE ZONE	FENESTRATION U-FACTOR	SKYLIGHT[b] U-FACTOR	GLAZED FENESTRATION SHGC	CEILING R-VALUE	WOOD FRAME WALL R-VALUE	MASS WALL R-VALUE	FLOOR R-VALUE	BASEMENT[c] WALL R-VALUE	SLAB[d] R-VALUE & DEPTH	CRAWL SPACE[c] WALL R-VALUE
1	1.2	0.75	0.40	30	13	3	13	0	0	0
2	0.75	0.75	0.40	30	13	4	13	0	0	0
3	0.65	0.65	0.40[e]	30	13	5	19	0	0	5/13
4 except Marine	0.40	0.60	NR	38	13	5	19	10 / 13	10, 2 ft	10 / 13
5 and Marine 4	0.35	0.60	NR	38	19 or 13+5[g]	13	30[f]	10 / 13	10, 2 ft	10 / 13
6	0.35	0.60	NR	49	19 or 13+5[g]	15	30[f]	10 / 13	10, 4 ft	10 / 13
7 and 8	0.35	0.60	NR	49	21	19	30[f]	10 / 13	10, 4 ft	10 / 13

a. R-values are minimums. U-factors and SHGC are maximums. R-19 insulation shall be permitted to be compressed into a 2x6 cavity.
b. The fenestration U-factor column excludes skylights. The solar heat gain coefficient (SHGC) column applies to all glazed fenestration.
c. The first R-value applies to continuous insulation, the second to framing cavity insulation; either insulation meets the requirement.
d. R-5 shall be added to the required slab edge R-values for heated slabs.
e. There are no solar heat gain coefficient (SHGC) requirements in the Marine Zone.
f. Or insulation sufficient to fill the framing cavity, R-19 minimum.
g. "13+5" means R-13 cavity insulation plus R-5 insulated sheathing. If structural sheathing covers 25% or less of the exterior, R-5 sheathing is not required where structural sheathing is used. If structural sheathing covers more than 25% of exterior, structural sheathing shall be supplemented with insulated sheathing of at least R-2.

TABLE N1102.1.2
EQUIVALENT U-factors[a]

CLIMATE ZONE	FENESTRATION U-FACTOR	SKYLIGHT U-FACTOR	CEILING U-FACTOR	FRAME WALL U-FACTOR	MASS WALL U-FACTOR	FLOOR U-FACTOR	BASEMENT WALL U-FACTOR	CRAWL SPACE WALL U-FACTOR
1	1.20	0.75	0.035	0.082	0.197	0.064	0.360	0.477
2	0.75	0.75	0.035	0.082	0.165	0.064	0.360	0.477
3	0.65	0.65	0.035	0.082	0.141	0.047	0.360	0.136
4 except Marine	0.40	0.60	0.030	0.082	0.141	0.047	0.059	0.065
5 and Marine 4	0.35	0.60	0.030	0.060	0.082	0.033	0.059	0.065
6	0.35	0.60	0.026	0.060	0.06	0.033	0.059	0.065
7 and 8	0.35	0.60	0.026	0.057	0.057	0.033	0.059	0.065

a. Nonfenestration U-factors shall be obtained from measurement, calculation or an approved source.

Table N1102.2.4.
Steel-Frame Ceiling, Wall and Floor Insulation (R-value)

WOOD FRAME R-VALUE REQUIREMENT	COLD-FORMED STEEL EQUIVALENT R-VALUE[A]
Steel Truss Ceilings[a]	
R-30	R-38 or R-30 + 3 or R-26 + 5
R-38	R-49 or R-38 + 3
R-49	R-38 + 5
Steel Joist Ceilings[b]	
R-30	R-38 in 2x4 or 2x6 or 2x8 R-49 in any framing
R-38	R-49 in 2x4 or 2x6 or 2x8 or 2x10
Steel Framed Wall	
R-13	R-13 + 5 or R-15 + 4 or R-21 + 3
R-19	R-13 + 9 or R-19 + 8 or R-25 + 7
R-21	R-13 +10 or R-19 + 9 or R-25 + 8
Steel Joist Floor	
R-13	R-19 in 2x6 R-19 + R6 in 2x8 or 2x10
R-19	R-19 +R-6 in 2x6 R-19 + R-12 in 2x8 or 2x10

For SI: 1 inch = 25.4mm.
Notes:
a. Cavity insulation R-value is listed first, followed by continuous insulation R-value.
b. Insulation exceeding the height of the framing shall cover the framing.

Part V—Mechanical

CHAPTER 13
GENERAL MECHANICAL SYSTEM REQUIREMENTS

Section M1303.1 Change to read as shown: (EC48-03/04)

M1303.1 Label information. A permanent factory-applied nameplate(s) shall be affixed to appliances on which shall appear, in legible lettering, the manufacturer's name or trademark, the model number, serial number, and the seal or mark of the testing agency. A label shall also include the following:

1. Electrical appliances. Electrical rating in volts, amperes and motor phase; identification of individual electrical components in volts, amperes or watts and motor phase; and in Btu/h (W) output and required clearances.

2. Absorption units. Hourly rating in Btu/h (W), minimum hourly rating for units having step or automatic modulating controls, type of fuel, type of refrigerant, cooling capacity in Btu/h (W) and required clearances.

3. Fuel-burning units. Hourly rating in Btu/h (W), type of fuel approved for use with the appliance and required clearances.

4. Electric comfort heating appliances. Name and trademark of the manufacturer; the model number or equivalent; the electric rating in volts, amperes and phase; Btu/h (W) output rating; individual marking for each electrical component in amperes or watts, volts and phase; required clearances from

2004 SUPPLEMENT TO THE IRC

combustibles and a seal indicating approval of the appliance by an approved agency.

5. Maintenance instructions. Required regular maintenance actions and title or publication number for the operation and maintenance manual for that particular model and type of product.

Section M1305.1 Change to read as shown: (RM1-03/04)

M1305.1 Appliance access for inspection service, repair and replacement. Appliances shall be accessible for inspection, service, repair and replacement without removing permanent construction. A level working space at least 30 inches deep and 30 inches wide shall be provided in front of the control side to service an appliance. Room heaters shall be permitted to be installed with at least an 18-inch (457 mm) working space. A platform shall not be required for room heaters.

Section M1305.1.3.1 Change to read as shown: (EL3-03/04)

M1305.1.3.1 Electrical requirements. A luminaire controlled by a switch located at the required passageway opening and a receptacle outlet shall be provided at or near the appliance location in accordance with Chapter 38.

Section M1305.1.4.3 Change to read as shown: (EL3-03/04)

M1305.1.4.3 Electrical requirements. A luminaire controlled by a switch located at the required passageway opening and a receptacle outlet shall be provided at or near the appliance location in accordance with Chapter 38.

Section M1308.3 Add new section to read as shown: (RM19-03/04)

M1308.3 Foundations and supports. Foundations and supports for outdoor mechanical systems shall be raised at least 3 inches (76 mm) above the finished grade, and shall also conform to the manufacturer's installation instructions.

CHAPTER 14
HEATING AND COOLING EQUIPMENT

Section M1406.2 Change to read as shown: (EL3-03/04)

M1406.2 Clearances. Clearances for radiant heating panels or elements to any wiring, outlet boxes and junction boxes used for installing electrical devices or mounting luminaires shall comply with Chapters 33 through 42 of this code.

Section M1411.3.1 Change to read as shown: (RM4-03/04 and M15-03/04)

M1411.3.1 Auxiliary and secondary drain systems. In addition to the requirements of Section M1411.3, a secondary drain or auxiliary drain pan shall be required for each cooling or evaporator coil where damage to any building components will occur as a result of overflow from the equipment drain pan or stoppage in the condensate drain piping. Such piping shall maintain a minimum horizontal slope in the direction of discharge of not less than one-eighth unit vertical in 12 units horizontal (1-percent slope). Drain piping shall be a minimum of 3/4-inch (19.1 mm) nominal pipe size. One of the following methods shall be used:

1. An auxiliary drain pan with a separate drain shall be provided under the coils on which condensation will occur. The auxiliary pan drain shall discharge to a conspicuous point of disposal to alert occupants in the event of a stoppage of the primary drain. The pan shall have a minimum depth of 1.5 inches (38 mm), shall not be less than 3 inches (76 mm) larger than the unit or the coil dimensions in width and length and shall be constructed of corrosion-resistant material. Metallic pans shall have a minimum thickness of not less than 0.0276-inch (0.7 mm) galvanized sheet metal. Nonmetallic pans shall have a minimum thickness of not less than 0.0625 inch (1.6 mm).

2. A separate overflow drain line shall be connected to the drain pan provided with the equipment. Such overflow drain shall discharge to a conspicuous point of disposal to alert occupants in the event of a stoppage of the primary drain. The overflow drain line shall connect to the drain pan at a higher level than the primary drain connection.

3. An auxiliary drain pan without a separate drain line shall be provided under the coils on which condensate will occur. Such pan shall be equipped with a water level detection device that will shut off the equipment served prior to overflow of the pan. The auxiliary drain pan shall be constructed in accordance with Item 1 of this section.

4. A water level detection device shall be provided that will shut off the equipment served in the event that the primary drain is blocked. The device shall be installed in the primary drain line, the overflow drain line or the equipment-supplied drain pan, located at a point higher than the primary drain line connection and below the overflow rim of such pan.

Sections M1411 and M1411.4 Change section heading and add new section to read as shown: (RM3-03/04)

M1411
HEATING AND COOLING EQUIPMENT

M1411.4 Auxiliary drain pan. Category IV condensing appliances shall be provided with an auxiliary drain pan where damage to any building component will occur as a result of stoppage in the condensate drainage system. Such pans shall be installed in accordance with the applicable provisions located in Section M1411.3.

> **Exception:** Fuel-fired appliances that automatically shut down operation in the event of a stoppage in the condensate drainage system.

(Renumber remaining sections)

CHAPTER 15
EXHAUST SYSTEMS

Sections M1501 and M1501.1 Add new section heading and section to read as shown: (RM5-03/04)

SECTION M1501
GENERAL

M1501.1 Outdoor discharge. The air removed by every mechanical exhaust system shall be discharged to the outdoors. Air shall not be exhausted into an attic, soffit, ridge vent or crawl space.

> **Exception:** Whole-house ventilation-type attic fans that discharge into the attic space of dwelling units having private attics shall not be prohibited.

(Renumber remaining sections)

Section M1501.3 Change to read as shown: (RM8-03/04)

M1501.3 Length limitation. The maximum length of a clothes dryer exhaust duct shall not exceed 25 feet (7620 mm) from the dryer location to the wall or roof termination. The maximum length of the duct shall be reduced 2.5 feet (762 mm) for each 45-degree (0.79 rad) bend and 5 feet (1524 mm) for each 90-degree (1.6 rad) bend. The maximum length of the exhaust duct does not include the transition duct.

> **Exception:** Where the make and model of the clothes dryer to be installed is known and the manufacturer's installation instructions for such dryer are provided to the building official, the maximum length of the exhaust duct, including any transition duct, shall be permitted to be in accordance with the dryer manufacturer's installation instructions.

Section M1506.2 Change to read as shown: (RM9-03/04)

M1506.2 Recirculation of air. Exhaust air from bathrooms and toilet rooms shall not be recirculated within a residence or to another dwelling unit and shall be exhausted directly to the outdoors. Exhaust air from bathrooms and toilet rooms shall not discharge into an attic, crawl space or other areas inside the building.

CHAPTER 16
DUCT SYSTEMS

Section M1601.1.1 Change to read as shown: (RB199-03/04)

M1601.1.1 Above-ground duct systems. Above-ground duct systems shall conform to the following:

1. Equipment connected to duct systems shall be designed to limit discharge air temperature to a maximum of 250 °F (121°C).

2. Factory-made air ducts shall be constructed of Class 0 or Class 1 materials as designated in Table M1601.1.1(1).

3. Fibrous duct construction shall conform to the SMACNA *Fibrous Glass Duct Construction Standards* or NAIMA *Fibrous Glass Duct Construction Standards*.

4. Minimum thickness of metal duct material shall be as listed in Table M1601.1.1(2). Galvanized steel shall conform to ASTM A 653.

5. Gypsum products are permitted to be used to construct return air ducts or plenums, provided that the air temperature does not exceed 125°F (52°C) and exposed surfaces are not subject to condensation.

6. Duct systems shall be constructed of materials having a flame spread index not greater than 200.

7. Stud wall cavities and the spaces between solid floor joists to be utilized as air plenums shall comply with the following conditions:

    7.1. Such cavities or spaces shall not be utilized as a plenum for supply air.

2004 SUPPLEMENT TO THE IRC

7.2. Such cavities or spaces shall not be part of a required fire-resistance-rated assembly.

7.3. Stud wall cavities shall not convey air from more than one floor level.

7.4. Stud wall cavities and joist-space plenums shall be isolated from adjacent concealed spaces by tight-fitting fire blocking in accordance with Section R602.8.

CHAPTER 17
COMBUSTION AIR

Section M1703.2.1 Change to read as shown: (RM12-03/04)

M1703.2.1 Size of openings. Where directly communicating with the outdoors, or where communicating with the outdoors by means of vertical ducts, each opening shall have a free area of at least 1 square inch per 4,000 Btu/per hour (0.550 mm^2/W) of total input rating of all appliances in the space. Where horizontal ducts are used, each opening shall have a free area of at least 1 square inch per 2,000 Btu/per hour (1.1 mm^2/W) of total input of all appliances in the space. Ducts shall be of the same minimum cross-sectional area as the required free area of the openings to which they connect. The minimum cross-sectional dimension of rectangular air ducts shall be 3 inches (76 mm).

CHAPTER 20
BOILERS AND WATER HEATERS

Section M2006.3 Change to read as shown: (RM13-03/04)

M2006.3 Temperature and pressure-limiting devices. Pool heaters shall have temperature relief valves.

CHAPTER 21
HYDRONIC PIPING

Table M2101.1 Change to read as shown: (RM14-03/04 and RM16-03/04)

TABLE M2101.1
HYDRONIC PIPING MATERIALS

MATERIAL	USE CODE [a]	STANDARD [b]	JOINTS	NOTES
Cross-linked Polyethylene(PEX)	1, 2, 3	ASTM F 876, F 877	(See PEX fittings)	Install in accordance with manufacturer's instructions
PEX Fittings		ASTM F 1807 ASTM F 1960 ASTM F 2098	Copper-crimp/ insert fittings, cold expansion fittings, Stainless steel clamp, insert fittings	Install in accordance with manufacturer's instructions
Polypropylene (PP)	1, 2, 3	ISO 15874	Heat fusion joints, mechanical fittings, threaded adapters, compression joints	

(Portions of table not shown do not change)

a. Use code
   1. Above ground.
   2. Embedded in radiant systems.
   3. Temperatures below 180°F only.
   4. Low temperature (below 130°F) only.
b. Standards as listed in Chapter 43

Table M2101.9 Change to read as shown: (RM16-03/04)

TABLE M2101.9
HANGER SPACING INTERVALS

PIPING MATERIAL	MAXIMUM HORIZONTAL SPACING (feet)	MAXIMUM VERTICAL SPACING (feet)
PP < 1 inch pipe or tubing	2.67	4
PP > 1 1/4 inch	4	5

(Portions of table not shown do not change)

Section M2103.1 Change to read as shown: (RM15-03/04 and RM16-03/04)

M2103.1 Piping Materials. Piping for embedment in concrete or gypsum materials shall be standard-weight steel pipe, copper tubing, cross-linked polyethylene/aluminum/ cross-linked polyethylene (PEX-AL-PEX) pressure pipe, chlorinated polyvinyl chloride (CPVC), polybutylene, cross-linked polyethylene (PEX) tubing or polypropylene (PP) with a minimum rating of 100 psi at 180°F (689 kPa at 82°C).

Section M2103.2 Change to read as shown: (RM16-03/04 and RM17-03/04)

M2103.2 Piping joints. Piping joints that are embedded shall be installed in accordance with the following requirements.

1. Steel pipe joints shall be welded.

2. Copper tubing shall be joined with brazing material having a melting point exceeding 1,000°F (538°C).

3. Polybutylene pipe and tubing joints shall be installed with socket-type heat-fused polybutylene fittings.

4. CPVC tubing shall be joined using solvent cement joints.

5. Polypropylene pipe and tubing joints shall be installed with socket-type heat-fusion polypropylene fittings.

6. Cross-linked polyethylene (PEX) tubing shall be joined using cold expansion, insert or compression fittings.

Section M2104.2 Change to read as shown: (RM16-03/04)

M2104.2 Piping joints. Piping joints (other than those in Section M2103.2) that are embedded shall be installed in accordance with the following requirements:

1. Cross-linked polyethylene (PEX) tubing shall follow manufacturer's instructions.

2. Polyethylene tubing shall be installed with heat fusion joints.

3. Polypropylene (PP) tubing shall be installed in accordance with the manufacturer's instructions.

**CHAPTER 22
SPECIAL PIPING AND STORAGE SYSTEMS**

Section M2201.7 Add new text to read as shown: (M76-03/04)

M2201.7 Tanks abandoned or removed. All exterior above-grade fill piping shall be removed when tanks are abandoned or removed. Tank abandonment and removal shall be in accordance with the *International Fire Code*.

Part VI—Fuel Gas

**CHAPTER 24
FUEL GAS**

**SECTION G2403 (202)
GENERAL DEFINITIONS**

Change the definition of "Connector" to read as shown: (FG3-03/04)

CONNECTOR, CHIMNEY OR VENT. The pipe that connects an appliance to a chimney or vent.

Add new definition to read as shown: (FG3-03/04)

CONNECTOR, APPLIANCE (fuel). Rigid metallic pipe and fittings, semi-rigid metallic tubing and fittings or a listed and labeled device that connects an appliance to the gas piping system.

Change the definition of "Fuel Gas" to read as shown: (FG4-03/04)

FUEL GAS. A natural gas, manufactured gas, liquefied petroleum gas or mixtures of these gases.

2004 SUPPLEMENT TO THE IRC

Delete the definition of "Mechanical Exhaust System" (FG5-03/04)

Change the definition of "Point of Delivery" to read as shown: (FG6-03/04)

POINT OF DELIVERY. For natural gas systems, the point of delivery is the outlet of the service meter assembly or the outlet of the service regulator or service shutoff valve where a meter is not provided. Where a valve is provided at the outlet of the service meter assembly, such valve shall be considered to be downstream of the point of delivery. For undiluted liquefied petroleum gas systems, the point of delivery shall be considered the outlet of the first regulator that reduces pressure to 2 pounds per square inch gauge (13.8 kPag) or less.

Add new definitions to read as shown: (FG30-03/04)

VENT PIPING

> **Breather.** Piping run from a pressure regulating device to the outdoors, designed to provide a reference to atmospheric pressure. If the device incorporates an integral pressure relief mechanism, a breather vent can also serve as a relief vent.
>
> **Relief.** Piping run from a pressure-regulating or pressure-limiting device to the outdoors, designed to provide for the safe venting of gas in the event of excessive pressure in the gas piping system.

SECTION G2406 (303)
APPLIANCE LOCATION

Section G2406.2 (303.3) Change to read as shown: (FG8-03/04)

G2406.2 (303.3) Prohibited locations. Appliances shall not be located in sleeping rooms, bathrooms, toilet rooms or storage closets, or in a space that opens only into such rooms or spaces, except where the installation complies with one of the following:

1. The appliance is a direct-vent appliance installed in accordance with the conditions of the listing and the manufacturer's instructions.

2. Vented room heaters, wall furnaces, vented decorative appliances, vented gas fireplaces, vented gas fireplace heaters and decorative appliances for installation in vented solid-fuel-burning fireplaces are installed in rooms that meet the required volume criteria of Section G2407.5.

3. A single wall-mounted unvented room heater is installed in a bathroom and such unvented room heater is equipped as specified in Section G2445.6 and has an input rating not greater than 6,000 Btu per hour (1.76 kW). The bathroom shall meet the required volume criteria of Section G2407.5.

4. A single wall-mounted unvented room heater is installed in a bedroom and such unvented room heater is equipped as specified in Section G2445.6 and has an input rating not greater than 10,000 Btu per hour (2.93 kW). The bedroom shall meet the required volume criteria of Section G2407.5.

5. The appliance is installed in a room or space that opens only into a bedroom or bathroom, such room or space is used for no other purpose, and is provided with a solid weather-stripped door equipped with an approved self-closing device. All combustion air shall be taken directly from the outdoors in accordance with Section G2407.6.

SECTION G2408 (305)
INSTALLATION

Section G2408.6(307.4) Add new section to read as shown: (FG16-03/04)

G2408.6(307.4) Auxiliary drain pan. Category IV condensing appliances shall be provided with an auxiliary drain pan where damage to any building component will occur as a result of stoppage in the condensate drainage system. Such pan shall be installed in accordance with the applicable provisions of Chapter 14.

> **Exception:** An auxiliary drain pan shall not be required for appliances that automatically shut down operation in the event of a stoppage in the condensate drainage system.

Section G2408.7(307.1) Add new section to read as shown: (FG15-03/04)

G2408.7(307.1) Evaporators and cooling coils. Condensate drainage systems shall be provided for equipment and appliances containing evaporators and cooling coils in accordance with Chapter 14.

SECTION G2412 (401)
GENERAL

Section G2412.5 (401.5) Change to read as shown: (FG17-03/04)

G2412.5 (401.5) Identification. For other than steel pipe, exposed piping shall be identified by a yellow label marked "Gas" in black letters. The marking shall be spaced at intervals not exceeding 5 feet (1524 mm). The marking shall not be required on pipe located in the same room as the equipment served.

SECTION G2415 (404)
PIPING SYSTEM INSTALLATION

Section G2415.6 (404.6) Change to read as shown: (FG21-03/04)

G2415.6 (404.6) Piping in solid floors. Piping in solid floors shall be laid in channels in the floor and covered in a manner that will allow access to the piping with a minimum amount of damage to the building. Where such piping is subject to exposure to excessive moisture or corrosive substances, the piping shall be protected in an approved manner. As an alternative to installation in channels, the piping shall be installed in a conduit of Schedule 40 steel, wrought iron, PVC or ABS pipe with tightly sealed ends and joints. Both ends of such conduit shall extend not less than 2 inches (51 mm) beyond the point where the pipe emerges from the floor. The sleeve shall be vented above-grade to the outdoors and shall be installed so as to prevent the entry of water and insects.

SECTION G2420 (409)
GAS SHUTOFF VALVES

Section G2420.1.1 (409.1.1) Change to read as shown: (FG25-03/04)

G2420.1.1 (409.1.1) Valve approval. Shutoff valves shall be of an approved type. Shutoff valves shall be constructed of materials compatible with the piping. Shutoff valves shall comply with the standard that is applicable for the pressure and application, in accordance with Table G2420.1.1.

Table G2420.1.1 (Table 409.1.1) Add new table as shown: (FG25-03/04)

TABLE G2420.1.1 (Table 409.1.1)
MANUAL GAS VALVE STANDARDS

VALVE STANDARDS	APPLIANCE SHUTOFF VALVE APPLICATION UP TO ½ psig PRESSURE	OTHER VALVE APPLICATIONS			
		Up To ½ psig PRESSURE	Up To 2 psig PRESSURE	Up To 5 psig PRESSURE	Up To 125 psig PRESSURE
ANSI Z21.15	X				
CSA Requirement 3-88	X	X	If labeled 2G	If labeled 5G	
ASME B16.44	X	X	If labeled 2G	If labeled 5G	
ASME B16.33	X	X	X	X	X

For SI: 1 psig = 6.89 kPa.

2004 SUPPLEMENT TO THE IRC

SECTION G2421 (410)
FLOW CONTROLS

Section G2421.3 (410.3) Change to read as shown: (FG30-03/04)

G2421.3 (410.3) Venting of regulators. Pressure regulators that require a vent shall be vented directly to the outdoors. The vent shall be designed to prevent the entry of insects, water and foreign objects.

> **Exception:** A vent to the outdoors is not required for regulators equipped with and labeled for utilization with an approved vent-limiting device installed in accordance with the manufacturer's instructions.

Section G2421.3.1 (410.3.1) Add new section to read as shown: (FG30-03/04)

G2421.3.1 (410.3.1) Vent piping. Vent piping shall be not smaller than the vent connection on the pressure regulating device. Vent piping serving relief vents and combination relief and breather vents shall be run independently to the outdoors and shall serve only a single device vent. Vent piping serving only breather vents is permitted to be connected in a manifold arrangement where sized in accordance with an approved design that minimizes back pressure in the event of diaphragm rupture.

SECTION G2422 (411)
APPLIANCE CONNECTIONS

Section G2422.1 (411.1) Change to read as shown: (FG31-03/04 and FG32-03/04)

G2422.1 (411.1) Connecting appliances. Appliances shall be connected to the piping system by one of the following:

1. Rigid metallic pipe and fittings.

2. Corrugated Stainless Steel Tubing (CSST) where installed in accordance with the manufacturer's instructions.

3. Listed and labeled appliance connectors in compliance with ANSI Z21.24 and installed in accordance with the manufacturer's installation instructions and located entirely in the same room as the appliance.

4. Listed and labeled quick-disconnect devices used in conjunction with listed and labeled appliance connectors.

5. Listed and labeled convenience outlets used in conjunction with listed and labeled appliance connectors.

Section G2422.1.2 (411.1.2) Delete and substitute new Sections G2422.1.2 (411.1.2) through G2422.1.2.4 (411.1.2.4) to read as shown: (FG26-03/04; FG34-03/04)

G2422.1.2 (411.1.2) Connector installation. Appliance fuel connectors shall be installed in accordance with the manufacturer's instructions and Sections G2422.1.2.1 through G2422.1.2.4.

G2422.1.2.1 (411.1.2.1) Maximum length. Connectors shall have an overall length not to exceed 3 feet (914 mm), except for range and domestic clothes dryer connectors, which shall not exceed 6 feet (1829 mm) in overall length. Measurement shall be made along the centerline of the connector. Only one connector shall be used for each appliance.

> **Exception:** Rigid metallic piping used to connect an appliance to the piping system shall be permitted to have a total length greater than 3 feet, provided that the connecting pipe is sized as part of the piping system in accordance with Section G2413, and the location of the equipment shutoff valve complies with Section G2420.5.

G2422.1.2.2 (411.1.2.2) Minimum size. Connectors shall have the capacity for the total demand of the connected appliance.

G2422.1.2.3 (411.1.2.3) Prohibited locations and penetrations. Connectors shall not be concealed within, or extended through, walls, floors, partitions, ceilings or appliance housings.

> **Exception:** Fireplace inserts factory-equipped with grommets, sleeves or other means of protection in accordance with the listing of the appliance.

G2422.1.2.4 (411.1.2.4) Shutoff valve. A shutoff valve not less than the nominal size of the connector shall be installed ahead of the connector in accordance with Section G2420.5.

SECTION G2423 (413)
CNG GAS-DISPENSING SYSTEMS

Section G2423.2 (413.2.3) Add new section to read as shown: (FG35-03/04)

G2423.2 (413.2.3) Refueling appliances. Residential fueling appliances shall be listed. The capacity of a residential fueling appliance shall not exceed 5 standard cubic feet per minute (0.14 standard cubic meter/min) of natural gas.

Section G2423.3 (413.3) Add new section to read as shown: (FG35-03/04)

G2423.3 (413.3) Location of dispensing operations and equipment. Compression, storage and dispensing equipment shall be located above ground outside.

Exceptions:

1. Compression, storage or dispensing equipment is allowed in buildings of noncombustible construction, as set forth in the *International Building Code*, which are unenclosed for three-quarters or more of the perimeter.
2. Compression, storage and dispensing equipment is allowed to be located indoors in accordance with the *International Fire Code*.
3. Residential fueling appliances and equipment shall be allowed to be installed indoors in accordance with the equipment manufacturer's instructions and Section G2423.4.3.

Sections G2423.4 through G2423.4.3 (413.4 through 413.4.3) Add new sections to read as shown: (FG35-03/04)

G2423.4 (413.4) Residential fueling appliance installation. Residential fueling appliances shall be installed in accordance with Sections G2423.4.1 through 2423.4.3.

G2423.4.1 (413.4.1) Gas connections. Residential fueling appliances shall be connected to the premises gas piping system without causing damage to the piping system or the connection to the internal appliance apparatus.

G2423.4.2 (413.4.2) Outdoor installation. Residential fueling appliances located outdoors shall be installed on a firm, noncombustible base.

G2423.4.3 (413.4.3) Indoor installation. Where located indoors, residential fueling appliances shall be vented to the outdoors. A gas detector set to operate at one-fifth of the lower limit of flammability of natural gas shall be installed in the room or space containing the appliance. The detector shall be located within 6 inches (152 mm) of the highest point in the room or space. The detector shall stop the operation of the appliance and activate an audible or visual alarm.

SECTION G2436 (607)
VENTED WALL FURNACES

Section G2436.1 (608.1) Change to read as shown: (FG40-03/04)

G2436.1 (608.1) General. Vented wall furnaces shall be tested in accordance with ANSI Z21.86/CSA 2.32 and shall be installed in accordance with the manufacturer's installation instructions.

2004 SUPPLEMENT TO THE IRC

SECTION G2437 (609)
FLOOR FURNACES

Section G2437.1 (609.1) Change to read as shown: (FG41-03/04)

G2437.1 (609.1) General. Floor furnaces shall be tested in accordance with ANSI Z21.86/CSA 2.32 and shall be installed in accordance with the manufacturer's installation instructions.

SECTION G2446 (622)
VENTED ROOM HEATERS

Section G2446.1 (622.1) Change to read as shown: (FG44-03/04)

G2446.1 (622.1) General. Vented room heaters shall be tested in accordance with ANSI Z21.86/CSA 2.32, shall be designed and equipped as specified in Section G2432.2 and shall be installed in accordance with the manufacturer's installation instructions.

CHAPTER 25
PLUMBING ADMINISTRATION

Section P2503.6 Change to read as shown: (P9-03/04)

P2503.6 Water-supply system testing. Upon completion of the water-supply system or a section thereof, the system, or portion completed, shall be tested and proved tight under a water pressure not less than the working pressure of the system or, for piping systems other than plastic, by an air test of not less than 50 psi (344 kPa). This pressure shall be held for not less than 15 minutes. The water used for tests shall be obtained from a potable water source.

CHAPTER 26
GENERAL PLUMBING REQUIREMENTS

Section P2601.1 Change to read as shown: (RP3-03/04)

P2601.1 Scope. The provisions of this chapter shall govern the installation of plumbing not specifically covered in other chapters applicable to plumbing systems. The installation of plumbing, appliances, equipment and systems not addressed by this code shall comply with the applicable provisions of the *International Plumbing Code*.

Section P2604.1 Change to read as shown: (RP6-03/04)

P2604.1 Trenching and bedding. Where trenches are excavated such that the bottom of the trench forms the

2004 SUPPLEMENT TO THE IRC

bed for the pipe, solid and continuous load-bearing support shall be provided between joints. Where over-excavated, the trench shall be backfilled to the proper grade with compacted earth, sand, fine gravel or similar granular material. Piping shall not be supported on rocks or blocks at any point. Rocky or unstable soil shall be over excavated by two or more pipe diameters and brought to the proper grade with suitable compacted granular material.

CHAPTER 27
PLUMBING FIXTURES

Table P2701.1 Change to read as shown: (RP7-03/04 & RB8-03/04 and P33-03/04)

TABLE P2701.1
PLUMBING FIXTURES, FAUCETS AND FIXTURE FITTINGS

MATERIAL	STANDARD
Enameled cast-iron plumbing fixtures	ASME A112.19.1M, CSA B45.2
Individual shower control valves anti-scald	ASSE 1016, CSA B125
Nonvitreous ceramic plumbing fixtures	ASME A112.19.9M, CSA B45.1
Plastic bathtub units	ANSI Z124.1, CSA B45.5
Plastic lavatories	ANSI Z124.3, CSA B45.5
Plastic shower receptors and shower stall	ANSI Z124.2, CSA B45.5
Plastic sinks	ANSI Z124.6, CSA B45.5
Plastic water closet bowls and tanks	ANSI Z124.4, CSA B45.5
Plumbing fixture fittings	ASME A112.18.1, CSA B125
Porcelain enameled formed steel plumbing fixtures	ASME A112.19.4M, CSA B45.3
Stainless steel plumbing fixtures (residential)	ASME A112.19.3M, CSA B45.4
Thermoplastic accessible and replaceable plastic tube and tubular fittings	ASTM F 409
Water closet flush tank fill valves	ASSE 1002, CSA B125
Plumbing fixture waste fittings	ASTM F 409, CSA B125, ASME A112.18.2

(Portions of table not shown do not change)

Section P2702.2 Change to read as shown: (P33-03/04)

P2702.2 Waste fittings. Waste fittings shall conform to ASME A112.18.2, ASTM F 409, CSA B125 or to one of the standards listed in Table P3002.1 for above-ground drainage and vent pipe and fittings.

Section P2705.1 Change to read as shown: (RP9-03/04)

P2705.1 General. The installation of fixtures shall conform to the following:

1. Floor-outlet or floor-mounted fixtures shall be secured to the drainage connection and to the floor, when so designed, by screws, bolts, washers, nuts and similar fasteners of copper, brass or other corrosion-resistant material.

2. Wall-hung fixtures shall be rigidly supported so that strain is not transmitted to the plumbing system.

3. Where fixtures come in contact with walls and floors, the contact area shall be water tight.

4. Plumbing fixtures shall be usable.

5. The centerline of water closets or bidets shall not be less than 15 inches (381 mm) from adjacent walls or partitions or not less than 15 inches (381mm) from the centerline of a bidet to the outermost rim of an adjacent water closet. There shall be at least 21 inches (533 mm) clearance in front of the water closet, bidet or lavatory to any wall, fixture or door.

6. The location of piping, fixtures or equipment shall not interfere with the operation of windows or doors.

7. In areas prone to flooding as established by Table R301.2(1), plumbing fixtures shall be located or installed in accordance with Section R323.1.5.

8. Integral fixture-fitting mounting surfaces on manufactured plumbing fixtures or plumbing fixtures constructed on site, shall meet the design requirements of ASME A112.19.2 or ASME A112.19.3.

Section P2708.3 Change to read as shown: (P37-03/04)

P2708.3 Shower control valves. Individual shower and tub/shower combination valves shall be equipped with control valves of the pressure balance, the thermostatic mixing or the combination pressure balance/thermostatic mixing valve types with high limit stop in accordance with

ASSE 1016 or CSA B125. The high limit stop shall be set to limit water temperature to a maximum of 120°F (49°C). In-line thermostatic valves shall not be utilized for compliance with this section.

Section P2719.1 Change to read as shown: (P25-03/04)

P2719.1 Floor drains. Floor drains shall have waste outlets not less than 2 inches (51mm) in diameter and shall be provided with a removable strainer. The floor drain shall be constructed so that the drain is capable of being cleaned. Access shall be provided to the drain inlet.

Section P2720.1 Delete text and substitute as shown: (P32-03/04)

P2720.1 Access to pump. Access shall be provided to circulation pumps in accordance with the fixture manufacturer's installation instructions. Where the manufacturer's instructions do not specify the location and minimum size of field-fabricated access openings, a 12-inch x 12-inch (304 mm x 304 mm) minimum size door or panel shall be installed to provide access to the circulation pump. Where pumps are located more than 2 feet (609 mm) from the access opening, a 18-inch x 18-inch (457 mm x 457 mm) minimum size door or panel shall be installed. In all cases, access panel and door openings shall be unobstructed and large enough to permit the removal of the circulation pump.

Section P2721.2 Add new section as shown: (P23-03/04)

P2721.2 Bidet water temperature. The discharge water temperature from a bidet fitting shall be limited to a maximum temperature of 110°F (43°C) by a water temperature limiting device conforming to ASSE 1070.

Section P2722.1 Change to read as shown: (RP8-03/04)

P2722.1 General. Fixture supply valves and faucets shall comply with ASME A112.18.1 or CSA B125 as listed in Table P2701.1. Faucets and fixture fittings that supply drinking water for human ingestion shall conform to the requirements of NSF 61, Section 9. Flexible water connectors shall conform to the requirements of Section P2904.7.

Section P2722.2 Change to read as shown: (P69-03/04)

P2722.2 Hot water. Fixture fittings and faucets that are supplied with both hot and cold water shall be installed and adjusted so that the left-hand side of the water temperature control represents the flow of hot water when facing the outlet.

> **Exception:** Shower and tub/shower mixing valves conforming to ASSE 1016 or CSA B125, where the water temperature control corresponds to the markings on the device.

Section P2722.3 Add new section to read as shown: (P41-03/04)

P2722.3 Hose-connected outlets. Faucets and fixture fittings with hose-connected outlets shall conform to ASME A112.18.3 or CSA B125.

(Renumber remaining sections)

CHAPTER 28
WATER HEATERS

Section P2801.4 Delete exceptions without substitution as shown: (RP11-03/04)

P2801.4 Prohibited locations. Water heaters shall be located in accordance with Chapter 20.

Section P2801.5.1 Change to read as shown: (P47-03/04)

P2801.5.1 Pan size and drain. The pan shall be not less than 1.5 inches (38 mm) deep and shall be of sufficient size and shape to receive all dripping or condensate from the tank or water heater. The pan shall be drained by an indirect waste pipe having a minimum diameter of 3/4 inch (19 mm). Piping for safety pan drains shall be of those materials listed in Table P2904.5

Section P2801.7 Add new section as shown: (RB41-03/04)

P2801.7 Water heater seismic bracing. In Seismic Design Categories D_0, D_1 and D_2 and townhouses in Seismic Design Category C, water heaters shall be anchored or strapped in the upper one-third and in the lower one-third of the appliance to resist a horizontal force equal to one-third the operating weight of the water heater, acting in any horizontal direction, or in accordance with the appliance manufacturer's recommendations.

Section P2803.6.1 Change to read as shown: (P45-03/04)

P2803.6.1 Requirements of discharge pipe. The discharge piping serving a pressure relief valve, temperature relief valve or combination thereof shall:

1. Not be directly connected to the drainage system.

2. Discharge through an air gap located in the same room as the water heater.

3. Not be smaller than the diameter of the outlet of the valve served and shall discharge full size to the air gap.

4. Serve a single relief device and shall not connect to piping serving any other relief device or equipment.

2004 SUPPLEMENT TO THE IRC

5. Discharge to the floor, to an indirect waste receptor or to the outdoors. Where discharging to the outdoors in areas subject to freezing, discharge piping shall be first piped to an indirect waste receptor through an air gap located in a conditioned area.

6. Discharge in a manner that does not cause personal injury or property damage.

7. Discharge to a termination point that is readily observable by the building occupants.

8. Not be trapped.

9. Be installed so as to flow by gravity.

10. Not terminate more than 6 inches (152 mm) above the floor or waste receptor.

11. Not have a threaded connection at the end of such piping.

12. Not have valves or tee fittings.

13. Be constructed of those materials listed in Section P2904.5 or materials tested, rated and approved for such use in accordance with ASME A112.4.1.

Section P2803.6.2 Relief valve drains. Delete without substitution.

CHAPTER 29
WATER SUPPLY AND DISTRIBUTION

Table P2902.2 Add new standards as shown: (P71-03/04)

TABLE P2902.2
APPLICATION FOR BACKFLOW PREVENTERS

DEVICE	DEGREE OF HAZARD[a]	APPLICATION[b]	APPLICABLE STANDARDS
Backflow preventer with intermediate atmospheric vents	Low hazard	Backpressure or backsiphonage Sizes 1/4"-3/4"	ASSE 1012, CSA B64.3
Double check backflow prevention assembly and double check fire protection backflow prevention assembly	Low hazard	Backpressure or backsiphonage Sizes 3/8"-16"	ASSE 1015, AWWA C510 CSA B64.5, CSA B64.5.1
Hose connection backflow preventer	High or low hazard	Low head backpressure, rated working pressure backpressure or backsiphonage Sizes 1/2"-1"	ASSE 1052, CSA B64.2.1.1
Hose-connection vacuum breaker	High or low hazard	Low head backpressure or backsiphonage Sizes 1/2", 3/4", 1"	ASSE 1011, CSA B64.2, CSA B64.2.1
Pipe-applied atmospheric-type vacuum breaker	High or low hazard	Backsiphonage only Sizes 1/4"-4"	ASSE 1001, CSA B64.1.1
Pressure vacuum breaker assembly	High or low hazard	Backsiphonage only Sizes 1/2"-2"	ASSE 1020, CSA B64.1.2
Reduced pressure principle backflow preventer and reduced pressure principle fire protection backflow preventer	High or low hazard	Backpressure or backsiphonage Sizes 3/8"-16"	ASSE 1013, AWWA C511 CSA B64.4, CSA B64.4.1
Vacuum breaker wall hydrants, frost-resistant, automatic draining type	High or low hazard	Low head backpressure or backsiphonage Sizes 3/4", 1"	ASSE 1019, CSA B64.2.2

(Portions of table not shown do not change)

Section P2902.2 Add new section to read as shown: (P72-03/04)

P2902.2 Plumbing fixtures. The supply lines and fittings for every plumbing fixture shall be installed as to prevent backflow. Plumbing fixture fittings shall provide backflow protection in accordance with ASME A112.18.1.

Section P2902.2.2 Change to read as shown: (P75-03/04)

P2902.2.2 Atmospheric-type vacuum breakers. Pipe-applied atmospheric-type vacuum breakers shall conform to ASSE 1001 or CSA B64.1.1. Hose-connection vacuum breakers shall conform to ASSE 1011, ASSE 1019, ASSE 1035, ASSE 1052, CSA B64.2, CSA B64.2.1, CSA B64.2.1.1, CSA B64.2.2 or CSA B64.7. These devices shall operate under normal atmospheric pressure when the critical level is installed at the required height.

Section P2902.2.4 Change to read as shown: (P74-03/04)

P2902.2.4 Pressure-type vacuum breakers. Pressure-type vacuum breakers shall conform to ASSE 1020 or CSA B64.1.2 and spillproof vacuum breakers shall comply with ASSE 1056. These devices are designed for installation under continuous pressure conditions when the critical level is installed at the required height. Pressure-type vacuum breakers shall not be installed in locations where spillage could cause damage to the structure.

2004 SUPPLEMENT TO THE IRC

Section P2902.2.5 Change to read as shown: (P73-03/04)

P2902.2.5 Reduced pressure principle backflow preventers. Reduced pressure principle backflow preventers shall conform to ASSE 1013, AWWA C511, CSA B64.4 or CSA B64.4.1. Reduced pressure detector assembly backflow preventers shall conform to ASSE 1047. These devices shall be permitted to be installed where subject to continuous pressure conditions. The relief opening shall discharge by air gap and shall be prevented from being submerged.

Section P2902.2.6 Change to read as shown: (P76-03/04)

P2902.2.6 Double check-valve assemblies. Double check-valve assemblies shall conform to ASSE 1015, CSA B64.5, CSA B64.5.1 or AWWA C510. Double-detector check-valve assemblies shall conform to ASSE 1048. These devices shall be capable of operating under continuous pressure conditions.

Section P2903.4 Change to read as shown: (RP12-03/04)

P2903.4 Thermal expansion control. A means for controlling increased pressure caused by thermal expansion shall be provided where required in accordance with Sections P2903.4.1 and P2903.4.2.

Sections P2903.4.1 and P2903.4.2 Add new sections to read as shown: (RP12-03/04)

P2903.4.1 Pressure-reducing valve. For water service system sizes up to and including 2 inches (51 mm), a device for controlling pressure shall be installed where, because of thermal expansion, the pressure on the downstream side of a pressure-reducing valve exceeds the pressure-reducing valve setting.

P2903.4.2 Backflow prevention device or check valve. Where a backflow prevention device, check valve or other device is installed on a water supply system utilizing storage water heating equipment such that thermal expansion causes an increase in pressure, a device for controlling pressure shall be installed.

Section P2903.8 Change to read as shown: (P51-03/04)

P2903.8 Gridded and parallel water distribution system manifolds. Hot water and cold water manifolds installed with gridded or parallel-connected individual distribution lines to each fixture or fixture fittings shall be designed in accordance with Sections P2903.8.1 through P2903.8.7.

Section P2903.8.2 Change to read as shown: (RP14-03/04)

P2903.8.2 Minimum size. Where the developed length of the distribution line is 60 feet (18 288 mm) or less, and the available pressure at the meter is a minimum of 40 pounds per square inch (275.8 kPa), the minimum size of individual distribution lines shall be 3/8 inch (9.5 mm). Certain fixtures such as one-piece water closets and whirlpool bathtubs shall require a larger size where specified by the manufacturer. If a water heater is fed from the end of a cold water manifold, the manifold shall be one size larger than the water heater feed.

Section P2903.8.3 Maximum length. Delete without substitution. (RP14-03/04)

Section P2904.3 Change to read as shown: (P86-03/04)

P2904.3 Polyethylene plastic piping installation. Polyethylene pipe shall be cut square using a cutter designed for plastic pipe. Except when joined by heat fusion, pipe ends shall be chamfered to remove sharp edges. Pipe that has been kinked shall not be installed. For bends, the installed radius of pipe curvature shall be greater than 30 pipe diameters or the coil radius when bending with the coil. Coiled pipe shall not be bent beyond straight. Bends shall not be permitted within 10 pipe diameters of any fitting or valve. Joints between polyethylene plastic pipe and fittings shall comply with Sections P2904.3.1 and P2904.3.2.

Section P2904.3.1 and P2904.3.2 Add new sections to read as shown: (P86-03/04)

P2904.3.1 Heat-fusion joints. Joint surfaces shall be clean and free from moisture. All joint surfaces shall be heated to melting temperature and joined. The joint shall be undisturbed until cool. Joints shall be made in accordance with ASTM D 2657.

P2904.3.2 Mechanical joints. Mechanical joints shall be installed in accordance with the manufacturer's instructions.

Section P2904.4 Change to read as shown: (P53-03/04)

P2904.4 Water service pipe. Water service pipe shall conform to NSF 61 and shall conform to one of the standards listed in Table P2904.4.1. All water service pipe or tubing, installed underground and outside of the structure, shall have a minimum working pressure rating of 160 pounds per square inch at 73°F (1103 kPa at 23°C). Where the water pressure exceeds 160 pounds per square inch (1103 kPa), piping material shall have a rated working pressure equal to or greater than the highest available pressure. Water service piping materials not third-party certified for water distribution shall terminate at or before the full open valve located at the entrance to the structure. All ductile iron water service piping shall be cement mortar lined in accordance with AWWA C104.

2004 SUPPLEMENT TO THE IRC

Table P2904.4.1 Add new entries to read as shown: (P56-03/04 and P57-03/04)

TABLE P2904.4.1
WATER SERVICE PIPE

MATERIAL	STANDARD
Polypropylene (PP-R) plastic pipe	CSA B137.11
Cross-linked polyethylene/aluminum/high density polyethylene (PEX-AL-HDPE)	ASTM F1986

(Portions of table not shown do not change)

Section P2904.5.1 Change to read as shown: (RP15-03/04, RP16-03/04)

P2904.5.1 Under concrete slabs. Inaccessible water distribution piping under slabs shall be copper water tube minimum Type M, brass, ductile iron pressure pipe, cross-linked polyethylene/aluminum/cross-linked polyethylene (PEX-AL-PEX) pressure pipe, polyethylene/aluminum/polyethylene (PE-AL-PE) pressure pipe, chlorinated polyvinyl chloride (CPVC), polybutylene (PB), cross-linked polyethylene (PEX) plastic pipe or tubing or polypropylene (PP) pipe or tubing, all to be installed with approved fittings or bends. The minimum pressure rating for plastic pipe or tubing installed under slabs shall be 100 pounds per square inch at 180°F (689 kPa at 82°C).

Table P2904.5 Add new entries to read as shown: (P56-03/04 and P57-03/04)

TABLE P2904.5
WATER DISTRIBUTION PIPE

MATERIAL	STANDARD
Polypropylene (PP-R) plastic pipe	CSA B137.11
Cross-linked polyethylene/aluminum/high density polyethylene (PEX-AL-HDPE)	ASTM F 1986

(Portions of table not shown do not change)

Table P2904.6 Revise reference standards and add entries (PB), (PP-R) and (PEX-AL-HDPE), (P56-03/04, P57-03/04)

TABLE P2904.6
PIPE FITTINGS

MATERIAL	STANDARD
Chlorinated polyvinyl chloride (CPVC) plastic	ASTM F 437, ASTM F 438, ASTM F 439, CSA B137.6
Fittings for cross-linked polyethylene (PEX) plastic tubing	ASTM F 1807, ASTM F 1960, ASTM F 2080, CSA B137.5
Polybutylene (PB) plastic	CSA B137.8
Polyethylene (PE) plastic	ASTM D 2609, CSA B137.1
Polypropylene (PP-R) plastic	CSA B137.11
Cross-linked polyethylene/aluminum/high density polyethylene (PEX-AL-HDPE)	ASTM F 1986

(Portions of table not shown do not change)

Section P2904.9.1.2 Change to read as shown: (RP17-03/04)

P2904.9.1.2 CPVC plastic pipe. Joint surfaces shall be clean and free from moisture and an approved primer shall be applied. Solvent cement for CPVC plastic pipe, orange in color and conforming to ASTM F 493, shall be applied to all joint surfaces. The joint shall be made while the cement is wet and in accordance with ASTM D 2846 or ASTM F 493. Solvent-cement joints shall be permitted above or below ground.

> **Exception:** A primer is not required where all of the following conditions apply:
>
> 1. The solvent cement used is third-party certified as conforming to ASTM F 493.
> 2. The solvent cement used is yellow in color.
> 3. The solvent cement is used only for joining 1/2-inch (12.7 mm) through 2-inch (51 mm) diameter CPVC pipe and fittings.
> 4. The CPVC pipe and fittings are manufactured in accordance with ASTM D 2846.

Section P2904.9.1.3 Change to read as shown: (RP18-03/04)

P2904.9.1.3 PVC plastic pipe. A purple primer that conforms to ASTM F 656 shall be applied to all PVC solvent cemented joints. Solvent cement for PVC plastic pipe conforming to ASTM D 2564 shall be applied to all joint surfaces.

CHAPTER 30
SANITARY DRAINAGE

Section P3002.1 Change to read as shown: (RP20-03/04)

P3002.1 Piping within buildings. Drain, waste and vent (DWV) piping in buildings shall be as shown in Tables P3002.1(1) and P3002.1(2) except that galvanized wrought-iron or galvanized steel pipe shall not be used underground and shall be maintained not less than 6 inches (152 mm) above ground. Allowance shall be made for the thermal expansion and contraction of plastic piping.

Section P3002.2 Change to read as shown: (RP19-03/04 & RP20-03/04)

P3002.2 Building sewer. Building sewer piping shall be as shown in Table P3002.2 Forced main sewer piping shall conform to one of the standards for ABS plastic pipe, copper or copper-alloy tubing, PVC plastic pipe or pressure-rated pipe listed in Table P3002.3.

Section P3002.3 Change to read as shown: (RP20-03/04)

P3002.3 Fittings. Fittings shall be approved and compatible with the type of piping being used and shall be of a sanitary or DWV design for drainage and venting as shown in Table P3002.3. Water pipe fittings shall be permitted for use in engineer designed systems where the design indicates compliance with Section P3101.2.1.

Table P3002.1 Drain, Waste and Vent Piping and Fitting Materials and Table P3002.2 Building Sewer Piping. Delete in entirety and replace with tables shown: (RP18-03/04, RP20-03/04 and P83-03/04)

TABLE P3002.1(1)
ABOVE-GROUND DRAINAGE AND VENT PIPE

MATERIAL	STANDARD
Acrylonitrile butadiene styrene (ABS) plastic pipe	ASTM D 2661; ASTM F 628; CSA B181.1
Brass pipe	ASTM B 43
Cast-iron pipe	ASTM A 74; CISPI 301; ASTM A 888
Coextruded composite ABS DWV schedule 40 IPS pipe (solid)	ASTM F 1488
Coextruded composite ABS DWV schedule 40 IPS pipe (cellular core)	ASTM F 1488
coextruded composite PVC DWV schedule 40 IPS pipe (solid)	ASTM F 1488
Coextruded composite PVC DWV schedule 40 IPS pipe (cellular core)	ASTM F 1488; ASTM F 891
Coextruded composite PVC IPS-DR, PS140, PS200 DWV	ASTM F 1488
Copper or copper-alloy pipe	ASTM B 42; ASTM B 302
Copper or copper-alloy tubing (Type K, L, M or DWV)	ASTM B 75; ASTM B 88; ASTM B 251; ASTM B 306
Galvanized steel pipe	ASTM A 53
Polyolefin pipe	CSA B181.3
Polyvinyl chloride (PVC) plastic pipe (Type DWV)	ASTM D 2665; ASTM D 2949; CSA B181.2; ASTM F 1488
Primers for solvent cemented PVC-DWV pipe and fittings (purple in color)	ASTM F 656
Stainless steel drainage systems, Types 304 and 316L	ASME A112.3.1

TABLE P3002.1(2)
UNDERGROUND BUILDING DRAINAGE AND VENT PIPE

MATERIAL	STANDARD
Acrylonitrile butadiene styrene (ABS) plastic pipe	ASTM D 2661; ASTM F 628; CSA B181.1
Asbestos-cement pipe	ASTM C428
Cast-iron pipe	ASTM A 74; CISPI 301; ASTM A 888
Coextruded composite ABS DWV schedule 40 IPS pipe (solid)	ASTM F 1488
Coextruded composite ABS DWV schedule 40 IPS pipe (cellular core)	ASTM F 1488
Coextruded composite PVC DWV schedule 40 IPS pipe (solid)	ASTM F 1488
Coextruded composite PVC DWV schedule 40 IPS pipe (cellular core)	ASTM F 891, ASTM F 1488
Coextruded composite PVC IPS-DR, PS140, PS200 DWV	ASTM F 1488
Copper or copper alloy tubing (Type K, L, M or DWV)	ASTM B 75; ASTM B 88; ASTM B 251; ASTM B 306
Polyolefin pipe	CSA B181.3; ASTM F1412
Polyvinyl chloride (PVC) plastic pipe (Type DWV),	ASTM D 2665; ASTM D 2949; CSA B181.2
Primers for solvent cemented PVC-DWV pipe and fittings (purple in color)	ASTM F 656
Stainless steel drainage systems, Type 316L	ASME A112.3.1

TABLE P3002.2
BUILDING SEWER PIPE

MATERIAL	STANDARD
Acrylonitrile butadiene styrene (ABS) plastic pipe	ASTM D 2661; ASTM D 2751; ASTM F 628
Asbestos-cement pipe	ASTM C 428
Cast-iron pipe	ASTM A 74; ASTM A 888; CISPI 301
Coextruded composite ABS DWV schedule 40 IPS pipe (solid)	ASTM F 1488
Coextruded composite ABS DWV schedule 40 IPS pipe (cellular core)	ASTM F 1488
Coextruded composite PVC DWV schedule 40 IPS pipe (solid)	ASTM F 1488
Coextruded composite PVC DWV schedule 40 IPS pipe (cellular core)	ASTM F 1488; ASTM F 891
Coextruded composite PVC IPS-DR-PS DWV, PS140, PS200	ASTM F 1488
Coextruded composite ABS sewer and drain DR-PS in PS35, PS50, PS100, PS140, PS200	ASTM F 1488
Coextruded composite PVC sewer and drain DR-PS in PS35, PS50, PS100, PS140, PS200	ASTM F 1488
Coextruded composite PVC sewer and drain-PS 25, PS 50, PS 100 (cellular core)	ASTM F 891
Concrete pipe	ASTM C14; ASTM C 76; CSA A257.1; CSA A257.2
Copper or copper-alloy tubing (Type K or L)	ASTM B 75; ASTM B 88; ASTM B 251
Polyethylene (PE) plastic pipe (SDR-PR)	ASTM F 714
Polyolefin pipe	CSA B181.3
Polyvinyl chloride (PVC) plastic pipe (Type DWV, SDR26, SDR35, SDR41, PS50 or PS100)	ASTM D 2665; ASTM D 2949; ASTM D 3034; ASTM F1412; CSA B182.2; CSA B182.4
Primers for solvent cemented PVC-DWV pipe and fittings (purple in color)	ASTM F 656
Stainless steel drainage systems, Types 304 and 316L	ASME A 112.3.1

TABLE P3002.3
PIPE FITTINGS

MATERIAL	STANDARD
Acrylonitrile butadiene styrene (ABS) plastic pipe	ASTM D 3311; CSA B181.1; ASTM D 2661
Cast iron	ASME B 16.12; ASTM A 74; ASTM A 888; CISPI 301
Coextruded composite ABS DWV schedule 40 IPS pipe (solid or cellular core)	ASTM D 2661; ASTM D 3311; ASTM F 628
Coextruded composite PVC DWV schedule 40 IPS-DR, PS 140, PS 200 (solid or cellular core)	ASTM D 2665; ASTM D 3311; ASTM F 891
Coextruded composite ABS sewer and drain DR-PS in PS 35, PS 50, PS 100, PS 140, PS 200	ASTM D 2751
Coextruded composite PVC sewer and drain DR-PS in PS 35, PS 50, PS 100, PS 140, PS 200	ASTM D 3034
Copper or copper alloy	ASME B 16.23; ASME B 16.29
Gray iron and ductile iron	AWWA C110
Polyolefin	CSA B 181.3
Polyvinyl chloride (PVC) plastic	ASTM D 3311; ASTM D 2665; ASTM F1412; ASTM F1866; CSA B 181.2; CSA B 182.1
Stainless steel drainage systems, Types 304 and 316L	ASME A 112.3.1

Section P3003.3 through P3003.4.5 Delete and substitute: (RP22-03/04, P85-03/04, P87-03/04)

P3003.3 ABS plastic. Joints between ABS plastic pipe or fittings shall comply with Sections P3003.3.1 through P3003.3.3.

P3003.3.1 Mechanical joints. Mechanical joints on drainage pipes shall be made with an elastomeric seal conforming to ASTM C 1173, ASTM D 3212 or CSA B602. Mechanical joints shall only be installed in underground systems unless otherwise approved. Joints shall be installed in accordance with the manufacturer's instructions.

P3003.3.2 Solvent cementing. Joint surfaces shall be clean and free from moisture. Solvent cement that conforms to ASTM D 2235 or CSA B181.1 shall be applied to all joint surfaces. The joint shall be made while the cement is wet. Joints shall be made in accordance with ASTM D 2235, ASTM D 2661, ASTM F 628 or CSA B181.1. Solvent-cement joints shall be permitted above or below ground.

P3003.3.3 Threaded joints. Threads shall conform to ASME B1.20.1. Schedule 80 or heavier pipe shall be permitted to be threaded with dies specifically designed for plastic pipe. Approved thread lubricant or tape shall be applied on the male threads only.

P3003.4 Asbestos-cement. Joints between asbestos-cement pipe or fittings shall be made with a sleeve coupling of the same composition as the pipe, sealed with an elastomeric ring conforming to ASTM D 1869.

P3003.5 Brass. Joints between brass pipe or fittings shall comply with Sections P3003.5.1 through P3003.5.3.

P3003.5.1 Brazed joints. All joint surfaces shall be cleaned. An approved flux shall be applied where required. The joint shall be brazed with a filler metal conforming to AWS A5.8.

P3003.5.2 Mechanical joints. Mechanical joints shall be installed in accordance with the manufacturer's instructions.

P3003.5.3 Threaded joints. Threads shall conform to ASME B1.20.1. Pipe-joint compound or tape shall be applied on the male threads only.

P3003.6 Cast iron. Joints between cast-iron pipe or fittings shall comply with Sections P3003.6.1 through P3003.6.3.

P3003.6.1 Caulked joints. Joints for hub and spigot pipe shall be firmly packed with oakum or hemp. Molten lead shall be poured in one operation to a depth of not less than 1 inch (25 mm). The lead shall not recede more than 0.125 inch (3.2 mm) below the rim of the hub and shall be caulked tight. Paint, varnish or other coatings shall not be permitted on the jointing material until after the joint has been tested and approved. Lead shall be run in one pouring and shall be caulked tight. Acid-resistant rope and acidproof cement shall be permitted.

P3003.6.2 Compression gasket joints. Compression gaskets for hub and spigot pipe and fittings shall conform to ASTM C 564. Gaskets shall be compressed when the pipe is fully inserted.

P3003.6.3 Mechanical joint coupling. Mechanical joint couplings for hubless pipe and fittings shall comply with CISPI 310 or ASTM C 1277. The elastomeric sealing sleeve shall conform to ASTM C 564 or CSA B602 and shall be provided with a center stop. Mechanical joint couplings shall be installed in accordance with the manufacturer's installation instructions.

P3003.7 Concrete joints. Joints between concrete pipe and fittings shall be made with an elastomeric seal conforming to ASTM C 443, ASTM C 1173, CSA A257.3M or CSA B602.

P3003.8 Coextruded composite ABS pipe, joints. Joints between coextruded composite pipe with an ABS outer layer or ABS fittings shall comply with Sections P3003.8.1 and P3003.8.2.

P3003.8.1 Mechanical joints. Mechanical joints on drainage pipe shall be made with an elastomeric seal conforming to ASTM C1173, ASTM D 3212 or CSA B602. Mechanical joints shall not be installed in above-ground systems, unless otherwise approved. Joints shall be installed in accordance with the manufacturer's instructions.

P3003.8.2 Solvent cementing. Joint surfaces shall be clean and free from moisture. Solvent cement that conforms to ASTM D 2235 or CSA B181.1 shall be applied to all joint surfaces. The joint shall be made while the cement is wet. Joints shall be made in accordance with ASTM D 2235, ASTM D 2661, ASTM F 628 or CSA B181.1. Solvent-cement joints shall be permitted above or below ground.

P3003.9 Coextruded composite PVC pipe. Joints between coextruded composite pipe with a PVC outer layer or PVC fittings shall comply with Sections P3003.9.1 and P3003.9.2.

P3003.9.1 Mechanical joints. Mechanical joints on drainage pipe shall be made with an elastomeric seal conforming to ASTM D 3212. Mechanical joints shall not be installed in above-ground systems, unless otherwise approved. Joints shall be installed in accordance with the manufacturer's instructions.

P3003.9.2 Solvent cementing. Joint surfaces shall be clean and free from moisture. A purple primer that conforms to ASTM F 656 shall be applied. Solvent cement not purple in color and conforming to ASTM D 2564, CSA B137.3 or CSA B181.2 shall be applied to all joint surfaces. The joint shall be made while the cement is wet and shall be in accordance with ASTM D 2855. Solvent-cement joints shall be permitted above or below ground.

P3003.10 Copper pipe. Joints between copper or copper-alloy pipe or fittings shall comply with Sections P3003.10.1 through P3003.10.4.

P3003.10.1 Brazed joints. All joint surfaces shall be cleaned. An approved flux shall be applied where required. The joint shall be brazed with a filler metal conforming to AWS A5.8.

P3003.10.2 Mechanical joints. Mechanical joints shall be installed in accordance with the manufacturer's instructions.

P3003.10.3 Soldered joints. Solder joints shall be made in accordance with the methods of ASTM B 828. All cut tube ends shall be reamed to the full inside diameter of the tube end. All joint surfaces shall be cleaned. A flux conforming to ASTM B 813 shall be applied. The joint shall be soldered with a solder conforming to ASTM B 32.

P3003.10.4 Threaded joints. Threads shall conform to ASME B1.20.1. Pipe-joint compound or tape shall be applied on the male threads only.

P3003.11 Copper tubing. Joints between copper or copper-alloy tubing or fittings shall comply with Sections P3003.11.1 through P3003.11.3.

P3003.11.1 Brazed joints. All joint surfaces shall be cleaned. An approved flux shall be applied where required. The joint shall be brazed with a filler metal conforming to AWS A5.8.

P3003.11.2 Mechanical joints. Mechanical joints shall be installed in accordance with the manufacturer's instructions.

P3003.11.3 Soldered joints. Solder joints shall be made in accordance with the methods of ASTM B 828. All cut tube ends shall be reamed to the full inside diameter of the tube end. All joint surfaces shall be cleaned. A flux conforming to ASTM B 813 shall be applied. The joint shall be soldered with a solder conforming to ASTM B 32.

P3003.12 Steel. Joints between galvanized steel pipe or fittings shall comply with Sections P3003.12.1 and P3003.12.2.

P3003.12.1 Threaded joints. Threads shall conform to ASME B1.20.1. Pipe-joint compound or tape shall be applied on the male threads only.

P3003.12.2 Mechanical joints. Joints shall be made with an approved elastomeric seal. Mechanical joints shall be installed in accordance with the manufacturer's instructions.

P3003.13 Lead. Joints between lead pipe or fittings shall comply with Sections P3003.13.1 and P3003.13.2.

P3003.13.1 Burned. Burned joints shall be uniformly fused together into one continuous piece. The thickness of the joint shall be at least as thick as the lead being joined. The filler metal shall be of the same material as the pipe.

P3003.13.2 Wiped. Joints shall be fully wiped, with an exposed surface on each side of the joint not less than 0.75 inch (19.1 mm). The joint shall be at least 0.325 inch (9.5 mm) thick at the thickest point.

P3003.14 PVC plastic. Joints between PVC plastic pipe or fittings shall comply with Sections P3003.14.1 through P3003.14.3.

P3003.14.1 Mechanical joints. Mechanical joints on drainage pipe shall be made with an elastomeric seal conforming to ASTM C 1173, ASTM D 3212 or CSA B602. Mechanical joints shall not be installed in above-ground systems, unless otherwise approved. Joints shall be installed in accordance with the manufacturer's instructions.

P3003.14.2 Solvent cementing. Joint surfaces shall be clean and free from moisture. A purple primer that conforms to ASTM F 656 shall be applied. Solvent cement not purple in color and conforming to ASTM D 2564, CSA B137.3 or CSA B181.2 shall be applied to all joint surfaces. The joint shall be made while the cement is wet and shall be in accordance with ASTM D 2855. Solvent-cement joints shall be permitted above or below ground.

P3003.14.3 Threaded joints. Threads shall conform to ASME B1.20.1. Schedule 80 or heavier pipe shall be permitted to be threaded with dies specifically designed for plastic pipe. Approved thread lubricant or tape shall be applied on the male threads only.

P3003.15 Vitrified clay. Joints between vitrified clay pipe or fittings shall be made with an elastomeric seal conforming to ASTM C 425, ASTM C 1173 or CSA B602.

P3003.16 Polyolefin plastic. Joints between polyolefin plastic pipe and fittings shall comply with Sections P3003.16.1 and P3003.16.2.

P3003.16.1 Heat-fusion joints. Heat-fusion joints for polyolefin pipe and tubing joints shall be installed with socket-type heat-fused polyolefin fittings or electrofusion polyolefin fittings. Joint surfaces shall be clean and free from moisture. The joint shall be undisturbed until cool. Joints shall be made in accordance with ASTM F 1412 or CSA B181.3.

P3003.16.2 Mechanical and compression sleeve joints. Mechanical and compression sleeve joints shall be installed in accordance with the manufacturer's instructions.

P3003.17 Joints between different materials. Joints between different piping materials shall be made with a mechanical joint of the compression or mechanical-sealing type conforming to ASTM C1173, ASTM C1460 or ASTM C1461. Connectors and adapters shall be approved for the application and such joints shall have an elastomeric seal conforming to ASTM C 425, ASTM C 443, ASTM C 564, ASTM C 1440, ASTM D 1869, ASTM F 477, CSA A257.3M or CSA B602, or as required in Sections P3003.17.1 through P3003.17.6. Joints between glass pipe and other types of materials shall be made with adapters having a TFE seal. Joints shall be installed in accordance with the manufacturer's instructions.

P3003.17.1 Copper or copper-alloy tubing to cast-iron hub pipe. Joints between copper or copper-alloy tubing and cast-iron hub pipe shall be made with a brass ferrule or compression joint. The copper or copper-alloy tubing shall be soldered to the ferrule in an approved manner, and the ferrule shall be joined to the cast-iron hub by a caulked joint or a mechanical compression joint.

P3003.17.2 Copper or copper-alloy tubing to galvanized steel pipe. Joints between copper or copper-alloy tubing and galvanized steel pipe shall be made with a brass converter fitting or dielectric fitting. The copper tubing shall be soldered to the fitting in an approved manner, and the fitting shall be screwed to the threaded pipe.

P3003.17.3 Cast-iron pipe to galvanized steel or brass pipe. Joints between cast-iron and galvanized steel or brass pipe shall be made by either caulked or threaded joints or with an approved adapter fitting.

P3003.17.4 Plastic pipe or tubing to other piping material. Joints between different grades of plastic pipe or between plastic pipe and other piping material shall be made with an approved adapter fitting. Joints between plastic pipe and cast-iron hub pipe shall be made by a caulked joint or a mechanical compression joint.

P3003.17.5 Lead pipe to other piping material. Joints between lead pipe and other piping material shall be made by a wiped joint to a caulking ferrule, soldering nipple, or bushing or shall be made with an approved adapter fitting.

P3003.17.6 Stainless steel drainage systems to other materials. Joints between stainless steel drainage systems and other piping materials shall be made with approved mechanical couplings.

P3003.18 Joints between drainage piping and water closets. Joints between drainage piping and water closets or similar fixtures shall be made by means of a closet flange compatible with the drainage system material, securely fastened to a structurally firm base. The inside diameter of the drainage pipe shall not be used as a socket fitting for a four by three closet flange. The joint shall be bolted, with an approved gasket, flange to fixture connection complying with ASME A112.4.3 or setting compound between the fixture and the closet flange.

Section P3005.2.2 Change to read as shown: (RP23-03/04)

P3005.2.2 Spacing. Cleanouts shall be installed not more than 100 feet (30 480 mm) apart in horizontal drainage lines measured from the upstream entrance of the cleanout.

Section P3005.2.7 Change to read as shown: (RP24-03/04)

P3005.2.7 Building drain and building sewer junction. There shall be a cleanout near the junction of the building drain and building sewer. This cleanout shall be either inside or outside the building wall, provided it is brought up to finish grade or to the lowest floor level. An approved two-way cleanout shall be permitted to serve as the required cleanout for both the building drain and the building sewer. The cleanout at the junction of the building drain and building sewer shall not be required where a cleanout on a 3-inch (76 mm) or larger diameter soil stack is located within a developed length of 10 feet (3048 mm) of the building drain and building sewer junction.

Section P3005.2.9 and Table P3005.2.9 Delete existing text and table and replace as shown: (RP25-03/04)

P3005.2.9 Cleanout size. Cleanouts shall be the same nominal size as the pipe they serve up to 4 inches (102 mm). For pipes larger than 4 inches (102 mm) nominal size, the minimum size of the cleanout shall be 4 inches (102 mm).

Exceptions:

1. "P" trap connections with slip joints or ground joint connections, or stack cleanouts that are not more than one pipe diameter smaller than the drain served, shall be permitted.

2. Cast-iron cleanout sizing shall be in accordance with referenced standards in Table P3002.1, ASTM A 74 for hub and spigot fittings or ASTM A 888 or CISPI 301 for hubless fittings.

CHAPTER 31
VENTS

Sections P3102.1, P3102.2 and P3102.3 Delete and substitute as shown: (P93-03/04)

P3102.1 Required vent extension. The vent system serving each building drain shall have at least one vent pipe that extends to the outdoors.

P3102.2 Installation. The required vent shall be a dry vent that connects to the building drain or an extension of a drain that connects to the building drain. Such vent shall not be an island fixture vent as allowed by Section P3112.

P3102.3 Size. The required vent shall be sized in accordance with Section P3113.1 based on the required size of the building drain.

(Renumber remaining sections)

Section P 3103.1 Change to read as shown: (RP27-03/04)

P3103.1 Roof extension. All open vent pipes which extend through a roof shall be terminated at least 6 inches (152 mm) above the roof or 6 inches (152 mm) above the anticipated snow accumulation, whichever is greater, except that where a roof is to be used for any purpose other than weather protection, the vent extension shall be run at least 7 feet (2134 mm) above the roof.

Section P3105.2 Change to read as shown: (RP28-03/04)

P3105.2 Fixture drains. The total fall in a fixture drain due to pipe slope shall not exceed one pipe diameter, nor shall the vent pipe connection to a fixture drain, except for water closets, be below the weir of the trap.

Section P3105.3 and Figure P3105.3 Delete without substitution: (RP28-03/04)

P3108.1 Change to read as shown: (RP30-03/04)

P3108.1 Wet vent permitted. Any combination of fixtures within two bathroom groups located on the same floor level are permitted to be vented by a wet vent. The wet vent shall be considered the vent for the fixtures and shall extend from the connection of the dry vent along the direction of flow in the drain pipe to the most downstream fixture drain connection to the horizontal branch drain. Only the fixtures within the bathroom groups shall connect to the wet vented horizontal branch drain. Any additional fixtures shall discharge downstream of the wet vent.

2004 SUPPLEMENT TO THE IRC

Figures P3108.1(1), P3108.1(2), P3108.1(3) Relocate figures to new Appendix J: (RP30-03/04):

P3108.4 Change to read as shown: (RP31-03/04)

P3108.4 Vertical wet vent. A combination of fixtures located on the same floor level are permitted to be vented by a vertical wet vent. The vertical wet vent shall extend from the connection to the dry vent down to the lowest fixture drain connection. Each fixture shall connect independently to the vertical wet vent. All water closet drains shall connect at the same elevation. Other fixture drains shall connect above or at the same elevation as the water closet fixture drains. The dry vent connection to the vertical wet vent shall be an individual or common vent serving one or two fixtures.

Figures P3108.2(1), P3108.2(2) Relocate figures to new Appendix J: (RP31-03/04):

P3109.2 Change to read as shown: (RP32-03/04)

P3109.2 Stack installation. The waste stack shall be vertical, and both horizontal and vertical offsets shall be prohibited. Every fixture drain shall connect separately to the waste stack. The stack shall not receive the discharge of water closets or urinals.

Figure P3109.2 Relocate figure to new Appendix J: (RP32-03/04)

P3110.4 Change to read as shown: (RP33-03/04)

P3110.4 Additional fixtures. Fixtures, other than the circuit vented fixtures, are permitted to discharge to the horizontal branch drain. Such fixtures shall be located on the same floor as the circuit vented fixtures and shall be either individually or common vented.

Figure P3110.4 Relocate figure to new Appendix J: (RP33-03/04)

CHAPTER 32
TRAPS

Section P3201.2 Change to read as shown: (RP36-03/04)

P3201.2 Trap seals and trap seal protection. Traps shall have a liquid seal not less than 2 inches (51 mm) and not more than 4 inches (102 mm). Traps for floor drains shall be fitted with a trap primer or shall be of the deep seal design.

Section P3201.5 Change to read as shown: (RP28-03/04)

P3201.5 Prohibited trap designs. The following types of traps are prohibited:

1. Bell traps.
2. Separate fixture traps with interior partitions, except those lavatory traps made of plastic, stainless steel or other corrosion-resistant material.
3. "S" traps.
4. Drum traps.
5. Trap designs with moving parts.

2004 SUPPLEMENT TO THE IRC

CHAPTER 43
REFERENCED STANDARDS

Change, delete or add the following referenced standards to read as shown: (RB98-03/04, RB99-03/04, RB102-03/04, RB124-03/04, RB154-03/04, RB157-03/04, RB199-03/04, RB202-03/04, RB207-03/04, RB208-03/04, RB211-03/04, RB212-03/04, RB213-03/04, RB222-03/04, RB234-03/04, RB239-03/04, RB240-03/04, RB241-03/04, RB242-03/04, RB243-03/04, RB244-03/04, RB245-03/04, RB246-03/04, RB247-03/04, RB249-03/04, RB250-03/04, RB251-03/04, RB253-03/04, RB254-03/04, RB255-03/04, RB256-03/04, RB265-03/04, RP7-03/04, RP8-03/04, RP22-03/04, RP38-03/04, RM18-03/04, FG25-03/04, FG40-03/04, FG49-03/04, EC48-03/04, P33-03/04, P37-03/04, P41-03/04, P53-03/04, P56-03/04, P69-03/04, P71-03/04, P72-03/04, P73-03/04, P74-03/04, P75-03/04, P76-03/04, P83-03/04, P85-03/04, P86-03/04, P87-03/04, S51-03/04) (STANDARDS NOT SHOWN DO NOT CHANGE)

AAMA
American Architectural Manufacturers Association
1827 Walden Office Square, Suite 550
Schaumburg, IL 60173

Standard reference number	Title	Referenced in code section number
450-00	Voluntary Performance Rating Method for Mulled Fenestration Assemblies	R613.6.1
506-00	Voluntary Specifications for Hurricane Impact and Cycle Testing of Fenestration Products	R613.4.1

ACCA
Air Conditioning Contractors of America
2800 Shirlington Road, Suite 300
Arlington, VA 22206

Standard reference number	Title	Referenced in code section number
Manual J-2002	Residential Load Calculation - Eight Edition	M1401.3

AITC
American Institute of Timber Construction
7012 S. Revere Parkway, Suite 140
Englewood, CO 80112

Standard reference number	Title	Referenced in code section number
A 190.1-02	Structural Glued Laminated Timber	R502.1.5, R602.1.2, R802.1.4

ANSI
American National Standards Institute
25 West 43rd Street - 4th Floor
New York, NY 10036

Standard reference number	Title	Referenced in code section number
LC 1-1997	Interior Fuel Gas Piping Systems Using Corrugated Stainless Steel Tubing - with Addenda LC 1a-1999 and LC 1b-2001	G2414.5.3
Z21.1-2000	Household Cooking Gas Appliances - with Addenda Z21.1a-2003	G2447.1
Z21.5.1-1999	Gas Clothes Dryers - Volume I - Type 1 Clothes Dryers - with Addenda Z21.5.1a-2003	G2438.1

2004 SUPPLEMENT TO THE IRC

ANSI (continued)

Z21.8-1994 (R2002)	Installation of Domestic Gas Conversion Burners	G2443.1
Z21.10.1-2001	Gas Water Heaters - Volume I Storage, Water Heaters with Input Ratings of 75,000 Btu per Hour or Less - with Addenda Z21.10.1a-2002	G2448.1
Z21.10.3-2001	Gas Water Heaters - Volume III - Storage Water Heaters with Input Ratings Above 75,000 Btu Per Hour, Circulating and Instantaneous	G2448.1
Z21.11.1-1991	DELETED	
Z21.11.2-2002	Gas-Fired Room Heaters - Volume II - Unvented Room Heaters	G2445.1
Z21.13-2000	Gas-Fired Low Pressure Steam and Hot Water Boilers - with Addenda Z21.13a-2002	G2452.1
Z21.15-1997 (R2003)	Manually Operated Gas Valves for Appliances, Appliance Connector Valves and Hose End Valves - with Addenda Z21.15a-2001 (R2003)	Table G2420.1.1
Z21.22-99	Relief Valves for Hot Water Supply Systems - with Addenda Z21.22a-2000 and Z21.22b-2001	P2803.2
Z21.24-97	Connectors for Gas Appliances	G2422.1
Z21.40.1-1996(R2002)	Gas-Fired, Heat Activated Air Conditioning and Heat Pump Appliances - with Addenda Z21.40.1a-1997 (R2002)	G2449.1
Z21.40.2-1996 (R2002)	Gas-Fired Work Activated Air Conditioning and Heat Pump Appliances (Internal Combustion) - with Addenda Z21.40.2a-1997 (R2002)	G2449.1
Z21.42-1993 (R2002)	Gas-Fired Illuminating Appliances	G2450.1
Z21.47-2001	Gas-Fired Central Furnaces - with Addenda Z21.47a-2001 and Z21.47b-2002	R617.1, G2442.1
Z21.48-92	DELETED	
Z21.49-92	DELETED	
Z21.50-2000	Vented Gas Fireplaces - with Addenda Z21.50a-2001 and Z21.50b-2002	G2434.1
Z21.56-2001	Gas-Fired Pool Heaters	R616.1, G2441.1
Z21.58-1995 (R2002)	Outdoor Cooking Gas Appliances - with Addenda Z21.58a-1998 (R2002) and Z21.58b-2002	R622.1, G2447.1
Z21.60-2003	Decorative Gas Appliances for Installation in Solid-Fuel Burning Fireplaces	G2432.1
Z21.69-2002	Connectors for Movable Gas Appliances	G2422.1
Z21.84-2002	Manually Lighted, Natural Gas Decorative Gas Appliances for Installation in Solid-Fuel Burning Fireplaces	G2432.1, G2432.2
Z21.86-2000	Gas-Fired Vented Space Heating Appliances - with Addenda Z21.86a-2002 and Z21.86b-2002	G2436.1, G2437.1, G2446.1
Z21.88-2002	Vented Gas Fireplace Heaters	G2435.1
Z83.8-2002	Gas Unit Heaters and Gas-Fired Duct Furnaces	G2444.1

APA

APA - The Engineered Wood Association
7011 South 19th Street
Tacoma, WA 98466

Standard reference number	Title	Referenced in code section number
APA E30-03	Engineered Wood Construction Guide	R803.2.3

ASME

American Society of Mechanical Engineers
Three Park Avenue
New York, NY 10016-5990

Standard reference number	Title	Referenced in code section number
A112.1.2-1991 (Reaffirmed 2002)	Air Gaps in Plumbing Systems	P2902.2.1, Table P2902.2

2004 SUPPLEMENT TO THE IRC

ASME (continued)

A112.4.1-1993 (Reaffirmed 2002)	Water Heater Relief Valve Drain Tubes	P2803.6.1
A112.4.3-1999	Plastic Fittings for Connecting Water Closets to the Sanitary Drainage System	P3003.18
A112.6.1M-1997 (Reaffirmed 2002)	Floor-Affixed Supports for Off-the-Floor Plumbing Fixtures for Public Use	P2702.4
A112.18.1-2000	Plumbing Fixture Fittings	Table P2701.1, P2722.1, P2902.2
A112.18.2-2002	Plumbing Fixture Waste Fittings	Table P2701.1, P2702.2
A112.18.3-1996	Performance Requirements for Backflow Protection Devices and Systems in Plumbing Fixture Fittings	P2722.3
A112.19.1M-1994 (Reaffirmed 1999)	Enameled Cast Iron Plumbing Fixtures - with 1998 and 2000 Supplements	Table P2701.1, P2711.1
A112.19.2M-1998 (Reaffirmed 2002)	Vitreous China Plumbing Fixtures - with 2000 Supplement	Table P2701.1, P2711.1 P2712.2
A112.19.3-2000	Stainless Steel Plumbing Fixtures (Designed for Residential Use) - with 2002 Supplement	Table P2701.1, P2711.1
A112.19.9M-1991 (Reaffirmed 1998)	Non-Vitreous Ceramic Plumbing Fixtures - with 2002 Supplement	Table P2701.1, P2712.1
B1.20.1-1983 (Reaffirmed 2001)	Pipe Threads, General Purpose (inch)	P3003.5.3, P3003.10.4, P3003.12.1, P3003.14.3
B16.9-2001	Factory-Made Wrought Steel Buttwelding Fittings	Table P2904.6
B16.11-2001	Forged Fittings, Socket-Welding and Threaded	Table P2904.6
B16.18-2001	Cast Copper Alloy Solder Joint Pressure Fittings	Table P2904.6
B16.22-2001	Wrought Copper and Copper Alloy Solder Joint Pressure Fittings	Table P2904.6
B16.23-2002	Cast Copper Alloy Solder Joint Drainage Fittings: DWV	Table P2904.6, Table P3002.4
B16.29-2001	Wrought Copper and Wrought Copper Alloy Solder Joint Drainage Fittings - DWV	Table P2904.6, Table P3002.4
B16.33-2002	Manually Operated Metallic Gas Valves for Use in Gas Piping Systems up to 125 psig (Sizes ½ through 2)	Table G2420.1.1
B16.44-01	Manually Operated Metallic Gas Valves For Use in House Piping Systems	Table G2420.1.1
BPVC-2001	ASME Boiler & Pressure Vessel Code (2001 Edition) (Sections I, II, IV, V & IX)	G2452.1
CSD-1-2002	Controls and Safety Devices for Automatically Fired Boilers	M2001.1.1, G2452.1

ASSE

American Society of Sanitary Engineering
28901 Clemens Road, Suite A
Westlake, OH 44145

Standard reference number	Title	Referenced in code section number
1001-2002	Performance Requirements for Atmospheric Type Vacuum Breakers	Table P2902.2, P2902.2.2
1003-2001	Performance Requirements for Water Pressure Reducing Valves	P2903.3.1
1010-1996	Performance Requirements for Water Hammer Arresters	P2903.5
1011-1993	Performance Requirements for Hose Connection Vacuum Breakers	Table P2902.2, P2902.2.2, P2902.3.3
1012-2002	Performance Requirements for Backflow Preventers with Intermediate Atmospheric Vent	Table P2902.2, P2902.2.3, P2902.3.3, P2902.4.1, P2902.4.5
1024-2003	Performance Requirements for Dual Check Valve Backflow Preventers	Table P2902.2
1035-2002	Performance Requirements for Laboratory Faucet Blackflow Preventers	Table P2902.2, P2902.2.2
1050-2002	Performance Requirements for Stack Air Admittance Valves for Sanitary Drainage Systems	P3114.1

2004 SUPPLEMENT TO THE IRC

ASSE (continued)

1051-2002	Performance Requirements for Individual and Branch Type Air Admittance Valves for Plumbing Drainage Systems	P3114.1
1052-1993	Performance Requirements for Hose Connection Backflow Preventers	Table P2701.1, Table P2902.2, P2902.2.2 P2902.3.3
1056-2001	Performance Requirements for Spill Resistant Vacuum Breaker	Table P2902.2, P2902.2.4, P2902.3.3
1070-2004	Performance Requirements for Water Temperature Limiting Devices	P2721.2

ASTM

ASTM International
100 Barr Harbor Drive
West Conshohocken, PA 19428-2859

Standard reference number	Title	Referenced in code section number
A 36/A 36M-02	Specification for Carbon Structural Steel	R606.14
A 53/A 53M-02	Specification for Pipe, Steel, Black and Hot Dipped, Zinc-Coated Welded and Seamless	Table M2101.1, G2414.4.2, Table 2904.4.1, Table P3002.1
A 74-03	Specification for Cast Iron Soil Pipe and Fittings	Table P3002.1, Table P3002.2, Table P3002.3, Table P3002.4, P3005.2.9
A 106-02a	Specification for Seamless Carbon Steel Pipe for High-Temperature Service	Table M2101.1, G2414.4.2
A 153-02	Specification for Zinc Coating (Hot Dip) on Iron and Steel Hardware	Table R606.14.1
A 254-97(2002)	Specification for Copper Brazed Steel Tubing	Table M2101.1, G2414.5.1
A 312/A 312M-02	Specification for Seamless and Welded Austenitic Stainless Steel Pipes	Table P2904.4.1, Table P2904.5, Table P2904.6, P2904.10.2
A 463/A 463M-02a	Standard Specification for Steel Sheet, Aluminum-coated by the Hot-Dip Process	Table R905.10.3
A 510-02	Specification for General Requirements for Wire Rods and Coarse Round Wire, Carbon Steel	R606.14
A525-93	DELETED	
A 615/A 615M-02	Specification for Deformed and Plain Billet-Steel Bars for Concrete Reinforcement	R404.4.6.1, R611.6.2
A 653/A 653M-02a	Specification for Steel Sheet, Zinc-Coated Galvanized or Zinc-Iron Alloy-Coated Galvanized by the Hot-Dip Process	R505.2.1, R505.2.3, R603.2.1, R603.2.3, Table R606.14.1, R804.2.1, R804.2.3, Table R905.10.3, M1601.1.1
A 706/A 706M-02	Specification for Low-Alloy Steel Deformed and Plain Bars for Concrete Reinforcement	R404.4.6.1, R611.6.2
A 755/A 755M-01 (2003)	Specification for Steel Sheet, Metallic-Coated by the Hot-Dip Process and Prepainted by the Coil-Coating Process for Exterior Exposed Building Products	Table R905.10.3
A 792/A 792M-02	Specification for Steel Sheet, 55% Aluminum-Zinc Alloy-Coated by the Hot-Dip Process	R505.2.1, R603.2.1, R603.2.3, R804.2.1, R804.2.3
A 875/A 875M-02a	Specification for Steel Sheet Zinc-5%, Aluminum Alloy-Coated by the Hot-Dip Process	R505.2.1, R505.2.3, R603.2.1, R603.2.3, R804.2.1, R804.2.3, Table R905.10.3
A888-98e1	Specification for Hubless Cast Iron Soil Pipe and Fittings for Sanitary and Storm Drain, Waste and Vent Piping Application	Table P3002.1, Table P3002.2, P3005.2.9
A 951-02	Specification for Masonry Joint Reinforcement	R606.14
A 996/A 996M-02	Specification for Rail-Steel and Axle-Steel Deformed Bars for Concrete Reinforcement	R404.4.6.1, R611.6.2
A 1003/A 1003M-00	Standard Specification for Steel Sheet, Carbon, Metallic- and Nonmetallic-Coated for Cold-formed Framing Members	R505.2.1, R603.2.1, R804.2.1

2004 SUPPLEMENT TO THE IRC

ASTM (continued)

B 32-00e01	Specification for Solder Metal	P3003.10.3, P3003.11.3
B 42-02	Specification for Seamless Copper Pipe, Standard Sizes	Table M2101.1, Table P2904.5, Table P3002.1
B 75-02	Specification for Seamless Copper Tube	Table M2101.1, Table P2904.4.1, Table P2904.5, Table P3002.1, Table P3002.2, Table P3002.3
B 88-02	Specification for Seamless Copper Water Tube	Table M2101.1, G2414.5.2, Table P2904.4.1, Table P2904.5, Table P3002.1, Table P3002.2, Table P3002.3
B 101-02	Specification for Lead-Coated Copper Sheet and Strip for Building Construction	R905.10.3
B 135-02	Specification for Seamless Brass Tube	Table M2101.1
B 209-02a	Specification for Aluminum and Aluminum-Alloy Steel and Plate	Table 905.10.3
B 227-02	Specification for Hard-Drawn Copper-Clad Steel Wire	R606.14
B 251-02	Specification for General Requirements for Wrought Seamless Copper and Copper-Alloy Tube	Table M2101.1, Table P2904.4.1.Table P2904.5, Table P3002.1, Table P3002.2, Table P3002.3
B 280-02	Specification for Seamless Copper Tube for Air Conditioning and Refrigeration Field Service	G2414.5.2
B 302-02	Specification for Threadless Copper Pipe, Standard Sizes	Table M2101.1, Table P2904.5, P2904.4, Table P3002.1
B 306-02	Specification for Copper Drainage Tube (DWV)	Table M2101.1, Table P3002.1, Table P3002.2
B 447-02	Specification for Welded Copper Tube	Table P2904.4.1, Table P2904.5
B813-00e01	Specification for Liquid and Paste Fluxes for Soldering Applications Of Copper and Copper Alloy Tube	M2101.1, P2904.11, P3003.3.4, P3003.10.3, P3003.11.3
B 828-02	Practice for Making Capillary Joints by Soldering of Copper and Copper Alloy Tube and Fittings	P2904.11, P3003.10.3, P3003.11.3
C 27-98(2002)	Specification for Standard Classification of Fireclay and High-Alumina Refractory Brick	R1003.5, R1003.8
C 28/C 28M-00e01	Specification for Gypsum Plasters	R702.2
C 67-03	Test Methods of Sampling and Testing Brick and Structural Clay Tile	R905.3.5
C 76-02	Specification for Reinforced Concrete Culvert, Storm Drain, and Sewer Pipe	Table P3002.3
C 90-02a	Specification for Loadbearing Concrete Masonry Units	Table R301.2(1)
C 140-02c	Test Method Sampling and Testing Concrete Masonry Units and Related Units	R905.3.5
C 236.89 (1993)e1	DELETED	
C 270-03	Specification for Mortar for Unit Masonry	R607.1
C 315-02	Specification for Clay Flue Linings	Table R1001.11(1),Table R1001.11(2), R1001.8.1, G2425.12
C 425-02	Specification for Compression Joints for Vitrified Clay Pipe and Fittings	Table P3002.2, P3003.15, P3003.17
C 428-97(2002)	Specification for Asbestos-Cement Nonpressure Sewer Pipe	Table P3002.2, Table P3002.3
C 443-02a	Specification for Joints for Concrete Pipe and Manholes, Using Rubber Gaskets	P3003.7, P3003.17
C 475/C 475M-02	Specification for Joint Compound and Joint Tape for Finishing Gypsum Wallboard	R702.3.1
C 476-02	Specification for Grout for Masonry	R609.1.1
C564-03	Specification for Rubber Gaskets for Cast Iron Soil Pipe and Fittings	P3003.6.2, P3003.6.3, P3003.17
C 587-02	Specification for Gypsum Veneer Plaster	R702.2
C 588/C 588M-01	Specification for Gypsum Base for Veneer Plasters	R702.2
C 631-95a (2002)	Specification for Bonding Compounds for Interior Gypsum Plastering	R702.2
C 700-02	Specification for Vitrified Clay Pipe, Extra Strength, Standard Strength, and Perforated	P3002.2
C 836-03	Specification for High-Solids Content, Cold Liquid-Applied Elastomeric Waterproofing Membrane for Use with Separate Wearing Course	R905.15.2

2004 SUPPLEMENT TO THE IRC

ASTM (continued)

C 887-79a (2001)	Specification for Packaged, Dry, Combined Materials for Surface Bonding Mortar	R406.1
C 926-98a	Specification for Application of Portland Cement Based-Plaster	R703.6
C 931/C 931M-02	Specification for Exterior Gypsum Soffit Board	R702.3.1
C 960/C 960M-01	Specification for Predecorated Gypsum Board	R702.3.1
C 1029-02	Specification for Spray-Applied Rigid Cellular Polyurethane Thermal Insulation	R905.14.2
C 1032-96 (2002)	Specification for Woven Wire Plaster Base	R702.2
C 1157-02	Performance Specification for Hydraulic Cement	R402.2
C 1167-03	Specification for Clay Roof Tiles	
C 1173-02	Specification for Flexible Transition Couplings for Underground Piping Systems	P3003.3, P3003.7.1, P3003.8.1, P3003.14.1, P3003.15, P3003.17
C 1178/C 1178M-01	Specification for Glass Mat Water-resistant Gypsum Backing Panel	R702.3.1, R702.3.7, R702.4.2, R702.4.4
C 1186-02	Specification for Flat Nonasbestos Fiber Cement Sheets	R703.4
C 1277-03	Specification for Shielded Couplings Joining Hubless Cast Iron Soil Pipe and Fittings	Table P3002.1, Table P3002.2
C 1283-02	Practice for Installing Clay Flue Lining	R1001.9
C 1288-01	Standard Specification For Discrete Non-asbestos Fiber-Cement Interior Substrate Sheets	R702.4.4
C 1325-01	Standard Specification for Non-Asbestos Fiber-Mat Reinforced Cement Interior Substrate Sheets	R702.4.4
C 1396M-02	Specification for Gypsum Board	R702.3.1
C1440-99e01	Specification for Thermoplastic Elastomeric (TPE) Gasket Materials for Drain, Waste, and Vent (DWV), Sewer, Sanitary and Storm Plumbing Systems	P3003.17
C1460-00	Specification for Shielded Transition Couplings for Use with Dissimilar DWV Pipe and Fittings Above Ground	P3003.17
C 1461-02	Specification for Mechanical Couplings Using Thermoplastic Elastomeric (TPE) Gaskets for Joining Drain, Waste, and Vent (DWV) Sewer, Sanitary, and Storm Plumbing Systems for Above and Below Ground Use	Table P3002.1, Table P3002.2
D 224-89(1996)	DELETED	
D 225-02	Specification for Asphalt Shingles (Organic Felt) Surfaced with Mineral Granules	R905.2.4
D 249-89(96)	DELETED	
D371-89(1996)	DELETED	
D 422-63(2002)	Test Method for Particle-Size Analysis of Soils	R403.1.7.5.1
D 1248-02	Specification for Polyethylene Plastics Extrusion Materials for Wire and Cable	M1601.1.2
D 1784-02	Specification for Rigid Poly (Vinyl Chloride) (PVC) Compounds and Chlorinated Poly (Vinyl Choloride) (CPVC) Compounds	M1601.1.2
D1869-95(2000)	Specification for Rubber Rings for Asbestos Cement Pipe	P2904.15, P3003.4, P3003.17
D2235-01	Specification for Solvent Cement for Acrylonitrile-Butadiene-Styrene (ABS) Plastic Pipe and Fittings	P2904.9.1.1, Table P3002.1, Table P3003.2, P3003.3.2, P3003.8.2
D 2412-02	Test Method for Determination of External Loading Characteristics of Plastic Pipe by Parallel-Plate Loading	M1601.1.2
D 2466-02	Specification for Poly (Vinyl Chloride) (PVC) Plastic Pipe Fittings, Schedule 40	Table P2904.6
D 2467-02	Specification for Poly (Vinyl Chloride) (PVC) Plastic Pipe Fittings, Schedule 80	Table P2904.6
D 2513-03	Specification for Thermoplastic Gas Pressure Pipe, Tubing, and Fittings	Table M2101.1, M2104.2.1.3, G2414.6, G2414.6.1, G2414.11, G2415.14.3
D 2564-02	Specification for Solvent Cements for Poly (Vinyl Chloride) (PVC) Plastic Piping Systems	P2904.9.1.3, Table P3002.1, Table P3002.2, P3003.9.2, P3003.14.2
D2657-97	Standard Practice for Heat Fusion-Joining of Polyolefin Pipe and Fittings	P2904.3.1

2004 SUPPLEMENT TO THE IRC

ASTM (continued)

Standard	Title	Referenced in
D 2661-02	Specification for Acrylonitrile-Butadiene-Styrene (ABS) Schedule 40 Plastic Drain, Waste, and Vent Pipe and Fittings	Table P3002.1, Table P3002.2, Table P3002.3, Table P3002.4, P3003.3.2, P3003.8.2
D 2665-02a	Specification for Poly (Vinyl Chloride) (PVC) Plastic Drain, Waste, Vent Pipe and Fittings	Table P3002.1, Table P3002.2, Table P3002.3, Table P3002.4
D 2751-96a	Specification for Acrylonitrile.Butadiene.Styrene (ABS) Sewer Pipe and Fittings	Table P3002.3
D 2824-02	Specification for Aluminum-Pigmented Asphalt Roof Coatings, Non-Fibered, Asbestos Fibered, and Fibered without Asbestos	Table R905.9.2, Table R905.11.2
D 2837-02	Test Method for Obtaining Hydrostatic Design Basis for Thermoplastic Pipe Materials	Table M2101.1
D 2855-96(2002)	Standard Practice for Making Solvent-Cemented Joints with Poly (Vinyl Chloride) (PVC) Pipe and Fittings	P3003.9.2, P3003.14.2
D 2949-01a	Specification for 3.25-in. Outside Diameter Poly (Vinyl Chloride) (PVC) Plastic Drain, Waste, and Vent Pipe and Fittings	Table P3002.1, Table P3002.2, Table P3002.3
D 3034-00	Specification for Type PSM Poly (Vinyl Chloride) (PVC) Sewer Pipe and Fittings	Table P3002.3, Table P3002.4
D 3161-03	Test Method for a Wind Resistance of Asphalt Shingles (Fan Induced Method)	R905.2.6
D 3212-96a	Specification for Joints for Drain and Sewer Plastic Pipes Using Flexible Elastomeric Seals	P3003.3.1, P3003.8.1, P3003.9.1, P3003.14.1
D 3311-02	Specification for Drain, Waste, and Vent (DWV) Plastic Fittings Patterns	Table P3002.4
D 3350-02a	Specification for Polyethylene Plastics Pipe and Fittings Materials	Table M2101.1
D 3462-03	Specification for Asphalt Shingles Made From Glass Felt and Surfaced with Mineral Granules	R905.2.4
D 3679-02a	Specification for Rigid Poly (Vinyl Chloride) (PVC) Siding	Table R703.4
D 3737-02	Practice for Establishing Allowable Properties for Structural Glued Laminated Timber (Glulam)	R502.1.5, R602.1.2, R802.1.4
D 3747-79 (2002)	Specification for Emulsified Asphalt Adhesive for Adhering Roof Insulation	Table R905.9.2, Table R905.11.2
D 3909-97b	Specification for Asphalt Roll Roofing (Glass Felt) Surfaced with Mineral Granules	R905.2.8.2, R905.3.3, R905.5.4, Table R905.9.2, Table R906.3.2
D 4869-02	Specification for Asphalt-Saturated (Organic Felt) Underlayment Used in Steep Slope Roofing	R905.2.3, R905.4.3, R905.5.3, R905.6.3, R905.7.3, R905.8.3
D 5019-96e01	Specification for Reinforced Non-Vulcanized Polymeric Sheet Used in Roofing Membrane	R905.12.2
D 5055-02	Specification for Establishing and Monitoring Structural Capacities of Prefabricated Wood I-Joists	R502.1.4
D 5516-02	Test Method of Evaluating the Flexural Properties of Fire-Retardant Treated Softwood Plywood Exposed to the Elevated Temperatures	R802.1.3.2.1
D 5664-02	Test Methods for Evaluating the Effects of Fire-Retardant Treatments and Elevated Temperatures on Strength Properties of Fire-Retardant Treated Lumber	R802.1.3.2.2
D 6222-02	Specification for Atactic Polypropylene (APP) Modified Bituminous Sheet Materials Using Polyester Reinforcement	Table R905.11.2
D 6223-02	Specification for Atactic Polypropylene (APP) Modified Bituminous Sheet Materials Using a Combination of Polyester and Glass Fiber Reinforcement	Table R905.11.2
D 6305-02e01	Practice for Calculating Bending Strength Design Adjustment Factors for Fire-Retardant-Treated Plywood Roof Sheathing	R802.1.3.2.1
D 6380-01[E1]	Standard Specification for Asphalt Roll Roofing (Organic Felt) [1]	R905.2.8.2,

IRC-93

2004 SUPPLEMENT TO THE IRC

ASTM (continued)

		R905.3.3, R905.5.4
D 6694-01	Standard Specification Liquid-Applied Silicone Coating Used in Spray Polyurethane Foam Roofing[1]	R905.15.2
D 6754-02	Standard Specification for Ketone Ethylene Ester Based Sheet Roofing [1]	R905.13.2
D 6757-02	Standard Specification for Inorganic Underlayment for Use with Steep Slope Roofing Products [1]	R905.2.3
D 6878 - 03	Standard Specification for Thermoplastic Polyolefin Based Sheet Roofing [1]	R905.13.2
E 84-03	Test Method for Surface Burning Characteristics of Building Materials	R202, R3141.1, R3142.6, R314.3, R315.3, R316.1, R316.2, R802.1.3, M1601.2.1, M1601.4.2
E 90-02	Test Method for Laboratory Measurement of Airborne Sound Transmission Loss of Building Partitions and Elements	AK102.1
E 96-00e01	Test Method for Water Vapor Transmission of Materials	R202, R806.4, M1411.,4, M1601.3.4, N1102.1.7
E 136-99e01	Test Method for Behavior of Materials in a Vertical Tube Furnace at 750 Degrees C	R202
E 152-95	DELETED	
E 283	Test Method for Determining the Rate of Air Leakage Through Exterior Windows, Curtain Walls and Doors Under Specified Pressure Differences Across the Specimen	R806.4, N1102.1.11, N1102.4.3
E 330-02	Test Method for Structural Performance of Exterior Windows, Doors, Skylights and Curtain Walls by Uniform Static Air Pressure Difference	R613.3
E 331-00	Test Method for Water Penetration of Exterior Windows, Skylights, Doors, and Curtain Walls by Uniform Static Air Pressure Difference	R703.1
E 814-02	Test Method of Fire Tests of Through-Penetration Firestops	R317.3.1.2
E 1886-02	Test Method for Performance of Exterior Windows, Curtain Walls, Doors and Storm Shutters Impacted by Missiles and Exposed to Cyclic Pressure Differentials	R301.2.1.2, R613.4.1, N1102.3.7
E 1996-02	Specification for Performance of Exterior Windows, Curtain Walls, Doors and Storm Shutters Impacted by Winborne Debris in Hurricanes	R301.2.1.2, R613.4.1, N1102.3.7
F 409-02	Specification for Thermoplastic Accessible and Replaceable Plastic Tube and Tubular Fittings	Table P2701.1, P2702.2, P2702.3
F 438-02	Specification for Socket-Type Chlorinated Poly (Vinyl Chloride) (CPVC) Plastic Pipe Fittings, Schedule 40	Table P2904.6
F 439-02	Specification for Socket-Type Chlorinated Poly (Vinyl Chloride) (CPVC) Plastic Pipe Fittings, Schedule 80	Table P2904.6
F 441/F 441M-02	Specification for Chlorinated Poly (Vinyl Chloride) (CPVC) Plastic Pipe, Schedules 40 and 80	Table P2904.4.1, Table P2904.5
F 477-02e01	Specification for Elastomeric Seals (Gaskets) for Joining Plastic Pipe	P2904.15, P3003.17
F 628-01	Specification for Acrylonitrile.Butadiene.Styrene (ABS) Schedule 40 Plastic Drain, Waste, and Vent Pipe with a Cellular Core	Table P3002.1, Table P3002.2, Table P3002.3, P3003.3.2, P3003.8.2
F 656-02	Specification for Primers for Use in Solvent Cement Joints of Poly (Vinyl Chloride) (PVC) Plastic Pipe and Fittings	P2904.9.1.3, Table P3002.1, Table P3002.2, P3003.9.2, P3003.14.2
F 714-01	Specification for Polyethylene (PE) Plastic Pipe (SDR-PR) Based on Outside Diameter	Table P3002.3
F 876-02e01	Specification for Crosslinked Polyethylene (PEX) Tubing	Table M2101.1, Table P2904.4.1
F 877-02e01	Specification for Crosslinked Polyethylene (PEX) Plastic Hot- and Cold-Water Distribution Systems	Table M2101.1, Table P2904.4.1, Table P2904.5
F 891-00e01	Specification for Coextruded Poly (Vinyl Chloride) (PVC) Plastic Pipe with a Cellular Core	Table P3002.1, Table P3002.2, Table P3002.3, Table P3002.4
F 1281-02e02	Specification for Crosslinked Polyethylene/Aluminum/Crosslinked Polyethylene (PEX-AL-PEX) Pressure Pipe	Table P2904.4.1, Table P2904.5, Table M2101.1
F 1282-02e02	Specification for Polyethylene/Aluminum/Polyethylene (PE-AL-PE) Composite Pressure Pipe	Table P2904.4.1, Table P2904.5

ASTM (continued)

F 1346-91(2003)	Performance Specification for Safety Covers and Labeling Requirements for All Covers for Swimming Pools, Spas and Hot Tubs	AG105.2, AG105.5
F 1412-01	Specification for Polyolefin Pipe and Fittings for Corrosive Waste Drainage	Table P3002.2, P3003.16.1
F 1488-00e01	Specification for Coextruded Composite Pipe	Table P3002.1, Table P3002.2, Table P3002.3
F 1667-03	Specification for Driven Fasteners: Nails, Spikes, and Staples	R905.2.5
F 1807-02a	Specifications for Metal Insert Fittings Utilizing a Copper Crimp Ring for SDR9 Cross-linked Polyethylene (PEX) Tubing	Table P2904.6, P2904.9.1.4.2, Table M2101.1
F 1866-98	Specification for Poly (Vinyl Chloride) (PVC) Plastic Schedule 40 Drainage and DWV Fabricated Fittings	Table P3002.4
F 1960-03	Specification for Cold Expansion Fittings with PEX Reinforcing Rings for Use with Cross-linked Polyethylene (PEX) Tubing	Table P2904.6
F 1974-02	Specification for Metal Insert Fittings for Polyethylene/Aluminum/ Polyethylene and Crosslinked Polyethylene/Aluminum/Crosslinked Polyethylene Composite Pressure Pipe	Table P2904.6
F 1986-00a	Multilayer Pipe Type 2, Compression Fittings and Compression Joints for Hot and Cold Drinking Water Systems	Table P2904.4.1, Table P2904.5, Table P2904.6
F 2080-02	Specification for Cold-Expansion Fittings with Metal Compression-Sleeves for Cross-linked Polyethylene (PEX) Pipe	P2904.6, P2904.9.1.4.2

AWS

American Welding Society
550 N.W. LeJeune Road
Miami, FL 33126

Standard reference number	Title	Referenced in code section number
A5.8-92	Specifications for Filler Metals for Brazing and Braze Welding	P3003.5.1, P3003.10.1, P3003.11.1

AWPA

American Wood-Preservers' Association
P.O. Box 5690
Granbury, Texas 76049

Standard reference number	Title	Referenced in code section number
C1-00	DELETED	
C2-01	DELETED	
C3-99	DELETED	
C4-99	DELETED	
C9-00	DELETED	
C15-00	DELETED	
C18-99	DELETED	
C22-96	DELETED	
C23-00	DELETED	
C24-96	DELETED	
C28-99	DELETED	
C31-01	DELETED	
U1-02	USE CATEGORY SYSTEM: User Specification for Treated Wood except Section 7 Commodity Specification H	R319.1, R402.1.2, R504.3, Table R905.8.5
P5-02	Standard for Waterborne Preservatives	R319.1, R323.1.7

2004 SUPPLEMENT TO THE IRC

AWWA

American Water Works Association
6666 West Quincy Avenue
Denver, CO 80235

Standard reference number	Title	Referenced in code section number
C104-98	Standard for Ductile-iron and Gray-iron Fittings, 3 inches through 48 inches, for Water	P2904.4
C151/A21.51-02	Ductile-Iron Pipe, Centrifugally Cast, for Water	Table P2904.4.1

CDA DELETED

Standard reference number	Title	Referenced in code section number
4050	DELETED	

CISPI

Cast Iron Soil Pipe Institute
5959 Shallowford Road, Suite 419
Chattanooga, TN 37421

Standard reference number	Title	Referenced in code section number
301-00	Specification for Hubless Cast Iron Soil Pipe and Fittings for Sanitary and Storm Drain, Waste and Vent Piping Applications	P3005.2.9, Table P3002.1, Table P3002.2, Table P3002.3, Table P3002.4
310-97	Specification for Coupling for Use in Connection with Hubless Cast Iron Soil Pipe and Fittings for Sanitary and Storm Drain, Waste, and Vent Piping Applications	Table P3002.1, Table P3002.2, P3003.6.3

CSA

CSA America, Inc.
3501 E. Pleasant Valley Road
Cleveland, OH 44131-5575

Standard reference number	Title	Referenced in code section number
CSA Requirement 3-88	Manually Operated Gas Valves For Use In House Piping Systems	Table G2420.1.1

CSA

Canadian Standards Association
5060 Spectrum Way, Suite 100
Mississauga, Ontario, Canada L4W 5N6

Standard reference number	Title	Referenced in code section number
A 257.1M-92	Circular Concrete Culvert, Storm Drain, Sewer Pipe and Fittings	Table P3002.3
A 257.2M-92	Reinforced Circular Concrete Culvert, Storm Drain, Sewer Pipe and Fittings	Table P3002.3

CSA (continued)

A257.3M-92	Joints for Circular Concrete Sewer and Culvert Pipe, Manhole Sections, and Fittings Using Rubber Gaskets	P3003.7, P3003.17
B 45.1-02	Ceramic Plumbing Fixtures	Table P2701.1, P2711.1, P2712.1
B 45.2-02	Enameled Cast-Iron Plumbing Fixtures	Table P2701.1, P2711.1
B 45.3-02	Porcelain Enameled Steel Plumbing Fixtures	Table P2701.1, P2711.1
B 45.4-02	Stainless-Steel Plumbing Fixtures	Table P2701.1, P2711.1, P2712.1
B 45.5-02	Plastic Plumbing Fixtures	Table P2701.1, P2711.2, P2712.1
B 45.9-02	Macerating Systems and Related Components	P3007.1
B 64.1.1-01	Vacuum Breakers, Atmospheric Type (AVB)	Table P2902.2, P2902.2
B64.1.2-01	Vacuum Breakers, Pressure Type (PVB)	Table P2902.2, P2902.2.4
B64.2-01	Vacuum Breakers, Hose Connection Type (HCVB)	Table P2902.2, P2902.2.2
B64.2.1-01	Vacuum Breakers, Hose Connection Type (HCVB) with Manual Draining Feature	Table P2902.2, P2902.2.2
B64.2.1.1-01	Vacuum Breakers, Hose Connection Dual Check Type (HCDVB)	Table P2902.2, P2902.2.2
B64.2.2-01	Vacuum Breakers, Hose Connection Type (HCVB) with Automatic Draining Feature	Table P2902.2, P2902.2
B64.3-01	Backflow Preventers, Dual Check Valve Type with Atmospheric Port (DCAP)	Table P2902.2, P2902.2.2, P2902.2.3, P2902.4.1
B64.4-01	Backflow Preventers, Reduced Pressure Principle Type (RP)	Table P2902.2, P2902.2.3, P2902.2.5, P2902.4.1
B64.4.1-01	Backflow Preventers, Reduced Pressure Principle Type for Fire Systems (RPF)	Table P2902.2, P2902.2.5
B64.5-01	Backflow Preventers, Double Check Valve Type (DCVA)	Table P2902.2, P2902.2.6
B64.5.1-01	Backflow Preventers, Double Check Valve Type for Fire Systems (DCVAF)	Table P2902.2, P2902.2.6
B 64.7-01	Vacuum Breakers, Laboratory Faucet Type (LFVB)	Table P2902.2, P2902.2.2
B125-01	Plumbing Fittings	Table P2701.1, P2702.2, P2708.3, P2722.1, P2722.2, P2722.3
B137.1-02	Polyethylene Pipe, Tubing and Fittings for Cold Water Pressure Services	Table P2904.4.1, Table P2904.6
B137.2-02	PVC Injection-Moulded Gasketed Fittings for Pressure Applications	Table P2904.6
B137.3-02	Rigid Poly (Vinyl Chloride) (PVC) Pipe for Pressure Applications	Table P2904.4.1, P3003.9.2, P3003.14.2
B137.5-02	Cross-Linked Polyethylene (PEX) Tubing Systems for Pressure Applications	Table P2904.4.1, Table P2904.5, Table P2904.6
B137.6-02	CPVC Pipe, Tubing and Fittings for Hot and Cold Water Distribution System	Table P2904.4.1, Table P2904.5, Table P2904.6
B137.8-02	Polybutylene (PB) Piping for Pressure Applications	Table P2904.4.1, Table P2904.5, Table P2904.6
B137.9-02	Polyethylene/Aluminum/Polyethylene Composite Pressure-Pipe Systems	Table P2904.4.1
B137.10-02	Crosslinked Polyethylene/Aluminum/Crosslinked Polyethylene Composite Pressure-Pipe Systems	Table P2904.4.1, Table P2904.5, Table M2101.1
B137.11-99	Polypropylene (PP-R) Pipe and Fittings for Pressure Applications	Table P2904.4.1, Table P2904.5, Table P2904.6
B181.1-02	ABS Drain, Waste, and Vent Pipe and Pipe Fittings	Table P3002.1, Table P3002.2, Table P3002.4, P3003.3.2, P3003.8.2
B181.2-02	PVC Drain, Waste, and Vent Pipe and Pipe Fittings	Table P3002.1, Table P3002.2, P3003.9.2, P3003.14.2
B181.3-02	Polyolefin Laboratory Drainage Systems	Table P3002.2, P3003.16.1
B182.2-02	PVC Sewer Pipe and Fittings (PSM Type)	Table P3002.3
B182.4-02	Profile PVC Sewer Pipe & Fittings	Table P3002.3
B602-02	Mechanical Couplings for Drain, Waste, and Vent Pipe and Sewer Pipe	P3003.3.1, P3003.6.3, P3003.7, P3003.8.1, P3003.14.1, P3003.15, P3003.17
CSA 6.19-01	Residential Carbon Monoxide Alarming Devices	[F]R313.2

2004 SUPPLEMENT TO THE IRC

NAIMA

North American Insulation Manufacturers
44 Canal Center Plaza, Suite 310
Alexandria, VA 22314

Standard reference number	Title	Referenced in code section number
AH 116-02	Fibrous Glass Duct Construction Standards, Fifth Edition	M1601.1.1

NCMA

National Concrete Masonry Association
2302 Horse Pen Road
Herndon, VA 20171-3499

Standard reference number	Title	Referenced in code section number
TR 68B (2001)	Basement Manual Design and Construction Using Concrete Masonry	R404.1

NFPA

National Fire Protection Association
1 Batterymarch Park
Quincy, MA 02269-9101

Standard reference number	Title	Referenced in code section number
13-02	Installation of Sprinkler Systems	R317.1
72-02	National Fire Alarm Code	R313.1
501-00	Manufactured Housing	R202

NFRC

National Fenestration Rating Council, Inc.
8484 Georgia Avenue, Suite 320
Silver Spring, MD 20910

Standard reference number	Title	Referenced in code section number
100-2001	Procedure for Determining Fenestration Product U-factors - Second Edition	N1101.3.2, N1101.3.2.1, N1101.5
200-2001	Procedure for Determining Fenestration Product Solar Heat Gain Coefficients and Visible Transmittance at Normal Incidence - Second Edition	N1101.3.2, N1101.3.2.1, N1101.5
400-2001	Procedure for Determining Fenestration Product Air Leakage - Second Edition	N1101.3.2.2, N1102.4.2

NSF

NSF International
789 North Dixboro Road
Ann Arbor, MI 48105

Standard reference number	Title	Referenced in code section number
14-2003	Plastic Piping System Components and Related Materials	P2608.3, P2907.3
42-2002	Drinking Water Treatment Units - Aesthetic Effects	P2907.1, P2907.3
44-2002	Residential Cation Exchange Water Softners	P2907.1, P2907.3

2004 SUPPLEMENT TO THE IRC

NSF (continued)

53-2002	Drinking Water Treatment Units - Health Effects	P2907.1, P2907.3
58-2002	Reverse Osmosis Drinking Water Treatment Systems	P2907.2, P2907.3
61-2002	Drinking Water System Components - Health Effects	P2608.5, P2722.1, P2904.4, P2904.5, P2907.3

NSPI

National Spa and Pool Institute
2111 Eisenhower Ave
Alexandria, VA 22314-4698

Standard reference number	Title	Referenced in code section number
5-2003	Standard for Residential Inground Swimming Pools	AG103.1

RMA — DELETED

Standard reference number	Title	Referenced in code section number
RP1-90	DELETED	
RP2-90	DELETED	
RP3-85	DELETED	

SMACNA

Sheet Metal & Air Conditioning Contractors
4021 Lafayette Center Road
Chantilly, VA 20151-1209

Standard reference number	Title	Referenced in code section number
SMACNA-03	Fibrous Glass Duct Construction Standards (2003)	M1601.1.1

UL

Underwriters Laboratories, Inc.
333 Pfingsten Road
Northbrook, IL 60062-2096

Standard reference number	Title	Referenced in code section number
80-1996	Steel Tanks for Oil-Burner Fuel - with revisions through July 2000	M2201.1
103-2001	Factory-Built Chimneys, for Residential Type and Building Heating Appliances	R202, R1002.3, G2430.1
181A-1998	Closure Systems for Use with Rigid Air Ducts and Air Connectors - with Revisions through December 1998	M1601.2, M1601.3.1
181B-1995	Closure Systems for Use with Flexible Air Ducts and Air Connectors - with Revisions through May 2000	M1601.2, M1601.3.1
217-1997	Single and Multiple Station Smoke Alarms-with Revisions Through January 1999	[F]R313.1
325-2002	Door, Drapery, Louver and Window Operators and Systems - with Revisions through March 2003	R309.6
343-1997	Pumps for Oil-Burning Appliances - with Revisions through May 2002	M2204.1
441-1996	Gas Vents - with Revisions through December 1999	G2426.1
726-1998	Oil-Fired Boiler Assemblies - with Revisions through January 2001	M2001.1.1, M2006.1, G2425.1

2004 SUPPLEMENT TO THE IRC

UL (continued)

Standard reference number	Title	Referenced in code section number
923-2002	Microwave Cooking Appliances - with Revisions through January 2003	M1503.1
1256-2002	Fire Test of Roof Deck Construction	R906.1
1261-2001	Electric Water Heaters for Pools and Tubs	M2006.1
1479-1994	Fire Tests of Through-Penetration Firestops - with Revisions through August 2000	R317.3.1.2
1715-1997	Fire Test of Interior Finish Material - with Revisions through October 2002	R314.3
2034-2002	Standard for Safety for Single and Multiple Station Carbon Monoxide Alarms	[F]R313.2

ULC

Underwriters Laboratories of Canada
7 Crouse Road
Toronto, Ontario, Canada   MIR 3A9

Standard reference number	Title	Referenced in code section number
S 102-1988	Standard Methods of Test for Surface Burning Characteristics of Building Materials and Assemblies - with 2000 Revisions	R316.2

2004 SUPPLEMENT TO THE IRC

Part X—Appendices

APPENDIX J
EXISTING BUILDINGS AND STRUCTURES

Appendix J102.4 Change to read as shown: (EC48-03/04)

AJ102.4 Replacement windows. Regardless of the category of work, when an entire existing window, including frame, sash and glazed portion is replaced, the replacement window shall comply with the requirements of Chapter 11.

Appendix J601.1.2 Change to read as shown: (EB40-03/04

AJ601.1.2 Handrails. Every required exit stairway that has four or more risers; is part of the means of egress for any work area; and is not provided with at least one handrail, or in which the existing handrails are judged to be in danger of collapsing, shall be provided with handrails designed and installed in accordance with Section R311 for the full length of the run of steps on at least one side.

APPENDIX J Add new appendix and figures as shown: (RP31-03/04, RP32-03/04, RP33-03/04, RP34-03/04)

APPENDIX J
VENTING METHODS

(This appendix is informative and is not part of the code. This appendix provides examples of various illustrations of venting methods.)

FIGURE J3108.1(1)
TYPICAL SINGLE BATH WET VENT ARRANGEMENTS
(Figure P3108.1(1) relocated to Appendix)

FIGURE J3108.1(2)
TYPICAL DOUBLE BATH WET VENT ARRANGEMENTS
(Figure P3108.1(2) relocated to Appendix)

FIGURE J3108.1(3)
TYPICAL HORIZONTAL WET VENTING
(Figure P3108.1(3) relocated to Appendix)

FIGURE J3108.2(1)
TYPICAL METHODS WET VENTING
(Figure P3108.2(1) relocated to Appendix)

FIGURE J3108.2(2)
SINGLE STACK SYSTEM FOR A TWO STORY DWELLING
(Figure P3108.2(2) relocated to Appendix)

FIGURE J3109.2
WASTE STACK VENTING
(Figure P3109.2 relocated to Appendix)

FIGURE J3110.4
CIRCUIT VENT WITH ADDITIONAL NONCIRCUIT VENTED BRANCHES
(Figure P3110.4 relocated to Appendix)

(Renumber remaining Appendices)

Appendix L Add new text as shown and rename current Appendix L: (G11-03/04)

APPENDIX L
PERMIT FEES

TOTAL VALUATION	FEE
$1 to $500	$24
$501 to $2,000	$24 for the first $500; plus $3 for each additional $100 or fraction thereof, to and including $2,000
$2,000 to $40,000	$69 for the first $2,000; plus $11 for each additional $1,000 or fraction thereof, to and including $40,000
$40,001 to $100,000	$487 for the first $40,000; plus $9 for each additional $1,000 or fraction thereof, to and including $100,000
$100,001 to $500,000	$1,027 for the first $100,000; plus $7 for each additional $1,000 or fraction thereof, to and including $500,000
$500,001 to $1,000,000	$3,827 for the first $500,000; plus $5 for each additional $1,000 or fraction thereof, to and including $1,000,000
$1,000,001 to $5,000,000	$6,327 for the first $1,000,000; plus $3 for each additional $1,000 or fraction thereof, to and including $5,000,000
$5,000,000 and over	$18,327 for the first $5,000,000; plus $1 for each additional $1,000 or fraction thereof

2004 SUPPLEMENT TO THE IRC

Appendix M Add new text as shown: (RB268-03/04)

APPENDIX M
HOME DAYCARE – R-3 OCCUPANCY

SECTION AM 101
GENERAL

AM101.1. General. This appendix shall apply to a home daycare operated within a dwelling. It is to include buildings and structures occupied by persons of any age who receive custodial care for less than 24 hours by individuals other than parents or guardians or relatives by blood, marriage, or adoption, and in a place other than the home of the person cared for.

SECTION AM102
DEFINITIONS

EXIT ACCESS. That portion of a means of egress system that leads from any occupied point in a building or structure to an exit.

SECTION AM103
MEANS OF EGRESS

AM103.1. Exits Required. If the occupant load of the residence is more than nine, including those who are residents, during the time of operation of the daycare, two exits are required from the ground-level story. Two exits are required from a home daycare operated in a manufactured home regardless of the occupant load. Exits shall comply with Section R311.

AM103.1.1. Exit Access Prohibited. An exit access from the area of daycare operation shall not pass through bathrooms, bedrooms, closets, garages, fenced rear yards or similar areas.

> **Exception:** An exit may discharge into a fenced yard if the gate or gates remain unlocked during daycare hours. The gates may be locked if there is an area of refuge located within the fenced yard and more than 50 feet (15 240 mm) from the dwelling. The area of refuge shall be large enough to allow 5 square feet (0.47 m²) per occupant.

AM103.1.2. Basements. If the basement of a dwelling is to be used in the daycare operation, two exits are required from the basement regardless of the occupant load. One of the exits may pass through the dwelling and the other must lead directly to the exterior of the dwelling.

> **Exception:** An emergency and escape window complying with Section R310, which does not conflict with Section AM103.1.1, Exit Access Prohibited, may be used as the second means of egress from a basement.

AM103.1.3. Yards. If the yard is to be used as part of the daycare operation it shall be fenced.

AM103.1.3.1. Type of Fence and Hardware. The fence shall be of durable materials at least 6 feet (1529 mm) in height completely enclosing the area used for the daycare operations. Each opening shall be a gate or door equipped with a self-closing and self-latching device to be installed at a minimum of 5 feet (1528 mm) above the ground.

> **Exception:** The door of any dwelling which forms part of such enclosure need not be equipped with self-closing and self-latching devices.

AM103.1.3.2. Construction of Fence. Openings in the fence, wall or enclosure required by this section shall have intermediate rails or an ornamental pattern such that a sphere 4 inches (102 mm) in diameter cannot pass through. In addition, the following criteria must be met:

1. The maximum vertical clearance between grade and the bottom of the fence, wall or enclosure shall be 2 inches (51 mm).

2. Solid walls or enclosures that do not have openings, such as masonry or stone walls, shall not contain indentations or protrusions except for tooled masonry joints.

3. Maximum mesh size for chain link fences shall be 1-1/4-inches (32 mm) square unless the fence is provided with slats at the top or bottom which reduce the opening to no more than 1-3/4 inches (44 mm). The wire shall not be less than 9 gage [(0.148 in.)(3.8 mm)].

AM103.1.3.3. Decks. Decks that are more than 12 inches (305 mm) above grade shall have a guard in compliance with Section R316.

AM103.2. Width and Height of an Exit. The minimum width of a required exit is 36 inches (914 mm) with a net clear width of 32 inches (813 mm). The minimum height of a required exit is 6 feet 8 inches (2032 mm).

AM103.3. Type of Lock and Latches for Exits. Regardless of the occupant load served, exit doors shall be openable from the inside without the use of a key or any special knowledge or effort. When the occupant load is 10 or less, a night latch, dead bolt or security chain may be used, provided such devices are openable from the inside without the use of a key or tool and mounted at a height not to exceed 48 inches (1219 mm) above the finished floor.

AM103.4. Landings. Landings for stairways and doors shall comply with Section R312 except that landings shall be required for the exterior side of a sliding door when a home daycare is being operated in a Group R-3 Occupancy.

SECTION AM104
SMOKE DETECTION

AM104.1. General. Dwelling units used for home daycare operations shall be provided with smoke detectors. Detectors shall be installed in accordance with the approved manufacturer's instructions. If the current smoke detection system in the dwelling is not in compliance with the currently adopted code for smoke detection, it shall be upgraded to meet the currently adopted code requirements and Section AM103 before daycare operations commence.

AM104.2. Power Source. Required smoke detectors shall receive their primary power from the building wiring when such wiring is served from a commercial source and shall be equipped with a battery backup. The detector shall emit a signal when the batteries are low. Wiring shall be permanent and without a disconnecting switch other than those required for over-current protection. All required smoke detectors shall be interconnected so if one detector is activated, all detectors are activated.

AM104.3. Location. A detector shall be located in each bedroom and any room that is to be used as a sleeping room and centrally located in the corridor, hallway or area giving access to each separate sleeping area. When the dwelling unit has more than one story and in dwellings with basements, a detector shall be installed on each story and in the basement. In dwelling units where a story or basement is split into two or more levels, the smoke detector shall be installed on the upper level, except that when the lower level contains a sleeping area, a detector shall be installed on each level. When sleeping rooms are on the upper level, the detector shall be placed at the ceiling of the upper level in close proximity to the stairway. In dwelling units where the ceiling height of a room open to the hallway serving the bedrooms or sleeping areas exceeds that of the hallway by 24 inches (610 mm) or more, smoke detectors shall be installed in the hallway and adjacent room. Detectors shall sound an alarm audible in all sleeping areas of the dwelling unit in which they are located.

International Urban-Wildland Interface Code

2004 Supplement

ICC
INTERNATIONAL CODE COUNCIL®

2004 SUPPLEMENT TO THE IUWIC

INTERNATIONAL URBAN-WILDLAND INTERFACE CODE 2004 SUPPLEMENT

CHAPTER 7
REFERENCED STANDARDS

Change the following standards to read as shown: (UWIC7-03/04)

(STANDARDS NOT SHOWN DO NOT CHANGE)

ASTM

ASTM International
100 Barr Harbor Drive
West Conshohocken, PA 19428-2859

Standard reference number	Title	Referenced In code section number
E 84-03	Test Method for Surface Burning Characteristics of Building Materials	202
E 136-99e01	Test Method for Behavior of Materials in a Vertical Tube Furnace at 750 Degrees C (*This update is to coordinate with reference in the IMC.)	202

2004 SUPPLEMENT TO THE IUWIC

INTERNATIONAL ZONING CODE®

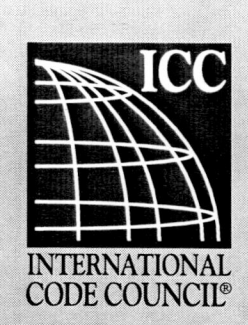

2004 SUPPLEMENT

INTERNATIONAL ZONING CODE
2004 SUPPLEMENT

CHAPTER 8
GENERAL REQUIREMENTS

Section 806.1 Change to read as shown: (Z3-03/04)

806.1 General. Loading spaces shall be provided on the same lot for every building in the C or FI zones. No loading space is required if prevented by an existing lawful building.

Section 807 Grading and Excavation Regulations Delete section in its entirety without substitution: (Z4-03/04)

ERRATA TO THE 2003 INTERNATIONAL CODES

The pages that follow include the errata to the 2003 International Codes, formatted by code and printing. This includes errata published on the ICC website (www.iccsafe.org) which has been reviewed and edited to include only technical errata.

Each code is identified, followed by the errata based on the respective printing. Each entry includes the section number of the code followed by the text. For example, Section 508.4 of the first printing of the *International Building Code* reads: "Section 508.4 Parking beneath Group R. Where a maximum ... not less than the mixed occupancy separation required in Section 302.3.3. The revised wording, as reflected in the errata, indicates that the text should read, "... separation required in Section 302.3.2" (underlining added for identification only).

INTERNATIONAL BUILDING CODE EDITORIAL CHANGES – FIRST PRINTING

CERAMIC FIBER BLANKET: Section reference now reads . . . 721.1.1
CONCRETE CARBONATE AGGREGATE: now reads . . . See Section 721.1.1.
CONCRETE , CELLULAR: now reads . . . See Section 721.1.1.
CONCRETE, LIGHTWEIGHT AGGREGATE: now reads . . . See Section 721.1.1.
CONCRETE, PERLITE: now reads . . . See Section 721.1.1.
CONCRETE, SAND-LIGHTWEIGHT: now reads . . . See Section 721.1.1.
CONCRETE, SILICEOUS AGGREGATE: now reads . . . See Section 721.1.1.
CONCRETE, VERMICULITE: now reads . . . See Section 721.1.1.
GLASS FIBERBOARD: now reads . . . See Section 721.1.1.
MINERAL BOARD: now reads . . . See Section 721.1.1.
302.2: Section reference in line 3 now reads . . . Section 302.3.2
307.2: Last line of COMBUSTIBLE FIBERS now reads . . . wastepaper, certain synthetic fibers or other like materials.
Table 307.7(1): Column 2, row 2, line 2 now reads . . . IIIA
Table 307.7(1): Column 4, row 12, line 1 now reads . . . $1^{e,g}$
402.4.1.4: Section reference in 1st sentence now reads . . . Section 1004.
402.7.3: Section reference in last sentence now reads . . . Section 705.
406.1.2: Section reference in last sentence now reads . . . Section 705.
406.2.7: Section reference in last sentence now reads . . . Section 302.3.2.
406.3.4: Last line now reads . . . of Sections 302.3, 402.7.1, 406.3.13, 508.3, 508.4 and 508.7.
408.3.3: Section reference now reads . . . Section 1009.9
408.3.6: Item 2, section reference now reads . . . Section 715.3.
412.1.3: Exception, section reference now reads . . . Section 1019.1.8
Table 415.3.2: Column 2, row 3, last line now reads . . . Division 1.6
415.7.3.4.1: Line 8 now reads . . . Section 715.
415.7.3.5.2: Section reference now reads . . . Section 715
415.9.2.2: Exception 2, last line now reads . . . Section 715
505.3: Line 3 now reads . . . Section 1013.3.
505.3: Exceptions 1 and 2 now read . . . Section 1014.1 and 1007, respectively.
507.6: Last line now reads . . . with Table 302.3.2
508.4: Last line now reads . . . Section 302.3.2.
508.7.1: Line 4 now reads . . . in Table 302.3.2 ...
603.1: Item 15 now reads . . . Nailing or furring strips as permitted by Section 803.4.
705.6: Exception 2, line 1 now reads . . . Two-hour fire-resistance-rated walls shall be ...
705.6: Exception 4.3. line 7 now reads . . . by a minimum of 2-inch (51 mm) nominal ledgers ...
706.7: Section references now read . . . Section 715 and 1019.1.1, respectively.
706.8.1: Section reference now reads . . . Section 1019.1.2
707.2: Exception 2.2, line 9 now reads . . . Section 907.10
707.8.1: Exception now reads . . . Section 1019.1.2
708.1: New item added now reads . . . 5. Elevator lobby separation as required by Section 707.14.1.
708.3: Last line now reads . . . wall shall be at least 1 hour.
710.5.2: Last line now reads . . . the ambient temperture test and the elevated temperature exposure test.
715.4: Last line now reads . . . Section 715.4.8.
716.5.3: Last line now reads . . . as permitted by Section 1019.1.2.
716.6.2: Last line now reads . . . with Section 712.4.2, where exhaust ducts are located with the cavity of a wall, and where exhaust ducts do not pass through another dwelling unit or tenant space.
Table 720.1(2): Column 3, row 2, line 4 now reads . . .with $2^1/_4$" Type S drywall screws, spaced 12^2 on center, wallboard joints covered with paper tape and joint compound, fastener heads covered with joint compound, ...
803.7: Section reference now reads . . . Section 803.6
803.7: Exception, last line now reads . . . with Sections 803.1 or 803.6.
805.1.2: Exception, last line now reads . . . with Section 803.4.
905.10: Line 2 now reads . . . during construction and demolition operations shall ...

ERR-2

909.5.2: Last 2 lines now read . . . Door openings shall be protected by fire door assemblies complying with Section 715.3.3.
909.5.2: Exception 1, last line now reads . . . in accordance with Section 907.10.
909.20.2.1: Line 5 now reads . . . in accordance with Section 715.3.7.
909.20.3.1: Last line now reads . . . in accordance with Section 715.3.
909.20.3.2: Section references now read . . . Section 715.3.
909.20.4.1: Section references now read . . . Section 715.3.
Table 1004.1.2: Column 2, row 6 now reads . . . See Section 1004.7
1021.3: Last line now reads . . . with Section 907.10.
1203.1: Section reference now reads . . . Section 1203.4
1203.4.3: Section reference now reads . . . Section 1206
Table 1507.2: Column 2, row 10, line 2 now reads . . . (0.105 inch)
1604.6: Section reference now reads . . . Section 1713
1615.1.4: Equation 16-43, section reference in notation T now reads . . . (see Section 9.5.5.3 of ASCE 7).
1617.2.2.2: Reference to r in notation r_{max} now reads . . . ρ
1617.6.1.1: Item 3, last line now reads . . . factor of 4.
1617.6.1: New subsection 1617.6.1.3 added.
1622.1.3: ASCE 7, Section changed to 9.14.7.9 and section reference in 5th line now reads 9.1.3
1704.5: Last line of Section and Exception 1 now read . . . Table 1604.5 and Section 1617.2).
1704.5.2: Last line now reads . . . Table 1604.5 and Section 1617.2).
Table 1704.5.3: Column 5, row 12 now reads . . . Sec. 1.2.2(e), 2.1.4, 3.1.6
1805.9: paragraph 2, line 2 and last line line of exception 2 now reads . . . provisions of ACI 318, Sections 21.10.1 to 21.10.3
1808.2.23.2: Line 6 now reads . . . Provisions of ACI 318, Section 21.10.4; Exception 2, last line now reads . . . Section 21.10.4; Exception 3 now reads . . . Section 21.10.4.4(a) of ACI 318...
1910.4.1: Last line now reads . . . as modified by Section 1908.1.7.
1910.5.2: Last line now reads . . . Section 1908.1.6.
2106.5.1: Section now reads . . . When calculating in-plane shear or diagonal tension stresses by the working stress design method, shear walls that resist seismic forces shall be designed to resist 1.5 times the seismic forces required by Chapter 16. The 1.5 multiplier need not be applied to the overturning moment.
2107.2.6: 2nd paragraph deleted
2113.3: line 5 now reads . . . in Sections 2113.3.1, 2113.3.2 and 2113.4.
2305.2.4.1: line 3 now reads . . . requirements in Section 1620.5 or Section 9.5.2.6.5 of ASCE 7
Table 2305.3.3: Note a now reads . . . For design to resist seismic forces, shear wall height-width ratios greater than 2:1, but not exceeding $3^1/_2$:1, are permitted provided the allowable shear resistance values in Table 2306.4.1 are multiplied by $2w/h$.
Table 2308.12.4: Column 5, row 1 now reads . . . 1.00 << S_{DS}
2406.2: Section reference in line 2 now reads . . . Section 2406.2.1

EDITORIAL CHANGES – SECOND PRINTING
302.3.2: Exception, line 4 now reads . . . the fire-resistance ratings in Table 302.3.2...
Table 302.3.2: Note d now reads . . . See Section 406.1.4.
Table 307.7(1): Note n added in title.
Table 307.7(1): Note n added in title.
Table 307.7(1): Note e now reads . . . e. Maximum allowable quantities shall be increased 100 percent when stored in approved storage cabinets, gas cabinets, exhausted enclosures or safety cans as specified in the *International Fire Code*. Where Note d also applies, the increase for both notes shall be applied accumulatively.
Table 307.7(1): Note m now reads . . . m. For gallons of liquids, divide the amount in pounds by 10 in accordance with Section 2703.1.2 of the *International Fire Code*.
Table 307.7(1): Note n now reads . . . n. For storage and display quantities in Group M and storage quantities in Group S occupancies complying with Section 414.2.4, see Table 414.2.4.
402.7.3: Exception, last line now reads . . . complying with Section 705.
410.3.1: Exception 1, last line now reads . . . with Section 410.3.4.
Section 414.1.2.2 now reads **414.1.2.1 Aerosols** and 414.1.2.2 deleted.
Table 414.2.4: Note i now reads . . . i. The permitted quantities shall not be limited in a building equipped throughout with an automatic sprinkler system in accordance with Section 903.3.1.1.

LIQUID STORAGE ROOM, last line now reads . . . liquids in a closed condition.
Table 503: Group I-3, Type IIB55 now reads . . . 10,000
Table 721.2.3(2): Title now reads . . . COVER THICKNESS FOR PRESTRESSED CONCRETE FLOOR OR ROOF SLABS (inches)
907.2.13: Line 4 now reads . . . be activated in accordance with Section 907.6.
1007.1: Exception 3, last line now reads . . . in Section 1024.8.
1008.1.2: Exception 6 now reads . . . Power-operated doors in accordance with Section 1008.1.3.2.
1016.4: now reads . . . **Air movement in corridors.** Exit access corridors shall not serve as supply, return, exhaust, relief or ventilation air ducts.
1021.3: Line 3 now reads . . . smoke detector installed in accordance with Section 907.10.
1405.9.1 has been deleted.
1405.9.1.1: now reads . . . **1405.9.1 Interior adhered masonry veneers.** Interior adhered masonry veneers shall have a maximum weight of 20 psf (0.958 kg/m^2) and shall be installed in accordance with Section 1405.9. Where the interior adhered masonry veneer is supported by wood construction, the supporting members shall be designed to limit deflection to 1/600 of the span of the supporting members.
2110.1.1: Exception 1, line 4 now reads . . . 715 in fire barriers and fire partitions that have a...
Table 2111.1: Row 2, column 3 now reads . . . 4-inch minimum thickness for hearth, 2-inch minimum thickness for hearth extension.
2211.2.2: Item 8, last line now reads . . . by Table 2211.2(1).
Table 2211.2(2): Title now reads . . . NOMINAL SHEAR VALUES FOR WIND AND SEISMIC FORCES IN POUNDS PER FOOT FOR SHEAR WALLS FRAMED WITH COLD-FORMED STEEL STUDS AND FACED WITH GYPSUM BOARD[a,b]
Table 2211.3: Table notes now read . . . a. See Section 2211.3.2, item 2. b. See Section 2211.3.2, item 1.
2211.3.2: Line 3 now reads . . . multiplied by the sum of the widths (SL_i) of the Type II shear...
2211.3.3: Line 1 now reads . . . **Anchorage and load path.** Design of Type II shear wall...
2211.3.3.3: Line 3 now reads . . . Section 2211.3.3.1, Type II shear wall bottom plates...
2211.3.3.4: Line 2 now reads . . . each end of each Type II shear wall segment shall be...
Table 2306.3.2: Note a, line 2 now reads . . . above for nail size of actual grade, and (3) Multiply value by the following adjustment factor = [1 - (0.5 - SG)], where SG = Specific gravity of the framing lumber.
Table 2902.1: Title now reads . . . MINIMUM NUMBER OF REQUIRED PLUMBING FIXTURES[a]

EDITORIAL CHANGES – THIRD PRINTING
[F] TABLE 307.7(2): Note j added in title.
[F] TABLE 307.7(2): Note j reads . . . For gallons of liquids, divide the amount in pounds by 10 in accordance with Section 2703.1.2 of the *International Fire Code*.
402.7.3: Exception, last two lines now read . . . building shall be separated by 2-hour fire-barriers complying with Section 706.
406.2.7: last 2 lines now read . . . rated from other occupancies in accordance with Section 302.3.
TABLE 601: Type II, column B, last row, add a table note c to 0.
1621.1.3: Modified ASCE 7, Section 9.6.3.13, line 11 now reads . . . system listed in Section 307, shall, itself, be designed to . . .
1623.1.1: Modified ASCE 7, Section 9.13.6.2.3, line 2 now reads . . . system shall comply with Section 714.7 of the . ..
1704.5: line 5 now reads . . . Table 1604.5 and Section 1616.2).
1704.5: exception 1, row 5 now reads . . . 1604.5 and Section 1616.2).
1704.5.2: line 6 now reads . . . 1616.2), shall comply with Table 1704.5.1.
1707.7.1: now reads . . . Special inspection is required for the installation of the following components, where the component has a Component Importance Factor of 1.0 or 1.5 in accordance with Section 9.6.1.5 of ASCE 7.
1808.2.10: line 3 now reads . . . for each pile type in Sections 1809 and 1810 are permitted...
1810.6.2: exception line 2 now reads . . . 1808.2.10, the allowable stresses are permitted to be in-...
1908.1: line 2 now reads . . . cated in Sections 1908.1.1 through 1908.1.7.
2205.2.2: line 4 now reads . . . 341, Part I or III.
2602.7.2: line 4 now reads . . . ing, at an ambient temperature of at least 200°F (111°C) be- ...

EDITORIAL CHANGES – FOURTH PRINTING
Table 307.7(1): Oxidizer row, column 3, 2nd line now reads . . . H-2 or H-3
415.2: FLAMMABLE VAPORS OR FUMES, line 2 now reads . . . flammable constituents in air that exceed 25 percent of their...

Table 415.3.1, Footnote c, line 4 now reads . . . explosive materials contained in detonator buildings or magazines shall govern in regard to the spacing of said detonator buildings or magazines from buildings or magazines containing other explosive materials. If any two or more buildings or magazines are separated from each other by less than the specified "Separation of Magazines" distances, then such two or more buildings or magazines, as a group, shall be considered as one building or magazine, and the total quantity of explosive materials stored in such group shall be treated as if the explosive were in a single building or magazine located on the site of any building or magazine of the group, and shall comply with the minimum distance specified from other magazines or inhabited buildings.

Table 720.1(3): Last sentence of item number 22 now reads . . . The wood structural panel thickness shall not be less than nominal $1/2$" nor less than required by Chapter 23.

801.2.2: line 2 now reads . . . interior finish or trim except as provided in Section 2603.8 or 2604.

909.5.1, line 2 now reads . . . is the product of the smoke barrier gross area multiplied by...

1015.2: last line now reads . . . feet (122 m) for occupancies in Group F-1 or S-1.

Table 1907.5.2.1: Row 1 under Depth now reads . . . $d £ 8$

FIGURE 2308.12.6(1) changed as shown

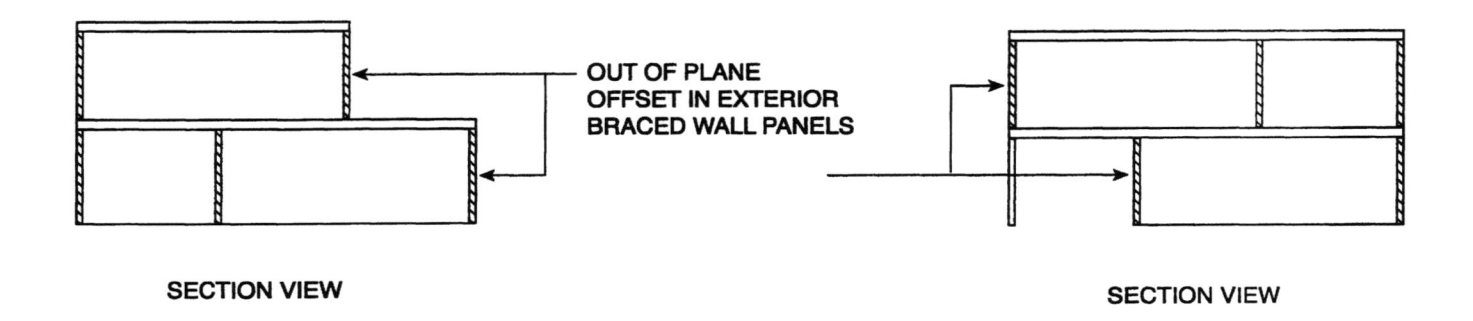

E104.3.2: Section reference now reads . . . E104.3.4.

INTERNATIONAL ENERGY CONSERVATION CODE EDITORIAL CHANGES – SECOND PRINTING

402.4.1: Last line now reads . . . of effect of factors specified in Section 402.5.
502.1.3: Item 3, line 5 now reads . . . tested at 1.57 pounds per square foot (psf) (75 Pa)
Table 502.2: Columns 3 and 4 now read . . . DETACHED ONE- AND TWO-FAMILY DWELLINGS and GROUP R-2, R-4 OR TOWNHOUSES respectively.
Table 602.1.1.1(1): Column 2, Zone 15 now reads . . . 7,000 - 8,499
802.3.2: Line 5 now reads . . . square foot (psf) (75 Pa)
802.3.7: Item 3, line 5 now reads . . . be tested at 1.57 psf (75 Pa)

INTERNATIONAL FIRE CODE EDITORIAL CHANGES – FIRST PRINTING

202: High-hazard Group H-3 list now reads ...Oxidizers, Class 3, that are used or stored in normally closed containers or systems pressurized at least 15 pounds per square inch gauge (103 kPa)

Section 308.6 is added to read ...**308.6 Flaming food and beverage preparation.** The preparation of flaming foods or beverages in places of assembly and drinking or dining establishments shall be in accordance with Section 308.6.

308.6.1 Dispensing. Flammable or combustible liquids used in the preparation of flaming foods or beverages shall be dispensed from one of the following:

1. A 1-ounce (29.6 ml) container; or
2. A container not exceeding 1-quart (946.5 ml) capacity with a controlled pouring device that will limit the flow to a 1-ounce (29.6 ml) serving.

308.6.2 Containers not in use. Containers shall be secured to prevent spillage when not in use.

308.6.3 Serving of flaming food. The serving of flaming foods or beverages shall be done in a safe manner and shall not create high flames. The pouring, ladling or spooning of liquids is restricted to a maximum height of 8 inches (203 mm) above the receiving receptacle.

308.6.4 Location. Flaming foods or beverages shall be prepared only in the immediate vicinity of the table being serviced. They shall not be transported or carried while burning.

308.6.5 Fire protection. The person preparing the flaming foods or beverages shall have a wet cloth towel immediately available for use in smothering the flames in the event of an emergency.

313.1: Exception 2 now reads ...Where allowed by Section 313 or 314.

Page 30, 313.2: line 4 now reads ...buildings.

804.1.1: Exception 2 now reads ...Trees shall be permitted within dwelling units in Group R-2 occupancies.

804.4: line 2 now reads ...Such flame retardance...

805.1.2: Exception, last line now reads ...with Section 803.4 of the *International Building Code.*

805.2: line 3 now reads ...approved agency and pass Test 1 or 2, as described...

Page 58, 806.2: Class C now reads ...Flame spread index 76-200

907.2.10.1: line 3 now reads ...in Sections 907.2.10.1.1 through 907.2.10.1.3.

Table 1004.1.2: Column 2, row 6 now reads ...See Section 1004.7

1007.5: line 5 now reads ...with Section 604 shall be installed in accordance with ASME...

1026.21: now reads ...**Stairway floor number signs.** Existing stairs shall be marked in accordance with Section 1019.1.7.

1417.3: Exception deleted

2206.2.4: line 5 now reads ...Section 3404.2.8 and shall comply with Sections 2206.2.4.1...

2209.5: line 3 now reads ...with Sections 2209.5.1 through 2209.5.4.3.

2209.5.4.2: Item 1, line 3 now reads ...2209.5.4.2 for the combinations of maximum...

Table 2209.5.4.1 is now ...TABLE 2209.5.4.2

Figure 2209.5.4.1 is now ...FIGURE 2209.5.4.2

2211.8.2: last line now reads ...with Section 2211.8.1.2.

2505.6: Item 1, line 2 now reads ...requirements in Sections 2505.1 through 2505.5.

2701.3.3.8: line 3 now reads ...immediate harm to persons or property, means of mitigating the dangerous effects of a release shall be provided.

Table 2703.8.2: last line of Explosives Class now reads ...Division 1.6

3204.3.1.1: Paragraph 2, line 3 now reads ...limits established by law as the limits of districts in

3404.2.9.5.1: line 4 now reads ...within the limits established by law as the limits of

3405.2.4: now reads ...Class I, II and III liquids. Class I and II liquids or Class III liquids that are heated up to or above their flash points shall be transferred by one of the following methods:

3406.2.4.4: line 4 now reads ...by law as the limits of districts in which such storage

Page 312, 3406.2.8: Item 2 now reads ...The dispensing hose does not exceed 100 feet (30 480 mm) in length.

3406.5.4: last line now reads ...through 3406.5.4.5.

3703.2.5: now reads ...**3703.2.5 Weather protection for highly toxic liquids and solids — outdoor storage or use.** Where overhead weather protection is provided for outdoor storage or use of highly toxic liquids or solids, and the weather protection is attached to a building, the storage or use area shall either be equipped throughout with an approved automatic sprinkler system in accordance with Section 903.3.1.1, or storage or use vessels shall be fire-resistance rated. Weather protection shall be provided in accordance with Section 2704.13 for storage and Section 2705.3.9 for use.

EDITORIAL CHANGES – SECOND PRINTING
313.1: exception 2 now reads ...Where allowed by Section 314.
TABLE 1304.1: row 11, column 2 now reads ...Manufacturing, Processing and Handling of Combustible Particulate Solids.
2211.3.2: second line now reads ...garages except in approved locations.
2701.3: last line now reads ...Chapters 28 through 44.
2701.5: second paragraph, last line now reads ...closure plan in accordance with Section 2701.6.3.
2701.6.2: last line now reads ... closure plan in accordance with Section 2701.6.3.
2701.6.3: second line now reads ...is required in accordance with Section 2701.5 to terminate ...
Table 2703.1.1(1): Column 3, row 2, second line now reads ... H-2 or H-3
TABLE 2703.1.1(1)—continued: footnote f now reads ...Quantities shall not be limited in a building equipped throughout with an approved automatic sprinkler system in accordance with Section 903.3.1.1.
TABLE 3304.5.2(2): line 19, column 4 now reads ...680.
3406.5.4.5: subsection 1, line 2 now reads ...been issued a permit to conduct mobile fueling.
3704.1.2: line 3 now reads ...2703.8.6 and the following requirements:

EDITORIAL CHANGES – THIRD PRINTING
202: Residential Group R, R-3, line 4 now reads ...two dwelling units as applicable in Section 101.2 of the *International Building Code*.
909.5.1: line 2 now reads ...is the product of the smoke barrier gross area multiplied by
1007.1: Exception 3, line 5 now reads ...in Section 1024.8.
1008.1.2: Exception 3, line 3 now reads ...1001.1.
1008.1.2: Exception 6, line 2 now reads ...1008.1.3.2.
1015.2: line 6 now reads ...feet (122 m) for occupancies in Group F-1 or S-1.
2206.2.3: line 3 now reads . . .of Class I, II or IIIA liquid motor fuels except as provided by
2206.2.4: line 3 now reads . . .of Class I, II or IIIA liquid motor fuels are allowed to be in-Page
2206.2.6: line 4 now reads . . .motor fuels are allowed to be installed in buildings in special
2206.3: line 2 now reads ...motor fuels shall be safeguarded from public access or unaut-Page
3404.2.9.5.1.6: line 7 now reads . . .3404.2.9.5.1.2 through 3404.2.9.5.1.5 assuming a . . .

INTERNATIONAL FUEL GAS CODE EDITORIAL CHANGES – FIRST PRINTING

Table 402.2: Column of Input Btu/H values adjusted as shown

TABLE 402.2
APPROXIMATE GAS INPUT FOR TYPICAL APPLIANCES

APPLIANCE	INPUT BTU/H (Approx.)

Space Heating Units
   Hydronic boiler
     Single family
     Multifamily, per unit
   Warm-air furnace
     Single family
     Multifamily, per unit
	
	100,000
	60,000
	
	100,000
	60,000

Space and Water Heating Units
   Hydronic boiler
     Single family
     Multifamily, per unit
	
	120,000
	75,000

Water Heating Appliances
   Water heater, automatic instantaneous
     Capacity at 2 gal./minute
     Capacity at 4 gal./minute
     Capacity at 6 gal./minute
   Water heater, automatic storage 30- to 40-gal. tank
   Water heater, automatic storage, 50-gal. tank
   Water heater, domestic, circulating or side-arm
	
	142,800
	285,000
	428,400
	35,000
	50,000
	35,000

Cooking Appliances
   Built-in oven or broiler unit, domestic
   Built-in top unit, domestic
   Range, free-standing, domestic
	
	25,000
	40,000
	65,000

Other Appliances
   Barbecue
   Clothes dryer, Type I (domestic)
   Gas fireplace, direct vent
   Gas light
   Gas log
   Refrigerator
	
	40,000
	35,000
	40,000
	2,500
	80,000
	3,000

For SI: 1 British thermal unit per hour = 0.293 W, 1 gallon = 3.785 L, 1 gallon per minute = 3.785 L/m.

Table 402.4(27): Title now reads . . . SEMI-RIGID COPPER TUBING
Table 503.4: Column 2, row 3 now reads . . . Type B-W gas vent (Sections 503.6, 608)
Table 504.2(5): Appliance Type now reads . . . Draft hood equipped
603.1: line 2 now reads . . . CSA 8 and installed in accordance with the
703.1.2: Section reference now reads . . . Section 502.16
Example 5C, Solution, paragraph 4, line 1 now reads . . . According to Section 504.3.19,
Example 5C, Solution, paragraph 5, line 9 now reads . . . by Section 504.3.21.

EDITORIAL CHANGES – SECOND PRINTING
202 BTU: last line now reads . . . of water 1°F (0.56°C) (1 Btu = 1055 J).
304.10: line 4 now reads . . . of louver, grille or screen is known, it shall be used in calculating the size opening required to provide the free area specified. Where the design and free area of louvers and grilles are not known, it shall be assumed that wood louvers will have 25-percent free area and metal louvers and grilles will have 75-percent free area. Screens shall have a mesh size not smaller than $^1/_4$ inch.
503.6.9.1: Listed items are no longer exceptions
614.2: line 6 now reads . . . thickness specified in Table 603.4 of the *International Mechanical Code*...
706.3.6: last line now reads . . . in accordance with Chapter 27 of the *International Building Code*.

EDITORIAL CHANGES – THIRD PRINTING
501.8: subsection 4, third line now reads . . . 614).

INTERNATIONAL MECHANICAL CODE EDITORIAL CHANGES – FIRST PRINTING

304.5: line 2 now reads . . . motor fuel-dispensing facilities, repair garages or other areas
506.3.1.2: line 4 now reads . . . with Sections 603.1, 603.3, 603.4, 603.9, 603.10 and 603.12.
607.5.4: line 4 now reads . . . barrier wall or a corridor enclosure required to have smoke

EDITORIAL CHANGES – SECOND PRINTING
Added definition: ENERGY RECOVERY VENTILATION SYSTEM. Systems that employ air-to-air heat exchangers to recover energy from or reject energy to exhaust air for the purpose of pre-heating, pre-cooling, humidifying or dehumidifying outdoor ventilation air prior to supplying such air to a space, either directly or as part of an HVAC system.
TABLE 403.3 — continued: row 2, column 3, line 8 now reads . . . intermittent
501.2: Add arrow above section
502.7.3.2, Exception 1.2: now reads . . . of the lower flammable limit
509.1, now reads . . . Commercial cooking appliances
607.6.2, Added sentence . . . Ceiling radiation dampers shall not be required where exhaust duct penetrations are protected in accordance with Section 712.4.2 of the *International Building Code* and the exhaust ducts are located within the cavity of a wall and do not pass through another dwelling unit or tenant space.

INTERNATIONAL PLUMBING CODE EDITORIAL CHANGES – FIRST PRINTING

504.6.2: line 2 now reads . . . those materials listed in Section 605.4 or shall be tested, rated ...
608.13.3: line 4 now reads . . . CAN/CSA-B64.3.
ASTM: Standard reference number now reads . . . D 2665—01
NSF: Standard reference number now reads . . . 1996a

EDITORIAL CHANGES – SECOND PRINTING

FLUSH TANK: line 1 now reads . . . A tank designed with a fill valve and flush . . .
TABLE 702.1: row 8, column 2 now reads . . . ASTM F 1488, ASTM F 891.
TABLE 702.1: row 15, column 2 now reads . . . ASTM D 2665; ASTM D 2949; CSA B181.2; ASTM F 1488.
TABLE 702.2: row 8, column 2 now reads . . . ASTM F 1488, ASTM F 891.
TABLE 702.2: row 12, column 2 now reads . . . ASTM D 2665; ASTM D 2949; CSA B181.2.
TABLE 702.3: row 3, column 1 now reads . . . Asbestos-cement pipe
TABLE 702.3: row 8, column 2 now reads . . . ASTM F 1488, ASTM F 891.
TABLE 702.3: inserted new row 12 which now reads . . . Coextruded PVC sewer and drain PS 25, PS 50, PS 100 (cellular core) ASTM F 891.
TABLE 702.3: row 16, column 2 now reads . . . ASTM D 2665; ASTM D 2949; ASTM D 3034; CSA B182.2; CAN/CSA B182.4.
Standard Reference Number 1055 now reads . . . 1055—97 Performance Requirements for Chemical Dispensing Systems . . .
TABLE E201.1—continued: row 4, column 8 now reads . . . 5.

EDITORIAL CHANGES – THIRD PRINTING

417.1: line 4 now reads . . . form to the requirements of Section 424.3.
ASSE: reference number 1002–99 now reads . . . Performance Requirements for Antisiphon Fill Valves (Ballcocks) for Gravity Water Closet Flush Tanks
ASSE: reference number 1004–90 now reads . . . Performance Requirements for Backflow Prevention Requirements for Commercial Dishwashing Machines
ASSE: reference number 1006–89 now reads . . . Performance Requirements for Residential Use Dishwashers
ASSE: reference number 1016–96 now reads . . . Performance Requirements for Individual Thermostatic, Pressure Balancing and Combination Control Valves for Individual Fixture Fittings
ASSE: reference number 1019–97 now reads . . . Performance Requirements for Vacuum Breaker Wall Hydrants, Freeze Resistant, Automatic Draining Type
ASSE: reference number 1066–97 now reads . . . Performance Requirements for Individual Pressure Balancing In-Line Valves for Individual Fixture Fittings
ASSE: reference number 5013–98 now reads . . . Performance Requirements for Testing Reduced Pressure Principle Backflow Prevention Assembly (RPA) and Reduced Pressure Fire Protection Principle Backflow Preventers (RFP)
ASSE: reference number 5015–98 now reads . . . Performance Requirements for Testing Double Check Valve Backflow Prevention Assembly (DCVA)
ASSE: reference number 5048–98 now reads . . . Performance Requirements for Testing Double Check Valve Detector Assembly (DCDA)

INTERNATIONAL PRIVATE SEWAGE DISPOSAL CODE EDITORIAL CHANGES – FIRST PRINTING

202: High-hazard Group H-3 list now reads ...Oxidizers, Class 3, that are used or stored in normally closed containers or systems pressurized at least 15 pounds per square inch gauge (103 kPa).

Section 308.6 is added to read ...**308.6 Flaming food and beverage preparation.** The preparation of flaming foods or beverages in places of assembly and drinking or dining establishments shall be in accordance with Section 308.6.

308.6.1 Dispensing. Flammable or combustible liquids used in the preparation of flaming foods or beverages shall be dispensed from one of the following:
1. A 1-ounce (29.6 ml) container; or
2. A container not exceeding 1-quart (946.5 ml) capacity with a controlled pouring device that will limit the flow to a 1-ounce (29.6 ml) serving.

308.6.2 Containers not in use. Containers shall be secured to prevent spillage when not in use.

308.6.3 Serving of flaming food. The serving of flaming foods or beverages shall be done in a safe manner and shall not create high flames. The pouring, ladling or spooning of liquids is restricted to a maximum height of 8 inches (203 mm) above the receiving receptacle.

308.6.4 Location. Flaming foods or beverages shall be prepared only in the immediate vicinity of the table being serviced. They shall not be transported or carried while burning.

308.6.5 Fire protection. The person preparing the flaming foods or beverages shall have a wet cloth towel immediately available for use in smothering the flames in the event of an emergency.

313.1 Exception 2 now reads ...Where allowed by Section 313 or 314.
Page 30, 313.2: line 4 now reads ...buildings.
804.1 1 Exception 2 now reads ...Trees shall be permitted within dwelling units in Group R-2 occupancies.
804.4: line 2 now reads ...Such flame retardance...
805.1.2: Exception, last line now reads ...with Section 803.4 of the *International Building Code.*
805.2: line 3 now reads ...approved agency and pass Test 1 or 2, as described...
806.2: Class C now reads ...Flame spread index 76-200
907.2.10.1 line 3 now reads ...in Sections 907.2.10.1 1 through 907.2.10.1.3.
Table 1004.1.2: Column 2, row 6 now reads ...See Section 1004.7
1007.5: line 5 now reads ...with Section 604 shall be installed in accordance with ASME...
1026.21 now reads ...**Stairway floor number signs.** Existing stairs shall be marked in accordance with Section 1019.1.7
1417.3: Exception deleted
2206.2.4: line 5 now reads ...Section 3404.2.8 and shall comply with Sections 2206.2.4.1...
2209.5: line 3 now reads ...with Sections 2209.5.1 through 2209.5.4.3.
2209.5.4.2: Item 1 line 3 now reads ...2209.5.4.2 for the combinations of maximum...
Table 2209.5.4.1 is now ...TABLE 2209.5.4.2
Figure 2209.5.4.1 is now ...FIGURE 2209.5.4.2
2211.8.2: last line now reads ...with Section 2211.8.1.2.
2505.6: Item 1 line 2 now reads ...requirements in Sections 2505.1 through 2505.5.
2701.3.3.8: line 3 now reads ...immediate harm to persons or property, means of mitigating the dangerous effects of a release shall be provided.
Table 2703.8.2: last line of Explosives Class now reads ...Division 1.6
3204.3.1 1: Paragraph 2, line 3 now reads ...limits established by law as the limits of districts in
3404.2.9.5.1 line 4 now reads ...within the limits established by law as the limits of
3405.2.4: now reads ...Class I, II and III liquids. Class I and II liquids or Class III liquids that are heated up to or above their flash points shall be transferred by one of the following methods:
3406.2.4.4: line 4 now reads ...by law as the limits of districts in which such storage
3406.2.8: Item 2 now reads ...The dispensing hose does not exceed 100 feet (30 480 mm) in length.
3406.5.4: last line now reads ...through 3406.5.4.5.
3703.2.5: now reads ...**3703.2.5 Weather protection for highly toxic liquids and solids — outdoor storage or use.** Where overhead weather protection is provided for outdoor storage or use of highly toxic liquids or solids, and the weather protection is attached to a building, the storage or use area shall either be equipped throughout with an approved automatic sprinkler system in accordance with Section 903.3.1 1 or storage or use vessels shall be fire-resistance rated. Weather protection shall be provided in accordance with Section 2704.13 for storage and Section 2705.3.9 for use.

ERR-13

EDITORIAL CHANGES – SECOND PRINTING
313.1: exception 2 now reads ...Where allowed by Section 314.
TABLE 1304.1: row 11, column 2 now reads ...Manufacturing, Processing and Handling of Combustible Particulate Solids.
2211.3.2: second line now reads ...garages except in approved locations.
2701.3: last line now reads ...Chapters 28 through 44.
2701.5: second paragraph, last line now reads ...closure plan in accordance with Section 2701.6.3.
2701.6.2: last line now reads ...ity closure plan in accordance with Section 2701.6.3.
2701.6.3: second line now reads ...is required in accordance with Section 2701.5 to terminate ...
Table 2703.1.1(1): Column 3, row 2, second line now reads ...H-2 or H-3
TABLE 2703.1.1(1)—continued: footnote f now reads ...Quantities shall not be limited in a building equipped throughout with an approved automatic sprinkler system in accordance with Section 903.3.1.1.
TABLE 3304.5.2(2): line 19, column 4 now reads ... 680.
3406.5.4.5: subsection 1, line 2 now reads ...been issued a permit to conduct mobile fueling.
3704.1.2: line 3 now reads ...2703.8.6 and the following requirements:

EDITORIAL CHANGES – THIRD PRINTING
202: Residential Group R, R-3, line 4 now reads ...two dwelling units as applicable in Section 101.2 of the *International Building Code*.
909.5.1: line 2 now reads ... is the product of the smoke barrier gross area multiplied by
1007.1: Exception 3, line 5 now reads ...in Section 1024.8.
1008.1.2: Exception 3, line 3 now reads ...1001.1.
1008.1.2: Exception 6, line 2 now reads ...1008.1.3.2.
1015.2: line 6 now reads ...feet (122 m) for occupancies in Group F-1 or S-1.
2206.2.3: line 3 now reads . . .of Class I, II or IIIA liquid motor fuels except as provided by
2206.2.4: line 3 now reads . . .of Class I, II or IIIA liquid motor fuels are allowed to be in-Page
2206.2.6: line 4 now reads . . .motor fuels are allowed to be installed in buildings in special
2206.3: line 2 now reads ...motor fuels shall be safeguarded from public access or unaut-Page
3404.2.9.5.1.6: line 7 now reads . . .3404.2.9.5.1.2 through 3404.2.9.5.1.5 assuming a

INTERNATIONAL RESIDENTIAL CODE EDITORIAL CHANGES – FIRST PRINTING

Table R301.5: 7th row of Live Loads now reads . . . 50
Figure R403.3(2): Revuse the "3000" contours in NY, VT, NH and ME to read "2000"
R602.3.1: Exception 2, line 2 now reads . . . in accordance with Table R602.3.1.
Figure R602.3(1): Section numbers changed as indicated
R602.11.1: line 6 now reads . . . ¼ inch by 3 inches by 3 inches (6.4 mm by 76 mm by 76...
Table G2409.2: In first column title, change figure references to G2409.2(1) and G2409.1(2). In first column change "0.024" to 0.024 inch (24 gage)" in all locations
Table G2413.2: Input btu/h values changed as indicated

TABLE G2413.2
APPROXIMATE GAS INPUT FOR TYPICAL APPLIANCES

APPLIANCE	INPUT BTU/H (Approx.)
Space Heating Units	
Hydronic boiler	
Single family	100,000
Multifamily, per unit	60,000
Warm-air furnace	
Single family	100,000
Multifamily, per unit	60,000
Space and Water Heating Units	
Hydronic boiler	
Single family	120,000
Multifamily, per unit	75,000
Water Heating Appliances	
Water heater, automatic instantaneous	
Capacity at 2 gal./minute	142,800
Capacity at 4 gal./minute	285,000
Capacity at 6 gal./minute	428,400
Water heater, automatic storage 30- to 40-gal. tank	35,000
Water heater, automatic storage, 50-gal. tank	50,000
Water heater, domestic, circulating or side-arm	35,000
Cooking Appliances	
Built-in oven or broiler unit, domestic	25,000
Built-in top unit, domestic	40,000
Range, free-standing, domestic	65,000
Other Appliances	
Barbecue	40,000
Clothes dryer, Type I (domestic)	35,000
Gas fireplace, direct vent	40,000
Gas light	2,500
Gas log	80,000
Refrigerator	3,000

For SI: 1 British thermal unit per hour = 0.293 W, 1 gallon = 3.785 L, 1 gallon per minute = 3.785 L/m.

G2414.11: item 3, line 7 now reads . . . the outside end of the compression fitting...
P2801.4: Exception 2, line 3 now reads . . . in accordance with Section M1703...

P2802.3: Deleted
P2902.2.6: line 2 now reads . . . shall conform to ASSE 1015 or AWWA C510.
Figure P3108.1(3): The stack vent shown was deleted
E3605.4.4: line 6 now reads . . . shall comply with Section E3605.1 and Table E3605.5.3.
AAMA: Reference to 101/I.S.2/NAFS—02 now reads . . . R308.6.9, R613.3, N1101.3.2.2
AWWA: Reference to C50—00 now reads . . . Table P2902.2, P2902.2.6
TPI: Standard reference number now reads . . . TPI 1—2002
WDMA: Reference to 101/I.S.2/NAFS—02 now reads . . . R308.6.9, R613.3, N1101.3.2.2
LINTEL: now reads . . . R606.9, R611.7.3, R703.7.3, R1003.7

EDITORIAL CHANGES – SECOND PRINTING

R110.2: line 3 now reads . . . Sections 3406 and 3407 of the *International Building Code*.
Table R301.2.2.4: Table deleted
R308.4: Exception 9.1, line 4 now reads . . . of Sections 1012 and 1607.7 of the *International Building Code*; and
R308.4: line 4 now reads . . . of Sections 1012 and 1607.7 of the *International Building Code*; and
R317.2.2: line 3 now reads . . . an extension of exterior walls or common walls in accordance with
R403.1.4: line 3 now reads . . . ground surface. Where applicable, the depth of footings shall also
R403.1.4.2: line 4 now reads . . . shall extend to a depth of not less than 12 inches (305 mm)
R403.1.4: line 3 now reads . . . ground surface. Where applicable, the depth of footings shall
R403.1.4.2: line 4 now reads . . . grade shall extend to a depth of not less than 12 inches (305 mm) below the top of slab.
Table R404.1.1(1): row 15, column 3 now reads . . . 8^g
R502.2.1: last line now reads . . . live load specified in Table R301.5 acting on the cantilevered portion of the deck.
Table R502.3.3(1): footnote f now reads . . . See Section R301.2.2.2.2, item 1, for additional limitations on cantilevered floor joists for detached one- and two-family dwellings in Seismic Design Categories D_1 and D_2 and townhouses in Seismic Design Categories C, D_1, and D_2.
Table R602.3(1): footnote e now reads . . . Spacing of fasteners not included in this table shall be based on Table R602.3(2).
R602.10.11: Exception, line 6 now reads . . . spaced greater or less than 25 feet (7620 mm) apart
R602.10.11: 2nd exception, line 3 now reads . . . to begin no more than 8 feet (2438 mm) from each
Table R703.4: footnote to Horizontal aluminum now reads . . . Horizontal aluminum^e
Figure R703.7.2.1 is now Figure R703.7.2.2
Figure R703.7.2.2 is now Figure R703.7.2.1
R808.1: line 9 now reads . . . shall meet the requirements of Section N1102.1.11.
G2407.10: now reads . . . The required size of openings for combustion, ventilation and dilution air shall be based on the net free area of each opening. Where the free area through a design of louver, grille or screen is known, it shall be used in calculating the size opening required to provide the free area specified. Where the design and free area of louvers and grilles are not known, it shall be assumed that wood louvers will have 25-percent free area and metal louvers and grilles will have 75-percent free area. Screens shall have a mesh size not smaller than $^1/_4$ inch (6 mm). Nonmotorized louvers and grilles shall be fixed in the open position. Motorized louvers shall be interlocked with the equipment so that they are proven to be in the full open position prior to main burner ignition and during main burner operation. Means shall be provided to prevent the main burner from igniting if the louvers fail to open during burner startup and to shut down the main burner if the louvers close during operation.
G2427.6.8.1: Items 1–4 are not exceptions
E3902.9 now reads . . . **E3902.10 Wet locations other than outdoors.**
E3902.9 now reads . . . **E3902.9 Outdoor wet locations.** Where installed outdoors in a wet location, 15- and 20-ampere, 125- and 250-volt receptacles shall have an enclosure that is weatherproof whether or not the attachment plug cap is inserted.
E3902.10 now reads . . . **E3902.11 Bathtub and shower space.**
E3902.11 now reads . . . **E3902.12 Flush mounting with faceplate.**
E3902.12 now reads . . . **E3902.13 Outdoor installation.**

EDITORIAL CHANGES – THIRD PRINTING

Definition UNUSUALLY TIGHT CONSTRUCTION now reads . . . UNUSUALLY TIGHT CONSTRUCTION. Construction in which:
Definition UNUSUALLY TIGHT CONSTRUCTION subsection 1, line 1 now reads . . . Walls and ceilings comprising the building thermal envelope have a . . .
Definition UNUSUALLY TIGHT CONSTRUCTION subsection 2 now reads . . . Storm windows or weatherstripping is applied around the threshold and jambs of opaque doors and openable windows.

R301.2.2.3.3: last line now . . . Section R611 or R612.

R319.1.4: exception 2, line 5 now reads . . . from exposed ground, and are separated there from . . .

TABLE R502.5(2): line 12, column 3 now reads . . . 9-0, column 5 now reads . . . 7-8, column 6 now reads . . . 1, column 7 now reads . . . 6-9, column 8 now reads . . . 1.

TABLE R502.5(2): line 23, column 3 now reads . . . 6-1, column 4 now reads . . . 1, column 5 now reads . . . 5-3, column 7 now reads . . . 4-8.

TABLE R602.10.5: row 1, column 1 now reads . . . MINIMUM.

R611.2: first paragraph, line 15 now reads . . . sidered irregular as defined in Section R301.2.2.2.2.

R905.3.8: line 18 now reads . . . to the roofing underlayment for slopes less than seven units vertical in 12 units horizontal (58-percent slope) . . .

P3108.1: line 10 now reads . . . the wet vent. [See Figures P3108.1(1), P3108.1(2), P3108.1(3) and P3108.1(4) for typical wet vent configurations.]

P3108.4: line now reads . . . serving one or two fixtures. [See Figures P3108.1(4) and P3108.1(5) for typical vertical wet vent configurations.]

FIGURE P3108.2(1); figure number now reads . . . FIGURE P3108.1(4)

FIGURE P3108.2(2); figure number now reads . . . FIGURE P3108.1(5)

ASTM—continued: last line, reference standard C 67—02 is deleted.

ASTM—continued: line 4, reference standard C 140—01ae01 is deleted.

ASTM—continued: line 50, reference standard C 1167—96 now reads . . . C 1167—96 Specification for Clay Roof Tiles R905.3, R905.3.4

ASTM—continued: line 11, reference standard D 2898—94(1999) now reads . . . D 2898—94(1999) Test Methods for Accelerated Weathering of Fire–retardant–treated Wood for Fire Testing R802.1.3.3

FIGURE B-10: See Section callout now reads . . . G2428.3.5.

AG103.1: last line now reads . . . Section AG108.

AG103.2: last line now reads . . . AG108.

AS104.1: last line now reads . . . Section AG108.

AS104.2: last line now reads . . . ANSI/NSPI-6 as listed in Section AG108.

AK101.1: now reads . . . Wall and floor-ceiling assemblies separating dwelling units including those separating adjacent townhouse units shall provide airborne sound insulation for walls, and both airborne and impact sound insulation for floor-ceiling assemblies.

AK102.1: line 4 now reads . . . ASTM E 90. Penetrations or openings in construction assemblies for piping; electrical devices; recessed cabinets; bathtubs; soffits; or heating, ventilating or exhaust ducts shall be sealed, lined, insulated or otherwise treated to maintain the required ratings. Dwelling unit entrance doors, which share a common space, shall be tight fitting to the frame and sill.

ERR-18

2004 SUPPLEMENT TO THE INTERNATIONAL CODES
EDITORIAL CHANGES - FIRST PRINTING

International Building Code

1405.12.2: Remove entry regarding stay of action on FS138-03/04. Add the following entry: Section 1405.12.2 Add new section to read as shown: (FS138-04/05) **1405.12.2 Window sills.** In occupancy Group R, one-and two family and multiple single family dwellings, where the rough opening for the sill portion of an operable window is located more than 72 inches above the grade or other surface below, the rough opening for the sill, or lowest part of the operable portion of the window, shall be a minimum of 24 inches above the finished floor of the room in which the window is located. **Exception.** Windows whose openings will not allow a 4 inch diameter sphere to pass through the opening when the opening is in its largest opened position.

1510.3: Exception 3 now reads: 3. The application of a new protective coating over an existing spray polyurethane foam roofing system shall be permitted without tear-off of existing roof coverings.

1609.7.2: Reference to 1609.7.2 in the Exception should read, 1609.7.3

1609.7.2: Paragraph after the exception and before the last paragraph reads: Asphalt shingles installed over a roof deck complying with IBC Section 1609.7.1 shall be tested to determine the resistance of the sealant to uplift forces using ASTM D6381.

Table 1617.6: Change footnote b to read, "... overstrength factor Ω_o is permitted ...

1704.10: Move after Section 1704.9.

1707.7.1, 1707.7.2 and 1707.7.3: Change to Sections 1707.8.1, 1707.8.2 and 1707.8.3.

Table 2306.3.2: Footnotes a, b and e now read:

a. For framing of other species: (1) Find specific gravity for species of framing lumber in AFPA and National Design Specification. (2) For staples, find shear value from table above for Structural I panels (regardless of actual grade) and multiply value by 0.82 for species with specific gravity of 0.42 or greater, or from 0.65 fro all other species. (3) For nails, find shear value from table above for nail size of actual grade and multiply value by the following adjustment factor: Specific Gravity Adjustment Factor = [1- (0.5 - SG)], where SG = Specific Gravity of the framing lumber. This adjustment factor shall not be greater than 1.

b. Fastening along intermediate framing members: Space fasteners a maximum of 12 inches on center, except 6 inches on center for spans greater than 32 inches.

c. and d. (No change to current text)

e. The minimum nominal depth of framing members shall be inches nominal. The minimum nominal width of framing members not located at boundaries or adjoining panel edges shall be 2 inches.

Table 2306.4.1: Footnote i: Change "plywood" to "wood structural panel"

2406.1.2: Remove entry regarding stay of action on S85-04/05 and replace with the following entry: Section 2406.1.2 Wired glass. Delete section without substitution.

International Energy Conservation Code

Table 301.1:
    (a) "Clear Creek" is a single county in Colorado. This is incorrectly identified as two counties: "Clear" and "Creek";
    (b) "Palo Pinto" is a single county in Texas. This is incorrectly identified as two counties: "Palo" and "Pinto";
    (c) "Red River" is a single county in Texas. This is incorrectly identified as two counties: "Red" and "River";
    (d) "Pend Oreille" is a single county in the state of Washington. This is incorrectly identified as two counties: "Pend" and "Oreille."

Table 301.2:
    (a) "Palo Pinto" is a single county in Texas. This is incorrectly identified as two counties: "Palo" and "Pinto";
    (b) "Red River" is a single county in Texas. This is incorrectly identified as two counties: "Red" and "River";

Table 402.1: Relocate the reference to footnote "e" from the Glazed Fenestration SHGC column heading to the Glazed Fenestration SHGC column for Climate Zone 3.

Table 802.2(1): In the section for "Walls, Above Grade" in the row for "Mass" walls under the column for climate zone 3, add a footnote reference to footnote 'c.' The contents of this cell will then read "R-5.7 ci^{c,e}"

802.3, Air leakage: Revise the sub-sections which are referenced to include Sections 802.3.3 through 802.3.7. The sentence will then read, "...as specified in Sections 802.3.1 through 802.3.7."

802.3.2, Curtain wall, storefront glazing and commercial entrance doors: Revise the units for pressure from "pounds per square inch (psi)" to "pounds per square foot (psf)." The sentence will then read, "...for air leakage at 1.57 pounds per square foot (psf) (75 Pa) in accordance..."

802.3.7, Recessed luminaires: Revise the units for pressure from "(psi)" to "(psf)." The sentence now reads, "...shall be tested at 1.57 (psf) (75 Pa) pressure difference..."

Table 803.2.2(2): In the row for Air cooled (Heating mode), under the column for Size Category, add the parenthetical term "(Cooling capacity)" under the entry for < 65,000 Btu/h.

Table 803.2.2(4): In the "Test Procedure" column, revise the referenced standard from "ANSI Z83.9" to "ANSI Z83.8." ANSI Standard Z83.8 is inclusive of gas fired duct furnaces and unit heaters.

803.2.3, Temperature and humidity controls: Revise the sub-sections which are referenced to include Section 803.2.3.3. The sentence now reads, "...as specified in Sections 803.2.3.1 through 803.2.3.3."

803.2.6, Cooling with outdoor air: Revise the table which is referenced from Table 703.2.6(1) to Table 803.2.6(1). The sentence now reads "....cooling system as shown in Table 803.2.6(1)."

Table 803.3.2(5): In the bottom left cell of the table, Under the Leaving Chilled Water Temperature, Entering Condenser Water Temperature and Lift columns, the last row now reads as follows: " Condenser ΔT^b " *[this will indicate that the condenser delta T is determined as indicated in footnote "b"].*

Table 803.3.2(5): In the row for the metric conversions which follows the table, revise the flow rate for 1 gallon per minute as follows: "1 gallon per minute = 3.785 L/min."

803.3.3, HVAC system controls: Revise the sub-sections which are referenced to include Sections 803.3.3.6 through 803.3.3.8. The sentence now reads, "...as required in Sections 803.3.3.1 through 803.3.3.8."

803.3.3.7.2, Two-pipe changeover system: Revise the metric conversion at the end of the last sentence of the paragraph now reads as follows: "....no more than 30°F (16.7°C) apart."

803.3.6, Duct and plenum insulation and sealing: Below the equation, in the description for the factor "F" revise the first word from "he" to "The." The description now reads "F = The measured leakage...."

Table 804.2: Under the Size Category column for "Instantaneous Water Heaters, Gas" in the last row for that equipment type, the entry should address items which are greater than *or equal to* 200,000 Btu/h and read as follows "≥ 200,000 Btu/h." (The entry currently only addresses items which are *greater* than 200,000 Btu/h and it needs to also address those which are "equal to" 200,000 Btu/h.)

Table 804.2: Under the Subcategory or Rating Condition column, in the row for "Instantaneous Water Heaters, Oil" which are ">210,000 Btu/h" in size, the entry should address items which are *greater than or equal to* 10 gallons and read as follows: "≥4,000 Btu/h/gal and ≥10 gal"

International Fire Code

910.2.2: Exception reads as follows: **Exception:** Buildings of noncombustible construction containing only noncombustible materials.

1805.2.3.5: Revise existing condition "3" to be "2.3", and add the following Condition 3 immediately before 3.1: 3. Cabinet exhaust ventilation system: An exhaust ventilation system shall be provided for cabinets and shall comply with the following:

Page IFC-37, before Table 2703.8.3.2, insert: Sections 2703.8.3.1 and 2703.8.3.3 Change to read as shown: (G 94-03/04)

> **[B] 2703.8.3.1 Construction requirements.** Control areas shall be separated from each other by fire barriers constructed in accordance with Section 706 of the *International Building Code*.
>
> **[B] 2703.8.3.3 Fire-resistance rating requirements.** The required fire-resistance rating for fire barrier assemblies shall be in accordance with Table 2703.8.3.2. The floor construction of the control area and construction supporting the floor of the control area shall have a minimum 2-hour fire-resistance rating.

International Fuel Gas Code

Section 101.2: After the section, insert "(Exceptions unchanged)"

International Residential Code

Table 505.2(3): Delete entry in this Supplement in its entirety.
R505.3.6: Delete entry in this Supplement in its entirety.
R505.3.6.1: Delete entry in this Supplement in its entirety.
Table R602.3.1: Remove superscript "b" from Category D_1 in title of Table R602.3.1
R602.10.11: In the second exception revise "12.5 feet (2438 mm)" to "8 feet (2438 mm)"
R603.3.5: Delete entry in this Supplement in its entirety.
R603.3.5.1: Delete entry in this Supplement in its entirety.
R606.4: Remove the text. Leave the heading, "Support conditions."
R606.4.1: Remove the paragraph and replace with the following text: "Each masonry wythe shall be supported by at least two-thirds of the wythe thickness."
R613.2: Remove entry regarding stay of action on RB205-3/04. Add the following entry: Section R613.2: Add new section to read as shown: (RB205-03/04) **R613.2 Window sills.** In dwelling units, where the rough opening for the sill portion of an operable window is located more than 72 inches above the ground or other surface below, the rough opening for the sill portion of the window shall be a minimum of 24 inches above the finished floor of the room in which the window is located. **Exception.** Windows whose openings will not allow a 4 inch diameter sphere to pass through the opening when the opening is in its largest opened position.
Section R702.3.7.1: Renumber to be Section R702.3.8.1 and relocate below Section R702.3.8
Section R804.3.6: Delete entry in this Supplement in its entirety.
Section R804.3.6.1: Delete entry in this Supplement in its entirety.
Table P3002.2: In row 6, under "Standard," delete CSA B181.5; In row 16, under "Standard," Add ASTM F1412.
Table P3002.3: In row 10, under "Standard," add ASTM F1412.
Section P3003.16.1: Renumber Section P3003.16.1 Polyolefin plastic to RP3003.16
Appendix J, Section AJ601.1.2: This section should be located as part of "Appendix J, Existing Buildings and Structures.